Biological and Medical Physics, Biomedical Engineering

More information about this series at http://www.springer.com/series/3740

BIOLOGICAL AND MEDICAL PHYSICS, BIOMEDICAL ENGINEERING

The fields of biological and medical physics and biomedical engineering are broad, multidisciplinary and dynamic. They lie at the crossroads of frontier research in physics, biology, chemistry, and medicine. The Biological and Medical Physics, Biomedical Engineering Series is intended to be comprehensive, covering a broad range of topics important to the study of the physical, chemical and biological sciences. Its goal is to provide scientists and engineers with textbooks, monographs, and reference works to address the growing need for information.

Books in the series emphasize established and emergent areas of science including molecular, membrane, and mathematical biophysics; photosynthetic energy harvesting and conversion; information processing; physical principles of genetics; sensory communications; automata networks, neural networks, and cellular automata. Equally important will be coverage of applied aspects of biological and medical physics and biomedical engineering such as molecular electronic components and devices, biosensors, medicine, imaging, physical principles of renewable energy production, advanced prostheses, and environmental control and engineering.

Tatiana K. Rostovtseva
Editor

Molecular Basis for Mitochondrial Signaling

Springer

Editors
Tatiana K. Rostovtseva
Section on Molecular Transport
Eunice Kennedy Shriver National Institute of Child Health
and Human Development, National Institutes of Health
Bethesda, MD, USA

ISSN 1618-7210 ISSN 2197-5647 (electronic)
Biological and Medical Physics, Biomedical Engineering
ISBN 978-3-319-55537-9 ISBN 978-3-319-55539-3 (eBook)
DOI 10.1007/978-3-319-55539-3

Library of Congress Control Number: 2017938879

Printed on acid-free paper

This Springer imprint is published by Springer Nature
The registered company is Springer International Publishing AG
The registered company address is: Gewerbestrasse 11, 6330 Cham, Switzerland

Preface

Mitochondria are fascinating cellular organelles that, even 100 years after their discovery, continue to puzzle researchers and challenge our experimental capabilities. Enormous progress in our understanding of these ancient organelles' function and their vital involvement in a plethora of cellular signaling pathways has been made in the last 20 years, and yet, there remain old riddles (as well as emerging new ones) waiting to be solved. The focus of this book, the molecular basis for mitochondrial signaling, treats just one part of the current surge in mitochondrial research but aims to inform readers on the vibrancy and versatility of modern mitochondrial research: how it actively embraces new disciplines and research areas and how it quickly assimilates modern technologies. A wide array of cutting-edge methods is covered in this book, ranging from electrophysiology and cell biology to structural and computational biology. It is hoped that readers will find this volume fulfilling and useful in their own investigations.

Traditionally, mitochondria were viewed as "the powerhouse of the cell," using oxidative phosphorylation to convert dietary calories into usable energy. While this remains true, it is now well recognized that mitochondria are involved in multiple crucial cellular functions, including Ca^{2+} signaling, programmed cell death or apoptosis, adaptation to stressful conditions, steroidogenesis, and aging. Mitochondrial dysfunction plays a central role in a wide range of age-related disorders, myopathies, neurodegenerations, and various forms of cancer. Mitochondrial channels and transporters are directly involved in the regulation of mitochondrial functions and in controlling metabolic response to nutritional conditions, energy demands, and developmental needs. They compose a molecularly diverse group of channels and transporters with the purpose of translocating ions, metabolites, and proteins across the two mitochondrial membranes, providing a dynamic exchange of energy and matter between mitochondria and the cytosol. The physiological importance of mitochondrial channels and transporters includes key roles in the regulation of the production of both mitochondrial energy in the form of ATP and toxic reactive oxygen species and in the regulation of cellular Ca^{2+} levels, apoptosis, and cellular metabolism. All these play crucial roles in normal cellular physiology and in pathological conditions.

Recent advances in the study of mitochondrial channels will be of particular interest to readers. After years of failed attempts as well as exciting discoveries, the molecular identity of two important channels in the mitochondrial inner membrane has finally been established. These are the mitochondrial Ca^{2+} uniporter (MCU) and the mitochondrial permeability transition pore (mPTP). Mitochondria play a pivotal role in Ca^{2+} homeostasis, as they represent a central hub for the complex network of Ca^{2+} signaling pathways and control of Ca^{2+} dynamics under physiological and pathological conditions. Recent progress in the development of fluorescently labeled and genetically encoded Ca^{2+} probes targeted to the mitochondrial matrix allowed the dissection of the physiology and molecular identity of the MCU. Furthermore, the so-called Ca^{2+} microdomains found on the surface of mitochondria are functionally and structurally coupled with endoplasmic reticular (ER) membranes and Ca^{2+}-releasing channels. Recent exciting findings confirmed that the c-subunit of the ATP synthase, generally required for ATP production, can form a large uncoupling channel in the mitochondrial inner membrane (mPTP) under certain conditions, such as excessive Ca^{2+} uptake by mitochondria. The persistent opening of the PTP produces osmotic dysregulation of the inner membrane and leads to the disruption of ATP production and consequently cell death. Another new promising research direction arises from recent discoveries that structural changes of mPTP are associated with its activity during cell development but also in aging and during stressful or degenerative events. These discoveries have reaffirmed that mitochondria are key organelles in the modulation of intracellular Ca^{2+} homeostasis.

The extensive functional studies on mitochondrial channels and transporters are now beginning to merge with structural information. After years of work on the physiological importance of the voltage-dependent anion channel (VDAC) of the mitochondrial outer membrane, we now know the VDAC structure and how this channel is regulated by cytosolic proteins. The groundbreaking research solving VDAC structure stimulated major recent findings regarding the function and regulation of this large mitochondrial transport channel. The impressive amount of data accumulated from structural, biochemical, and biophysical studies makes feasible at last the deciphering of the exact mechanisms governing selective metabolite transport through VDAC and its signature gating. Modeling studies using modern powerful computational approaches offer insights at the molecular level on VDAC's function in mammalian and plant cells. While the potential across the inner membrane (the mitochondrial potential) has been successfully measured since the 1950s, the existence of a potential across the outer membrane is still debated. Conventional thinking holds that this potential is essentially zero due to the high abundance of VDAC in the outer membrane. An intriguing theoretical model of the outer membrane potential identifies the VDAC-hexokinase complex as a potential-generating "battery." The model incorporates the well-known ability of VDAC to gate under applied potential, thus giving a new physiological relevance not only to the VDAC voltage gating but also to the channel's interaction with hexokinases and other cytosolic regulators, such as tubulin and alpha-synuclein. Investigations of this mitochondrial channel from different tissues and species present a good example of how a combination of structural, functional, and modeling studies delivers results inaccessible by a single approach.

Mitochondrial steroidogenesis, the process by which mitochondria maintain an effective exchange of sterols between mitochondria and other cellular compartments, has taken on new importance. The 18 kDa protein translocator protein (TSPO), known to shuttle cholesterol into the mitochondria for pregnenolone synthesis, possesses the specific function of mediating cholesterol transport across the mitochondrial membranes. Thus, the importance of the recently discovered high-resolution TSPO structure is difficult to overestimate. TSPO is another impressive example of how a breakthrough in structural studies stimulates research leading to progress in our understanding of mitochondrial function and provides essential clues for defining therapeutic strategies against a wide range of diseases.

Mitochondria communicate with other cellular compartments by exchanging information in the form of ions, metabolites, proteins, amino acids, and nucleic acids. Without exception, each exchange molecule must cross one or both mitochondrial membranes. Considering that almost all mitochondrial proteins have to be delivered from the cytoplasm, the protein import machinery spanning both mitochondrial membranes is vital for the control of not only mitochondrial but also whole-cell metabolism. With the recent development of tracing techniques, interest in the field of mitochondrial protein import has surged, leading to the identification of the regulatory mechanisms of protein import pathways.

The pro- and anti-apoptotic Bcl-2 members have remained a focus of intensive research for 20 years, not only because they orchestrate and execute apoptosis by congregating on mitochondrial membranes but also because they form a new, fascinating class of large oligomeric channels in the outer membrane. By forming multidomain channels in the mitochondrial outer membrane, pro-apoptotic proteins, such as Bax and Bak, irreversibly trigger apoptosis. In parallel with progress in our understanding of endogenous mitochondrial channel structure and functions, recent advancements in structural and imaging methods have led to breakthroughs in understanding the complex relationships between mitochondria and Bcl-2 family proteins. The notion that mitochondria are "passive" players in programmed cell death or apoptosis has been put to rest; modern research considers mitochondria and the Bcl-2 family of proteins as equal players and co-regulators of both cell death and energy metabolism.

This book illustrates a surprising aspect of many mitochondrial endogenous and associated proteins: they possess multiple functions. Sometimes these functions are quite opposite like the "pro-" and "antilife" activities of cytochrome c, ATP synthase, or Bcl-2 proteins. Mechanistically, these multiple functions most likely arise from protein-protein, protein-lipid, and protein-ion interactions between mitochondrial proteins (such as interactions between membrane proteins TSPO and VDAC or the interactions of pyruvate dehydrogenase, the key enzymatic complex, with other players of the tricarboxylic acid cycle) as well as between mitochondrial and cytosolic proteins (such as interactions with Bcl-2 proteins or interaction of VDAC with cytosolic proteins) and ions (such as Ca^{2+}). This complex array of interacting proteins, lipids, and signaling pathways has resulted, from time to time, in understandable confusion in the interpretation of results, which has occasionally moved the field in unproductive directions. With the increasing involvement of modern computational and systems biology approaches, future confusion will hopefully be minimized or at least short-lived. At the same time, the overwhelming complexity of mitochondrial signaling pathways challenges our curiosity and attracts fearless new researchers into the field.

Bethesda, MD, USA Tatiana K. Rostovtseva

Contents

Contents

Contributors

Jeff Abramson Department of Physiology, David Geffen School of Medicine, University of California, Los Angeles, CA, USA

Institute for Stem Cell Biology and Regenerative Medicine (inStem), National Centre for Biological Sciences–Tata Institute of Fundamental Research, Bangalore, Karnataka, India

Kambiz N. Alavian Division of Brain Sciences, Department of Medicine, Imperial College London, London, UK

Miguel A. Aon Laboratory of Cardiovascular Science, National Institute on Aging, National Institutes of Health, Baltimore, MD, USA

Ma Su Su Aung Macaulay Honors College at Baruch College of City University of New York, New York, NY, USA

Lucie Bergdoll Department of Physiology, David Geffen School of Medicine, University of California, Los Angeles, CA, USA

Gisela Beutner Department of Pediatrics (Cardiology), University of Rochester Medical Center, Rochester, NY, USA

Sergey M. Bezrukov Section on Molecular Transport, Eunice Kennedy Shriver National Institute of Child Health and Human Development, National Institutes of Health, Bethesda, MD, USA

Michelangelo Campanella Department of Comparative Biomedical Sciences, The Royal Veterinary College, University of London, London, UK

University College London Consortium for Mitochondrial Research, London, UK

Vanessa Checchetto Department of Biology and CNR Institute of Neurosciences, University of Padova, Padova, Italy

Rongmin Chen Department Internal Medicine, Section of Endocrinology, Yale University School of Medicine, New Haven, CT, USA

Sonia Cortassa Laboratory of Cardiovascular Science, National Institute on Aging, National Institutes of Health, Baltimore, MD, USA

Diego De Stefani Department of Biomedical Sciences, University of Padova, Padova, Italy

David N. DeHart Departments of Drug Discovery and Biomedical Sciences and the Hollings Cancer Center, Medical University of South Carolina, Charleston, SC, USA

Laurent Dejean Department of Chemistry, California State University, Fresno, CA, USA

Chiara Fedeli Department of Biomedical Sciences, University of Padova, Padova, Italy

Riccardo Filadi Department of Biomedical Sciences, University of Padova, Padova, Italy

Michele Frison Department of Comparative Biomedical Sciences, The Royal Veterinary College, University of London, London, UK

Michael Grabe Cardiovascular Research Institute, Department of Pharmaceutical Chemistry, University of California, San Francisco, CA, USA

Ruth Hartke Graduate Center and Baruch College of City University of New York, New York, NY, USA

Fabrice Homblé Structure et Fonction des Membranes Biologiques, Université Libre de Bruxelles (ULB), Brussels, Belgium

David P. Hoogerheide Center for Neutron Research, National Institute of Standards and Technology, Gaithersburg, MD, USA

Elizabeth A. Jonas Department Internal Medicine, Section of Endocrinology, Yale University School of Medicine, New Haven, CT, USA

Emma Kennedy Department of Comparative Biomedical Sciences, The Royal Veterinary College, University of London, London, UK

Hana Kmita Laboratory of Bioenergetics, Institute of Molecular Biology and Biotechnology, Faculty of Biology, Adam Mickiewicz University, Poznan, Poland

John J. Lemasters Department of Biochemistry & Molecular Biology, Medical University of South Carolina, Charleston, SC, USA

Institute of Theoretical and Experimental Biophysics, Russian Academy of Sciences, Pushchino, Russian Federation

Department of Drug Discovery & Biomedical Sciences, Medical University of South Carolina, Charleston, SC, USA

Hollings Cancer Center, Medical University of South Carolina, Charleston, SC, USA

Center for Cell Death, Injury & Regeneration, Medical University of South Carolina, Charleston, SC, USA

Victor V. Lemeshko Escuela de Física, Facultad de Ciencias, Universidad Nacional de Colombia, Medellín, Colombia

Marc Léonetti I.R.P.H.E, Aix-Marseille Université, CNRS, Technopôle de Château-Gombert, Marseille Cedex 13, France

Stephen Madamba Graduate Center and Baruch College of City University of New York, New York, NY, USA

Eduardo N. Maldonado Department of Drug Discovery & Biomedical Sciences, Medical University of South Carolina, Charleston, SC, USA

Hollings Cancer Center, Medical University of South Carolina, Charleston, SC, USA

Center for Cell Death, Injury & Regeneration, Medical University of South Carolina, Charleston, SC, USA

Anna Katherina Mallach Department of Comparative Biomedical Sciences, The Royal Veterinary College, University of London, London, UK

Stéphen Manon CNRS, Université de Bordeaux, UMR5095, Bordeaux, France

Nikita Mehta Department Internal Medicine, Section of Endocrinology, Yale University School of Medicine, New Haven, CT, USA

Oygul Mirzalieva Baruch College of City University of New York, New York, NY, USA

Nelli Mnatsakanyan Department Internal Medicine, Section of Endocrinology, Yale University School of Medicine, New Haven, CT, USA

Han-A. Park Department Internal Medicine, Section of Endocrinology, Yale University School of Medicine, New Haven, CT, USA

Evgeny V. Pavlov Department of Basic Sciences, New York University, College of Dentistry, New York, NY, USA

Pablo M. Peixoto Graduate Center and Baruch College of City University of New York, New York, NY, USA

Paola Pizzo Department of Biomedical Sciences, University of Padova, Padova, Italy

Neuroscience Institute, National Research Council (CNR), Padova, Italy

George A. Porter Jr Department of Pediatrics (Cardiology), University of Rochester Medical Center, Rochester, NY, USA

Anna Raffaello Department of Biomedical Sciences, University of Padova, Padova, Italy

Rosario Rizzuto Department of Biomedical Sciences, University of Padova, Padova, Italy

Alice Rossi Department of Biomedical Sciences, University of Padova, Padova, Italy

Tatiana K. Rostovtseva Section on Molecular Transport, Eunice Kennedy Shriver National Institute of Child Health and Human Development, National Institutes of Health, Bethesda, MD, USA

Amandine Rovini Section on Molecular Transport, Eunice Kennedy Shriver National Institute of Child Health and Human Development, National Institutes of Health, Bethesda, MD, USA

Hayet Saidani Structure et Fonction des Membranes Biologiques, Université Libre de Bruxelles (ULB), Brussels, Belgium

Maria E. Solesio Department of Basic Sciences, New York University, College of Dentistry, New York, NY, USA

Steven J. Sollott Laboratory of Cardiovascular Science, National Institute on Aging, National Institutes of Health, Baltimore, MD, USA

Ildiko Szabo Department of Biology and CNR Institute of Neurosciences, University of Padova, Padova, Italy

Pierre Theurey Department of Biomedical Sciences, University of Padova, Padova, Italy

Part I
Mitochondrial Calcium Signaling

Chapter 1
Mitochondrial Ca^{2+} Handling and Behind: The Importance of Being in Contact with Other Organelles

Riccardo Filadi, Pierre Theurey, Alice Rossi, Chiara Fedeli, and Paola Pizzo

1.1 The Mitochondrial Network as a Fundamental Structure in Cellular Ca^{2+} Homeostasis

The capacity of mitochondria to take up Ca^{2+} was first documented in the 1960s (Deluca and Engstrom 1961). Mitchell's chemiosmotic theory, proposed in the same years (Mitchell and Moyle 1967), early provided the ultimate thermodynamic basis for the entry of Ca^{2+} (i.e., a positively charged ion) into the mitochondrial matrix, thanks to the generation, by the respiratory chain, of an electrochemical gradient ($\Delta\psi$) across the inner mitochondrial membrane (IMM), negative on the side of the matrix (−180 mV). However, the molecular mechanisms and the significance of this accumulation have started to be elucidated much more recently. Since the pivotal studies on isolated mitochondria, it was clear that the outer mitochondrial membrane (OMM) is largely permeable to ions and small solutes (at least in part due to the presence of different isoforms of the voltage-dependent anion channels, VDACs; see Chaps. 7, 8, 9, 10, and 11), and thus it is not a limiting step for mitochondrial Ca^{2+} accumulation. On the contrary, the inner mitochondrial membrane (IMM) is highly impermeable to ions and requires specialized transport mechanisms (Bragadin et al. 1979). The molecular identity of the underlying proteins has been only recently revealed (Baughman et al. 2011) (see also Chap. 2), but their activity was functionally characterized in the 1970s. The mitochondrial high-capacity mechanism to take up Ca^{2+} was called "mitochondrial Ca^{2+} uniporter"

R. Filadi • P. Theurey • A. Rossi • C. Fedeli
Department of Biomedical Sciences, University of Padova, Padova, Italy

P. Pizzo (✉)
Department of Biomedical Sciences, University of Padova, Padova, Italy

Neuroscience Institute, National Research Council (CNR), Padova, Italy
e-mail: paola.pizzo@unipd.it

T.K. Rostovtseva (ed.), *Molecular Basis for Mitochondrial Signaling*,
Biological and Medical Physics, Biomedical Engineering,
DOI 10.1007/978-3-319-55539-3_1

(MCU) and demonstrated to exhibit dependence on $\Delta\psi$, sensitivity to the inhibitor ruthenium red, low affinity for Ca^{2+} ($K_d \sim 20\ \mu M$), and Ca^{2+} cooperativity (Carafoli 2003). In addition, the existence of antiporters that export Ca^{2+} from the matrix in exchange for Na^+ (especially in excitable tissues, such as the brain and heart) or H^+ (in non-excitable tissues, such as liver) further controls and limits mitochondrial Ca^{2+} accumulation (Nicholls and Crompton 1980).

The initial enthusiasm for mitochondria in the cellular Ca^{2+} scenario temporarily vanished in the 1980s. Indeed, the development of fluorescent indicators and the discovery that inositol 1,4,5-triphosphate (IP3, generated upon stimulation of plasma-membrane (PM) receptors coupled to phospholipase C (PLC)) induces the release of Ca^{2+} into the cytosol from a "non-mitochondrial intracellular store" (Streb et al. 1983) clearly demonstrated that physiological cytosolic $[Ca^{2+}]$ oscillations (ranging from 50 to 100 nM in basal conditions, to peaks of 1–3 μM) are not compatible with the activation of MCU, whose K_d is around 20 μM (Patron et al. 2013). Thus, the initial idea of mitochondria as key organelles in the modulation of intracellular Ca^{2+} homeostasis was overcome by the general consensus that, only under pathological conditions with a massive cytosolic Ca^{2+} overload, an appreciable mitochondrial Ca^{2+} uptake would occur.

The introduction of genetically encoded Ca^{2+} probes targeted to the mitochondrial matrix (firstly aequorin (Rizzuto et al. 1992) and then GFP-based fluorescent probes (Miyawaki et al. 1997)) showed, however, that, upon cell stimulation, mitochondria promptly take up Ca^{2+} in a fashion that largely exceeds what is expected on the basis of the K_d of MCU for Ca^{2+}, thus actively taking part in the regulation of the intracellular Ca^{2+} dynamics. The contradiction between these data was only apparent and was solved by the demonstration that mitochondria are strategically positioned close to the regions of Ca^{2+} release from the intracellular stores (mainly the endoplasmic reticulum, ER (Rizzuto et al. 1998)), or Ca^{2+} entry from the PM. Indeed, in these areas, the opening of specific Ca^{2+} channels generates, close to their mouths, transient microdomains of high $[Ca^{2+}]$ that are experienced by nearby mitochondria, allowing the overcoming of the low affinity of MCU and resulting in a rapid mitochondrial Ca^{2+} accumulation that follows the cytosolic Ca^{2+} rise. Below, a brief summary on Ca^{2+} microdomain generation is provided.

1.1.1 Molecular Determinants of Ca^{2+} Microdomains

A microdomain can be defined as a localized region, within a cell, that differs in composition from the surrounding areas. Usually, the term refers to the presence, for specific molecules, of appreciable concentration gradients in a given environment, which can be relatively long lasting or, on the contrary, promptly drop out within a few ms. The importance of the existence of microdomains is immediately clear when referred to those of Ca^{2+}. Indeed, the versatility of Ca^{2+} as a key intracellular second messenger, regulating different physiological processes, is achieved by combining the possibility to transmit Ca^{2+} signals as temporally distinct oscillations and by confining these events within spatially defined regions (not necessarily membrane enclosed), allowing a fine and localized regulation of specific activities (Berridge et al. 2000). The generation of a Ca^{2+} microdomain typically begins with the opening of few Ca^{2+} channels located in the membrane of intracellular Ca^{2+} stores (endoplasmic/sarcoplasmic reticulum or Golgi apparatus) or in the PM. In the simpler model, immediately after their opening, Ca^{2+} flows from the mouth of the channels and spreads into the cytosol (where its concentration is lower), following Fick's diffusion laws (for a recent review, see Filadi and Pozzan (2015)). Accordingly, its flux in a given point is inversely proportional to the distance from the channel. However, the presence of different Ca^{2+} buffers (typically proteins, such as parvalbumin/calbindin or the Ca^{2+}-modulated calmodulin/calcineurin, but also other molecules, such as ATP and negatively charged phospholipids), the high viscosity of the cytosol, the existence of organelles endowed with pumps/exchangers/channels that can promptly remove or further release Ca^{2+}, and the possibility to tune the frequency and the time of channel opening are all players that actively shape the microdomain (Filadi and Pozzan 2015).

Elegant mathematical models (Naraghi and Neher 1997) and, more recently, Ca^{2+}-imaging experiments with sufficient temporal and spatial resolutions (Tadross et al. 2013) allowed to define the shape of a Ca^{2+} microdomain. Upon opening of a channel, within a few hundreds of nm, a Ca^{2+} microdomain is formed and reaches the steady state in less than 1 ms. When the channel closes, the microdomain vanishes immediately. A Ca^{2+} gradient was calculated to extend up to 50–70 nm from the mouth of a channel (Naraghi and Neher 1997), but clearly the amplitude of a Ca^{2+} microdomain in the cellular context depends on a multitude of parameters. Among them, particularly important are the flux of Ca^{2+} from a given channel determined by its current (the higher the flux, the higher the $[Ca^{2+}]$ reached in the microdomain), the concentration/affinity/diffusion constant of the buffers that surround the channel (the higher these parameters are, the higher the capacity of a given buffer to damp the microdomain), the presence of isolated or clustered channels (the more channels opening, the higher the flux of Ca^{2+}), and the abundance of the Ca^{2+} reservoir (virtually infinite in the case of the extracellular milieu, depletable in the case of intracellular Ca^{2+} stores).

The generation of Ca^{2+} microdomains on the surface of mitochondria has variable and important consequences for the cell fate, from activation of mitochondrial metabolism to control of cell death, from regulating autophagy to sustain tumor growth (see below).

1.1.2 *Generation of Ca^{2+} Microdomains on the Mitochondria Surface: The Importance of Being Coupled with Other Cell Compartments*

As discussed above (and see also Chap. 2), the process of mitochondrial Ca^{2+} uptake largely takes advantage from the exposure of mitochondria to high $[Ca^{2+}]$ microdomains on their surface, thanks to the location of these organelles (or, more precisely, of part of them) near the mouths of Ca^{2+} channels. On this aspect, the close proximity of mitochondria to the ER and, to a lesser extent, to PM, has been extensively investigated (Table 1.1 and see also next section).

As to the former, the first evidence that, upon an IP3-dependent Ca^{2+} release from the ER, mitochondria experienced on their surface an averaged Ca^{2+} concentration higher than that of the bulk cytosol was obtained thanks to the targeting of the photoprotein aequorin to the mitochondrial inter-membrane space (MIMS) (Rizzuto et al. 1998). Regions of close proximity between mitochondria and ER were observed in living cells, supporting the idea that mitochondrial Ca^{2+} uptake may be favored by their juxtaposition to the sites of Ca^{2+} release from the ER (Rizzuto et al. 1998). A multitude of evidence in support of this notion were provided, for instance, an increased heterogeneity in the mitochondrial Ca^{2+} peaks upon a dynamin-related protein 1 (Drp1)-overexpression-dependent mitochondria fragmentation (Szabadkai et al. 2004). These results, obtained in single-cell Ca^{2+} imaging experiments employing mitochondrial GFP-based Ca^{2+} probes, further suggested a physiological origin of the mitochondrial Ca^{2+} rises from localized regions that then spread along the mitochondrial network. Finally, using FRET-based Ca^{2+} probes spanning on the cytosolic side of the OMM, it was directly demonstrated that, during an IP3-dependent ER Ca^{2+} release, microdomains of high $[Ca^{2+}]$ (10–30 μM, compared to peaks of up to 3 μM in the bulk cytosol) are promptly generated on discrete sites on the OMM (Csordas et al. 2010; Giacomello et al. 2010) (Fig. 1.1).

Regarding the distance between ER and mitochondria for an efficient Ca^{2+} transfer, it has been demonstrated, by the use of artificial linkers, that a decrease in the distance, or an increase in the surface of apposition, between the two organelles, correlates with an increase in the efficiency of Ca^{2+} transfer (Csordas et al. 2006, 2010). However, below 7 nm in the distance between the two opposing membranes, the process resulted to be substantially impaired, likely because the IP3Rs span for ~10 nm from the ER membrane and thus cannot be accommodated when the width of the cleft is lower. However, the physiological thickness of ER-mitochondria appositions observed by EM (10–15 up to 25–30 nm in the case of smooth ER and 25–30 up to 50–80 nm in the case of rough ER) seems to be largely compatible with the process. Moreover, it must be stressed that, as discussed above, a Ca^{2+} microdomain is predicted to extend from the mouth of a channel for at least 50 nm (Naraghi and Neher 1997). In addition, given that IP3Rs form clusters (Foskett et al. 2007) and that they can open together, potentially engaging also the opening of Ryanodine receptors (RyRs) in the process of the Ca^{2+}-induced Ca^{2+} release (CICR), further increasing the amount of Ca^{2+} locally released from the ER, it appears reasonable that, under certain conditions, microdomains of high $[Ca^{2+}]$ can reach an extension of up to 100–150 nm. Thus, even mitochondria located at distances of ~100 nm may be exposed to $[Ca^{2+}]$ sufficiently high to induce an efficient uptake of Ca^{2+}. Importantly, regions in which ER membranes (especially those of the rough ER) are juxtaposed and follow the profile of the OMM at distances of 50–100 nm have been reported (Filadi et al. 2015; Giacomello and Pellegrini 2016); whether they are physiologically relevant is not clear. Recently, a mathematical model proposed a distance between IP3Rs and MCU of ~30–85 nm for an

Table 1.1 Techniques for investigating physical/functional interactions between mitochondria and other organelles

	Technique		Principle	Advantages	Disadvantages	References
Imaging	Electron microscopy		Observation and calculation of membrane proximity between mitochondria and partners	High resolution Quantification (fraction of membranes in proximity) 3D observation of organelles' interface when coupled with tomography	Fixation and treatment of samples: no live imaging; delicate in certain tissues Not trivial and time-consuming analysis	Mito apposition to ER (Cosson et al. 2012), peroxisomes (Hicks and Fahimi 1977), melanosomes (Daniele et al. 2014), PM (Perkins et al. 2010), and 3D tomography (Perkins et al. 2010; Friedman et al. 2011)
	Fluorescence microscopy	Fluorescent probes	Observation and quantification of co-localization of mitochondrial and partner fluorescent signal	Live imaging	Expression of recombinant proteins requires the use of viruses in primary cells Can lead to artefactual results if the morphology of organelles is modified	Rizzuto et al. (1998), Csordas et al. (2010), Zampese et al. (2011), Arruda et al. (2014), and Filadi et al. (2015)
		IF		Usable in cell lines, primary cells, tissue	Fixation of the samples: no live imaging Relying on specificity and distribution of the chosen antibodies Relying on the behavior of the chosen interacting proteins (recruitment at the interface, etc.)	Hedskog et al. (2013) and Tubbs et al. (2014)
	In situ PLA		Observation of physical interactions between two proteins or two organelles (mitochondria and partners)	Highly sensitive: DNA amplification-mediated detection Specific: dual recognition with two antibodies Quantitative: number of fluorescent dots		Hedskog et al. (2013) and Tubbs et al. (2014)

(continued)

Table 1.1 (continued)

	Technique		Principle	Advantages	Disadvantages	References
Biochemistry	Co-immunoprecipitation/ co-purification			Proof of real physical interaction of the partners	Poorly quantitative In part unspecific	de Brito and Scorrano (2008), Filadi et al. (2015), and Theurey et al. (2016)
	Purification of interface fractions		Subcellular fractionation/ ultracentrifugation on a gradient to purify the interaction interface, proteins quantification to evaluate purification yields	Estimation of interaction at the scale of the entire organ or tissue Purification of the fraction for subsequent analysis (proteomic, lipidomic, etc.)	No pure fraction, only enriched Poorly quantitative Experimental conditions can affect purification, rigorous normalization necessary Real quantification of fraction only for MAM	MAM (Vance 1990; Wieckowski et al. 2009; Filadi et al. 2016), Mito-peroxisome (Islinger et al. 2006), and Mito-PM (Suski et al. 2014)
Ca^{2+} transfer	Aequorin	Matrix aequorin	Measurement of mitochondrial Ca^{2+} uptake	Cell population assay; robust technique; rapid and simple evaluation of the averaged Ca^{2+} concentrations	Highly efficient transfection; Probe reconstitution with coelenterazine; -Only for relatively short experiments;	wt aequorin(Rizzuto et al. 1992) and mutated aequorin (Rapizzi et al. 2002; Zampese et al. 2011)
		IMS aequorin	Indirect measurement of mitochondria exposure to Ca^{2+} microdomains			Rizzuto et al. (1998)

	Fluorescence microscopy	Genetically encoded Ca^{2+} probes (i.e., cameleons, Pericam)	Direct measurement of mitochondrial Ca^{2+} uptake or Ca^{2+} microdomains on mitochondrial surface	Single cell measurements of both mitochondrial and cytosolic Ca^{2+} rises at the same time (by using a nuclear probe as a surrogate of the cytosolic one); pixel by pixel analysis allows the spatial/temporal definition of Ca^{2+} microdomains	Transfection of the probes is necessary; intrinsic variability of single cell data; calibrations of the probes are not always available; time-consuming data analysis	Csordas et al. (2010), Giacomello et al. (2010), and Zampese et al. (2011)
		Chemical dyes	Measurement of mitochondrial Ca^{2+} uptake	Measurements of mitochondrial Ca^{2+} uptake in single cells; possibility to measure both cytosolic/mitochondrial Ca^{2+} rises by using two dyes (Rhod-2 and Fluo-3); transfection is not necessary	High Ca^{2+} buffering capacity of the chemical dyes; they potentially alter the dynamics of Ca^{2+} diffusion; Rhod-2 is not suitable for repetitive Ca^{2+} oscillations, has a very high affinity for Ca^{2+} ($K_d \sim 1\ \mu M$), and may be easily saturated into mitochondrial matrix	Csordas et al. (2010) and Fonteriz et al. (2010)
Lipid transfer	Lipid synthesis assays		Measurements of specific lipid synthesis/modifications in different sub-cellular compartments	Possibility to follow step-by-step the process of lipid synthesis/modification in different sub-cellular fractions (i.e., MAMs); PE synthesis into mitochondria upon PS transfer from ER, as well as cholesteryl esters synthesis in MAMs, are indirect measurements of inter-organellar coupling	Most of the assays are based on radioactive molecules; the process is affected by the intrinsic activity/abundance of the underlying enzymes; many steps/players of the process are currently unknown	Vance (1990, 2014) and Area-Gomez et al. (2012)

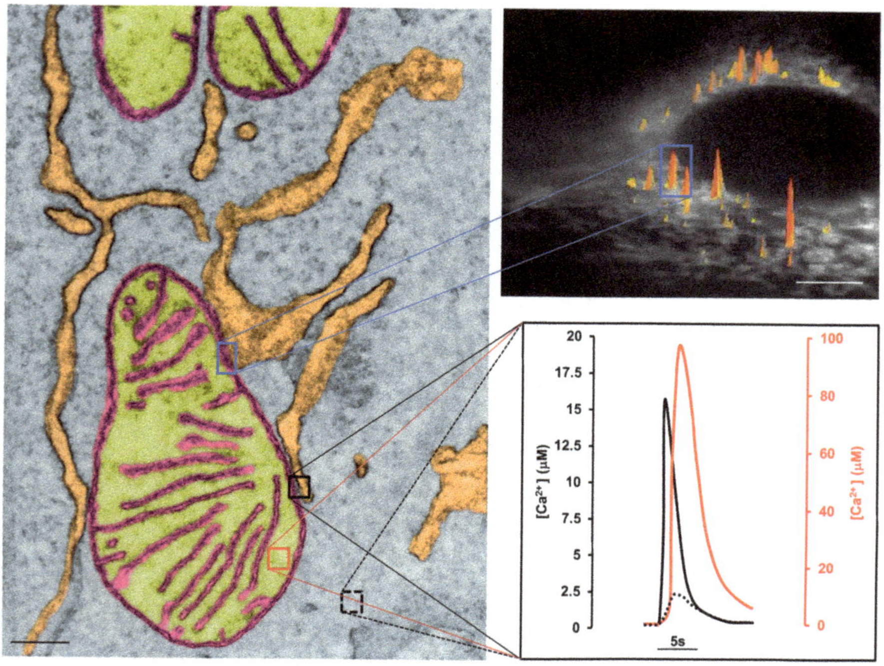

Fig. 1.1 Mitochondria-ER tethering is fundamental for their Ca^{2+} crosstalk. *Left* panel: EM image of a MEF cell showing several close appositions between ER (*orange*) and OMM (*purple*) (The EM image was acquired by G. Turacchio, Institute of Protein Biochemistry (CNR), Naples). Scale bar: 100 nm. The juxtaposition between the two opposing membranes allows the generations of high $[Ca^{2+}]$ on OMM upon Ca^{2+} release from ER Ca^{2+} channels. *Upper right* panel: an SH-SY5Y cell transiently expressing the OMM-targeted cameleon Ca^{2+} probe N33-D1cpv. *Yellow*-to-*red* spikes indicate the generation of high $[Ca^{2+}]$ microdomains in discrete regions of OMM upon bradykinin cell stimulation (100 nM) and IP_3-mediated ER Ca^{2+} release. Scale bar: 5 μm (Image modified from Filadi et al. 2012). *Lower right* panel: simulations of $[Ca^{2+}]$ changes over time upon IP3-mediated ER Ca^{2+} release in the bulk cytoplasm (*black, dotted trace*), at regions of ER-mitochondria interface (*black, continuous trace*) and in mitochondrial matrix (*red trace*). Note the different scales and that mitochondrial Ca^{2+} uptake is slightly delayed compared to the almost immediate generation of Ca^{2+} hot spots on mitochondrial surface

optimal ER-mitochondria Ca^{2+} transfer (Qi et al. 2015). Considering that the distance between the OMM and the IMM (where MCU is located) is at least of 10–20 nm (Reichert and Neupert 2002), the optimal gap between the ER and the OMM is predicted to be in the range of 10–60 nm. Surely, the fact that the thickness of the cleft is dynamically regulated may suggest that different types of contact could have different functions or represent dormant states of contact, ready to be recruited and reactivated upon a change in the metabolic state of the cells (Sood et al. 2014; Giacomello and Pellegrini 2016).

In addition to IP3Rs, the ER of different cell types and the sarcoplasmic reticulum (SR) of the striated muscles are endowed with RyRs. Though in principle the situation is similar to that of IP3Rs (i.e., a Ca^{2+} microdomain near the mouth of

RyRs can be formed upon their opening and be experienced by juxtaposed mitochondria), the peculiar organization and ordered distribution of these Ca^{2+} channels in striated muscles make the case of SR-mitochondria coupling unique. Indeed, in this tissue, the situation is peculiar; while regions of close contact (mediated by proteinaceous tethers of 10–15 nm (Franzini-Armstrong 2007; Boncompagni et al. 2009)) between SR and mitochondria have been clearly documented, RyRs are specifically clustered in the dyadic cleft, i.e., the portion of SR (junctional SR) that faces the T-tubules (for a recent review, see Filadi and Pozzan (2015)). Importantly, the space between T-tubules and SR is too narrow (~10 nm) to allow the accommodation of mitochondria, which have been reported to be at a distance of ~130 nm in skeletal muscle (Boncompagni et al. 2009) and ~35 nm in cardiomyocytes (Sharma et al. 2000), where the SR cisternae are flatter. However, as discussed above, the presence of RyR clusters and the fact that the two opposing membranes in the dyadic cleft represent a barrier (that only allows the lateral diffusion of Ca^{2+}) are factors that shape the Ca^{2+} microdomain and laterally spread it, so that mitochondria surrounding the Ca^{2+} releasing unit (CRU) are exposed to $[Ca^{2+}]$ sufficiently high to induce an appreciable uptake, as measured in different experimental models (Pacher et al. 2000, 2002; Szalai et al. 2000; Robert et al. 2001; Rudolf et al. 2004; Drago et al. 2012). Though in adult cardiomyocytes it is still a matter of debate whether mitochondria could efficiently take up Ca^{2+} during the fast physiological Ca^{2+} transients in a beat-to-beat manner, in neonatal cardiomyocytes, such oscillations in mitochondrial Ca^{2+} have been measured (Robert et al. 2001; Pacher et al. 2002; Drago et al. 2012) and demonstrated to actively impact the amplitude of cytosolic Ca^{2+} rises (Drago et al. 2012).

Finally, as far as Ca^{2+} crosstalk is concerned, it is important to stress that the existence of regions of close apposition between mitochondria and ER/SR is physiologically relevant not only for the process of mitochondrial Ca^{2+} uptake but also for ER Ca^{2+} handling. First of all, the physical presence of mitochondria near ER membranes limits the free Ca^{2+} diffusion upon release from the ER Ca^{2+} channels, potentially sustaining the local $[Ca^{2+}]$ and favoring its reuptake by SERCA pumps. Moreover, the local high $[Ca^{2+}]$ may have opposite effects: on the one hand, affecting the process of CICR (recruiting and opening more Ca^{2+} releasing channels, RyRs and IP3Rs), and, on the other, negatively regulating the opening of IP3Rs when the local $[Ca^{2+}]$ reaches a certain threshold (Landolfi et al. 1998; Foskett et al. 2007). On the contrary, upon MCU activation and rapid mitochondrial Ca^{2+} accumulation, mitochondria can act as Ca^{2+} buffers that dampen not only bulk cytosolic Ca^{2+} peaks but also, locally, Ca^{2+} microdomains (Qi et al. 2015). However, given that mitochondria do not store Ca^{2+} (with the only exception of Ca^{2+} phosphates precipitation, favored by the alkaline pH; (Nicholls and Chalmers 2004)) and, after Ca^{2+} uptake, they release it back to the cytosol, the latter phenomenon could facilitate Ca^{2+} reuptake into the ER by the SERCA. In addition, mitochondria have been reported to locally sustain SERCA activity by fueling it with ATP (De Marchi et al. 2011), further indicating the existence of a functional, bidirectional crosstalk.

Much less investigated has been the role of mitochondria-PM contacts in the process of mitochondrial Ca^{2+} uptake. In many excitable cell types, a fast mitochondrial

Ca^{2+} rise has been measured upon voltage-operated channel (VOC)'s opening and Ca^{2+} entry from the extracellular environment, subsequent to PM depolarization (see Giacomello et al. 2010 and, for a review, Rizzuto and Pozzan 2006). Thus, regions of high [Ca^{2+}] are likely to be generated on the surface of sub-PM located mitochondria when Ca^{2+} enters through VOC's. On the contrary, more debated is the generation of such Ca^{2+} microdomains on the OMM upon capacitative Ca^{2+} entry (CCE)). Indeed, evidence that mitochondria are exposed to CCE-generated Ca^{2+} microdomains has been provided (Quintana et al. 2006; Watson and Parekh 2012), and mitochondrial membrane potential, as well as functional mitochondrial Na^+/Ca^{2+} exchanger, has been shown to be necessary to sustain CCE, likely because of mitochondria capacity to buffer sub-PM local Ca^{2+} increases and maintain CCE activation (Naghdi et al. 2010). However, mitochondrial depolarization has also been described to negatively impact on CCE by a different Mfn2-dependent mechanism, impairing migration of the ER Ca^{2+} sensor stromal interaction molecule 1 (STIM-1) protein to form PM-located ER *punctae*, critical for CCE induction (Singaravelu et al. 2011). Notably, during CEE, the close apposition between ER and PM is narrow enough to exclude mitochondria from the mouth of PM-located Orai1 channels (calcium release-activated calcium channel protein 1), thus arguing against the possibility that mitochondria could be exposed to high [Ca^{2+}] during CCE. Moreover, data recently obtained in our lab (Giacomello et al. 2010; Filadi et al. unpublished results) did not support an appreciable contribution of CCE to mitochondrial Ca^{2+} uptake. Probably, these contrasting results depend on the specific cell type employed in different studies and, consequently, on the amplitude of the Ca^{2+} influx and on the presence, or not, of mitochondria in the regions surrounding the ER-PM platform generated during CCE. Further investigations will be required to address this issue. Interestingly, however, in HeLa cells, the presence of sub-PM mitochondria has been suggested to indirectly modulate the activity of both PM Ca^{2+} ATPases (PMCA) and store-operated Ca^{2+} channels (Frieden et al. 2005). The phenomenon was proposed to be due, again, to a possible Ca^{2+} buffering effect exerted by sub-PM mitochondria on the entry of the ion through the PM located channels (Frieden et al. 2005). Thus, it appears clear how the interrelationship between mitochondria and PM, as for the ER, may be bidirectional.

1.2 Functional Consequences of Mitochondria-ER Ca^{2+} Cross-Talk

Ca^{2+} rises on the cytosolic surface of both the ER and mitochondria can have important regulatory roles. For example, [Ca^{2+}] in the mitochondrial inter-membrane space modulates the activity of metabolite carriers or dehydrogenases located on the outer surface of the IMM (see below); furthermore, as mentioned above, local mitochondrial Ca^{2+} sequestration has profound effects on the allosteric modulation of ER Ca^{2+}-releasing channels. Regarding the [Ca^{2+}] within mitochondria, two possible and opposite effects can be achieved: activation of mitochondrial metabolism by stimulating the Krebs cycle, NADH formation, and the respiratory chain activity,

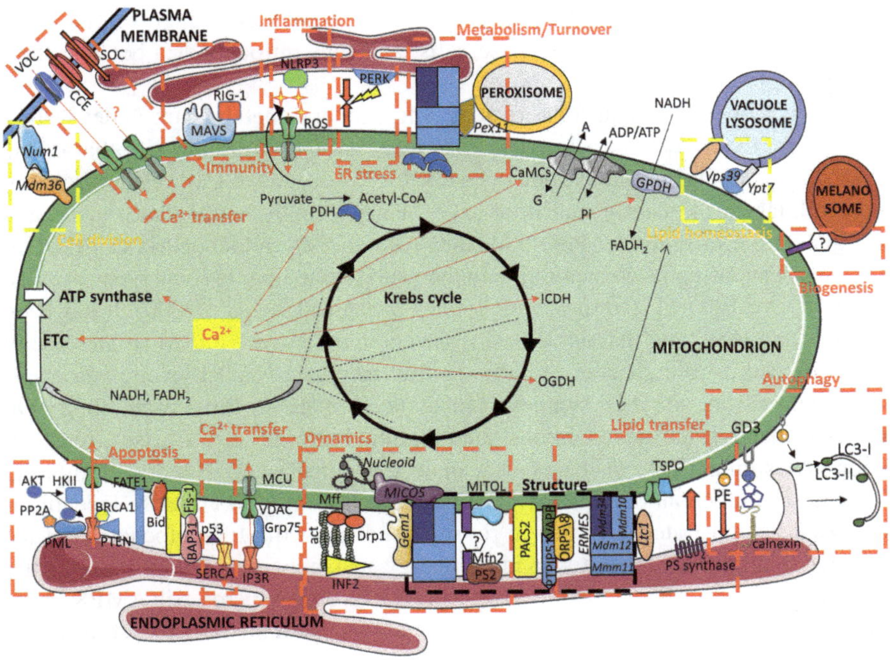

Fig. 1.2 Mitochondria interact with different organelles/compartments within the cell. The cartoon represents multiple mitochondria interactions with other organelles, as detailed in the text. *Red-dotted squares* underline specific functions sustained by the indicated inter-organelle interaction, with the representation of the main proteins involved; *yellow-dotted squares* are those only identified in yeast; in the *black-dotted square* are represented proteins that have been implicated in the structure of the inter-organelle interface. See text for all the abbreviations used and the description of the specific functions. In italic, yeast proteins

with the result of an increase in ATP production, or sensitization toward cell death pathways, due to mitochondrial Ca^{2+} overload, opening of the mitochondrial permeability transition pore (mPTP), and release of cytochrome C. Moreover, an important role for ER-mitochondria Ca^{2+} transfer in modulating autophagy/mitophagy and tumor progression has been recently shown. While a more detailed discussion on the role of ER-mitochondria Ca^{2+} transfer in regulating cell death is provided elsewhere (see Chap. 6 and Orrenius et al. 2015 for a recent review), below a brief summary on activation of mitochondrial metabolism, autophagy, and cancer growth modulation by ER-mitochondria Ca^{2+} crosstalk is provided (Fig. 1.2).

1.2.1 Mitochondrial Metabolism

Bioenergetics homeostasis and mitochondrial functionality are two essential features that provide the correct energy to the cell accordingly to its requirements; in particular, in excitable cells, such as neurons, cardiac, and skeletal muscle cells, ATP produced through the glycolysis process is not enough to guarantee the multiplicity of activities of these cells, and thus mitochondria become their energy suppliers.

Ca^{2+} is a specific signal for activating mitochondria metabolism: under resting conditions, the mitochondrial matrix Ca^{2+} level is in equilibrium with that of the cytosol, i.e., around 100 nM. Upon cytosolic Ca^{2+} rises, due to Ca^{2+} release from the intracellular stores, mainly the ER, or to Ca^{2+} entry from the extracellular space, and the generation of high Ca^{2+} microdomains (see above), mitochondria can take up Ca^{2+} thus increasing its concentration in the matrix, reaching, in some types of cells, values above 100 μM (Rizzuto et al. 1992; Pozzan and Rizzuto 2000). It has been known for a long time that a rise in Ca^{2+} levels in the matrix, due to an increase in the workload or to a specific cell stimulation, increases NADH levels, mitochondrial metabolism, and ATP synthesis (Jouaville et al. 1999; Pitter et al. 2002). On the other hand, it has been demonstrated that the buffering of Ca^{2+} rises, obtained upon cell stimulations, in the cytosol or in the mitochondrial matrix, limits the increase in mitochondrial metabolism (Wiederkehr et al. 2011). More recently, it has been demonstrated that also at resting conditions, the ER-mitochondria Ca^{2+} transfer is important for mitochondrial energy metabolism: the inhibition, or the absence, of the transfer causes a decrease in ATP levels, an increase in AMPK phosphorylation, and, eventually, a strong autophagy activation (Cardenas et al. 2010) (see below).

Mitochondria produce ATP for the cell through two main processes, the tricarboxylic acid (TCA) or Krebs cycle and the respiration by the electron transport chain (ETC). These two processes need pyruvate, the product of glycolysis, to move in the mitochondrial matrix where it is converted in acetyl-CoA, which enters in the TCA cycle. From each molecule of acetyl-CoA, mitochondria produce three NADH and one $FADH_2$. These are two high-energy molecules that are essential for the ETC activity. Since 1970, it is known that four mitochondrial dehydrogenases are regulated by Ca^{2+}. One of these, the FAD-glycerol phosphate dehydrogenase (GPDH), is located in the IMM and senses the cation concentration in the intermembrane space, while the other three enzymes catalyze reactions of the TCA cycle (or immediately upstream) and sense mitochondrial matrix [Ca^{2+}]: pyruvate dehydrogenase (PDH), isocitrate dehydrogenase (ICDH), and oxoglutarate dehydrogenase (OGDH) (Denton 2009). GPDH transfers, through a redox reaction, reducing equivalents from NADH, produced by glycolysis, to the ETC as $FADH_2$; it presents a Ca^{2+}-binding motif that lies in the intermembrane space and has a K_d for Ca^{2+} of 0.1 μM; moreover, its activation increases cellular ATP levels (Garrib and McMurray 1986). PDH plays a crucial role in mitochondrial metabolism since it converts pyruvate into acetyl-CoA, allowing the TCA cycle activation. Its activity is regulated by a kinase/phosphatase cycle: when phosphorylated it is inactive, while a dephosphorylation event activates it. Ca^{2+} binds ($K_d \sim 1$ μM) both the dehydrogenase itself and its phosphatase, increasing the active form of the enzyme (Turkan et al. 2004). On the other hand, ICDH and OGDH catalyze two reactions of the TCA cycle: they are both regulated by Ca^{2+} through the direct binding of the cation; the latter event induces an increase in the affinity of the enzymes for their substrates (K_d of 20–30 μM and 1 μM, respectively; (Denton et al. 1978)).

More recently, in isolated mitochondria, it has been reported that Ca^{2+} can regulate directly the ETC and the F1F0 ATP synthase activity, indicating that mitochondrial matrix Ca^{2+} concentration is a key factor not only for TCA cycle enzymes but it can directly stimulate, independently from dehydrogenase functionalities, ATP production, and the activity of the respiratory chain complexes (Territo et al. 2000; Glancy et al. 2013).

As mentioned above, Ca^{2+} can regulate mitochondrial functions without reaching their matrix, modulating the shuttle of nucleotides, metabolites, and cofactors inside the organelles. Specific mitochondrial carriers (MCs), localized in the IMM, exchange nucleotides, substrates, and metabolites between cytosol and mitochondria. Among them, there are two Ca^{2+}-binding MCs (CaMCs): the L-CaMCs (long Ca^{2+}-dependent MCs) and the S-CaMCs (short Ca^{2+}-dependent MCs). Both molecules sense Ca^{2+} in the intermembrane space by the presence of EF-hand Ca^{2+}-binding domains localized in their N-terminal fragments facing the intermembrane space (del Arco and Satrustegui 2004; Satrustegui et al. 2007).

L-CaMCs are the aspartate/glutamate carriers (AGC), and they belong to the malate-aspartate shuttle (MAS). They catalyze the exchange of a glutamate and an H^+ (from the cytosol) for an aspartate (from mitochondria) and, being part of the MAS, the entry into mitochondria of a NADH molecule that contributes to mitochondrial metabolism (Palmieri et al. 2001). The role of AGC in substrates transport has been known for several years but only recently, thank to structural studies, it has been shown its Ca^{2+} dependence by the presence of an EF-hand domain in its N-terminal fragment. Moreover, AGC is activated by low [Ca^{2+}] (its K_d is around 300 nM), well below the concentration needed for MCU activation. Thus, also low cytosolic Ca^{2+} rises can induce the entrance into mitochondria of NADH and metabolites, stimulating ETC activity and ATP production (Pardo et al. 2006; Thangaratnarajah et al. 2014).

S-CaMCs, instead, are ATP-Mg/Pi carriers that catalyze the exchange of ATP and ADP for one phosphate across the IMM, modulating the levels of adenine nucleotides (AdN: AMP+ADP+ATP) inside mitochondria; due to this function, these carriers can modulate several cell functionalities, such as mitochondrial metabolism and oxidative phosphorylation. A defect in these carriers, or their absence, in different tissues, can induce an impairment in energy production required for several functionalities (Anunciado-Koza et al. 2011; Amigo et al. 2013). The activity of ATP-Mg/Pi carriers is regulated by Ca^{2+}, and they also present EF-hand Ca^{2+} binding domains, homologous to calmodulin, in their N-terminal parts facing the intermembrane space; their K_d is around 1.5–3 μM, requiring a higher increase in cytosolic [Ca^{2+}], than L-CaMCs, for their activation and the regulation of AdN levels within mitochondria (Haynes et al. 1986; Nosek et al. 1990).

1.2.2 Autophagy/Mitophagy

Ca^{2+} plays a complex role in autophagy regulation and the first link between autophagy, and intracellularly stored Ca^{2+} was described in rat hepatocytes in 1993 (Gordon et al. 1993). More recently, both inhibitory and activatory effects of Ca^{2+} on autophagy induction have been reported (see La Rovere et al. (2016) for a recent review).

As far as Ca^{2+} transfer from the ER to mitochondria is concerned, it has been shown that, under basal conditions, the constitutive, low level, IP3R-mediated Ca^{2+} release and subsequent Ca^{2+} uptake by mitochondria, throughout the MCU complex, are essential for the maintenance of optimal cellular bioenergetics (see above). As a consequence of this energetic balance, autophagy is maintained at low levels in healthy cells. On the other hand, disturbances in ER-mitochondria Ca^{2+} transfer trigger autophagy by increasing the AMP/ATP ratio (due to a decrease in mitochondria bioenergetics) and activating AMPK, a highly sensitive indicator of cellular energy status whose activity increases under conditions of metabolic stress (Cardenas et al. 2010). The importance of the ER-mitochondria Ca^{2+} crosstalk in autophagy induction was also confirmed by molecular (upon MCU overexpression) or pharmacological (upon treatments with drugs that increase the uptake of Ca^{2+} by mitochondria) approaches finalized to correct dysfunctional mitochondrial Ca^{2+} uptake in human fibroblasts from mitochondrial disorder patients, resulting, in these cells, in a rescue of autophagy levels similar to those observed in control cells (Granatiero et al. 2016). Consistently, stable knockdown of MCU and MCUR1, key components of the mitochondrial Ca^{2+} uptake machinery (Murgia and Rizzuto 2015) (and see also Chap. 2), reduced cellular oxygen consumption rate, activated AMPK, and induced autophagy (Mallilankaraman et al. 2012).

Finally, other alterations in mitochondrial Ca^{2+} signaling can lead to mitophagy. PINK-1, a protein involved in mitophagy initiation, regulates Ca^{2+} efflux from mitochondria through the Na^+/Ca^{2+} exchanger, and its loss may lead to mitochondrial Ca^{2+} accumulation (Gandhi et al. 2009), leading to mPTP opening and release of pro-apoptotic factors. In addition, elevation of local Ca^{2+} concentrations in the vicinity of mitochondria, a phenomenon very much depending on ER-mitochondria tethering (see above), with impaired membrane potential, promotes the activation of the mitochondrial shaping molecule Drp1, triggering mitochondrial fission and subsequent mitophagy (Sandebring et al. 2009).

More recently, it has been demonstrated that the overexpression of Parkin, the cytosolic E3 ubiquitin ligase involved in mitophagy, favors, in normal conditions, ER-mitochondria tethering and their Ca^{2+} crosstalk, sustaining mitochondria morphology and ATP production and consequently limiting ubiquitination of mitochondrial targets and mitophagy (Cali et al. 2013a). Upon mitochondrial damage (such as organelle depolarization), the protective effect of Parkin is however insufficient, and, after its recruitment to mitochondria, the mitophagic process is activated.

1.2.3 Cancer

Among multiple mechanisms, cancer growth relies mainly on dysregulation on cell proliferation and cell death, and Ca^{2+} and mitochondria are crucial actors in modulating both events. In particular, while the correct mitochondrial Ca^{2+}uptake sustains the increased energy demand of proliferating cells, mitochondrial Ca^{2+} overload induces organelles morphology alterations, permeabilization, and swelling, with the subsequent release of pro-apoptotic factors. Both Ca^{2+} -dependent pathways are very much linked to mitochondria-ER coupling.

Indeed, enhanced resistance to apoptosis of cancer cells involves also dysregulation of the ER-mitochondria Ca^{2+} axis and several tumor suppressor and oncogenic proteins, mutated or deleted in various types of human cancers, act at the interface between the two organelles and modify their Ca^{2+} crosstalk. The oncogene AKT, for example, has been reported to phosphorylate two key targets, hexokinase 2 (HK2), promoting its binding with the mitochondrial channel VDAC1 (Majewski et al. 2004), and the ER Ca^{2+}-releasing channel IP3R3 (Marchi et al. 2012), which in turn negatively affects the Ca^{2+}-dependent apoptotic response. On the contrary, the tumor suppressor protein phosphatase and tensin homolog (PTEN), commonly lost or mutated in human cancers, have been shown to directly interact with IP3R, counteracting, by its enzymatic activity, the reduced IP3R-dependent Ca^{2+} release mediated by AKT phosphorylation (Bononi et al. 2013). The tumor suppressor PML has been shown to act on this ER-mitochondria Ca^{2+} pathway as well: it promotes the formation of a multiprotein complex containing IP3R3, AKT, and the protein phosphatase PP2a, which regulates ER-mitochondria Ca^{2+} transfer (Giorgi et al. 2010). Finally, other two examples that further indicate the ER-mitochondria Ca^{2+} crosstalk as a critical hub in cancer growth are the tumor suppressor proteins BRCA1 and p53: the first was found to be recruited to the ER during apoptosis in an IP3R-dependent manner, sensitizing the IP3R to its ligand, thus favoring the Ca^{2+}-dependent apoptotic response (Hedgepeth et al. 2015); the second, in an extra-nuclear fraction, has been shown to interact with the ER Ca^{2+}-ATPase SERCA, modulating ER Ca^{2+} content and mitochondrial Ca^{2+} uptake, organelle swelling, and apoptosis induction (Giorgi et al. 2015). Of note, recently, the cancer-testis antigen FATE1 has been reported to uncouple the two organelles and to protect from Ca^{2+}- and drug-induced cell death (Doghman-Bouguerra et al. 2016), highlighting the importance of the modulation of ER-mitochondria interface in cancer cells.

1.3 Molecular Components Involved in Mitochondria-Organelles Contacts

The first electron microscopy (EM) studies of cellular ultrastructure reporting that mitochondria are in physical contact with other organelles, in particular with the ER, dated back to the 1950s (Robertson 1960). In these identified regions of juxtaposition, the two opposing organelles do not fuse but maintain their identity (Fig. 1.1 and Table 1.1).

Contacts with the ER are conserved between different species (Rowland and Voeltz 2012) and have been well documented in yeast, mammals, and, recently, also in plants (Mueller and Reski 2015). Different types of contact have been reported, depending on the fact that mitochondria can be engaged in juxtaposition with both smooth ER and rough ER and that distances between ER membranes and OMM can be extremely variable, ranging from ~10 nm up to 80–100 nm (for a recent review, see (Giacomello and Pellegrini 2016)). However, while in the case of the classical close contacts (below 25–30 nm), OMM and ER membranes appear to be clearly tethered by electron-dense filamentous structures, proteinaceous in their nature (Csordas et al. 2006); such structures, to the best of our knowledge, have never been observed for the more distant areas of apposition. A complete list of the proteins that has been involved in the modulation of ER-mitochondria coupling is far away from the scope of the present chapter. However, below, a brief summary of both yeast and mammalian tethers is provided (Fig. 1.2).

In yeast, the ER-mitochondria encounter structure (ERMES), formed by the cytosolic protein Mdm12, the ER membrane protein Mmm1, and the OMM proteins Mdm34 and Mdm10, was identified by a genetic screening (Kornmann et al. 2009). ERMES has been implicated in ER-mitochondria lipid transport (AhYoung et al. 2015), though different groups failed to find defects in lipid metabolism in cells lacking the complex (reviewed in Murley and Nunnari 2016). Importantly, ERMES homologs have not been identified in mammals.

Recently, a proteomic analysis identified the ER transmembrane protein Ltc1/Lam6 as a potential additional tether in yeast, thanks to its interaction with the OMM proteins TOM70/TOM71 (Murley et al. 2015).

ER-mitochondria contacts, in both yeast and mammals, mark sites of mitochondrial division (Friedman et al. 2011), and, in yeast, the additional ERMES-associated subunit Gem1 (a GTPase whose mammalian homolog is Miro-1) regulates ER-mitochondria connections (Kornmann et al. 2011) and ER-dependent mitochondrial division (Murley et al. 2013).

In metazoan cells, while the functional significance of the ER-mitochondria juxtaposition has been established in a number of different studies, the nature of the proteins involved in the physical tethering between the two organelles remains much less understood. Different proteins have been demonstrated to modulate this parameter, but the formal demonstration that the lack of a given protein abolish close contacts has never been provided. Notably, the possibility that different and independent tethering complexes may exist, and compensate one for the lack of the others, increases the complexity of the above analysis.

Among the proteins that have been associated with ER-mitochondria tethering, the OMM resident protein Mitofusin-2 (Mfn2) has received a lot of attention. Mfn2 was initially proposed to be a tether, given the presence of a fraction of the protein in ER membranes (particularly in the mitochondria associated membranes, MAMs) and its ability to form homotypic interactions with the OMM resident counterpart (de Brito and Scorrano 2008). The demonstration that the lack of this protein deeply reduces ER-mitochondria co-localization, visualized by confocal microscopy techniques, as well as a number of indirect evidence involving Mfn2 in organelles' coupling (Schneeberger et al. 2013), initially created a consensus on the fact that Mfn2 could be a tether. However, we (Filadi et al. 2015, 2016; Leal et al. 2016) and others (Cosson et al. 2012; Wang et al. 2015) have shown, by a number of different approaches, that Mfn2 depletion is actually associated to an increased ER-mitochondria physical and functional coupling, challenging the initial idea of Mfn2 as a tether. Interestingly, a Gp78 (an E3-ubiquitin ligase)-dependent Mfn2 degradation has been correlated with an increased association of rough ER to mitochondria, while Mitofusin-1 has been shown to negatively affect the tethering with the smooth ER, thus suggesting that the coupling with mitochondria of rough and smooth ER may be differently regulated (Wang et al. 2015).

In addition to Mfn2, others proteins have been reported to be involved in the regulation of ER-mitochondria juxtaposition. PACS-2, a cytosolic multi-sorting protein, have been shown to control their apposition in a BAP31 (an ER cargo receptor)-dependent manner (Simmen et al. 2005), though its role as a molecular scaffold or as a simple regulator is not fully clarified.

Interestingly, a physical interaction between the ER resident IP3Rs (especially the MAM-enriched IP3R3), the cytosolic fraction of the mitochondrial chaperone Grp75, and the OMM located VDAC1 has been reported and demonstrated to be functionally involved in the efficacy of mitochondrial Ca^{2+} uptake (Szabadkai et al. 2006). However, the interpretation that sees the IP3Rs-Grp75-VDAC1 complex as a potential tether between the two opposing organelles is, in our opinion, unlikely, because in DT40 cells knockout (KO) for the three IP3Rs isoforms, it has been demonstrated, by EM analysis, that ER-mitochondria physical association is not modified (Csordas et al. 2006), thus arguing against a role of the IP3Rs as tethers.

The familial Alzheimer's disease (FAD)-related protein Presenilin-2 (PS2) has been shown by us to favor the physical and functional ER-mitochondria coupling, with FAD-linked PS2 mutants more effective than the wt counterpart in the modulation of these parameters (Zampese et al. 2011; Kipanyula et al. 2012). Recently, we have demonstrated that PS2 is able to increase the number of ER-mitochondria close contacts in a Mfn2-dependent manner, by sequestering the latter protein and thus removing its negative effect on organelles tethering (Filadi et al. 2015, 2016) (see also below).

An interaction between the ER resident protein VAPB and the OMM protein PTPIP51 has also been demonstrated to positively correlate with the physical and functional ER-mitochondrial coupling (De Vos et al. 2012).

Finally, the ER-stress related protein kinase RNA-like ER kinase (PERK; see also below) has also been shown to modulate ER-mitochondria contact sites, independently from its enzymatic activity and canonical role during ER stress (Verfaillie et al. 2012; van Vliet and Agostinis 2016). As to the contacts with the PM, their functions and molecular composition have been much less investigated, and contrasting results have been obtained. For instance, in HeLa cells, it has been reported that ~10% of the PM surface co-localized with mitochondria and that sub-PM mitochondria are important for the modulation of PM Ca^{2+} ATPases (Frieden et al. 2005).

However, given the optical resolution of confocal microscopy (at best, ~0.5 μm along the *z*-axis), this value may be overestimated and include also mitochondria that are in proximity but not really associated to PM. Indeed, in another study performed in RBL-2H3 and H9c2 cells, just a few mitochondria-PM contact points were observed, and a large fraction of peri-PM mitochondria was found to lack a direct contact with the PM (Csordas et al. 2010). Often, an interleaving ER-stack was observed to be located between PM and mitochondria, hindering a direct association between them (Csordas et al. 2010). Moreover, it has been shown that, upon Drp1/Fis1 overexpression-induced mitochondrial fragmentation, the majority of mitochondria loose the co-localization with the PM, suggesting that just few tethering points between the two structures may exist (Frieden et al. 2005).

The molecular nature of the junctions between mitochondria and the PM is mysterious. Recently, by employing a proteomic approach in murine liver, it has been reported that the connexin protein Cx32, enriched at gap junctions, but retrieved also in the IMM, physically interacts with the OMM resident fraction of syderoflexin-1 (SFXN-1), thus suggesting a putative role for the Cx32-SFXN1 axis as a PM-mitochondria tether (Fowler et al. 2013). In yeast, the interaction between the PM protein Num1 (Klecker et al. 2013), the OMM-associated adaptor protein Mdm36 (Lackner et al. 2013), and still unknown additional partners has been demonstrated to be important for the association of mitochondria with the cell cortex and for the retention of part of them in the mother cell after cell division (reviewed in Klecker et al. 2014). Thus, PM-mitochondria contacts may be essential for the partitioning of mitochondria between the mother cell and the bud.

Recently, different types of contact have been also described between mitochondria and other cellular organelles, such as peroxisomes, through the binding of the peroxisomal protein Pex11 and the ERMES subunit Mdm34 in yeast (Mattiazzi Usaj et al. 2015), and melanosomes, through undefined Mfn2-dependent molecular tether structures (Daniele et al. 2014) (see also below). Moreover, tight contacts (vCLAMPs, ~10 nm the distance between the opposing membranes and ~100 nm the length of the juxtaposition) have been reported between vacuoles and mitochondria in yeast (Elbaz-Alon et al. 2014). These contacts depend on the vacuolar proteins Vps39 and the Rab GTPase Ypt7 and have a role in lipid metabolism/exchange between these organelles (see below).

1.4 Mitochondria Inter-organelle Contacts: Not Only Matter of Ca^{2+} Signaling

1.4.1 Lipid Synthesis and Trafficking

The maintenance of the lipid composition of cellular membranes involves a tight communication between organelles (Lebiedzinska et al. 2009). The more extensively studied, and probably the more important interorganellar interaction for this function, is between mitochondria and the ER. Indeed, lipid transfer is one of the most prominent feature of ER-mitochondria contact domains, even though the underlying molecular mechanisms are mostly unknown (Murley and Nunnari 2016).

The ER is the master organelle for lipid synthesis. Nevertheless, the specific domains associated to mitochondria (MAMs) have been shown to be enriched, compared to the bulk ER membranes, in enzymes involved in lipid metabolism, such as membrane-anchoring proteins, cholesterol metabolism, and triglycerides/phospholipids synthesis enzymes, including the two phosphatidylserine (PS) synthases (Stone and Vance 2000). Interestingly, certain phospholipids synthetic pathways require enzymes that are located at both the ER and the mitochondria: for example, biosynthesis of phosphatidylcholine (PC) and phosphatidylethanolamine (PE) follows a complete circuit, starting from PS synthesis at the ER side of MAM by PS synthase (Zborowski et al. 1983; Stone and Vance 2000) to its decarboxylation into PE at the IMM and back to the ER for conversion into PC (Vance 1990; Voelker 2000; Osman et al. 2011).

Beyond synthesis, there is a bidirectional and extensive lipid exchange between membranes of the two organelles, although the factors involved in lipid transport remain elusive (Rowland and Voeltz 2012). PC is the most abundant mitochondrial phospholipid, but it is synthesized only in elements of the ER. PS and PC must then be imported from the ER to the mitochondria, through yet uncharacterized mechanisms but likely involving ER membranes and OMM close juxtapositions (Vance 2014).

The importance of mitochondria-ER lipid exchanges is also illustrated by the fact that the precursor for the mitochondrial synthesis of cardiolipin, the phosphatidic acid, comes from the ER, through a yet unknown mechanism (Vance 2014). Moreover, sterols and sphingolipids are minor but potentially functionally significant constituents of mitochondrial membranes. MAMs have been reported to be enriched in cholesterol, compared to the bulk ER, and studies have suggested that depletion of cholesterol from MAMs promotes the association of the two organelles (Fujimoto et al. 2012). Thus, mitochondria-ER contacts could contain microdomains that are enriched in cholesterol and gangliosides, modified sphingolipids, similarly to the detergent-resistant lipid rafts at the PM (Hayashi and Fujimoto 2010). Interestingly, in yeast, sterols have been suggested to be directly transported from the ER to the mitochondria through Ltc1 (also known as Lam6), an ER-located sterol transporter enriched in MAMs (Murley et al. 2015) which interacts with the ERMES complex (Gatta et al. 2015). Finally, ceramides, like other sphingolipids, constitute a quantitatively minor, but functionally significant, constituent of mitochondria, since they can induce OMM permeabilization-mediated apoptosis. Interestingly, ceramides synthesis has been shown to occur at MAMs (Stiban et al. 2008).

Even though the interplay between mitochondria and ER seems undeniable in the maintenance of cellular lipid homeostasis, a clear identification of molecular actors implicated in the process in mammalian cells is still missing. Insights for molecular mechanisms come almost exclusively from yeast. Indeed, an exciting clue arised from the identification in yeast of the members of the ERMES complex (see above) as highly hydrophobic proteins involved in lipid trafficking (Kopec et al. 2010). More precisely, recent studies showed that synaptotagmin-like mitochondrial lipid-binding protein (SMP) domains, present in the ERMES complex, are able to assemble in heterotetramer domains to form tubular structures, as hydrophobic tunnels for lipid transfer (AhYoung et al. 2015). Nevertheless, the fact that these SMP domains display a strong preference for PC (AhYoung et al. 2015) suggests that other glycerol-phospholipids transfer might be mediated by different parts of the ERMES complex or by other molecules (Murley and Nunnari 2016). Evidence has pointed toward a possible implication of some yeast members of the oxysterol-binding homology (Osh) protein family that display high affinity for PS and are crucial for PS homeostasis (Maeda et al. 2013), as well as for the ER membrane complex (EMC) proteins (Lahiri et al. 2014), for increasing the rate of transfer of PS from the ER to mitochondria. All these elements suggest that defining the molecular mechanism of lipid transfer between ER and mitochondria could imply the discovery of mammalian homologues of ERMES to allow phospholipids transfer between organelles.

Recent studies, however, have provided the first possible molecular identity in a different protein implicated in ER-mitochondria steroid transfer. Indeed, the translocator protein (TSPO), located in the OMM, is implicated in the import and processing of steroids into mitochondria (Flis and Daum 2013), and many clues point toward a possible presence of TSPO in MAMs, particularly through its interaction with VDAC (Guilarte et al. 2016).

The ER is not the only partner with which mitochondria interact to maintain cellular lipid homeostasis. Indeed, experimental disruption of ERMES has been shown to be compensated by the expansion of mitochondrial contacts with vacuoles/lysosomes, called vacuolar and mitochondrial patches (vCLAMPs), to maintain lipid homeostasis, and vice versa. Moreover, the molecular ratio between ERMES and vCLAMP components in yeast is dependent on the metabolic state of the cell. Ltc1 (see above) could also be expressed at vCLAMPs, in conditions of expansion, and is necessary for the compensation of ERMES loss (Honscher et al. 2014 and reviewed in Murley and Nunnari 2016).

Peroxisomes have been also shown to interact with mitochondria (see also below). Interestingly, in yeast, they make contacts at sites of mitochondria-ER juxtapositions, through the interaction between the peroxisomal protein Pex11 and the ERMES subunit Mdm34 (Mattiazzi Usaj et al. 2015). Interestingly, in these areas, the mitochondrial matrix enzyme PDH, responsible for the production of acetyl-CoA (see above), is also always present. Thus, acetyl-coA, which is a metabolite at the crossroad between glucose and lipids metabolism, is produced at a cellular hub of organelles' interactions (Cohen et al. 2014). These observations could lead to exciting speculations about a nutrient-sensing system based on the distribution of mitochondria/ER/vacuoles and peroxisome contacts, as well as on lipid transporters composition, to adapt lipid homeostasis and metabolism to cellular environment (Murley and Nunnari 2016).

It is important to underlie, however, that even if close contacts have the strongest relevance in inter-organelle lipid trafficking, other mechanisms might exist. In particular, soluble cytosolic lipid transfer proteins have been identified, especially for PC and PI (Vance 2015).

1.4.2 Autophagy

The process of autophagy is considerably dependent on intracellular organelles' membranes and their possible interactions (Yang and Klionsky 2009).

The major intracellular source from which the initial isolation membrane (also known as "phagophore") can form has been reported to be the ER, both in yeast (Suzuki and Ohsumi 2010) and mammals (Axe et al. 2008; Hayashi-Nishino et al. 2009; Itakura and Mizushima 2010; Matsunaga et al. 2010). More recently, multiple pieces of evidence have been reported for mitochondria-ER contact sites as sources of membranes for autophagosome biogenesis. Depletion of ER-mitochondria tethering regulators, such as Mfn2, has been shown to impair autophagosome formation (Hailey et al. 2010).

In addition, a milestone following study (Hamasaki et al. 2013) clearly demonstrated that autophagosomes can form at mitochondria-ER contact sites: not only key components of the autophagic process (Atg5, Atg14, Beclin-1, Vps15, Vps34) were found to be localized in MAMs under starved conditions (and this association at mitochondria-ER contact sites was stable throughout the autophagosomes formation process) but also alterations of this association led to decreased autophagy. Moreover, it has been shown that MAM raft-like microdomains, containing the raft marker ganglioside-3 (GD3), were pivotal in autophagosome formation: under starved conditions, calnexin, a Ca^{2+}-binding chaperone protein enriched in MAMs, strongly associates with GD3 and crucial upstream regulators of autophagy, such as AMBRA1 and WIPI1, allowing autophagy to proceed (Garofalo et al. 2016). The reduction in GD3 levels, by knocking down a member of the glycosyltransferase family responsible for its synthesis, led to impaired starvation-induced association of core complex molecules at MAMs and defective autophagy, highlighting the key role played by the MAM platform in driving the process.

Finally, by scanning electron microscopy and electron tomography, a morphological analysis confirmed that phagophores can assemble at the mitochondria-ER contact sites (Biazik et al. 2015). This observation, together with the finding that lipids transported from the ER to the mitochondria are subsequently translocated into autophagosomes under starved conditions (Hailey et al. 2010), strongly indicates mitochondria-ER communications as fundamental for resupplying lipids to mitochondria to compensate for their transport into autophagosomes.

Very recently, FUNDC1, the integral OMM protein that functions as LC3 receptor during hypoxia-induced mitophagy (Liu et al. 2012), has been found to be enriched in MAMs and interact with calnexin, DRP1 and OPA1, mediating either mitochondria elongation in response to hypoxic stimuli or mitochondria fragmentation and mitophagy, depending on its phosphorylation pattern (Wu et al. 2016). Thus, FUNDC1, localized at mitochondria-ER contact sites, functions as an adaptor protein coordinating mitochondrial dynamics and quality control in response to different signaling pathways and mitochondrial stresses.

1.4.3 The Unfolded Protein Response (UPR)

Several physiological and pathological stress conditions can alter ER homeostasis, decreasing the amount of Ca^{2+} accumulated within the ER or compromising the folding ER capacity, leading to accumulation of unfolded proteins in the ER lumen, and causing the ER stress response known as "unfolded proteins response" (UPR) (Verfaillie et al. 2013).

Within the cell, the mitochondria-ER axis is a critical player in the complex modulation of the UPR; indeed, both organelles have a key role during UPR, and they are both modulated by it in their functionality and interrelationship. For example, the early phase of UPR is linked to changes in ER morphology and redistribution of mitochondria, which move toward the ER structure, strengthening ER-mitochondria contact sites. The increased ER-mitochondria interaction sustained a higher Ca^{2+} transfer between them, rising mitochondrial metabolism, oxygen consumption, and ATP production. The outcome can help the cell to rescue the stress state, supplying ATP to chaperone proteins, such as Bip, involved during UPR in the folding process (Bravo et al. 2011).

As mentioned above, if, however, the stress and the damage are too severe, the pro-survival pathway, mediated by the three sensors, turns into a pro-death response through the regulation of pro-apoptotic proteins. In this case, the stronger ER-mitochondria interaction favors a deregulated Ca^{2+} transfer, leading to mitochondrial Ca^{2+} overload, mPTP opening, and apoptotic protein activation. In particular, among the proteins that can trigger apoptosis, the truncated form of SERCA (S1T) has been shown to be involved in the late ER stress response: highly expressed during ER stress by the PERK-ATF4 pathway, the truncated pump increases ER-mitochondria tethering and blocks mitochondrial movements, causing an increased ER-mitochondrial Ca^{2+} transfer, mitochondrial Ca^{2+} overload, and eventually cell death (Chami et al. 2008).

1.4.4 Organelle Activity and Biogenesis

A peculiar molecular and functional relationship between mitochondria and peroxisomes has been known for a long time. Indeed, these organelles not only share some components of their fission machinery (such as Drp1, Fis1, and Mff in mammals and Dnm1 and Fis1 in yeast) but also cooperate in some metabolic pathways, such as β-oxidation of fatty acids (reviewed in Schrader et al. 2015). Moreover, the mitochondrial biogenesis transcriptional co-activator PGC-1α has been found to directly increase peroxisomes number (Bagattin et al. 2010).

Mitochondria and peroxisomes have been reported to be closely associated in ultrastructural studies in mammalian cells (Hicks and Fahimi 1977), and multiple evidences indicate physical connections between mitochondria and peroxisomes in yeast (Rosenberger et al. 2009), although their contribution to metabolite exchange between the two organelles is unclear. Instead, mitochondrion-derived vesicles (MDVs) were suggested to allow the latter feature (Neuspiel et al. 2008) and appear, together with vesicles derived from the ER, to be important for peroxisomal biogenesis (Mohanty and McBride 2013). As to the functions of mitochondria-peroxisomes contacts, they have been suggested to be relevant for peroxisome fission in yeast (Mao et al. 2014), as well as for the innate immune response against viral and bacterial infections (reviewed in Schrader et al. 2015 and see also Horner et al. 2011, 2015) and for redox homeostasis (Nordgren and Fransen 2014) (reviewed in Schrader et al. 2015). Moreover, the same molecular fission complex (see above) is reported to be recruited during the autophagic degradation of the two organelles, mitophagy (Mao et al. 2013), and pexophagy (Mao et al. 2014), respectively. Moreover, peroxisome division, occurring early in the process, has been shown to take place at mitochondria-peroxisome contact sites (Mao et al. 2014), highlighting a functional interplay of the two organelles in autophagy.

Mitochondria make also contacts with other organelles, although the significance and molecular characterization of these sites are not defined. For example, in yeast the acidic vacuole is functionally linked to mitochondria, and any dysregulation of each organelle is negatively reflected to the other (Hughes and Gottschling 2012). In mammals, a vesicular transport pathway has been described from mitochondria to lysosomes, as an early response to oxidative stress, independent from mitophagy and preventing mitochondrial dysfunctionality (Soubannier et al. 2012).

Finally, it has been recently demonstrated that ~7% of melanosomes are in close contact with mitochondria, thanks to the presence of proteinaceous bridges similar to those observed in ER-mitochondria regions of apposition (Daniele et al. 2014). This association has been reported to be important for melanosomes biogenesis, likely through a local ATP supply from mitochondria to melanosomes or through an exchange of other metabolites, such as Ca^{2+}. Interestingly, though its abundance in the sites of interorganellar apposition is strongly reduced compared to the bulk OMM (as revealed by immunogold analysis), Mfn2 has been suggested to be important for these contacts, because its downregulation decreases their number (Daniele et al. 2014).

1.4.5 Other Cell Functions

1.4.5.1 Mitochondria Inheritance

In yeast, it has been demonstrated that active nucleoids (the nucleoprotein structures responsible for mtDNA inheritance) are associated with a protein complex that spans the OMM and the IMM (Meeusen and Nunnari 2003). Components of the ERMES complex (involved in mitochondria-ER contacts in yeast, see above) have been observed to be spatially associated to active nucleoids (Murley et al. 2013). Moreover, Mmr1, an OMM protein specific of bud's but not of mother cell's mitochondria, physically interacts with Myo2 (Itoh et al. 2004), a class V myosin that mediates the transport of mitochondria to the bud; of note, Mmr1 has been observed to be enriched at mitochondria-ER contact sites (Swayne et al. 2011), thus suggesting that these regions of contact may be important for mitochondria inheritance.

Very recently, in mammalian cells, it has been reported that mtDNA synthesis is spatially associated to a subset of mitochondria-ER contacts involved in mitochondrial division (Lewis et al. 2016). Importantly, mitochondria-ER tether formation precedes and is necessary (but not sufficient) for mtDNA replication and mitochondrial division, thus playing a crucial role in the proper inheritance of mtDNA between dividing mitochondria. It is tempting to speculate that the existence of a contact with the ER creates a specialized platform on OMM and IMM, characterized by a specific lipid/protein composition (lipid rafts-like), important for the successive recruitment of proteins involved in specific activities, such as mtDNA replication and mitochondrial division.

1.4.5.2 Inflammation and Antiviral Response

The activation of innate immune response is another pathway in which the mitochondria-ER axis seems to be actively involved (reviewed in Marchi et al. 2014; van Vliet et al. 2014; Schrader et al. 2015). The cytosolic pathogen receptor RIG-1, responsible for the production of pro-inflammatory cytokines, has been demonstrated to be recruited for proper assembly by the OMM adaptor protein MAVS (mitochondria antiviral signaling protein; Belgnaoui et al. 2011), which has been recently reported to be specifically located at MAMs, where it regulates mitochondrial and peroxisomal signaling events during viral infections (Horner et al. 2011). A proteomic analysis revealed, in these latter conditions, a dynamic change in MAM protein composition, with an increase in MAVS interactors (Horner et al. 2015).

The inflammasome NOD-like receptor NLRP3 is a multiprotein complex that works like a sensor of damage (intracellular pattern-recognition receptors that recognize pathogen-associated molecular patterns, PAMPs) activating the pro-inflammatory response through specific pathways. Normally located at ER membranes, after mitochondrial-ROS-induced inflammasome activation, NLRP3 redistributes to MAMs (Zhou et al. 2011), thus highlighting the importance of this cell sub-compartment in ROS signaling and activation of inflammatory response. Importantly, the activity of VDACs, which are OMM proteins and important regulators of mitochondrial metabolic activity (see Chaps. 7, 8, 9, 10, 11, and 12 for extensive descriptions), is crucial for NLRP3 activation. Moreover, in cystic fibrosis cell models, it has been recently shown that MCU-dependent mitochondrial Ca^{2+} uptake integrates pro-inflammatory signals initiated by pathogen-associated molecules and relays the information to NLRP3 (Rimessi et al. 2015), acting as a synergic stress stimulus in triggering exacerbated inflammatory response. Thus, not only mitochondria play a central role in the orchestration of the inflammatory response but also constitute the signal-integrating organelle for inflammasome activation.

1.4.5.3 Mitochondria Dynamics and Transport

Mitochondria are dynamic organelles, continuously undergoing fusion and fission. These opposite processes maintain the shape, size, and number of mitochondria as well as their physiological functionality (Youle and van der Bliek 2012). Numerous studies have suggested a role for mitochondria-ER interactions in mitochondrial fission and fusion (see Marchi et al. 2014 for a recent review), through mechanisms involving physical and functional (Ca^{2+}) coupling. Moreover, many of the proteins involved in MAM structural integrity are mitochondria dynamics-related proteins.

Mitochondrial fission, in particular, seems to be highly dependent on ER proximity. The fragmentation process relies on the recruitment, by specific receptors at the mitochondrial surface, of dynamin-related proteins, forming helical oligomers that wrap around the organelles and divide them in a GTP-dependent manner (Klecker et al. 2014). Interestingly, these fission complexes assemble at specific sites where ER tubules circumscribe mitochondria to form constriction sites both in yeast and metazoan cells (Smirnova et al. 2001). In these latter, mitochondrial division mediated by ER constrictions involves actin polymerization to tighten the noose formed by Drp1, through the action of the ER protein inverted formin-2 (INF2), a promoter of actin polymerization and depolarization (Prudent and McBride 2016). Moreover, the mitochondrial protein Miro1, at the OMM, has been shown to sense [Ca^{2+}], stop mitochondria motility, and induce fission in a Drp1-dependent manner (Saotome et al. 2008). Interestingly, its yeast homolog Gem1, which presents similar features (Frederick et al. 2004), interacts with the MAM complex ERMES at sites of mitochondrial division (Nguyen et al. 2012). This interaction is relying on the Gem1 EF-hand Ca^{2+}-binding domain, suggesting that the association of the protein with ERMES could be Ca^{2+} dependent (Kornmann et al. 2011). Moreover, reduced mitochondria-ER interactions have been associated with mitochondrial fragmentation during glucose sensing in the liver (Theurey et al. 2016).

Apart from fission, mitochondria-ER interactions have been shown to be crucial for fusion processes as well. The main actors in the control of mitochondrial fusion are Mfn1 and Mfn2, two homologous GTPase proteins present, both, at OMM, but, the second one, also at ER membranes. In particular, Mfn2 coordinates the interactions between different mitochondria, leading to the stabilization of the whole mitochondrial network, but is also crucial for tuning their tethering to the ER, modulating MAM formation (see above) (Koshiba et al. 2004).

Only mitochondria-ER contacts have been clearly demonstrated to have a central role, with the described mechanisms, in mitochondrial shaping events. Nevertheless, clues suggest that mitochondria-peroxisome interactions might also play a part. For instance, induction of oxidative stress by expression of KillerRed in peroxisomes induces mitochondrial fragmentation, even though only functional interplay through ROS has been implicated in the phenomenon (Ivashchenko et al. 2011). Interestingly, both organelles undergo similar dynamics processes and share components of their division machinery, such as Drp1 and Fis1 (Schrader et al. 2016) (see above).

To this day, the only link observed between mitochondrial dynamics and mitochondria-PM interaction is the observation that organelles' fragmentation mediated by Fis1 or Drp1 overexpression is associated with their decreased interactions with the cell membrane. However, it is more likely that this element underlines the possibility that a small number of anchor points exist at the mitochondria-PM interface rather than a real functional interplay between mitochondrial dynamics and PM (Frieden et al. 2005).

1.5 Mitochondria-ER Contacts and Diseases

As detailed above, the communication between mitochondria and ER regulates a multitude of physiological processes, ranging from intracellular Ca^{2+} homeostasis to lipid metabolism and control of cell death. Thus, the emergence of alterations in organelles' physical and functional tethering as a common hallmark in different pathologies, such as neurodegenerative diseases, diabetes, obesity, and cancer is not surprising. A comprehensive and detailed review of all these findings is far away from the scope of the present chapter. Here, we limit ourselves to a brief summary of the main studies involving alterations in the mitochondria-ER axis as a key event in three neurodegenerative diseases: Alzheimer's disease (AD), Parkinson's disease (PD), and amyotrophic lateral sclerosis (ALS). The interested reader is also referred to additional reviews on this topic (Cali et al. 2013b).

1.5.1 Alzheimer's Disease

In Alzheimer's disease (AD), the pathogenicity of the altered production/accumulation of amyloid β peptide (Aβ) has been well established (Goedert and Spillantini 2006). Additionally, however, AD patients display early intracellular alterations (such as those in Ca^{2+} homeostasis, lipid metabolism, axonal transport, ROS production, autophagy, and mitochondrial dynamics) that have been much less investigated as potential therapeutic targets (Agostini and Fasolato 2016). Importantly, many of these altered processes are known to take place at MAMs. Furthermore, Presenilin 1 and 2 (PS1 and PS2), the two alternative core proteins of the γ-secretase complex responsible for Aβ production, mutated in the familial forms of the disease, as well as γ-secretase activity, have also been found to be enriched in MAMs (Area-Gomez et al. 2009; Schreiner et al. 2015). In addition, it has been originally demonstrated that wt PS2 and, more potently, familial AD (FAD)-PS2 mutants, but not PS1 neither FAD-PS1, favor ER-mitochondrial Ca^{2+} transfer by increasing the physical association between the two organelles (Zampese et al. 2011). Though an increased interorganellar apposition in the presence of FAD-PS2 mutations has been later confirmed by others in human fibroblasts from FAD patients (Area-Gomez et al. 2012), an increased tethering with an impact on phospholipids and cholesterol metabolism has been additionally observed in fibroblasts from FAD-PS1 and sporadic AD patients (Area-Gomez et al. 2012). In this paper, however, the authors reported that both endogenous PS1 and PS2 negatively affect ER-mitochondria coupling, while their FAD mutants favor the apposition. It is, however, unclear how sporadic AD cases may converge on MAMs alterations similarly to those mechanistically caused by PSs mutations, as it has been proposed. Moreover, in another independent study, FAD-PS1 expression in a neuronal cell model was found to be associated to a decreased, and not to an increased, ER-mitochondria physical and functional coupling (Sepulveda-Falla et al. 2014).

Very recently, we confirmed that only PS2 increases ER-mitochondria communications, by binding and sequestering Mfn2 (Filadi et al. 2016) (see also above). It is tempting to speculate that, while PS2 directly affect ER-mitochondria tethering, in the cases of FAD-PS1 and sporadic AD, other pathways may indirectly, and over a longer period of time, converge on ER-mitochondria association, though the molecular mechanism is not clear. On this line, an altered coupling between these two organelles has been reported in different AD models (Kipanyula et al. 2012; Hedskog et al. 2013), and an acute Aβ oligomer exposure has been demonstrated to increase ER-mitochondria Ca^{2+} transfer (Hedskog et al. 2013). Finally, it has been observed that increasing ER-mitochondria contacts (through Mfn2 down-regulation) deeply affects the process of Aβ production, by altering the maturation of γ-secretase (Leal et al. 2016), further suggesting that the ER-mitochondria platform may represent a novel target for AD pharmacological research.

1.5.1.1 Parkinson's Disease

The association between Parkinson's disease (PD) and mitochondrial alterations is particularly immediate. Indeed, some of the genes mutated in the familial form of the disease encode for proteins that are targeted or are transiently associated, to mitochondria, such as PINK-1 and Parkin (reviewed in Cali et al. 2011). As to the contacts with the ER, overexpression of wt Parkin has been reported to favor ER-mitochondria physical and functional coupling, increasing mitochondrial Ca^{2+} uptake and sustaining ATP production, a function exerted also by endogenous Parkin, whose downregulation results in mitochondrial fragmentation (Cali et al. 2013). However, a recent report described an increased ER-mitochondria apposition in fibroblasts from Parkin (PARK-2) KO mice, as well as from PD patients harboring PARK-2 mutations (Gautier et al. 2016). This increase has been associated to higher levels of the Parkin target Mfn2, proposed as an ER-mitochondria tether. However, the complete list of the possible Parkin targets still lacks, and thus whether the effects of Parkin-KO on the tethering depend or not from its effects on Mfn2 levels is not clear. Overexpression of wt DJ-1 (another protein whose mutations are associated with PD familial forms) has been also shown to increase ER-mitochondria communication, by counteracting the negative effects of the tumor suppressor protein p53 on this parameter (Ottolini et al. 2013). Interestingly, a small fraction of DJ-1 has been recently demonstrated to translocate into mitochondrial matrix upon nutrient deprivation, a process important to sustain ATP levels, which is lost in the presence of pathogenic DJ-1 mutants (Cali et al. 2015). α-Synuclein has been also reported to be involved in the modulation of ER-mitochondrial physical tethering, enhancing mitochondrial Ca^{2+} uptake (Cali et al. 2012), and has been recently demonstrated to be localized at MAMs (Guardia-Laguarta et al. 2014), with its disease-linked mutants less associated to MAMs and resulting in mitochondria-ER uncoupling. Finally, in drosophila PINK1/Parkin mutants, it has been found that defective mitochondria activate the PERK branch of UPR, a neurotoxic process that seems to be potentially modulated by Mfn2 (Celardo et al. 2016).

1.5.1.2 Amyotrophic Lateral Sclerosis

Amyotrophic lateral sclerosis (ALS) is the most common form of motor neuron disease and is clinically and genetically associated to frontotemporal dementia (FTD) (Ling et al. 2013). In the nucleus and cytosol of neurons and glia of patients, inclusions of TDP-43-, FUS-, and C9ORF72-derived protein can be retrieved. Importantly, mutations in the genes encoding for these proteins are responsible for ~50% of the familial forms of the disease (Ling et al. 2013). Recently, increasing evidence has been provided for a key role of the mitochondria-ER axis in ALS/FTD onset/progression (reviewed in Manfredi and Kawamata 2016). In particular, overexpression of both wt and mutated TDP-43 decreases ER-mitochondria contacts and Ca^{2+} transfer (Stoica et al. 2014), by disrupting the VAPB-PTPIP51 tethering complex through activation of GSK-3β (see above). Similarly, FUS, wt, and ALS-linked mutant overexpression has been reported to disrupt ER-mitochondria juxtaposition, Ca^{2+} shuttling, and ATP production (Stoica et al. 2016), by affecting, in a GSK-3β-dependent manner, the VAPB-PTPIP51 interaction. Importantly, VAPB mutations are responsible for some familial forms of ALS type-8, and VAPB-P56S expression increases mitochondrial Ca^{2+} uptake after its release from the ER (De Vos et al. 2012). In mouse embryonic motor neurons, overexpression of G93A hSOD1, as an ALS animal model, has been shown to affect ER and mitochondrial Ca^{2+} homeostasis (Lautenschlager et al. 2013). Recently, the loss of the MAM protein SIGMA1-R (a condition associated to some familial forms of ALS/FTD) has been reported to impair mitochondria-ER crosstalk (Bernard-Marissal et al. 2015), and, similarly, SIGMA1-R loss-of-function mutations (associated to distal hereditary motor neuropathies, dHMNs) have been shown to disrupt this interorganellar communication (Gregianin et al. 2016). Thus, different studies convincingly suggested that ER-mitochondria axis could be an interesting platform to exploit as a therapeutic target in ALS/FTD.

Acknowledgments The authors thank grants from the University of Padova, the Italian Ministry of University and Scientific Research, Fondazione Cassa di Risparmio di Padova e Rovigo (CARIPARO Foundation; Progetti d'eccellenza 2011/2012), and EU Joint Programme in Neurodegenerative Disease, 2015–2018, "Cellular Bioenergetics in Neurodegenerative Diseases: A system-based pathway and target analysis," for their support of the research work performed in their lab. P.T. is research fellow of the EU Joint Programme – Neurodegenerative Disease.

References

Agostini M, Fasolato C (2016) When, where and how? Focus on neuronal calcium dysfunctions in Alzheimer's disease. Cell Calcium 60(5):289–298

AhYoung AP, Jiang J, Zhang J et al (2015) Conserved SMP domains of the ERMES complex bind phospholipids and mediate tether assembly. Proc Natl Acad Sci U S A 112(25):E3179–E3188

Amigo I, Traba J, Gonzalez-Barroso MM et al (2013) Glucagon regulation of oxidative phosphorylation requires an increase in matrix adenine nucleotide content through Ca2+ activation of the mitochondrial ATP-Mg/Pi carrier SCaMC-3. J Biol Chem 288(11):7791–7802

Anunciado-Koza RP, Zhang J, Ukropec J et al (2011) Inactivation of the mitochondrial carrier SLC25A25 (ATP-Mg2+/Pi transporter) reduces physical endurance and metabolic efficiency in mice. J Biol Chem 286(13):11659–11671

Area-Gomez E, de Groof AJ, Boldogh I et al (2009) Presenilins are enriched in endoplasmic reticulum membranes associated with mitochondria. Am J Pathol 175(5):1810–1816

Area-Gomez E, Del Carmen Lara Castillo M, Tambini MD et al (2012) Upregulated function of mitochondria-associated ER membranes in Alzheimer disease. EMBO J 31(21):4106–4123

Arruda AP, Pers BM, Parlakgul G et al (2014) Chronic enrichment of hepatic endoplasmic reticulum-mitochondria contact leads to mitochondrial dysfunction in obesity. Nat Med 20(12):1427–1435

Axe EL, Walker SA, Manifava M et al (2008) Autophagosome formation from membrane compartments enriched in phosphatidylinositol 3-phosphate and dynamically connected to the endoplasmic reticulum. J Cell Biol 182(4):685–701

Bagattin A, Hugendubler L, Mueller E (2010) Transcriptional coactivator PGC-1alpha promotes peroxisomal remodeling and biogenesis. Proc Natl Acad Sci U S A 107(47):20376–20381

Baughman JM, Perocchi F, Girgis HS et al (2011) Integrative genomics identifies MCU as an essential component of the mitochondrial calcium uniporter. Nature 476(7360):341–345

Belgnaoui SM, Paz S, Hiscott J (2011) Orchestrating the interferon antiviral response through the mitochondrial antiviral signaling (MAVS) adapter. Curr Opin Immunol 23(5):564–572

Bernard-Marissal N, Medard JJ, Azzedine H, Chrast R (2015) Dysfunction in endoplasmic reticulum-mitochondria crosstalk underlies SIGMAR1 loss of function mediated motor neuron degeneration. Brain 138(Pt 4):875–890

Berridge MJ, Lipp P, Bootman MD (2000) The versatility and universality of calcium signalling. Nat Rev Mol Cell Biol 1(1):11–21

Biazik J, Yla-Anttila P, Vihinen H et al (2015) Ultrastructural relationship of the phagophore with surrounding organelles. Autophagy 11(3):439–451

Boncompagni S, Rossi AE, Micaroni M et al (2009) Mitochondria are linked to calcium stores in striated muscle by developmentally regulated tethering structures. Mol Biol Cell 20(3): 1058–1067

Bononi A, Bonora M, Marchi S et al (2013) Identification of PTEN at the ER and MAMs and its regulation of Ca(2+) signaling and apoptosis in a protein phosphatase-dependent manner. Cell Death Differ 20(12):1631–1643

Bragadin M, Pozzan T, Azzone GF (1979) Kinetics of Ca^{2+} carrier in rat liver mitochondria. Biochemistry 18(26):5972–5978

Bravo R, Vicencio JM, Parra V et al (2011) Increased ER-mitochondrial coupling promotes mitochondrial respiration and bioenergetics during early phases of ER stress. J Cell Sci 124(Pt 13):2143–2152

Cali T, Ottolini D, Brini M (2011) Mitochondria, calcium, and endoplasmic reticulum stress in Parkinson's disease. Biofactors 37(3):228–240

Cali T, Ottolini D, Negro A, Brini M (2012) Alpha-synuclein controls mitochondrial calcium homeostasis by enhancing endoplasmic reticulum-mitochondria interactions. J Biol Chem 287(22):17914–17929

Cali T, Ottolini D, Negro A, Brini M (2013a) Enhanced parkin levels favor ER-mitochondria crosstalk and guarantee Ca(2+) transfer to sustain cell bioenergetics. Biochim Biophys Acta 1832(4):495–508

Cali T, Ottolini D, Brini M (2013b) Calcium and endoplasmic reticulum-mitochondria tethering in neurodegeneration. DNA Cell Biol 32(4):140–146

Cali T, Ottolini D, Soriano ME, Brini M (2015) A new split-GFP-based probe reveals DJ-1 translocation into the mitochondrial matrix to sustain ATP synthesis upon nutrient deprivation. Hum Mol Genet 24(4):1045–1060

Carafoli E (2003) Historical review: mitochondria and calcium: ups and downs of an unusual relationship. Trends Biochem Sci 28(4):175–181

Cardenas C, Miller RA, Smith I et al (2010) Essential regulation of cell bioenergetics by constitutive InsP3 receptor Ca2+ transfer to mitochondria. Cell 142(2):270–283

Celardo I, Costa AC, Lehmann S et al (2016) Mitofusin-mediated ER stress triggers neurodegeneration in pink1/parkin models of Parkinson's disease. Cell Death Dis 7(6):e2271

Chami M, Oules B, Szabadkai G et al (2008) Role of SERCA1 truncated isoform in the proapoptotic calcium transfer from ER to mitochondria during ER stress. Mol Cell 32(5):641–651

Cohen Y, Klug YA, Dimitrov L et al (2014) Peroxisomes are juxtaposed to strategic sites on mitochondria. Mol BioSyst 10(7):1742–1748

Cosson P, Marchetti A, Ravazzola M, Orci L (2012) Mitofusin-2 independent juxtaposition of endoplasmic reticulum and mitochondria: an ultrastructural study. PLoS One 7(9):e46293

Csordas G, Renken C, Varnai P et al (2006) Structural and functional features and significance of the physical linkage between ER and mitochondria. J Cell Biol 174(7):915–921

Csordas G, Varnai P, Golenar T et al (2010) Imaging interorganelle contacts and local calcium dynamics at the ER-mitochondrial interface. Mol Cell 39(1):121–132

Daniele T, Hurbain I, Vago R et al (2014) Mitochondria and melanosomes establish physical contacts modulated by Mfn2 and involved in organelle biogenesis. Curr Biol 24(4):393–403

de Brito OM, Scorrano L (2008) Mitofusin 2 tethers endoplasmic reticulum to mitochondria. Nature 456(7222):605–610

De Marchi U, Castelbou C, Demaurex N (2011) Uncoupling protein 3 (UCP3) modulates the activity of Sarco/endoplasmic reticulum Ca^{2+}-ATPase (SERCA) by decreasing mitochondrial ATP production. J Biol Chem 286(37):32533–32541

De Vos KJ, Morotz GM, Stoica R et al (2012) VAPB interacts with the mitochondrial protein PTPIP51 to regulate calcium homeostasis. Hum Mol Genet 21(6):1299–1311

del Arco A, Satrustegui J (2004) Identification of a novel human subfamily of mitochondrial carriers with calcium-binding domains. J Biol Chem 279(23):24701–24713

Deluca HF, Engstrom GW (1961) Calcium uptake by rat kidney mitochondria. Proc Natl Acad Sci U S A 47:1744–1750

Denton RM (2009) Regulation of mitochondrial dehydrogenases by calcium ions. Biochim Biophys Acta 1787(11):1309–1316

Denton RM, Richards DA, Chin JG (1978) Calcium ions and the regulation of NAD+-linked isocitrate dehydrogenase from the mitochondria of rat heart and other tissues. Biochem J 176(3):899–906

Doghman-Bouguerra M, Granatiero V, Sbiera S et al (2016) FATE1 antagonizes calcium- and drug-induced apoptosis by uncoupling ER and mitochondria. EMBO Rep 17(9):1264–1280

Drago I, De Stefani D, Rizzuto R, Pozzan T (2012) Mitochondrial Ca^{2+} uptake contributes to buffering cytoplasmic Ca^{2+} peaks in cardiomyocytes. Proc Natl Acad Sci U S A 109(32): 12986–12991

Elbaz-Alon Y, Rosenfeld-Gur E, Shinder V et al (2014) A dynamic interface between vacuoles and mitochondria in yeast. Dev Cell 30(1):95–102

Filadi R, Pozzan T (2015) Generation and functions of second messengers microdomains. Cell Calcium 58(4):405–414

Filadi et al (2012) In: Agostinis P, Samali A (eds) Endoplasmic reticulum stress in health and disease. Springer Science+Business Media, Dordrecht

Filadi R, Greotti E, Turacchio G et al (2015) Mitofusin 2 ablation increases endoplasmic reticulum-mitochondria coupling. Proc Natl Acad Sci U S A 112(17):E2174–E2181

Filadi R, Greotti E, Turacchio G et al (2016) Presenilin 2 modulates endoplasmic reticulum-mitochondria coupling by tuning the antagonistic effect of mitofusin 2. Cell Rep 15(10): 2226–2238

Flis VV, Daum G (2013) Lipid transport between the endoplasmic reticulum and mitochondria. Cold Spring Harb Perspect Biol 5(6):a013235

Fonteriz RI, de la Fuente S, Moreno A et al (2010) Monitoring mitochondrial [Ca^{2+}] dynamics with rhod-2, ratiometric pericam and aequorin. Cell Calcium 48(1):61–69

Foskett JK, White C, Cheung KH, Mak DO (2007) Inositol trisphosphate receptor Ca^{2+} release channels. Physiol Rev 87(2):593–658

Fowler SL, Akins M, Zhou H et al (2013) The liver connexin32 interactome is a novel plasma membrane-mitochondrial signaling nexus. J Proteome Res 12(6):2597–2610

Franzini-Armstrong C (2007) ER-mitochondria communication. How privileged? Physiology (Bethesda) 22:261–268

Frederick RL, McCaffery JM, Cunningham KW et al (2004) Yeast Miro GTPase, Gem1p, regulates mitochondrial morphology via a novel pathway. J Cell Biol 167(1):87–98

Frieden M, Arnaudeau S, Castelbou C, Demaurex N (2005) Subplasmalemmal mitochondria modulate the activity of plasma membrane Ca2+-ATPases. J Biol Chem 280(52):43198–43208

Friedman JR, Lackner LL, West M et al (2011) ER tubules mark sites of mitochondrial division. Science 334(6054):358–362

Fujimoto M, Hayashi T, Su TP (2012) The role of cholesterol in the association of endoplasmic reticulum membranes with mitochondria. Biochem Biophys Res Commun 417(1):635–639

Gandhi S, Wood-Kaczmar A, Yao Z et al (2009) PINK1-associated Parkinson's disease is caused by neuronal vulnerability to calcium-induced cell death. Mol Cell 33(5):627–638

Garofalo T, Matarrese P, Manganelli V et al (2016) Evidence for the involvement of lipid rafts localized at the ER-mitochondria associated membranes in autophagosome formation. Autophagy 12(6):917–935

Garrib A, McMurray WC (1986) Purification and characterization of glycerol-3-phosphate dehydrogenase (flavin-linked) from rat liver mitochondria. J Biol Chem 261(17):8042–8048

Gatta AT, Wong LH, Sere YY et al (2015) A new family of StART domain proteins at membrane contact sites has a role in ER-PM sterol transport. elife 4:e07253

Gautier CA, Erpapazoglou Z, Mouton-Liger F et al (2016) The endoplasmic reticulum-mitochondria interface is perturbed in PARK2 knockout mice and patients with PARK2 mutations. Hum Mol Genet 25:2972. pii: ddw148

Giacomello M, Pellegrini L (2016) The coming of age of the mitochondria-ER contact: a matter of thickness. Cell Death Differ 23(9):1417–1427

Giacomello M, Drago I, Bortolozzi M et al (2010) Ca^{2+} hot spots on the mitochondrial surface are generated by Ca^{2+} mobilization from stores, but not by activation of store-operated Ca^{2+} channels. Mol Cell 38(2):280–290

Giorgi C, Ito K, Lin HK et al (2010) PML regulates apoptosis at endoplasmic reticulum by modulating calcium release. Science 330(6008):1247–1251

Giorgi C, Bonora M, Sorrentino G et al (2015) p53 at the endoplasmic reticulum regulates apoptosis in a Ca2+-dependent manner. Proc Natl Acad Sci U S A 112(6):1779–1784

Glancy B, Willis WT, Chess DJ, Balaban RS (2013) Effect of calcium on the oxidative phosphorylation cascade in skeletal muscle mitochondria. Biochemistry 52(16):2793–2809

Goedert M, Spillantini MG (2006) A century of Alzheimer's disease. Science 314(5800): 777–781

Gordon PB, Holen I, Fosse M et al (1993) Dependence of hepatocytic autophagy on intracellularly sequestered calcium. J Biol Chem 268(35):26107–26112

Granatiero V, Giorgio V, Cali T et al (2016) Reduced mitochondrial Ca(2+) transients stimulate autophagy in human fibroblasts carrying the 13514A>G mutation of the ND5 subunit of NADH dehydrogenase. Cell Death Differ 23(2):231–241

Gregianin E, Pallafacchina G, Zanin S et al (2016) Loss-of-function mutations in the SIGMAR1 gene cause distal hereditary motor neuropathy by impairing ER-mitochondria tethering and Ca2+ signalling. Hum Mol Genet 25:3741

Guardia-Laguarta C, Area-Gomez E, Rub C et al (2014) Alpha-synuclein is localized to mitochondria-associated ER membranes. J Neurosci 34(1):249–259

Guilarte TR, Loth MK, Guariglia SR (2016) TSPO finds NOX2 in microglia for redox homeostasis. Trends Pharmacol Sci 37(5):334–343

Hailey DW, Rambold AS, Satpute-Krishnan P et al (2010) Mitochondria supply membranes for autophagosome biogenesis during starvation. Cell 141(4):656–667

Hamasaki M, Furuta N, Matsuda A et al (2013) Autophagosomes form at ER-mitochondria contact sites. Nature 495(7441):389–393

Hayashi T, Fujimoto M (2010) Detergent-resistant microdomains determine the localization of sigma-1 receptors to the endoplasmic reticulum-mitochondria junction. Mol Pharmacol 77(4):517–528

Hayashi-Nishino M, Fujita N, Noda T et al (2009) A subdomain of the endoplasmic reticulum forms a cradle for autophagosome formation. Nat Cell Biol 11(12):1433–1437

Haynes RC Jr, Picking RA, Zaks WJ (1986) Control of mitochondrial content of adenine nucleotides by submicromolar calcium concentrations and its relationship to hormonal effects. J Biol Chem 261(34):16121–16125

Hedgepeth SC, Garcia MI, Wagner LE 2nd et al (2015) The BRCA1 tumor suppressor binds to inositol 1,4,5-trisphosphate receptors to stimulate apoptotic calcium release. J Biol Chem 290(11):7304–7313

Hedskog L, Pinho CM, Filadi R et al (2013) Modulation of the endoplasmic reticulum-mitochondria interface in Alzheimer's disease and related models. Proc Natl Acad Sci U S A 110(19):7916–7921

Hicks L, Fahimi HD (1977) Peroxisomes (microbodies) in the myocardium of rodents and primates. A comparative ultrastructural cytochemical study. Cell Tissue Res 175(4):467–481

Honscher C, Mari M, Auffarth K et al (2014) Cellular metabolism regulates contact sites between vacuoles and mitochondria. Dev Cell 30(1):86–94

Horner SM, Liu HM, Park HS et al (2011) Mitochondrial-associated endoplasmic reticulum membranes (MAM) form innate immune synapses and are targeted by hepatitis C virus. Proc Natl Acad Sci U S A 108(35):14590–14595

Horner SM, Wilkins C, Badil S et al (2015) Proteomic analysis of mitochondrial-associated ER membranes (MAM) during RNA virus infection reveals dynamic changes in protein and organelle trafficking. PLoS One 10(3):e0117963

Hughes AL, Gottschling DE (2012) An early age increase in vacuolar pH limits mitochondrial function and lifespan in yeast. Nature 492(7428):261–265

Islinger M, Luers GH, Zischka H et al (2006) Insights into the membrane proteome of rat liver peroxisomes: microsomal glutathione-S-transferase is shared by both subcellular compartments. Proteomics 6(3):804–816

Itakura E, Mizushima N (2010) Characterization of autophagosome formation site by a hierarchical analysis of mammalian Atg proteins. Autophagy 6(6):764–776

Itoh T, Toh EA, Matsui Y (2004) Mmr1p is a mitochondrial factor for Myo2p-dependent inheritance of mitochondria in the budding yeast. EMBO J 23(13):2520–2530

Ivashchenko O, Van Veldhoven PP, Brees C et al (2011) Intraperoxisomal redox balance in mammalian cells: oxidative stress and interorganellar cross-talk. Mol Biol Cell 22(9):1440–1451

Jouaville LS, Pinton P, Bastianutto C et al (1999) Regulation of mitochondrial ATP synthesis by calcium: evidence for a long-term metabolic priming. Proc Natl Acad Sci U S A 96(24):13807–13812

Kipanyula MJ, Contreras L, Zampese E et al (2012) Ca2+ dysregulation in neurons from transgenic mice expressing mutant presenilin 2. Aging Cell 11(5):885–893

Klecker T, Scholz D, Fortsch J, Westermann B (2013) The yeast cell cortical protein Num1 integrates mitochondrial dynamics into cellular architecture. J Cell Sci 126(Pt 13):2924–2930

Klecker T, Bockler S, Westermann B (2014) Making connections: interorganelle contacts orchestrate mitochondrial behavior. Trends Cell Biol 24(9):537–545
Kopec KO, Alva V, Lupas AN (2010) Homology of SMP domains to the TULIP superfamily of lipid-binding proteins provides a structural basis for lipid exchange between ER and mitochondria. Bioinformatics 26(16):1927–1931
Kornmann B, Currie E, Collins SR et al (2009) An ER-mitochondria tethering complex revealed by a synthetic biology screen. Science 325(5939):477–481
Kornmann B, Osman C, Walter P (2011) The conserved GTPase Gem1 regulates endoplasmic reticulum-mitochondria connections. Proc Natl Acad Sci U S A 108(34):14151–14156
Koshiba T, Detmer SA, Kaiser JT et al (2004) Structural basis of mitochondrial tethering by mitofusin complexes. Science 305(5685):858–862
La Rovere RM, Roest G, Bultynck G, Parys JB (2016) Intracellular Ca(2+) signaling and Ca(2+) microdomains in the control of cell survival, apoptosis and autophagy. Cell Calcium 60(2):74–87
Lackner LL, Ping H, Graef M et al (2013) Endoplasmic reticulum-associated mitochondria-cortex tether functions in the distribution and inheritance of mitochondria. Proc Natl Acad Sci U S A 110(6):E458–E467
Lahiri S, Chao JT, Tavassoli S et al (2014) A conserved endoplasmic reticulum membrane protein complex (EMC) facilitates phospholipid transfer from the ER to mitochondria. PLoS Biol 12(10):e1001969
Landolfi B, Curci S, Debellis L et al (1998) Ca^{2+} homeostasis in the agonist-sensitive internal store: functional interactions between mitochondria and the ER measured in situ in intact cells. J Cell Biol 142:1235–1243
Lautenschlager J, Prell T, Ruhmer J et al (2013) Overexpression of human mutated G93A SOD1 changes dynamics of the ER mitochondria calcium cycle specifically in mouse embryonic motor neurons. Exp Neurol 247:91–100
Leal NS, Schreiner B, Pinho CM et al (2016) Mitofusin-2 knockdown increases ER-mitochondria contact and decreases amyloid beta-peptide production. J Cell Mol Med 20(9):1686–1695
Lebiedzinska M, Szabadkai G, Jones AW et al (2009) Interactions between the endoplasmic reticulum, mitochondria, plasma membrane and other subcellular organelles. Int J Biochem Cell Biol 41(10):1805–1816
Lewis SC, Uchiyama LF, Nunnari J (2016) ER-mitochondria contacts couple mtDNA synthesis with mitochondrial division in human cells. Science 353(6296):aaf5549
Ling SC, Polymenidou M, Cleveland DW (2013) Converging mechanisms in ALS and FTD: disrupted RNA and protein homeostasis. Neuron 79(3):416–438
Liu L, Feng D, Chen G et al (2012) Mitochondrial outer-membrane protein FUNDC1 mediates hypoxia-induced mitophagy in mammalian cells. Nat Cell Biol 14(2):177–185
Maeda K, Anand K, Chiapparino A et al (2013) Interactome map uncovers phosphatidylserine transport by oxysterol-binding proteins. Nature 501(7466):257–261
Majewski N, Nogueira V, Bhaskar P et al (2004) Hexokinase-mitochondria interaction mediated by Akt is required to inhibit apoptosis in the presence or absence of Bax and Bak. Mol Cell 16(5):819–830
Mallilankaraman K, Cardenas C, Doonan PJ et al (2012) MCUR1 is an essential component of mitochondrial Ca2+ uptake that regulates cellular metabolism. Nat Cell Biol 14(12): 1336–1343
Manfredi G, Kawamata H (2016) Mitochondria and endoplasmic reticulum crosstalk in amyotrophic lateral sclerosis. Neurobiol Dis 90:35–42
Mao K, Wang K, Liu X, Klionsky DJ (2013) The scaffold protein Atg11 recruits fission machinery to drive selective mitochondria degradation by autophagy. Dev Cell 26(1):9–18
Mao K, Liu X, Feng Y, Klionsky DJ (2014) The progression of peroxisomal degradation through autophagy requires peroxisomal division. Autophagy 10(4):652–661
Marchi S, Marinello M, Bononi A et al (2012) Selective modulation of subtype III IP(3)R by Akt regulates ER Ca(2)(+) release and apoptosis. Cell Death Dis 3:e304

Marchi S, Patergnani S, Pinton P (2014) The endoplasmic reticulum-mitochondria connection: one touch, multiple functions. Biochim Biophys Acta 1837(4):461–469

Matsunaga K, Morita E, Saitoh T et al (2010) Autophagy requires endoplasmic reticulum targeting of the PI3-kinase complex via Atg14L. J Cell Biol 190(4):511–521

Mattiazzi Usaj M, Brloznik M, Kaferle P et al (2015) Genome-wide localization study of yeast Pex11 identifies peroxisome-mitochondria interactions through the ERMES complex. J Mol Biol 427(11):2072–2087

Meeusen S, Nunnari J (2003) Evidence for a two membrane-spanning autonomous mitochondrial DNA replisome. J Cell Biol 163(3):503–510

Mitchell P, Moyle J (1967) Chemiosmotic hypothesis of oxidative phosphorylation. Nature 213(72):137–139

Miyawaki A, Llopis J, Heim R et al (1997) Fluorescent indicators for Ca^{2+} based on green fluorescent proteins and calmodulin. Nature 388:882–887

Mohanty A, McBride HM (2013) Emerging roles of mitochondria in the evolution, biogenesis, and function of peroxisomes. Front Physiol 4:268

Mueller SJ, Reski R (2015) Mitochondrial dynamics and the ER: the plant perspective. Front Cell Dev Biol 3:78

Murgia M, Rizzuto R (2015) Molecular diversity and pleiotropic role of the mitochondrial calcium uniporter. Cell Calcium 58(1):11–17

Murley A, Nunnari J (2016) The emerging network of mitochondria-organelle contacts. Mol Cell 61(5):648–653

Murley A, Lackner LL, Osman C et al (2013) ER-associated mitochondrial division links the distribution of mitochondria and mitochondrial DNA in yeast. elife 2:e00422

Murley A, Sarsam RD, Toulmay A et al (2015) Ltc1 is an ER-localized sterol transporter and a component of ER-mitochondria and ER-vacuole contacts. J Cell Biol 209(4):539–548

Naghdi S, Waldeck-Weiermair M, Fertschai I et al (2010) Mitochondrial Ca2+ uptake and not mitochondrial motility is required for STIM1-Orai1-dependent store-operated Ca2+ entry. J Cell Sci 123(Pt 15):2553–2564

Naraghi M, Neher E (1997) Linearized buffered Ca^{2+} diffusion in microdomains and its implications for calculation of $[Ca^{2+}]$ at the mouth of a calcium channel. J Neurosci 17(18): 6961–6973

Neuspiel M, Schauss AC, Braschi E et al (2008) Cargo-selected transport from the mitochondria to peroxisomes is mediated by vesicular carriers. Curr Biol 18(2):102–108

Nguyen TT, Lewandowska A, Choi JY et al (2012) Gem1 and ERMES do not directly affect phosphatidylserine transport from ER to mitochondria or mitochondrial inheritance. Traffic 13(6):880–890

Nicholls DG, Chalmers S (2004) The integration of mitochondrial calcium transport and storage. J Bioenerg Biomembr 36(4):277–281

Nicholls DG, Crompton M (1980) Mitochondrial calcium transport. FEBS Lett 111(2):261–268

Nordgren M, Fransen M (2014) Peroxisomal metabolism and oxidative stress. Biochimie 98:56–62

Nosek MT, Dransfield DT, Aprille JR (1990) Calcium stimulates ATP-Mg/Pi carrier activity in rat liver mitochondria. J Biol Chem 265(15):8444–8450

Orrenius S, Gogvadze V, Zhivotovsky B (2015) Calcium and mitochondria in the regulation of cell death. Biochem Biophys Res Commun 460(1):72–81

Osman C, Voelker DR, Langer T (2011) Making heads or tails of phospholipids in mitochondria. J Cell Biol 192(1):7–16

Ottolini D, Cali T, Negro A, Brini M (2013) The Parkinson disease-related protein DJ-1 counteracts mitochondrial impairment induced by the tumour suppressor protein p53 by enhancing endoplasmic reticulum-mitochondria tethering. Hum Mol Genet 22(11):2152–2168

Pacher P, Csordas P, Schneider T, Hajnoczky G (2000) Quantification of calcium signal transmission from sarco-endoplasmic reticulum to the mitochondria. J Physiol 529(Pt 3):553–564

Pacher P, Thomas AP, Hajnoczky G (2002) Ca^{2+} marks: miniature calcium signals in single mitochondria driven by ryanodine receptors. Proc Natl Acad Sci U S A 99(4):2380–2385

Palmieri L, Pardo B, Lasorsa FM et al (2001) Citrin and aralar1 are Ca^{2+}-stimulated aspartate/glutamate transporters in mitochondria. EMBO J 20(18):5060–5069

Pardo B, Contreras L, Serrano A et al (2006) Essential role of aralar in the transduction of small Ca^{2+} signals to neuronal mitochondria. J Biol Chem 281(2):1039–1047

Patron M, Raffaello A, Granatiero V et al (2013) The mitochondrial calcium uniporter (MCU): molecular identity and physiological roles. J Biol Chem 288(15):10750–10758

Perkins GA, Tjong J, Brown JM et al (2010) The micro-architecture of mitochondria at active zones: electron tomography reveals novel anchoring scaffolds and cristae structured for high-rate metabolism. J Neurosci 30(3):1015–1026

Pitter JG, Maechler P, Wollheim CB, Spat A (2002) Mitochondria respond to Ca^{2+} already in the submicromolar range: correlation with redox state. Cell Calcium 31(2):97–104

Pozzan T, Rizzuto R (2000) The renaissance of mitochondrial calcium transport. Eur J Biochem 267:5269–5273

Prudent J, McBride HM (2016) Mitochondrial dynamics: ER actin tightens the Drp1 noose. Curr Biol 26(5):R207–R209

Qi H, Li L, Shuai J (2015) Optimal microdomain crosstalk between endoplasmic reticulum and mitochondria for Ca2+ oscillations. Sci Rep 5:7984

Quintana A, Schwarz EC, Schwindling C et al (2006) Sustained activity of calcium release-activated calcium channels requires translocation of mitochondria to the plasma membrane. J Biol Chem 281(52):40302–40309

Rapizzi E, Pinton P, Szabadkai G et al (2002) Recombinant expression of the voltage-dependent anion channel enhances the transfer of Ca^{2+} microdomains to mitochondria. J Cell Biol 159(4):613–624

Reichert A, Neupert W (2002) Contact sites between the outer and inner membrane of mitochondria-role in protein transport. Biochim Biophys Acta 1592(1):41–49

Rimessi A, Bezzerri V, Patergnani S et al (2015) Mitochondrial Ca2+-dependent NLRP3 activation exacerbates the Pseudomonas aeruginosa-driven inflammatory response in cystic fibrosis. Nat Commun 6:6201

Rizzuto R, Pozzan T (2006) Microdomains of intracellular Ca^{2+}: molecular determinants and functional consequences. Physiol Rev 86(1):369–408

Rizzuto R, Simpson AW, Brini M, Pozzan T (1992) Rapid changes of mitochondrial Ca^{2+} revealed by specifically targeted recombinant aequorin. Nature 358(6384):325–327

Rizzuto R, Pinton P, Carrington W et al (1998) Close contacts with the endoplasmic reticulum as determinants of mitochondrial Ca^{2+} responses. Science 280(5370):1763–1766

Robert V, Gurlini P, Tosello V et al (2001) Beat-to-beat oscillations of mitochondrial [Ca^{2+}] in cardiac cells. EMBO J 20(17):4998–5007

Robertson JD (1960) The molecular structure and contact relationships of cell membranes. Prog Biophys Mol Biol 10:343–418

Rosenberger S, Connerth M, Zellnig G, Daum G (2009) Phosphatidylethanolamine synthesized by three different pathways is supplied to peroxisomes of the yeast *Saccharomyces cerevisiae*. Biochim Biophys Acta 1791(5):379–387

Rowland AA, Voeltz GK (2012) Endoplasmic reticulum-mitochondria contacts: function of the junction. Nature reviews. Mol Cell Biol 13(10):607–625

Rudolf R, Mongillo M, Magalhães PJ, Pozzan T (2004) In vivo monitoring of Ca^{2+} uptake into mitochondria of mouse skeletal muscle during contraction. J Cell Biol 166(4):527–536

Sandebring A, Thomas KJ, Beilina A et al (2009) Mitochondrial alterations in PINK1 deficient cells are influenced by calcineurin-dependent dephosphorylation of dynamin-related protein 1. PLoS One 4(5):e5701

Saotome M, Safiulina D, Szabadkai G et al (2008) Bidirectional Ca^{2+}-dependent control of mitochondrial dynamics by the Miro GTPase. Proc Natl Acad Sci U S A 105(52):20728–20733

Satrustegui J, Pardo B, Del Arco A (2007) Mitochondrial transporters as novel targets for intracellular calcium signaling. Physiol Rev 87(1):29–67

Schneeberger M, Dietrich MO, Sebastian D et al (2013) Mitofusin 2 in POMC neurons connects ER stress with leptin resistance and energy imbalance. Cell 155(1):172–187

Schrader M, Godinho LF, Costello JL, Islinger M (2015) The different facets of organelle interplay-an overview of organelle interactions. Front Cell Dev Biol 3:56

Schrader M, Costello JL, Godinho LF et al (2016) Proliferation and fission of peroxisomes – an update. Biochim Biophys Acta 1863(5):971–983

Schreiner B, Hedskog L, Wiehager B, Ankarcrona M (2015) Amyloid-beta peptides are generated in mitochondria-associated endoplasmic reticulum membranes. J Alzheimers Dis 43(2): 369–374

Sepulveda-Falla D, Barrera-Ocampo A, Hagel C et al (2014) Familial Alzheimer's disease-associated presenilin-1 alters cerebellar activity and calcium homeostasis. J Clin Invest 124(4):1552–1567

Sharma VK, Ramesh V, Franzini-Armstrong C, Sheu SS (2000) Transport of Ca^{2+} from sarcoplasmic reticulum to mitochondria in rat ventricular myocytes. J Bioenerg Biomembr 32(1): 97–104

Simmen T, Aslan JE, Blagoveshchenskaya AD et al (2005) PACS-2 controls endoplasmic reticulum-mitochondria communication and Bid-mediated apoptosis. EMBO J 24(4): 717–729

Singaravelu K, Nelson C, Bakowski D et al (2011) Mitofusin 2 regulates STIM1 migration from the Ca2+ store to the plasma membrane in cells with depolarized mitochondria. J Biol Chem 286(14):12189–12201

Smirnova E, Griparic L, Shurland DL, van der Bliek AM (2001) Dynamin-related protein Drp1 is required for mitochondrial division in mammalian cells. Mol Biol Cell 12(8):2245–2256

Sood A, Jeyaraju DV, Prudent J et al (2014) A mitofusin-2-dependent inactivating cleavage of Opa1 links changes in mitochondria cristae and ER contacts in the postprandial liver. Proc Natl Acad Sci U S A 111(45):16017–16022

Soubannier V, McLelland GL, Zunino R et al (2012) A vesicular transport pathway shuttles cargo from mitochondria to lysosomes. Curr Biol 22(2):135–141

Stiban J, Caputo L, Colombini M (2008) Ceramide synthesis in the endoplasmic reticulum can permeabilize mitochondria to proapoptotic proteins. J Lipid Res 49(3):625–634

Stoica R, De Vos KJ, Paillusson S et al (2014) ER-mitochondria associations are regulated by the VAPB-PTPIP51 interaction and are disrupted by ALS/FTD-associated TDP-43. Nat Commun 5:3996

Stoica R, Paillusson S, Gomez-Suaga P et al. (2016) ALS/FTD-associated FUS activates GSK-3beta to disrupt the VAPB-PTPIP51 interaction and ER-mitochondria associations. EMBO Rep (17)9: 1326–1342.

Stone SJ, Vance JE (2000) Phosphatidylserine synthase-1 and -2 are localized to mitochondria-associated membranes. J Biol Chem 275(44):34534–34540

Streb H, Irvine RF, Berridge MJ, Schulz I (1983) Release of Ca^{2+} from a nonmitochondrial intracellular store in pancreatic acinar cells by inositol-1,4,5-trisphosphate. Nature 306:67–68

Suski JM, Lebiedzinska M, Wojtala A et al (2014) Isolation of plasma membrane-associated membranes from rat liver. Nat Protoc 9(2):312–322

Suzuki K, Ohsumi Y (2010) Current knowledge of the pre-autophagosomal structure (PAS). FEBS Lett 584(7):1280–1286

Swayne TC, Zhou C, Boldogh IR et al (2011) Role for cER and Mmr1p in anchorage of mitochondria at sites of polarized surface growth in budding yeast. Curr Biol 21(23):1994–1999

Szabadkai G, Simoni AM, Chami M et al (2004) Drp-1-dependent division of the mitochondrial network blocks intraorganellar Ca^{2+} waves and protects against Ca^{2+}-mediated apoptosis. Mol Cell 16(1):59–68

Szabadkai G, Bianchi K, Varnai P et al (2006) Chaperone-mediated coupling of endoplasmic reticulum and mitochondrial Ca^{2+} channels. J Cell Biol 175(6):901–911

Szalai G, Csordas G, Hantash BM et al (2000) Calcium signal transmission between ryanodine receptors and mitochondria. J Biol Chem 275(20):15305–15313

Tadross MR, Tsien RW, Yue DT (2013) Ca^{2+} channel nanodomains boost local Ca^{2+} amplitude. Proc Natl Acad Sci U S A 110(39):15794–15799

Territo PR, Mootha VK, French SA, Balaban RS (2000) Ca(2+) activation of heart mitochondrial oxidative phosphorylation: role of the F(0)/F(1)-ATPase. Am J Physiol Cell Physiol 278(2): C423–C435

Thangaratnarajah C, Ruprecht JJ, Kunji ER (2014) Calcium-induced conformational changes of the regulatory domain of human mitochondrial aspartate/glutamate carriers. Nat Commun 5:5491

Theurey P, Tubbs E, Vial G et al (2016) Mitochondria-associated endoplasmic reticulum membranes allow adaptation of mitochondrial metabolism to glucose availability in the liver. J Mol Cell Biol 8(2):129–143

Tubbs E, Theurey P, Vial G et al (2014) Mitochondria-associated endoplasmic reticulum membrane (MAM) integrity is required for insulin signaling and is implicated in hepatic insulin resistance. Diabetes 63(10):3279–3294

Turkan A, Hiromasa Y, Roche TE (2004) Formation of a complex of the catalytic subunit of pyruvate dehydrogenase phosphatase isoform 1 (PDP1c) and the L2 domain forms a Ca2+ binding site and captures PDP1c as a monomer. Biochemistry 43(47):15073–15085

van Vliet AR, Agostinis P (2016) When under pressure, get closer: PERKing up membrane contact sites during ER stress. Biochem Soc Trans 44(2):499–504

van Vliet AR, Verfaillie T, Agostinis P (2014) New functions of mitochondria associated membranes in cellular signaling. Biochim Biophys Acta 1843(10):2253–2262

Vance JE (1990) Phospholipid synthesis in a membrane fraction associated with mitochondria. J Biol Chem 265(13):7248–7256

Vance JE (2014) MAM (mitochondria-associated membranes) in mammalian cells: lipids and beyond. Biochim Biophys Acta 1841(4):595–609

Vance JE (2015) Phospholipid synthesis and transport in mammalian cells. Traffic 16(1):1–18

Verfaillie T, Rubio N, Garg AD et al (2012) PERK is required at the ER-mitochondrial contact sites to convey apoptosis after ROS-based ER stress. Cell Death Differ 19(11):1880–1891

Verfaillie T, Garg AD, Agostinis P (2013) Targeting ER stress induced apoptosis and inflammation in cancer. Cancer Lett 332(2):249–264

Voelker DR (2000) Interorganelle transport of aminoglycerophospholipids. Biochim Biophys Acta 1486(1):97–107

Wang PT, Garcin PO, Fu M et al (2015) Distinct mechanisms controlling rough and smooth endoplasmic reticulum contacts with mitochondria. J Cell Sci 128(15):2759–2765

Watson R, Parekh AB (2012) Mitochondrial regulation of CRAC channel-driven cellular responses. Cell Calcium 52(1):52–56

Wieckowski MR, Giorgi C, Lebiedzinska M et al (2009) Isolation of mitochondria-associated membranes and mitochondria from animal tissues and cells. Nat Protoc 4(11):1582–1590

Wiederkehr A, Szanda G, Akhmedov D et al (2011) Mitochondrial matrix calcium is an activating signal for hormone secretion. Cell Metab 13(5):601–611

Wu W, Lin C, Wu K et al (2016) FUNDC1 regulates mitochondrial dynamics at the ER-mitochondrial contact site under hypoxic conditions. EMBO J 35(13):1368–1384

Yang Z, Klionsky DJ (2009) An overview of the molecular mechanism of autophagy. Curr Top Microbiol Immunol 335:1–32

Youle RJ, van der Bliek AM (2012) Mitochondrial fission, fusion, and stress. Science 337(6098): 1062–1065

Zampese E, Fasolato C, Kipanyula MJ et al (2011) Presenilin 2 modulates endoplasmic reticulum (ER)-mitochondria interactions and Ca^{2+} cross-talk. Proc Natl Acad Sci U S A 108(7): 2777–2782

Zborowski J, Dygas A, Wojtczak L (1983) Phosphatidylserine decarboxylase is located on the external side of the inner mitochondrial membrane. FEBS Lett 157(1):179–182

Zhou R, Yazdi AS, Menu P, Tschopp J (2011) A role for mitochondria in NLRP3 inflammasome activation. Nature 469(7329):221–225

Chapter 2
Molecular Players of Mitochondrial Calcium Signaling: Similarities and Different Aspects in Various Organisms

Vanessa Checchetto, Diego De Stefani, Anna Raffaello, Rosario Rizzuto, and Ildiko Szabo

2.1 Introduction

Ca^{2+} acts as a second messenger in every cell type, controlling processes as diverse as secretion, cell death, and survival. The versatile and universal nature of calcium as intracellular messenger is guaranteed by a cell-specific Ca^{2+} signaling toolkit: several components (e.g., Ca^{2+} channels, pumps, and Ca^{2+}-binding proteins) can cooperate and generate a wide range of signals, where changes in intracellular Ca^{2+} concentration ($[Ca^{2+}]$i) vary in both spatial and temporal patterns. These specific patterns can then be decoded into specific cellular events (Berridge et al. 2003). Compartmentalization of $[Ca^{2+}]$ dynamics in the different organelles represents a further level of complexity. Mitochondria are thought to play an integral part that goes beyond acting as passive supporters by providing the ATP required for cellular readjustment of $[Ca^{2+}]$i following stimulation. These organelles are able to quickly and transiently accumulate Ca^{2+} upon cytosolic transients, and thus, they can a priori contribute to shaping cytosolic Ca^{2+} transients.

Only during the last decade, significant advances have been achieved regarding the identification of the molecular players of the mitochondrial Ca^{2+}-handling machinery. Here, we summarize our current knowledge on the main player, i.e., the

V. Checchetto • I. Szabo (✉)
Department of Biology and CNR Institute of Neurosciences, University of Padova, Viale G. Colombo 3, 35121 Padova, Italy
e-mail: ildi@civ.bio.unipd.it

D. De Stefani • A. Raffaello • R. Rizzuto
Department of Biomedical Sciences, University of Padova, Viale G. Colombo 3, 35121 Padova, Italy

T.K. Rostovtseva (ed.), *Molecular Basis for Mitochondrial Signaling*, Biological and Medical Physics, Biomedical Engineering,
DOI 10.1007/978-3-319-55539-3_2

MCU complex (MCUC) in different organisms including protozoa (*Trypanosoma*), fungi, plants, and animals. In addition, the physiology of mitochondrial calcium homeostasis will be discussed also in light of what we learned from studies in organisms where MCU complex components have been genetically targeted.

2.2 The Role of Mitochondrial Calcium Within the Organelle

During the last three decades, considerable experimental work has been carried out using either pharmacological or indirect genetic tools to alter mitochondrial calcium homeostasis and to dissect its pathophysiological role. Ca^{2+} inside mitochondria plays a pleiotropic role, with different cellular outcomes that depend on the investigated cell type, the metabolic state, and the concomitant presence of other stress signals. Not only calcium plays a regulatory role within the organelle itself ranging from the regulation of ATP production to the release of apoptotic cofactors with consequent cell death, but in most organisms, mitochondrial calcium can importantly impact cation homeostasis, both locally and globally (Rizzuto et al. 2012).

In respiring mitochondria, the major component of the electrochemical proton gradient ($\Delta\mu H^+$), the membrane potential difference ($\Delta\Psi$), represents a very large driving force for Ca^{2+} accumulation. The inner mitochondrial membrane (IMM) is impermeable for cations (including Ca^{2+}), and passage strictly requires channels/transporters. Early Ca^{2+} uptake studies with mammalian mitochondria revealed that transport of calcium required respiration (Vasington and Murphy 1962; Deluca and Engstrom 1961) and was accompanied by P_i transport (Greenawalt et al. 1964). The underlying transporter was proposed to be an electrophoretic Ca^{2+} uniporter that does not require ATP hydrolysis but is driven by the steep electrochemical gradient across the IMM (Rottenberg and Scarpa 1974). Similarly, studies using isolated mitochondria from different plant species evidenced that these organelles take up Ca^{2+} (Akerman and Moore 1983; Dieter and Marme 1980). Uptake strictly required energization (Dieter and Marme 1980). Ca^{2+} import in most cases was shown to require inorganic phosphate (P_i) (Hodges and Hanson 1965; Chen and Lehninger 1973), leading to a hypothesis that Ca^{2+} is imported by a symport mechanism together with P_i (Silva et al. 1992; Day et al. 1978). However, for example, in some studies, isolated plant mitochondria were not found to accumulate calcium (Moore and Bonner 1977; Martins and Vercesi 1985), casting doubt on the existence of a general mechanism of calcium handling by these organelles in plants. In mammals, once mitochondrial Ca^{2+} uptake could be monitored directly in intact cells using mitochondria-targeted aequorin as calcium-sensing probe (Rizzuto et al. 1992), it became evident that Ca^{2+} concentrations in mitochondria can reach up to hundred μM in some cell types. The speed and amplitude of Ca^{2+} uptake was shown to exceed the values that had been predicted from classical bioenergetic experiments in isolated mitochondria. Subsequent work in mammalian cells suggested an interaction of mitochondria with microdomains of high Ca^{2+} concentrations generated by localized release from the ER and the extracellular space, allowing highly efficient uptake (see, e.g., Rizzuto et al. (2012) also for historical overview).

In mammals, Ca^{2+} elevations in the mitochondrial matrix stimulate respiration and ATP synthesis to cover temporarily high-energy needs of cells, e.g. (Denton 2009). Increased biosynthesis rates of ATP rely on the activation of mitochondrial dehydrogenases by Ca^{2+} (McCormack et al. 1990). In addition, electron transfer chain (ETC) complexes as well as the ATP synthase are positively regulated by Ca^{2+}. Among the dehydrogenases, pyruvate dehydrogenase (PDH) (Denton et al. 1972), NAD-isocitrate dehydrogenase (NAD-ICDH) (Denton et al. 1978), and α-ketoglutarate (oxoglutarate) dehydrogenase (McCormack and Denton 1979) are activated by physiologically relevant Ca^{2+} concentrations (between 100 nM and 1 μM) in mitochondria isolated from mammalian tissues (Denton and McCormack 1980; Denton et al. 1980). The latter two enzymes do not contain any typical Ca^{2+}-binding motifs, such as EF-hands, but are directly, allosterically, regulated by Ca^{2+} (McCormack et al. 1990). Instead, both in animal and plants, the former enzyme is activated through Ca^{2+}-controlled PDH phosphatase: PDH activity is regulated through reversible phosphorylation (Holness and Sugden 2003; Tovar-Mendez et al. 2003) with activity being enhanced through a dephosphorylation step. These events in turn increase NADH availability and consequently the electron flow through the respiratory chain. In addition to matrix dehydrogenases, aspartate/glutamate exchangers of the inner membrane (aralar1, citrin, and the ATP-Mg/Pi carrier SCaMC-3) also seem to be regulated by Ca^{2+} via EF-hand Ca^{2+}-binding sites which are exposed to the intermembrane space (Rueda et al. 2015). Direct evidences in favor of regulation of metabolism by calcium are multiple: (i) an increased resting state level of Ca^{2+} in the mitochondrial matrix was shown to alter the PDH phosphorylation state in cultured cells (Mallilankaraman et al. 2012b) and (ii) an increase in mitochondrial and then cytosolic ATP was reported to occur upon cell stimulation and to depend on the $[Ca^{2+}]_{mt}$ rise (Jouaville et al. 1995), in order to match ATP synthesis to the increased demand of a stimulated cell. However, several findings argue against an universal conservation of Ca^{2+} regulation of mitochondrial metabolism in all organisms: (i) while the activity of TCA cycle enzymes NAD-ICDH and α-ketoglutarate dehydrogenase from various vertebrates is increased in the presence of Ca^{2+}, their homologs from *Escherichia coli*, yeast, insect flight muscle, and potato are insensitive to calcium (Nichols et al. 1994; McCormack and Denton 1981) and (ii) PDH phosphatase is not activated by Ca^{2+} in vitro or in intact mitochondria in all organisms (Miernyk and Randall 1987; Budde et al. 1988).

In addition to the regulation of mitochondrial metabolism, calcium plays an important role also in the context of cell death. Indeed, an excessive increase in mitochondrial Ca^{2+} concentration under certain stimuli may be harmful: when a certain threshold level is exceeded, it may result in long-lasting opening of the permeability transition pore (Bernardi and von Stockum 2012). The general consensus is that mitochondrial Ca^{2+} loading has a permissive role, allowing a variety of toxic challenges to cause the release of caspase cofactors from the organelle and thereby trigger apoptotic cell death. In turn, alteration of this cellular response has a role in the pathogenesis of human disorders such as neurodegenerative disorders and cancer (Bernardi et al. 2015; Szabo and Zoratti 2014). Ca^{2+} transfer from intracellular stores to mitochondria is also emerging as a site of pivotal importance in the regulation of both cell death and cell survival pathways as well as for autophagy (for details, see Filadi et al. in this book). Interestingly, a recent work highlights that a novel, cardiac mitochondrial cAMP-dependent pathway controls mitochondrial Ca^{2+} entry through the MCUC in order to prevent PTP opening (Wang et al. 2016).

2.3 The Role of Mitochondrial Calcium in Shaping the Cytosolic Calcium Signaling

Mitochondria play a pivotal role in shaping cytosolic Ca^{2+} signals. This has been demonstrated experimentally, since mitochondrial Ca^{2+} buffering was shown to influence cellular Ca^{2+} signals and consequently cell function in many different cell types (Rizzuto and Pozzan 2006). Indeed, the observation that mitochondria rapidly accumulate Ca^{2+} upon stimulation allowed to design experiments to prove that these organelles contribute to the buffering of either the whole cytoplasm or of specific cellular domains. Mitochondria were shown to be able to rapidly remove Ca^{2+} from the mouth of the ER-located Ca^{2+} release channel inositol trisphosphate receptor (IP_3R) and hence to modify the total amount of Ca^{2+} released from intracellular stores to the cytosol. Ca^{2+} released across this channel exerts a feedback regulatory action, either activating, inhibitory, or biphasic, depending on the local [Ca^{2+}]. This effect is a consequence of the bell-shaped relationship between cytosolic calcium concentration ($[Ca^{2+}]_c$) and IP_3R opening, where low and high $[Ca^{2+}]_c$ inhibit channel activity, whereas intermediate $[Ca^{2+}]_c$ increases cation release (Bezprozvanny et al. 1991). Accordingly, removal of Ca^{2+} from the proximity of IP_3Rs on one hand reduces the stimulus to opening, and on the other, it relieves Ca^{2+}-dependent inhibition of open channels, thus promoting Ca^{2+} release. This was first demonstrated in *Xenopus* oocytes, where mitochondrial energization and the resulting increase in mitochondrial Ca^{2+} uptake was shown to coordinate IP_3-induced $[Ca^{2+}]_c$ rises into single propagating waves of low frequency and high amplitude (Jouaville et al. 1995). Regulation of the spatiotemporal patterning of cytosolic Ca^{2+} waves by mitochondria has been observed then in numerous different cell types. Instead, in plants, although our current knowledge points to mitochondrial calcium uptake occurring in vivo, this event does not seem to correlate with shaping of the cytosolic calcium signaling (Wagner et al. 2015).

Mitochondrial Ca^{2+} buffering could also participate to the local accumulation of a large amount of cations in a defined cell region, thanks to the precise positioning of the organelles. For example, redistribution of mitochondria to the immunological synapse was shown to be necessary to maintain Ca^{2+} influx across the plasma membrane and for Ca^{2+}-dependent helper T cell activation (Quintana et al. 2007). Further work highlighted that calcium-dependent inactivation of the calcium influx-mediating ORAI channels was prevented by localizing mitochondria close to ORAI channels. Thus, the redistribution of mitochondria following the formation of immunological synapse maximized the efficiency of calcium influx through ORAI channels, but it also decreased calcium clearance by the exit pathway (calcium ATPase of the plasma membrane), resulting in a more sustained NFAT activity and subsequent activation of T cells (Quintana et al. 2011).

Overall, mitochondria appear to be efficient, high capacity Ca^{2+} buffers that shape cytosolic Ca^{2+} transients by either regulating the properties of Ca^{2+}-releasing channels or limiting the wide spreading of $[Ca^{2+}]_c$ rises, at least in animals. The precise positioning of the organelle is critical for shielding defined cell portions in specific cells or for regulating calcium-mediated feedback mechanisms. While these statements are in general valid for most mammalian cell types, the role of mitochondrial calcium buffering is much less clear in other cell types, for example, in plants (Wagner et al. 2016).

2.4 Ca^{2+} Import into Mitochondria

2.4.1 Calcium Flux Across the Outer Mitochondrial Membrane

Similarly, to other small molecules, Ca^{2+} is thought to freely pass the outer mitochondrial membrane (OMM) through VDACs (voltage-dependent anion channels, also called porins). VDACs allow flux of metabolites and ions including Ca^{2+}, for which mammalian VDAC also possesses binding sites, as demonstrated both in vitro and in vivo (Bathori et al. 2006; De Stefani et al. 2012; Gincel et al. 2001; Israelson et al. 2007; Rapizzi et al. 2002; Shoshan-Barmatz et al. 2010; Rizzuto et al. 2009). In mammals and plants, functionally distinct protein isoforms have been found in the OMM (for recent reviews, see Shoshan-Barmatz et al. 2010; Rostovtseva 2012; Takahashi and Tateda 2013). VDACs, although defined as anion channels, can conduct a substantial flow of Ca^{2+}, as demonstrated both in vitro and in vivo for the mammalian protein. The importance of calcium flux across VDACs is highlighted by recent studies addressing different aspects. For example, efsevin was shown to bind to VDAC2, to potentiate mitochondrial Ca^{2+} uptake, and to accelerate the transfer of Ca^{2+} from intracellular stores into mitochondria. In cardiomyocytes, efsevin inhibited Ca^{2+} overload-induced erratic calcium waves, demonstrating that VDAC2-dependent mitochondrial Ca^{2+} uptake plays a critical role in the regulation of cardiac rhythmicity (Shimizu et al. 2015). Uptake of calcium via VDAC1 seems to be required for inhibition of apoptosis by anti-apoptotic proteins. In particular, the BH4 domain of Bcl-XL, but not that of Bcl-2, was shown to selectively target VDAC1 and to inhibit apoptosis by decreasing VDAC1-mediated Ca^{2+} uptake into the mitochondria (Monaco et al. 2015). Not only uptake of calcium via VDACs have profound effect on pathophysiological processes but also its release: recent data indicate that mitochondrial calcium, released through VDAC1, triggers Schwann cell demyelination via a signaling pathway including ERK1/ERK2, p38, JNK, and c-JUN activation following sciatic nerve injury. Importantly, reduction of mitochondrial calcium release, either by VDAC1 silencing or pharmacological inhibition, prevented demyelination (Gonzalez et al. 2016).

2.4.2 Calcium Flux Across the Inner Mitochondrial Membrane

Mitochondrial calcium uptake mostly occurs via the mitochondrial calcium uniporter complex MCUC, but other uptake modes, differing from MCUC-mediated Ca^{2+} uptake in terms of Ca^{2+} affinity, uptake kinetics, and pharmacological control, seem to coexist at least in the mammalian system. Here below, we provide an updated overview of the MCUC components and function in different systems and briefly mention the main characteristics of the other uptake modes as well.

2.4.2.1 The Mitochondrial Calcium Uniporter Complex (MCUC)

The main bioenergetic properties of the uniporter have been characterized in fine detail, and low concentrations of ruthenium red and Ru360 were shown to lead to a direct inhibition of the uniporter (Reed and Bygrave 1974; Vasington et al. 1972; Moore 1971). The finding that a highly Ca^{2+}-selective ion channel recorded in the inner mitochondrial membrane (IMM) (in mitoplasts) by patch clamping (Kirichok et al. 2004) recapitulated the key characteristics observed for the mammalian mitochondrial uniporter in classical bioenergetic experiments, together with the establishment of the MitoCarta database, containing more than 1,000 mitochondrial proteins as identified by subtractive proteomics and GFP-fusion localization studies (Pagliarini et al. 2008), finally led to the molecular identification of the pore-forming component of MCUC. In parallel, regulatory subunits have been shown to affect channel activity and/or mitochondrial calcium uptake in several cell types. While the presence of the pore-forming subunit and of at least one regulatory subunit is a recurrent feature throughout different kingdoms, the actual composition of MCUC greatly varies in different organisms. Interestingly, complexity of MCUC does not necessarily reflect evolutionary order. At the current stage, the mammalian MCUC appears to consist of at least of the pore-forming protein MCU, an MCU paralog (MCUb), the essential MCU regulator (EMRE), the regulatory MICU proteins, and, possibly, the mitochondrial calcium uniport regulator 1 (MCUR1), as discussed below.

The Pore-Forming MCU Component, CCDC109A

In 2011, the Mootha group and some of us independently identified the 40 kDa coiled-coiled protein CCDC109A that gives rise to Ca^{2+}-permeable channel activity. This protein called mitochondrial calcium uniporter (MCU) was proposed to be the core component of the calcium uniporter of the inner mitochondrial membrane (Baughman et al. 2011; De Stefani et al. 2011) as it is a phylogenetically ancient molecule present in most eukaryotic taxa (Bick et al. 2012). The channel-forming and regulatory components of MCUC are currently studied by classical electrophysiological techniques, either using recombinant proteins or by direct patch clamping of mitoplasts (Fig. 2.1).

In the first paper where CCDC109A was shown to give rise to MCU channel activity, the protein was studied in planar lipid bilayers and displayed electrophysiological properties and inhibitor sensitivity of the uniporter (De Stefani et al. 2011), previously identified as a Ca^{2+}-permeable ion channel in patch-clamp experiments on mammalian mitoplasts (Kirichok et al. 2004). The channel was recorded in both works in 100 mM calcium gluconate solution displaying a conductance of 6–7 pS, an increased open probability with increasing negative voltages (on matrix side), and sensitivity to ruthenium red (RR) and gadolinium. The pore-forming nature of MCU was further proven by the following observations: (i) siRNA against the MCU

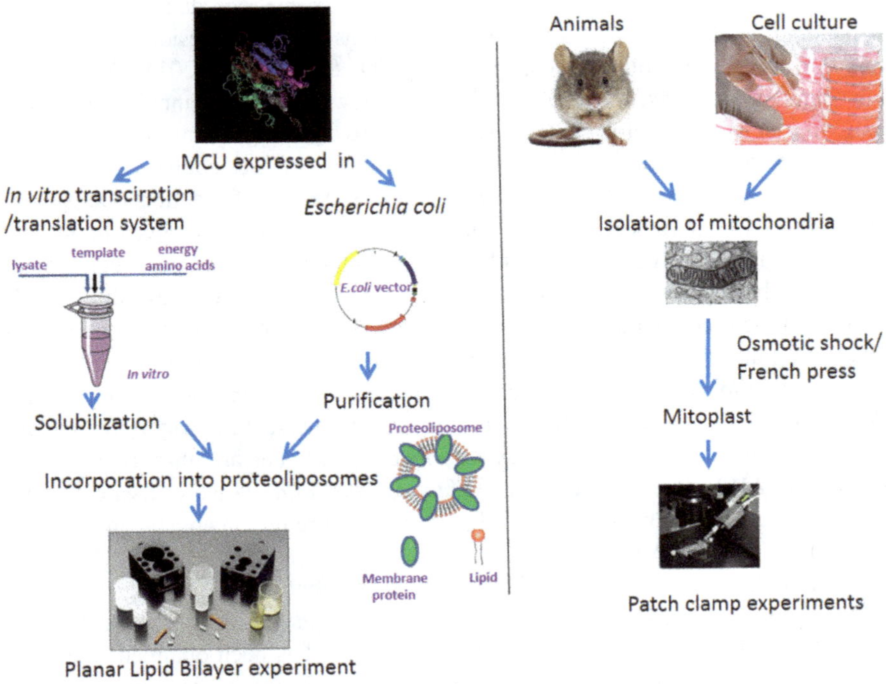

Fig. 2.1 Electrophysiological studies on the components of the MCU complex. The pore-forming protein MCU has been investigated in various studies either using the bilayer system with recombinant proteins (*left part*) or by direct patch clamping of mitoplasts (*right part*). The effect of regulators can be evaluated with both techniques, e.g., by co-incorporation of the regulator and the channel into proteoliposomes or by patch clamping of mitoplasts from knockout cells/animals lacking regulatory proteins. Shown structure of MCU is from PDB database (5ID3)

```
                 TM1          PORE        TM2
At4g36820   LWAGLGYLIIQTAGFMRLTFWDLTWDVMEPICFYVSSVYFMAGYTFFLKT
At2g23790   LWAGLGYLILQTAGFMRLTFWELSWDVMEPICFYVTSVYFMAGYAFFLRT
At5g66650   LWAGLGLIMAQTVGFFRLTFWELSWDVMEPICFYVTSTYFMAGYAFFLRT
At5g42610   LYCGLGFLAVQTIGFMRLTFWELSWDVMEPICFFVTTIHFILGYIFFLRT
At1g09575   LWCGLATSMVQIGLFFRLTFWEFSWDVMEPITFFATATGIIVGYAYFLMT
At1g57610   LWGGLGYSVVQIGIFVRLTFWEFSWDVMEPITFFTTATGIIVGYAYFLMT
hMCU        LWGGLAYMATQFGILARLTWWEYSWDIMEPVTYFITYGSAMAMYAYFVMT
cMCU        MWAGFAAMGVQTGLFARLTWWEYSWDIMEPVTYFATYSTVCATFGYYLYT
Dicty       IWTGLGYCFAQAAILARLTWWDLSWDIIEPVSYFLTFGSVLIGYTYFTMT
            :: *:.    *   : ***:*: :**::**: :: :       : ::  *
                                    ↑
```

Fig. 2.2 Amino acid sequence alignment of the transmembrane segments (*TM*) and the pore region of MCU from different species. Sequences for the six *Arabidopsis* isoforms and MCU from human (*hMCU*), from *C. elegans* (*cMCU*), and from *Dictyostelium* (*Dicty*) are shown. *Asterisks* show identical residues, while : indicates conservative substitutions. *Arrow* indicates the serine residue responsible for ruthenium red sensitivity

protein abolished the mitochondrial calcium uptake (De Stefani et al. 2011) (ii) siRNA against MCU abolished the calcium current recorded in mitoplasts (Chaudhuri et al. 2013). (iii) single-point mutations of specific residues in the pore region (see below) abolished channel activity (De Stefani et al. 2011). MCU does not share amino acid sequence similarity with known calcium channels in plants or animals, but its pore region contains several negatively charged amino acids that are crucial for calcium transport, since mutation of these residues abolished mitochondrial calcium uptake as well as calcium-carrying channel formation (De Stefani et al. 2011). In this region, a highly conserved serine residue is involved in binding of the inhibitor RR. These amino acids are highly conserved in the pore region of all species harboring homologs of the mammalian MCU, while much less similarity is present in the predicted transmembrane regions. MCU homologs are found in protozoa from diverse clades including kinetoplasts (*Trypanosoma cruzi*), heterolobosea (*Naegleria gruberi*), oomycetes (*Phytophthora infestans*), and ciliates (*Tetrahymena thermophila*). MCU homologs are also present in many fungi, including many basidiomycetes and *Allomyces macrogynus*, but are absent from yeast (Bick et al. 2012). In the model plant *Arabidopsis thaliana*, six isoforms are present. The amino acid sequences of the highly conserved pore region are shown for some of the MCUs in Fig. 2.2.

Despite the discovery that MCU homologs are present in different organisms and that the primary structure can be very different among species (except for the pore region), direct biophysical characterization of MCUs has been obtained solely for mammalian and plant proteins (Teardo et al. 2017). These studies reveal that the single channel conductance in calcium and the kinetic behavior are similar for mammalian and plant MCUs and that in both cases, sodium can permeate the channel when divalent cations are not present in the recording medium (Fig. 2.3). This aspect is typical of classical calcium channels (Talavera and Nilius 2006).

The consensus view concerning the topology of MCU is that both N- and C-terminal domains face the mitochondrial matrix, with the two membrane-

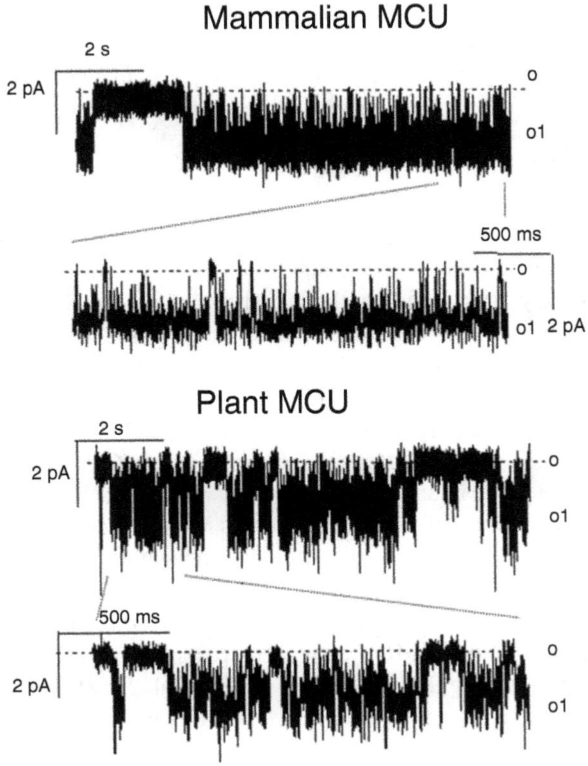

Fig. 2.3 Animal and plant MCU display similar activities in bilayer experiments. As mentioned in the text, MCU can conduct sodium in the absence of divalent cations. Current traces were recorded for all cases in 100 mM sodium gluconate, 10 mM Hepes, 5 mM NaEDTA, pH 7.3/NaOH. Applied voltage (on cis side of the chamber) is −40 mV. Current traces on expanded time scale are also shown

spanning domains connected in the intermembrane space (IMS) by the short pore region. The structure of the N-terminal domain has been resolved first, revealing that the N-terminal domain preceding the first coiled domain is essential for the modulation of MCU function: overexpression of MCU lacking this domain had a dominant-negative effect on mitochondrial Ca^{2+} uptake (Lee et al. 2015). A more recent structural study suggests that similarly to some other classical ion channels, pentamerization of the two transmembrane helix-containing subunits is required for formation of a functional channel (Oxenoid et al. 2016).

SILAC-based quantitative proteomics showed varying expression level of MCU in different mouse tissues, with cerebellum and gut showing highest level (Murgia and Rizzuto 2015). Several ion channels (including mitochondrial ones) display an altered expression in cancer cells, but this is apparently not the case of MCU (Peruzzo et al. 2016). Interestingly, while only one isoform of MCU is present in most organisms, six homologs of *MCU* were identified in the genomes of maize and *Arabidopsis* (Stael et al. 2012; Meng et al. 2015) (see also Fig. 2.2). The first proteomic evidence from *Arabidopsis* and potato suggests the presence of specific MCU homologs in mitochondrial fractions at low relative abundance, in accordance with MCU being and organellar ion channel (Wagner et al. 2015). The diversification of *MCU* genes in plants may provide regulatory flexibility on the different levels of gene expression, transcription, translation, and posttranslational organization/regulation. This idea is supported by differential tissue-dependent expression of *MCU* genes (Stael et al. 2012; Meng et al. 2015; Teardo et al. 2017).

The Dominant-Negative Pore-Forming MCU Component, CCDC109B

Mammalian MCU activity can be regulated through its paralog, CCDC109B/MCUb. MCUb is a 33 kDa protein with a very similar structure to MCU but devoid of calcium-permeable channel activity in bilayer experiments due to the substitution of key amino acids in the pore-forming region. The protein however is still able to form sodium-permeable channel in the absence of divalent cations (Raffaello et al. 2013, Raffaello et al. 2016). MCUb and MCU have been shown to interact and to form hetero-oligomers. When proteoliposomes contained both MCU and MCUb, the presence of MCUb decreased the likelihood of observing calcium-permeable channel activity, strongly indicating that MCUb is a dominant-negative regulatory subunit. In intact cells, overexpression of MCUb reduced the amplitude of calcium uptake into mitochondria, whereas MCUb silencing had the opposite effect, further proving that MCUb incorporates into the uniporter channel oligomer and reduces its activity. The ratio between MCU and MCUb might define the stoichiometry of channel assembly, thus setting a cell-specific baseline of MCU activity in various tissues. Indeed, MCU activity, as recorded by patch clamping of mitoplast, greatly varies among different tissues (Fieni et al. 2012). The molecular basis of this difference may lay in the ratio between MCU and MCUb that, at least at the mRNA level, has been shown to vary in different tissues. Interestingly, tissues characterized by low-amplitude mitochondrial calcium transients, such as the heart, show a relative abundance of MCUb, compared, for example, to skeletal muscle.

MICU Proteins: The EF-Hand Containing Regulatory Subunits

The mammalian MICU (mitochondrial calcium uptake) protein family consists of three members that share more than 40% sequence identity: (i) MICU1, (ii) MICU2, and (iii) MICU3. MICU1 was the first of the components of the complex to be described: it is a 50 kDa protein with two functional and two pseudo EF-hands that resides in the mitochondrial intermembrane space (Csordas et al. 2013; Patron et al. 2014; Hung et al. 2014; Petrungaro et al. 2015; Wang et al. 2014). MICU2 is a paralog of MICU1 with 27% sequence identity. It was first described as a protein whose silencing resulted in reduced mitochondrial calcium clearance in response to large extramitochondrial calcium pulses. Finally, comparative genomics analyses revealed also the presence of a third protein, MICU3. Whereas MICU1 and MICU2 had wide tissue expression, MICU3 was found to be almost exclusively expressed in neural tissues (Plovanich et al. 2013), and at present, the precise molecular function MICU3 remains unclear. All three proteins as well as their homologs in other organism harbor calcium-binding EF-hand helix-loop-helix motifs. The presence of EF-hands is a typical feature of Ca^{2+} sensors in animals in plants, but not every

Ca^{2+}-binding protein carries an EF-hand, and not every EF-hand binds Ca^{2+} (e.g., Gelhaye et al. 2004). Recent evidence indicates that increases in cytosolic Ca^{2+} are sufficient to induce rearrangement of MICU1 multimers and to trigger activation of mitochondrial Ca^{2+} uptake (EC_{50} of 4.4 μM) (Waldeck-Weiermair et al. 2015), in agreement with the emerging hypothesis that at low extramitochondrial Ca^{2+}concentrations MICU1 exerts a gatekeeping function, while it activates the channel when surrounding Ca^{2+} concentrations are high (Csordas et al. 2013; Mallilankaraman et al. 2012b). According to this hypothesis, MICU1 is sufficient to control calcium flux into mitochondria (Garg and Kirichok 2016; Tsai et al. 2016). Recombinant MICU1 was shown to directly increase MCU activity in planar lipid bilayer experiments in the presence of calcium (Patron et al. 2014). Multiple experimental evidence indicates that in resting conditions, MICU1-MICU2 heterodimers act as the MCU gatekeeper, while increases in calcium concentration, by inducing a conformational change in the dimer, would release MICU2-dependent inhibition and trigger MICU1-mediated enhancement of MCU channel activity (Patron et al. 2014). MICU2 indeed forms a heterodimer with MICU1 through an intermolecular disulfide bond and closes the channel at low extramitochondrial Ca^{2+} concentrations (Patron et al. 2014; Petrungaro et al. 2015). The stability of MICU2 depends on MICU1 (Plovanich et al. 2013; Patron et al. 2014), and loss of MICU2 in MICU1-silenced cells renders the difficult assignment of individual MICU1 and MICU2 functions. In summary, currently two models propose either MICU1 (1) to act exclusively as a uniporter activator at high cytosolic Ca^{2+} concentrations (Patron et al. 2014) or (2) to gradually disinhibit the uniporter with increasing Ca^{2+} concentrations in the cytosol (Csordas et al. 2013). However, hetero-dimerization with the ubiquitous MICU2 should be taken into account, at least in mammals when trying to describe models that best reflect the in vivo situation.

In addition to MICU1, skeletal muscle-specific alternative splice isoform of MICU1, MICU1.1, characterized by the addition of a micro-exon has recently been described (Vecellio Reane et al. 2016). MICU1.1 was shown to bind Ca^{2+} one order of magnitude more efficiently than MICU1 and activated MCU-mediated calcium uptake at lower Ca^{2+} concentrations than MICU1-MICU2 heterodimers.

MICU protein is conserved also in plants, where typically one or two homologs can be found depending on species (Wagner et al. 2015). *Arabidopsis* possesses only a single *MICU* gene, and knockout strongly affects mitochondrial Ca^{2+} dynamics (Wagner et al. 2015). *Arabidopsis* MICU contains an additional, third canonical EF-hand motif, which is conserved among plants and protists but is absent in mammalian MICUs and may open additional complexity of the regulation of MCUC activity by calcium.

Essential MCU Regulator EMRE

Another proposed core component of the mammalian MCUC is EMRE, a 10 kDa metazoan-specific protein that spans the inner mitochondrial membrane with only one transmembrane motif. Although the recombinant MCU protein when inserted

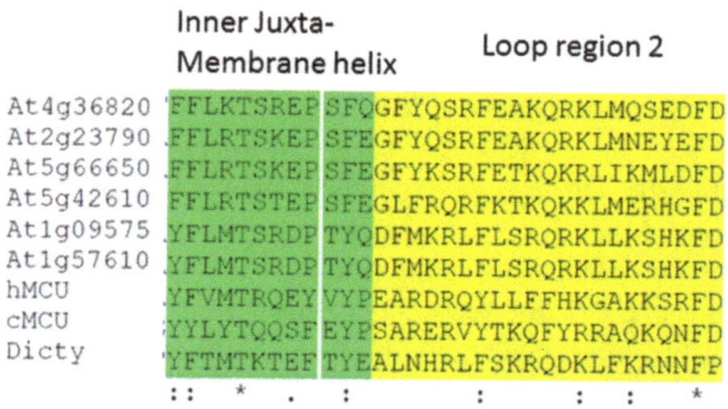

Fig. 2.4 Amino acid sequence alignment of the inner juxta membrane helix and of the loop region. Only few amino acids are conserved among different species in the regions proposed to be important for calcium conduction within the channel. Sequences for the six *Arabidopsis* isoforms and MCU from human (*hMCU*), from *C. elegans* (*cMCU*), and from *Dictyostelium* (*Dicty*) are shown. *Asterisks* show identical residues, while : indicates conservative substitutions. See text for further details

into a bilayer membrane in electrophysiological experiments alone is able to form a functional channel (De Stefani et al. 2011), in vivo EMRE seems to be required for channel formation, at least in mammals. A physical interaction between transmembrane helices between mammalian MCU and EMRE has been shown to take place (Tsai et al. 2016). Homologs of EMRE are not present in plants, fungi, or protozoa, and it has been recently shown that EMRE is required for Ca^{2+} uptake in the case of mammalian MCU, but not of MCU from the slime mold *Dictyostelium discoideum*: while expression of MCU from *Dictyostelium* alone was sufficient to import Ca^{2+} into yeast mitochondria (which lacks MCUC), human EMRE needed to be expressed alongside MCU to form an active Ca^{2+} uniporter system (Kovacs-Bogdan et al. 2014). Similarly to *Dictyostelium*, plants possess a minimal genetic uniporter configuration that lacks EMRE (Wagner et al. 2015, 2016).

Information about the structure of *C. elegans* MCU has recently become available: it was hypothesized that the outer and inner juxtamembrane helices as well as the loop region L2 are unstable regions which may undergo conformational changes upon activation by EMRE in order to create the lateral exit path for Ca^{2+} (Oxenoid et al. 2016). In plant and *Dictyostelium* MCUs, relatively few amino acids are conserved in the regions proposed to be important for the regulation of the *C. elegans* channel (Fig. 2.4).

Further research is needed to understand whether differences in these regions might account for the differential participation of EMRE in channel activity, and in general, mutations of the very few highly conserved amino acids in these regions might bring to the fine elucidation of ion permeation through this novel type of calcium channel. EMRE has been proposed to have another role as well, i.e., to bridge MCU and its regulators MICU1/MICU2 and thus to be indispensable for the activity of the mammalian uniporter in vivo (Sancak et al. 2013). However, when the binding properties of MCU/EMRE with MICU1 and MICU1.1 were investigated by surface plasmon resonance analysis, at least under the used experimental conditions (in vitro), EMRE was not found to be not involved in MCU-MICU1 interaction, in accordance with the electrophysiological data obtained regarding the effect of MICU1 on MCU activity (Vecellio Reane et al. 2016). Recently, EMRE was proposed to regulate MCU channel activity depending on the matrix Ca^{2+} concentration (Vais et al. 2016). Altogether, the role of EMRE is far from being clarified, even though MCU together with EMRE and MICU1 has been proposed to correspond to the minimal configuration of MCUC (Tsai et al. 2016). This statement however cannot be true for organisms where EMRE or a homolog is not present in the genome (like in plants and slime mold). It cannot be a priori excluded that a still unidentified protein fulfills the same function in plants and other EMRE-lacking organisms. Likewise, the possibility that EMRE helps the correct membrane insertion/folding of MCU cannot be dismissed as to date yet.

MCU Regulator 1 (MCUR1)

MCUR1 (mitochondrial calcium uniporter regulator 1)/CCDC90A is a 39 kDa protein with two predicted transmembrane domains that is supposed to interact with MCU (Mallilankaraman et al. 2012a) although later studies were unable to support this interaction (Sancak et al. 2013; Paupe et al. 2015). Paupe et al. (2015) provided evidence that MCUR1 is in fact an assembly factor of cytochrome *c* oxidase and argued that genetic manipulation modulates mitochondrial membrane potential, imposing only a secondary effect on Ca^{2+} transport. In support of this notion, MCUR1 has an orthologue in budding yeast which lacks core MCUC components. Although Vais et al. (2015) recently showed that MCUR1 affects MCU activity in patch-clamp experiments, direct regulation of Ca^{2+} uniport through MCUR1 is still debated. *Arabidopsis* possesses two MCUR1 homologs that lack functional characterization. Interestingly one of them has been identified as a plant-specific subunit of complex IV by proteome analysis (Millar et al. 2004; Klodmann et al. 2011).

Altogether, functional MCUC has different components in different organisms, with MCU and MICU family members being the only highly conserved constituents. Figure 2.5 shows the composition of MCUC in organisms where characteristics of this complex have been studied in detail.

2.4.2.2 Alternative Calcium Uptake Pathways

Additional Ca^{2+} uptake mechanisms in mammalian mitochondria were proposed by several groups still before the discovery of MCU. Ca^{2+} transients in mammalian cell culture where MCU expression is knocked down (De Stefani et al. 2011; Baughman

MCUC component	*Trypanosoma cruzi*	*Dictyostelium discoideum*	*Arabidopsis thaliana*	*Mus musculus*
MCU	+	+	+ (6 isoforms)	+
MCUb	-	-	-	+
EMRE	-	-	-	+
MICU	+	+	+	+ (3 isoforms)
MCUR	-	-	+ (2 isoforms)	+

Fig. 2.5 The presence of MCUC components in different organisms. See text for further details

et al. 2011; Bondarenko et al. 2014) and in the liver of *mcu* animals (Pan et al. 2013) are very efficiently abolished indicating that the MCUC has a dominating role among uptake mechanisms. This of course does not rule out the possibility that other mechanisms make major contributions to Ca^{2+} uptake, for example, during a specific developmental stage or in a specific tissue. Indeed, specific mitochondrial Ca^{2+} uptake modes (e.g., Ca^{2+}-selective conductance (mCa) 1 and 2 and rapid mode of uptake (RaM)) have been observed in animals, which however currently cannot be ascribed to well-defined molecules. These uptake modes apparently differ from MCUC-mediated Ca^{2+} uptake in terms of uptake kinetics, pharmacology, and Ca^{2+} affinity (Sparagna et al. 1995; Michels et al. 2009). Potential candidates for these distinct uptake modes include uncoupling proteins 2 and 3 (UCP2/UCP3), the transient receptor potential channel TRPC3, and ryanodine receptor RyR1. UCP2/UCP3 were originally proposed to be an essential components of mitochondrial Ca^{2+} uniport (Trenker et al. 2007), but currently it seems more likely that it has indirect effects on Ca^{2+} uptake into mitochondria (Brookes et al. 2008; De Marchi et al. 2011; Bondarenko et al. 2015). The mitochondrial ryanodine receptor (mRyR1) belongs to the RyR family that exists as three isoforms (RyR1–3) in animals but has no homologs in plants (Krinke et al. 2007). A low level of RyR1 is detectable in the IMM of heart mitochondria and provides rapid transport of Ca^{2+} that is insensitive to ruthenium red (Beutner et al. 2001, 2005). A small fraction of TRPC3 was found to be localized to mitochondria. It was then proven by genetic means that a significant fraction of mitochondrial Ca^{2+} uptake relies on TRPC3 expression (Wang et al. 2015; Feng et al. 2013). In summary, the abovementioned alternative pathways should be taken into account when interpreting the phenotypes observed in MCU knockdown systems. Finally, the hypothesis that MCUC is responsible also for the different uptake modes, at least in some cell types, cannot be formally excluded.

2.5 Mitochondrial Ca^{2+} Export

Following the transient accumulation of calcium in the matrix, a part of this ion is exported, and a part remains inside as non-free calcium. The exact chemical states of bound Ca^{2+} inside the matrix of the living cell and the relative contributions of proteins, metabolites, and Pi are largely unclear in both animals and plants. Ca^{2+} can be extruded from mitochondria by an antiport mechanisms giving rise to the so-called Ca^{2+} cycle (Carafoli 1979). This way matrix Ca^{2+} concentrations are regulated in order to avoid overload, which can be deleterious for mitochondrial function (see above). Two known Ca^{2+} export systems are two exchangers, namely, the cation/cation exchanger family member Na^+/Ca^{2+} exchanger (Crompton et al. 1977, 1978) and a H^+/Ca^{2+} exchanger (Akerman 1978; Fiskum and Lehninger 1979) of the cation/proton exchanger family.

2.5.1 *The Sodium-Calcium Exchanger NCLX and the Proton-Calcium Exchanger CAX*

The mammalian protein NCLX (Na/Li/Ca exchanger) (Palty et al. 2010) has been proposed to underlie molecular entity of electrogenic Ca^{2+} transport against Na^+. De Marchi et al. (2014) have provided further relevant evidence that NCLX represents the long-sought mediator of Ca^{2+} export from the mitochondrial matrix. As to CAX, this protein was located to mitochondria in *Plasmodium falciparum*, where it mediates Ca^{2+} efflux from the mitochondrial matrix (Rotmann et al. 2010). In other organisms the mechanisms of calcium release from mitochondria is less clear.

2.5.2 *Leucine Zipper-EF-Hand-Containing Transmembrane Protein1 (LETM1)*

LETM1 is a one-transmembrane segment-containing protein that is located in the mitochondrial inner membrane and is defective in Wolf-Hirschhorn syndrome (Zollino et al. 2003; Endele et al. 1999; Dimmer et al. 2008). Initially proposed to act as an a K^+/H^+-exchanger (Nowikovsky et al. 2004; Dimmer et al. 2008), a genome-wide RNAi screen for proteins mediating mitochondrial Ca^{2+} dynamics identified LETM1 as a Ca^{2+}/H^+ antiporter (Jiang et al. 2009; Waldeck-Weiermair et al. 2011; Tsai et al. 2014; Doonan et al. 2014). In vitro, LETM1 has been proposed to function as electroneutral Ca^{2+}/H^+ antiporter (Tsai et al. 2014). Recent electron microscopy studies reveal a hexameric structure with a central cavity and with two different conformational states under alkaline and acidic conditions (Shao et al. 2016). While a H^+-driven Ca^{2+} export by LETM1 is plausible, whether

LETM1 in vivo functions as Ca^{2+}/H^+ antiporter or as a K^+/H^+ exchanger, as proposed by Nowikovsky and Bernardi (2014), is still a highly debated issue. Several arguments point to LETM1 working as potassium/proton antiporter in intact mitochondria. For example, changes in mitochondrial morphology with altered LETM1 expression could be reverted through the ionophore nigericin that specifically mediates K^+/H^+-exchange (Nowikovsky et al. 2004). Interestingly, high-level expression of LETM1 was found to be an independent poor prognostic factor of breast cancer (Li et al. 2015). In plants, the *Arabidopsis* genome contains two genes with homology to *LETM1* (Zhang et al. 2012). Partial depletion of LETM did not affect mitochondrial morphology. Instead, mitochondrial protein translation was altered, possibly as a secondary effect of disrupted K^+ homeostasis (Hashimi et al. 2013), based on the observation that nigericin rescued the translation phenotype in cultured yeast cells.

2.5.3 Permeability Transition Pore

Transient opening of the mitochondrial permeability transition pore (PTP) has been proposed to cause release of Ca^{2+} from mammalian mitochondria (Bernardi and von Stockum 2012). At partial loss of membrane potential due to the opening of PTP, a large Ca^{2+} gradient (expected only at Ca^{2+} overload) would allow Ca^{2+} extrusion in a passive way. However, a partially or fully dissipated electrochemical gradient would not only allow Ca^{2+} extrusion but also severely interfere with matrix physiology, including ATP/ADP exchange, P_i uptake, and metabolite shuttling, which strictly depend on the proton motive force. Thus, it seems likely that such Ca^{2+} release via PTP occurs under specific, pathological conditions. Experimental evidence is still missing to either prove or disprove the above hypotheses.

2.6 Pathophysiological In Vivo Consequences of Alteration of Mitochondrial Calcium Homeostasis by Genetic Tools

Following the identification of proteins playing fundamental roles in the calcium uptake and exit pathways, the field of mitochondrial calcium signaling experienced a period of "Renaissance." Finally, fine dissection of the molecular pathways governing mitochondrial calcium homeostasis has become feasible using genetic tools. However, when interpreting the final outcome of knockout or knockdown experiments in terms of calcium levels, of metabolism, and of cell fate, one has to keep in mind that genetic manipulation of one MCUC component might lead to altered expression of (another) component(s) as well (e.g., when MICU1 is silenced, a dramatic reduction also of MICU2 protein occurs (Patron et al. 2014).

A few recent in vivo studies demonstrate that mitochondrial calcium homeostasis is crucial for regulation of metabolism, and its alterations are linked to pathologies. Genetic manipulation of MCU in lower organisms such as zebra fish (Prudent et al. 2013) and *Trypanosoma brucei* (Huang et al. 2013) resulted in major developmental and energetic defects, although such effect was not accentuated in the knockout mouse model, possibly due to compensatory mechanisms. Low levels of basal matrix calcium in the $MCU^{-/-}$ mice led to markedly increased levels of PDH phosphorylation (Pan et al. 2013). In another work, postnatal manipulation of MCU levels in mice (by using adeno-associated virus-mediated gene transfer) demonstrated the contribution of MCUC to the regulation of skeletal muscle tropism. MCU overexpression or downregulation caused muscular hypertrophy or atrophy, respectively, likely independent of metabolic alterations but dependent on a novel Ca^{2+}-dependent mitochondria-to-nucleus signaling pathway via transcriptional regulators (Mammucari et al. 2015). In mice with myocardial MCU inhibition, obtained by transgenic expression of a dominant-negative (DN) MCU, a strong correlation between MCU function, MCU-enhanced oxidative phosphorylation, and correct pacemaker cell function was observed (Wu et al. 2015). In addition, in vivo evidence exists in favor of a serine/threonine kinase LKB1-mediated regulation of MCU expression that controls mitochondrial calcium uptake and neurotransmitter release properties in a bouton-specific way through presynaptic Ca^{2+} clearance (Kwon et al. 2016). Changes of mitochondrial calcium level in neurons activated by insulin-like growth factor-1 receptor signaling also constitute a critical regulator of information processing in hippocampal neurons by maintaining evoked-to-spontaneous transmission ratio as assessed in vivo (Gazit et al. 2016). Furthermore, inhibition of MCU in *Drosophila*, during development in a brain region that is critical for olfactory memory formation, caused memory impairment in adults without altering the capacity to learn (Drago and Davis 2016). Lack of one of the MCU isoforms of *Arabidopsis* with prevalent expression in roots caused a profoundly altered mitochondrial ultrastructure and shortened root length in intact plants (Teardo et al. 2017).

As to the regulator, MICU1, mitochondria in a mouse model of MICU1 deficiency showed altered calcium uptake. Deletion of MICU1 resulted in significant perinatal mortality. MICU1 knockout animals displayed increased resting mitochondrial calcium levels, altered mitochondrial morphology, and reduced ATP. Deletion of one allele of EMRE helped to normalize calcium uptake while simultaneously rescuing the high perinatal mortality observed in young MICU1 $^{-/-}$ mice (Liu et al. 2016). In humans, homozygous patients carrying a loss-of-function mutation of MICU1 are characterized by myopathy, cognitive impairment, and extrapyramidal movement disorder (Logan et al. 2014), along with an increased agonist-induced mitochondrial Ca^{2+} uptake at low cytosolic Ca^{2+} concentrations and a decreased cytosolic Ca^{2+} signal. However, at least under resting conditions, the fibroblasts from affected individuals do not display defects in overall cellular metabolic function, but chronic elevation of the mitochondrial matrix Ca^{2+} load seems to lead to moderate mitochondrial stress, resulting in fragmentation of the mitochondrial network. In addition, homozygous deletion of exon 1 of MICU1 was shown to be associated with fatigue and lethargy in children with normal mitochondrial oxidative phosphorylation enzyme activities in muscle (Lewis-Smith et al. 2016).

2.7 Conclusion

As witnessed by the above-reported data, a considerable advancement has been achieved in the field of mitochondrial calcium handling in the last few years. Now, an even more exciting and stimulating era is expected to come. Following the identification of the calcium uptake and exit machineries, research will most probably focus on understanding the fine regulation of these components, e.g., by posttranslational modifications, and on elucidation of their role in different physiological and pathologic situations. Hopefully, the recently obtained information regarding the structure of several components will also prompt smart drug design in order to fully exploit the information arising in the field, in the context of pathologies linked to altered mitochondrial calcium handling.

Acknowledgments The authors are grateful for financial support by the Italian Association for Cancer Research (15544 to I.S. and 10016 to R.R.), the Italian Ministry (Progetti di Rilevanza Nazionale PRIN 2015795S5W to I.S.), the French Muscular Dystrophy Association (19471 to A.R.), and the European Research Council (ERC mitoCalcium no. 294777 to R.R.).

References

Akerman KE (1978) Effect of pH and Ca2+ on the retention of Ca2+ by rat liver mitochondria. Arch Biochem Biophys 189:256–262

Akerman KE, Moore AL (1983) Phosphate dependent, ruthenium red insensitive CA2+ uptake in mung bean mitochondria. Biochem Biophys Res Commun 114:1176–1181

Bathori G, Csordas G, Garcia-Perez C, Davies E, Hajnoczky G (2006) Ca2+–dependent control of the permeability properties of the mitochondrial outer membrane and voltage-dependent anion-selective channel (VDAC). J Biol Chem 281:17347–17358

Baughman JM, Perocchi F, Girgis HS, Plovanich M, Belcher-Timme CA, Sancak Y, Bao XR, Strittmatter L, Goldberger O, Bogorad RL, Koteliansky V, Mootha VK (2011) Integrative genomics identifies MCU as an essential component of the mitochondrial calcium uniporter. Nature 476:341–345

Bernardi P, Von Stockum S (2012) The permeability transition pore as a Ca(2+) release channel: new answers to an old question. Cell Calcium 52:22–27

Bernardi P, Rasola A, Forte M, Lippe G (2015) The mitochondrial permeability transition pore: channel formation by F-ATP synthase, integration in signal transduction, and role in pathophysiology. Physiol Rev 95:1111–1155

Berridge MJ, Bootman MD, Roderick HL (2003) Calcium signalling: dynamics, homeostasis and remodelling. Nat Rev Mol Cell Biol 4:517–529

Beutner G, Sharma VK, Giovannucci DR, Yule DI, Sheu SS (2001) Identification of a ryanodine receptor in rat heart mitochondria. J Biol Chem 276:21482–21488

Beutner G, Sharma VK, Lin L, Ryu SY, Dirksen RT, Sheu SS (2005) Type 1 ryanodine receptor in cardiac mitochondria: transducer of excitation-metabolism coupling. Biochim Biophys Acta 1717:1–10

Bezprozvanny I, Watras J, Ehrlich BE (1991) Bell-shaped calcium-response curves of Ins(1,4,5) P3- and calcium-gated channels from endoplasmic reticulum of cerebellum. Nature 351:751–754

Bick AG, Calvo SE, Mootha VK (2012) Evolutionary diversity of the mitochondrial calcium uniporter. Science 336:886

Bondarenko AI, Jean-Quartier C, Parichatikanond W, Alam MR, Waldeck-Weiermair M, Malli R, Graier WF (2014) Mitochondrial Ca(2+) uniporter (MCU)-dependent and MCU-independent Ca(2+) channels coexist in the inner mitochondrial membrane. Pflugers Arch 466:1411–1420
Bondarenko AI, Parichatikanond W, Madreiter CT, Rost R, Waldeck-Weiermair M, Malli R, Graier WF (2015) UCP2 modulates single-channel properties of a MCU-dependent Ca(2+) inward current in mitochondria. Pflugers Arch 467:2509–2518
Brookes PS, Parker N, Buckingham JA, Vidal-Puig A, Halestrap AP, Gunter TE, Nicholls DG, Bernardi P, Lemasters JJ, Brand MD (2008) UCPs – unlikely calcium porters. Nat Cell Biol 10:1235–1237. author reply 1237-40
Budde RJ, Fang TK, Randall DD (1988) Regulation of the phosphorylation of mitochondrial pyruvate dehydrogenase complex in situ: effects of respiratory substrates and calcium. Plant Physiol 88:1031–1036
Carafoli E (1979) The calcium cycle of mitochondria. FEBS Lett 104:1–5
Chaudhuri D, Sancak Y, Mootha VK, Clapham DE (2013) MCU encodes the pore conducting mitochondrial calcium currents. elife 2:e00704
Chen CH, Lehninger AL (1973) Ca 2+ transport activity in mitochondria from some plant tissues. Arch Biochem Biophys 157:183–196
Crompton M, Kunzi M, Carafoli E (1977) The calcium-induced and sodium-induced effluxes of calcium from heart mitochondria. Evidence for a sodium-calcium carrier. Eur J Biochem 79:549–558
Crompton M, Moser R, Ludi H, Carafoli E (1978) The interrelations between the transport of sodium and calcium in mitochondria of various mammalian tissues. Eur J Biochem 82:25–31
Csordas G, Golenar T, Seifert EL, Kamer KJ, Sancak Y, Perocchi F, Moffat C, Weaver D, de la Fuente Perez S, Bogorad R, Koteliansky V, Adijanto J, Mootha VK, Hajnoczky G (2013) MICU1 controls both the threshold and cooperative activation of the mitochondrial Ca(2)(+) uniporter. Cell Metab 17:976–987
Day DA, Bertagnolli BL, Hanson JB (1978) The effect of calcium on the respiratory responses of corn mitochondria. Biochim Biophys Acta 502:289–297
De Marchi U, Castelbou C, Demaurex N (2011) Uncoupling protein 3 (UCP3) modulates the activity of Sarco/endoplasmic reticulum Ca2+-ATPase (SERCA) by decreasing mitochondrial ATP production. J Biol Chem 286:32533–32541
De Marchi U, Santo-Domingo J, Castelbou C, Sekler I, Wiederkehr A, Demaurex N (2014) NCLX protein, but not LETM1, mediates mitochondrial Ca2+ extrusion, thereby limiting Ca2+-induced NAD(P)H production and modulating matrix redox state. J Biol Chem 289:20377–20385
De Stefani D, Raffaello A, Teardo E, Szabo I, Rizzuto R (2011) A forty-kilodalton protein of the inner membrane is the mitochondrial calcium uniporter. Nature 476:336–340
De Stefani D, Bononi A, Romagnoli A, Messina A, De Pinto V, Pinton P, Rizzuto R (2012) VDAC1 selectively transfers apoptotic Ca2+ signals to mitochondria. Cell Death Differ 19:267–273
Deluca HF, Engstrom GW (1961) Calcium uptake by rat kidney mitochondria. Proc Natl Acad Sci U S A 47:1744–1750
Denton RM (2009) Regulation of mitochondrial dehydrogenases by calcium ions. Biochim Biophys Acta 1787:1309–1316
Denton RM, Mccormack JG (1980) On the role of the calcium transport cycle in heart and other mammalian mitochondria. FEBS Lett 119:1–8
Denton RM, Randle PJ, Martin BR (1972) Stimulation by calcium ions of pyruvate dehydrogenase phosphate phosphatase. Biochem J 128:161–163
Denton RM, Richards DA, Chin JG (1978) Calcium ions and the regulation of NAD+-linked isocitrate dehydrogenase from the mitochondria of rat heart and other tissues. Biochem J 176:899–906
Denton RM, Mccormack JG, Edgell NJ (1980) Role of calcium ions in the regulation of intramitochondrial metabolism. Effects of Na+, Mg2+ and ruthenium red on the Ca2+-stimulated oxidation of oxoglutarate and on pyruvate dehydrogenase activity in intact rat heart mitochondria. Biochem J 190:107–117

Dieter P, Marme D (1980) Ca(2+) transport in mitochondrial and microsomal fractions from higher plants. Planta 150:1–8

Dimmer KS, Navoni F, Casarin A, Trevisson E, Endele S, Winterpacht A, Salviati L, Scorrano L (2008) LETM1, deleted in Wolf-Hirschhorn syndrome is required for normal mitochondrial morphology and cellular viability. Hum Mol Genet 17:201–214

Doonan PJ, Chandramoorthy HC, Hoffman NE, Zhang X, Cardenas C, Shanmughapriya S, Rajan S, Vallem S, Chen X, Foskett JK, Cheung JY, Houser SR, Madesh M (2014) LETM1-dependent mitochondrial Ca2+ flux modulates cellular bioenergetics and proliferation. FASEB J 28:4936–4949

Drago I, Davis RL (2016) Inhibiting the mitochondrial calcium uniporter during development impairs memory in adult drosophila. Cell Rep 16:2763–2776

Endele S, Fuhry M, Pak SJ, Zabel BU, Winterpacht A (1999) LETM1, a novel gene encoding a putative EF-hand Ca(2+)-binding protein, flanks the Wolf-Hirschhorn syndrome (WHS) critical region and is deleted in most WHS patients. Genomics 60:218–225

Feng S, Li H, Tai Y, Huang J, Su Y, Abramowitz J, Zhu MX, Birnbaumer L, Wang Y (2013) Canonical transient receptor potential 3 channels regulate mitochondrial calcium uptake. Proc Natl Acad Sci U S A 110:11011–11016

Fieni F, Lee SB, Jan YN, Kirichok Y (2012) Activity of the mitochondrial calcium uniporter varies greatly between tissues. Nat Commun 3:1317

Fiskum G, Lehninger AL (1979) Regulated release of Ca2+ from respiring mitochondria by Ca2+/2H+ antiport. J Biol Chem 254:6236–6239

Garg V, Kirichok Y (2016) Keeping a lid on calcium uptake. elife 5:e17293

Gazit N, Vertkin I, Shapira I, Helm M, Slomowitz E, Sheiba M, Mor Y, Rizzoli S, Slutsky I (2016) IGF-1 receptor differentially regulates spontaneous and evoked transmission via mitochondria at hippocampal synapses. Neuron 89:583–597

Gelhaye E, Rouhier N, Gerard J, Jolivet Y, Gualberto J, NAVROT N, Ohlsson PI, Wingsle G, Hirasawa M, Knaff DB, Wang H, Dizengremel P, Meyer Y, Jacquot JP (2004) A specific form of thioredoxin h occurs in plant mitochondria and regulates the alternative oxidase. Proc Natl Acad Sci U S A 101:14545–14550

Gincel D, Zaid H, Shoshan-Barmatz V (2001) Calcium binding and translocation by the voltage-dependent anion channel: a possible regulatory mechanism in mitochondrial function. Biochem J 358:147–155

Gonzalez S, Berthelot J, Jiner J, Perrin-Tricaud C, Fernando R, Chrast R, Lenaers G, Tricaud N (2016) Blocking mitochondrial calcium release in Schwann cells prevents demyelinating neuropathies. J Clin Invest 126:1023–1038

Greenawalt JW, Rossi CS, Lehninger AL (1964) Effect of active accumulation of calcium and phosphate ions on the structure of rat liver mitochondria. J Cell Biol 23:21–38

Hashimi H, Mcdonald L, Stribrna E, Lukes J (2013) Trypanosome Letm1 protein is essential for mitochondrial potassium homeostasis. J Biol Chem 288:26914–26925

Hodges TK, Hanson JB (1965) Calcium accumulation by maize mitochondria. Plant Physiol 40:101–109

Holness MJ, Sugden MC (2003) Regulation of pyruvate dehydrogenase complex activity by reversible phosphorylation. Biochem Soc Trans 31:1143–1151

Huang G, Vercesi AE, Docampo R (2013) Essential regulation of cell bioenergetics in Trypanosoma brucei by the mitochondrial calcium uniporter. Nat Commun 4:2865

Hung V, Zou P, Rhee HW, Udeshi ND, Cracan V, Svinkina T, Carr SA, Mootha VK, Ting AY (2014) Proteomic mapping of the human mitochondrial intermembrane space in live cells via ratiometric APEX tagging. Mol Cell 55:332–341

Israelson A, Abu-Hamad S, Zaid H, Nahon E, Shoshan-Barmatz V (2007) Localization of the voltage-dependent anion channel-1 Ca2+–binding sites. Cell Calcium 41:235–244

Jiang D, Zhao L, Clapham DE (2009) Genome-wide RNAi screen identifies Letm1 as a mitochondrial Ca2+/H+ antiporter. Science 326:144–147

Jouaville LS, Ichas F, Holmuhamedov EL, Camacho P, Lechleiter JD (1995) Synchronization of calcium waves by mitochondrial substrates in *Xenopus laevis* oocytes. Nature 377:438–441

Kirichok Y, Krapivinsky G, Clapham DE (2004) The mitochondrial calcium uniporter is a highly selective ion channel. Nature 427:360–364

Klodmann J, Senkler M, Rode C, Braun HP (2011) Defining the protein complex proteome of plant mitochondria. Plant Physiol 157:587–598

Kovacs-Bogdan E, Sancak Y, Kamer KJ, Plovanich M, Jambhekar A, Huber RJ, Myre MA, Blower MD, Mootha VK (2014) Reconstitution of the mitochondrial calcium uniporter in yeast. Proc Natl Acad Sci U S A 111:8985–8990

Krinke O, Novotna Z, Valentova O, Martinec J (2007) Inositol trisphosphate receptor in higher plants: is it real? J Exp Bot 58:361–376

Kwon SK, Sando R 3rd, Lewis TL, Hirabayashi Y, Maximov A, Polleux F (2016) LKB1 regulates mitochondria-dependent presynaptic calcium clearance and neurotransmitter release properties at excitatory synapses along cortical axons. PLoS Biol 14:e1002516

Lee Y, Min CK, Kim TG, Song HK, Lim Y, Kim D, Shin K, Kang M, Kang JY, Youn HS, Lee JG, An JY, Park KR, Lim JJ, Kim JH, Kim JH, Park ZY, Kim YS, Wang J, Kim Do H, Eom SH (2015) Structure and function of the N-terminal domain of the human mitochondrial calcium uniporter. EMBO Rep 16:1318–1333

Lewis-Smith D, Kamer KJ, Griffin H, Childs AM, Pysden K, Titov D, Duff J, Pyle A, Taylor RW, Yu-Wai-Man P, Ramesh V, Horvath R, Mootha VK, Chinnery PF (2016) Homozygous deletion in MICU1 presenting with fatigue and lethargy in childhood. Neurol Genet 2:e59

Li N, Zheng Y, Xuan C, Lin Z, Piao L, Liu S (2015) LETM1 overexpression is correlated with the clinical features and survival outcome of breast cancer. Int J Clin Exp Pathol 8:12893–12900

Liu JC, Liu J, Holmstrom KM, Menazza S, Parks RJ, Fergusson MM, Yu ZX, Springer DA, Halsey C, Liu C, Murphy E, Finkel T (2016) MICU1 serves as a molecular gatekeeper to prevent in vivo mitochondrial calcium overload. Cell Rep 16:1561–1573

Logan CV, Szabadkai G, Sharpe JA, Parry DA, Torelli S, Childs AM, Kriek M, Phadke R, Johnson CA, Roberts NY, Bonthron DT, Pysden KA, Whyte T, Munteanu I, Foley AR, Wheway G, Szymanska K, Natarajan S, Abdelhamed ZA, Morgan JE, Roper H, Santen GW, Niks EH, Van Der Pol WL, Lindhout D, Raffaello A, De Stefani D, Den Dunnen JT, Sun Y, Ginjaar I, Sewry CA, Hurles M, Rizzuto R, Duchen MR, Muntoni F, Sheridan E (2014) Loss-of-function mutations in MICU1 cause a brain and muscle disorder linked to primary alterations in mitochondrial calcium signaling. Nat Genet 46:188–193

Mallilankaraman K, Cardenas C, Doonan PJ, Chandramoorthy HC, Irrinki KM, Golenar T, Csordas G, Madireddi P, Yang J, Muller M, Miller R, Kolesar JE, Molgo J, Kaufman B, Hajnoczky G, Foskett JK, Madesh M (2012a) MCUR1 is an essential component of mitochondrial Ca2+ uptake that regulates cellular metabolism. Nat Cell Biol 14:1336–1343

Mallilankaraman K, Doonan P, Cardenas C, Chandramoorthy HC, Muller M, Miller R, Hoffman NE, Gandhirajan RK, Molgo J, Birnbaum MJ, Rothberg BS, Mak DO, Foskett JK, Madesh M (2012b) MICU1 is an essential gatekeeper for MCU-mediated mitochondrial Ca(2+) uptake that regulates cell survival. Cell 151:630–644

Mammucari C, Gherardi G, Zamparo I, Raffaello A, Boncompagni S, Chemello F, Cagnin S, Braga A, Zanin S, Pallafacchina G, Zentilin L, Sandri M, De Stefani D, Protasi F, Lanfranchi G, Rizzuto R (2015) The mitochondrial calcium uniporter controls skeletal muscle trophism in vivo. Cell Rep 10:1269–1279

Martins IS, Vercesi AE (1985) Some characteristics of Ca2+ transport in plant mitochondria. Biochem Biophys Res Commun 129:943–948

Mccormack JG, Denton RM (1979) The effects of calcium ions and adenine nucleotides on the activity of pig heart 2-oxoglutarate dehydrogenase complex. Biochem J 180:533–544

Mccormack JG, Denton RM (1981) A comparative study of the regulation of Ca2+ of the activities of the 2-oxoglutarate dehydrogenase complex and NAD+–isocitrate dehydrogenase from a variety of sources. Biochem J 196:619–624

Mccormack JG, Halestrap AP, Denton RM (1990) Role of calcium ions in regulation of mammalian intramitochondrial metabolism. Physiol Rev 70:391–425

Meng Q, Chen Y, Zhang M, Chen Y, Yuan J, Murray SC (2015) Molecular characterization and phylogenetic analysis of ZmMCUs in maize. Biologia 70:599–605

Michels G, Khan IF, Endres-Becker J, Rottlaender D, Herzig S, Ruhparwar A, Wahlers T, Hoppe UC (2009) Regulation of the human cardiac mitochondrial Ca2+ uptake by 2 different voltage-gated Ca2+ channels. Circulation 119:2435–2443

Miernyk JA, Randall DD (1987) Some properties of pea mitochondrial phospho-pyruvate dehydrogenase-phosphatase. Plant Physiol 83:311–315

Millar AH, Eubel H, Jansch L, Kruft V, Heazlewood JL, Braun HP (2004) Mitochondrial cytochrome c oxidase and succinate dehydrogenase complexes contain plant specific subunits. Plant Mol Biol 56:77–90

Monaco G, Decrock E, Arbel N, Van Vliet AR, La Rovere RM, De Smedt H, Parys JB, Agostinis P, Leybaert L, Shoshan-Barmatz V, Bultynck G (2015) The BH4 domain of anti-apoptotic Bcl-XL, but not that of the related Bcl-2, limits the voltage-dependent anion channel 1 (VDAC1)-mediated transfer of pro-apoptotic Ca2+ signals to mitochondria. J Biol Chem 290:9150–9161

Moore CL (1971) Specific inhibition of mitochondrial Ca++ transport by ruthenium red. Biochem Biophys Res Commun 42:298–305

Moore AL, Bonner WD Jr (1977) The effect of calcium on the respiratory responses of mung bean mitochondria. Biochim Biophys Acta 460:455–466

Murgia M, Rizzuto R (2015) Molecular diversity and pleiotropic role of the mitochondrial calcium uniporter. Cell Calcium 58:11–17

Nichols BJ, Rigoulet M, Denton RM (1994) Comparison of the effects of Ca2+, adenine nucleotides and pH on the kinetic properties of mitochondrial NAD(+)-isocitrate dehydrogenase and oxoglutarate dehydrogenase from the yeast *Saccharomyces cerevisiae* and rat heart. Biochem J 303(Pt 2):461–465

Nowikovsky K, Bernardi P (2014) LETM1 in mitochondrial cation transport. Front Physiol 5:83

Nowikovsky K, Froschauer EM, Zsurka G, Samaj J, Reipert S, Kolisek M, Wiesenberger G, Schweyen RJ (2004) The LETM1/YOL027 gene family encodes a factor of the mitochondrial K+ homeostasis with a potential role in the Wolf-Hirschhorn syndrome. J Biol Chem 279:30307–30315

Oxenoid K, Dong Y, Cao C, Cui T, Sancak Y, Markhard AL, Grabarek Z, Kong L, Liu Z, Ouyang B, Cong Y, Mootha VK, Chou JJ (2016) Architecture of the mitochondrial calcium uniporter. Nature 533:269–273

Pagliarini DJ, Calvo SE, Chang B, Sheth SA, Vafai SB, Ong SE, Walford GA, Sugiana C, Boneh A, Chen WK, Hill DE, Vidal M, Evans JG, Thorburn DR, Carr SA, Mootha VK (2008) A mitochondrial protein compendium elucidates complex I disease biology. Cell 134:112–123

Palty R, Silverman WF, Hershfinkel M, Caporale T, Sensi SL, Parnis J, Nolte C, Fishman D, Shoshan-Barmatz V, Herrmann S, Khananshvili D, Sekler I (2010) NCLX is an essential component of mitochondrial Na+/Ca2+ exchange. Proc Natl Acad Sci U S A 107:436–441

Pan X, Liu J, Nguyen T, Liu C, Sun J, Teng Y, Fergusson MM, Rovira II, Allen M, Springer DA, Aponte AM, Gucek M, Balaban RS, Murphy E, Finkel T (2013) The physiological role of mitochondrial calcium revealed by mice lacking the mitochondrial calcium uniporter. Nat Cell Biol 15:1464–1472

Patron M, Checchetto V, Raffaello A, Teardo E, Vecellio Reane D, Mantoan M, Granatiero V, Szabo I, De Stefani D, Rizzuto R (2014) MICU1 and MICU2 finely tune the mitochondrial Ca2+ uniporter by exerting opposite effects on MCU activity. Mol Cell 53:726–737

Paupe V, Prudent J, Dassa EP, Rendon OZ, Shoubridge EA (2015) CCDC90A (MCUR1) is a cytochrome c oxidase assembly factor and not a regulator of the mitochondrial calcium uniporter. Cell Metab 21:109–116

Peruzzo R, Biasutto L, Szabo I, Leanza L (2016) Impact of intracellular ion channels on cancer development and progression. Eur Biophys J 45:685–707

Petrungaro C, Zimmermann KM, Kuttner V, Fischer M, Dengjel J, Bogeski I, Riemer J (2015) The Ca(2+)-dependent release of the Mia40-induced MICU1-MICU2 Dimer from MCU regulates mitochondrial Ca(2+) uptake. Cell Metab 22:721–733

Plovanich M, Bogorad RL, Sancak Y, Kamer KJ, Strittmatter L, Li AA, Girgis HS, Kuchimanchi S, De Groot J, Speciner L, Taneja N, Oshea J, Koteliansky V, Mootha VK (2013) MICU2, a paralog of MICU1, resides within the mitochondrial uniporter complex to regulate calcium handling. PLoS One 8:e55785

Prudent J, Popgeorgiev N, Bonneau B, Thibaut J, Gadet R, Lopez J, Gonzalo P, Rimokh R, Manon S, Houart C, Herbomel P, Aouacheria A, Gillet G (2013) Bcl-wav and the mitochondrial calcium uniporter drive gastrula morphogenesis in zebrafish. Nat Commun 4:2330

Quintana A, Schwindling C, Wenning AS, Becherer U, Rettig J, Schwarz EC, Hoth M (2007) T cell activation requires mitochondrial translocation to the immunological synapse. Proc Natl Acad Sci U S A 104:14418–14423

Quintana A, Pasche M, Junker C, Al-Ansary D, Rieger H, Kummerow C, Nunez L, Villalobos C, Meraner P, Becherer U, Rettig J, Niemeyer BA, Hoth M (2011) Calcium microdomains at the immunological synapse: how ORAI channels, mitochondria and calcium pumps generate local calcium signals for efficient T-cell activation. EMBO J 30:3895–3912

Raffaello A, De Stefani D, Sabbadin D, Teardo E, Merli G, Picard A, Checchetto V, Moro S, Szabò I, Rizzuto R (2013) The mitochondrial calcium uniporter is a multimer that can include a dominant-negative pore-forming subunit. EMBO J 32:2362–2376.

Raffaello A, Mammucari C, Gherardi G, Rizzuto R (2016) Calcium at the center of cell signaling: interplay between endoplasmic reticulum, mitochondria, and lysosomes. Trends Biochem Sci 41:1035–1049

Rapizzi E, Pinton P, Szabadkai G, Wieckowski MR, Vandecasteele G, Baird G, Tuft RA, Fogarty KE, Rizzuto R (2002) Recombinant expression of the voltage-dependent anion channel enhances the transfer of Ca2+ microdomains to mitochondria. J Cell Biol 159:613–624

Reed KC, Bygrave FL (1974) The inhibition of mitochondrial calcium transport by lanthanides and ruthenium red. Biochem J 140:143–155

Rizzuto R, Pozzan T (2006) Microdomains of intracellular Ca2+: molecular determinants and functional consequences. Physiol Rev 86:369–408

Rizzuto R, Simpson AW, Brini M, Pozzan T (1992) Rapid changes of mitochondrial Ca2+ revealed by specifically targeted recombinant aequorin. Nature 358:325–327

Rizzuto R, Marchi S, Bonora M, Aguiari P, Bononi A, De Stefani D, Giorgi C, Leo S, Rimessi A, Siviero R, Zecchini E, Pinton P (2009) Ca(2+) transfer from the ER to mitochondria: when, how and why. Biochim Biophys Acta 1787:1342–1351

Rizzuto R, De Stefani D, Raffaello A, Mammucari C (2012) Mitochondria as sensors and regulators of calcium signalling. Nat Rev Mol Cell Biol 13:566–578

Rostovtseva TK (2012) VDAC structure, function, and regulation of mitochondrial and cellular metabolism. Biochim Biophys Acta 1818:1437

Rotmann A, Sanchez C, Guiguemde A, Rohrbach P, Dave A, Bakouh N, Planelles G, Lanzer M (2010) PfCHA is a mitochondrial divalent cation/H+ antiporter in Plasmodium falciparum. Mol Microbiol 76:1591–1606

Rottenberg H, Scarpa A (1974) Calcium uptake and membrane potential in mitochondria. Biochemistry 13:4811–4817

Rueda CB, Traba J, Amigo I, Llorente-Folch I, Gonzalez-Sanchez P, Pardo B, Esteban JA, Del Arco A, Satrustegui J (2015) Mitochondrial ATP-Mg/Pi carrier SCaMC-3/Slc25a23 counteracts PARP-1-dependent fall in mitochondrial ATP caused by excitotoxic insults in neurons. J Neurosci 35:3566–3581

Sancak Y, Markhard AL, Kitami T, Kovacs-Bogdan E, Kamer KJ, Udeshi ND, Carr SA, Chaudhuri D, Clapham DE, Li AA, Calvo SE, Goldberger O, Mootha VK (2013) EMRE is an essential component of the mitochondrial calcium uniporter complex. Science 342:1379–1382

Shao J, Fu Z, Ji Y, Guan X, Guo S, Ding Z, Yang X, Cong Y, Shen Y (2016) Leucine zipper-EF-hand containing transmembrane protein 1 (LETM1) forms a Ca2+/H+ antiporter. Sci Rep 6:34174

Shimizu H, Schredelseker J, Huang J, Lu K, Naghdi S, Lu F, Franklin S, Fiji HD, Wang K, Zhu H, Tian C, Lin B, Nakano H, Ehrlich A, Nakai J, Stieg AZ, Gimzewski JK, Nakano A, Goldhaber JI, Vondriska TM, Hajnoczky G, Kwon O, Chen JN (2015) Mitochondrial Ca(2+) uptake by the voltage-dependent anion channel 2 regulates cardiac rhythmicity. Elife 4 doi: 10.7554/eLife.04801

Shoshan-Barmatz V, De Pinto V, Zweckstetter M, Raviv Z, Keinan N, Arbel N (2010) VDAC, a multi-functional mitochondrial protein regulating cell life and death. Mol Asp Med 31:227–285

Silva MA, Carnieri EG, Vercesi AE (1992) Calcium transport by corn mitochondria: evaluation of the role of phosphate. Plant Physiol 98:452–457

Sparagna GC, Gunter KK, Sheu SS, Gunter TE (1995) Mitochondrial calcium uptake from physiological-type pulses of calcium. A description of the rapid uptake mode. J Biol Chem 270:27510–27515

Stael S, Wurzinger B, Mair A, Mehlmer N, Vothknecht UC, Teige M (2012) Plant organellar calcium signalling: an emerging field. J Exp Bot 63:1525–1542

Szabo I, Zoratti M (2014) Mitochondrial channels: ion fluxes and more. Physiol Rev 94:519–608

Takahashi Y, Tateda C (2013) The functions of voltage-dependent anion channels in plants. Apoptosis 18:917–924

Talavera K, Nilius B (2006) Biophysics and structure-function relationship of T-type Ca2+ channels. Cell Calcium 40:97–114

Teardo E, Carraretto L, Wagner S, Formentin E, Behera S, De Bortoli S, Larosa V, Fuschs P, Lo Schiavo F, Raffaello A, Rizzuto R, Costa A, Schwarzlander M, Szabo I (2017) Physiological characterization of a plant mitochondrial calcium uniporter in vitro and in vivo. Plant Physiol 173:1355–1370

Tovar-Mendez A, Miernyk JA, Randall DD (2003) Regulation of pyruvate dehydrogenase complex activity in plant cells. Eur J Biochem 270:1043–1049

Trenker M, Malli R, Fertschai I, Levak-Frank S, Graier WF (2007) Uncoupling proteins 2 and 3 are fundamental for mitochondrial Ca2+ uniport. Nat Cell Biol 9:445–452

Tsai MF, Jiang D, Zhao L, Clapham D, Miller C (2014) Functional reconstitution of the mitochondrial Ca2+/H+ antiporter Letm1. J Gen Physiol 143:67–73

Tsai MF, Phillips CB, Ranaghan M, Tsai CW, Wu Y, Willliams C, Miller C (2016) Dual functions of a small regulatory subunit in the mitochondrial calcium uniporter complex. Elife 5 doi: 10.7554/eLife.15545

Vais H, Tanis JE, Muller M, Payne R, Mallilankaraman K, Foskett JK (2015) MCUR1, CCDC90A, is a regulator of the mitochondrial calcium uniporter. Cell Metab 22:533–535

Vais H, Mallilankaraman K, Mak DO, Hoff H, Payne R, Tanis JE, Foskett JK (2016) EMRE is a matrix Ca(2+) sensor that governs gatekeeping of the mitochondrial Ca(2+) uniporter. Cell Rep 14:403–410

Vasington FD, Murphy JV (1962) Ca ion uptake by rat kidney mitochondria and its dependence on respiration and phosphorylation. J Biol Chem 237:2670–2677

Vasington FD, Gazzotti P, Tiozzo R, Carafoli E (1972) The effect of ruthenium red on Ca 2+ transport and respiration in rat liver mitochondria. Biochim Biophys Acta 256:43–54

Vecellio Reane D, Vallese F, Checchetto V, Acquasaliente L, Butera G, De Filippis V, Szabo I, Zanotti G, Rizzuto R, Raffaello A (2016) A MICU1 splice variant confers high sensitivity to the mitochondrial Ca2+ uptake machinery of skeletal muscle. Mol Cell 64:760–773

Wagner S, Behera S, De Bortoli S, Logan DC, Fuchs P, Carraretto L, Teardo E, Cendron L, Nietzel T, Fussl M, Doccula FG, Navazio L, Fricker MD, Van Aken O, Finkemeier I, Meyer AJ, Szabo I, Costa A, Schwarzlander M (2015) The EF-hand Ca2+ binding protein MICU choreographs mitochondrial Ca2+ dynamics in arabidopsis. Plant Cell 27:3190–3212

Wagner S, De Bortoli S, Schwarzlander M, Szabo I (2016) Regulation of mitochondrial calcium in plants versus animals. J Exp Bot 67:3809–3829

Waldeck-Weiermair M, Jean-Quartier C, Rost R, Khan MJ, Vishnu N, Bondarenko AI, Imamura H, Malli R, Graier WF (2011) Leucine zipper EF hand-containing transmembrane protein 1 (Letm1) and uncoupling proteins 2 and 3 (UCP2/3) contribute to two distinct mitochondrial Ca2+ uptake pathways. J Biol Chem 286:28444–28455

Waldeck-Weiermair M, Malli R, Parichatikanond W, Gottschalk B, Madreiter-Sokolowski CT, Klec C, Rost R, Graier WF (2015) Rearrangement of MICU1 multimers for activation of MCU is solely controlled by cytosolic Ca(2.) Sci Rep 5:15602

Wang L, Yang X, Li S, Wang Z, Liu Y, Feng J, Zhu Y, Shen Y (2014b) Structural and mechanistic insights into MICU1 regulation of mitochondrial calcium uptake. EMBO J 33:594–604

Wang L, Yang X, Shen Y (2015) Molecular mechanism of mitochondrial calcium uptake. Cell Mol Life Sci 72:1489–1498

Wang Z, Liu D, Varin A, Nicolas V, Courilleau D, Mateo P, Caubere C, Rouet P, Gomez AM, Vandecasteele G, Fischmeister R, Brenner C (2016) A cardiac mitochondrial cAMP signaling pathway regulates calcium accumulation, permeability transition and cell death. Cell Death Dis 7:e2198

Wu Y, Rasmussen TP, Koval OM, Joiner ML, Hall DD, Chen B, Luczak ED, Wang Q, Rokita AG, Wehrens XH, Song LS, Anderson ME (2015) The mitochondrial uniporter controls fight or flight heart rate increases. Nat Commun 6:6081

Zhang B, Carrie C, Ivanova A, Narsai R, Murcha MW, Duncan O, Wang Y, Law SR, Albrecht V, Pogson B, Giraud E, Van Aken O, Whelan J (2012) LETM proteins play a role in the accumulation of mitochondrially encoded proteins in *Arabidopsis thaliana* and AtLETM2 displays parent of origin effects. J Biol Chem 287:41757–41773

Zollino M, Lecce R, Fischetto R, Murdolo M, Faravelli F, Selicorni A, Butte C, Memo L, Capovilla G, Neri G (2003) Mapping the Wolf-Hirschhorn syndrome phenotype outside the currently accepted WHS critical region and defining a new critical region, WHSCR-2. Am J Hum Genet 72:590–597

Part II
The Mitochondrial Permeability Transition Pore: Structure and Function

Chapter 3
The Mitochondrial Permeability Transition Pore: Molecular Structure and Function in Health and Disease

Elizabeth A. Jonas, George A. Porter Jr., Gisela Beutner, Nelli Mnatsakanyan, Han-A. Park, Nikita Mehta, Rongmin Chen, and Kambiz N. Alavian

3.1 Mitochondria, Cell Metabolism and Death

Mitochondria are complex organelles responsible for producing energy in the form of ATP for most eukaryotic cells. They are also at the center of several other essential processes including Ca^{2+} homeostasis, heme and steroid biosynthesis. In addition, the mitochondrion regulates cellular responses to stress and controls cell death.

To produce ATP, mitochondria use substrates produced in the cytosol by carbohydrate, lipid, and protein metabolic pathways. Products such as acetyl co-enzyme A enter the tricarboxylic acid (TCA) cycle. The TCA cycle synthesizes NADH and $FADH_2$, which donate their electrons to the electron transport chain and their H^+ ions to be pumped out of the matrix by the NADH dehydrogenase and other electron transport complexes. This creates a proton motive force that in turn drives the F_1F_O ATP synthase (Mitchell 1961). Upon kinetic repositioning of the ATP synthase rotor, ATP is synthesized from ADP and P_i (Watt et al. 2010). The machinery required for ADP/ATP exchange between the cytoplasm and the matrix includes the outer membrane voltage-dependent anion channel (VDAC) and the adenine nucleotide transporter (ANT) at the inner membrane. These proteins are intimately linked to the ATP synthase (Chen et al. 2004).

Elizabeth A. Jonas and George A. Porter, Jr. contributed equally to the work.

E.A. Jonas (✉) • N. Mnatsakanyan • H.-A. Park • N. Mehta • R. Chen
Department Internal Medicine, Section of Endocrinology, Yale University School of Medicine, PO Box 208020, New Haven, CT 06520, USA
e-mail: Elizabeth.jonas@yale.edu

G.A. Porter Jr. • G. Beutner
Department of Pediatrics (Cardiology), University of Rochester Medical Center, Rochester, NY, USA

K.N. Alavian
Division of Brain Sciences, Department of Medicine, Imperial College London, London, UK

T.K. Rostovtseva (ed.), *Molecular Basis for Mitochondrial Signaling*, Biological and Medical Physics, Biomedical Engineering,
DOI 10.1007/978-3-319-55539-3_3

3.2 Mitochondrial Inner Membrane Leak: Regulator of Metabolic Rate and Uncoupling

Two currents complete the current loop of the proton pumping activity of the electron transport complexes. First, hydrogen ion (H^+) translocation through the ATP synthase in the opposite direction to that of the electron transport complexes forms a current during the synthesis of ATP. Second, an apparently wasteful leak in the inner mitochondrial membrane uncouples oxidation from phosphorylation as H^+ ions enter the matrix through channels independent of ATP production. Classically, uncoupling proteins generate heat for organisms with large surface area to volume ratio and may depolarize mitochondria in order to temper oxidative damage and regulate metabolic rate during hibernation and at other times (Nicholls and Rial 1999; Andrews et al. 2005; Divakaruni and Brand 2011; Fedorenko et al. 2012). In addition to uncoupling proteins, however, intrinsic uncoupling exists within several inner mitochondrial membrane channels and transporters including the F_1F_O ATP synthase (Caviston et al. 1998; D'Alessandro et al. 2008).

3.3 Mitochondrial Inner Membrane Channels and Exchangers Are Necessary for Ca^{2+} Cycling and Cellular Ca^{2+} Dynamics

Mitochondrial inner membrane depolarization occurs not only through proton movement but also via the flux of other ions including Ca^{2+} across mitochondrial membranes. Ca^{2+} movement into the mitochondrial matrix is a physiological event that takes place in response to increased cytosolic Ca^{2+} levels. Ca^{2+} buffering is frequently employed by mitochondria in cells that experience rapidly changing cytosolic Ca^{2+} levels such as those of excitable tissues.

Mitochondria regulate cytosolic levels of Ca^{2+} and the release of Ca^{2+} and metabolites using several ion channels and exchangers. The Ca^{2+} uniporter ion channel (MCU) located at the mitochondrial inner membrane participates in mitochondrial Ca^{2+} uptake within the cell body of many types of cells and also in the presynaptic terminals of neurons (Billups and Forsythe 2002; Kang et al. 2008; Chouhan et al. 2012; Rizzuto et al. 2012). Isoforms of MCU and its helper MICU confer tissue specificity and add complexity to the mechanisms of activity-dependent energy production by mitochondria (Raffaello et al. 2012; De Stefani and Rizzuto 2014). Mitochondrial Ca^{2+} release also appears to be highly regulated, but, unlike the MCU, the molecular components of a Ca^{2+} release channel were only recently discovered and form the main focus of this chapter.

Ca^{2+} rerelease from mitochondria prolongs the elevation of cytosolic Ca^{2+} after influx across the plasma membrane (Zucker and Regehr 2002). During neuronal activity, high cytosolic Ca^{2+} levels are cleared by the actions of Ca^{2+} ATPases at the

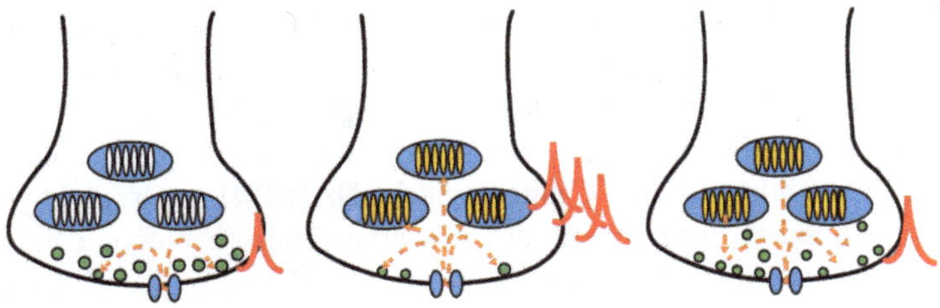

Fig. 3.1 Calcium rerelease from mitochondria is required for certain forms of presynaptic plasticity. *Left*: During an action potential invasion of the presynaptic nerve terminal, Ca^{2+} enters the terminal. A few synaptic vesicles fuse, releasing neurotransmitter. *Middle*: During high-frequency activity of the presynaptic nerve, Ca^{2+} fills the terminal and is taken up into mitochondria by the mitochondrial Ca^{2+} uniporter (MCU). Many vesicles fuse with the plasma membrane, and synaptic depression ensues. *Right*: After recovery of vesicle pools from the high-frequency event, mitochondria rerelease Ca^{2+} via Ca^{2+}-dependent exchangers and Ca^{2+}-sensitive channels in the inner mitochondrial membrane (mPTP-like channel). The increased Ca^{2+} level in the terminal provides for enhanced vesicle fusion upon a single action potential invasion of the terminal compared to the *left panel*, accounting for presynaptic plasticity of neurotransmitter release

plasma membrane and by buffering through uptake by intracellular stores including ER and mitochondria (Rizzuto et al. 2012; Lopreiato et al. 2014); these processes reset the normally low Ca^{2+} levels present in resting cells. The buffered Ca^{2+} is eventually rereleased. In synaptic endings, residual Ca^{2+} increases the Ca^{2+} available for synaptic vesicle fusion, enhancing the amount of neurotransmitter released for a given stimulus (Fig. 3.1) (Jonas 2006; Neher and Sakaba 2008).

Ca^{2+} influx into cells and into mitochondria also regulates TCA cycle enzymes (Denton 2009) and enzymes in the electron transport chain (Gellerich et al. 2010) to speed the process of ATP synthesis. As cytosolic Ca^{2+} increases during cellular activity, mitochondrial TCA cycle enzymatic activity is enhanced, increasing the levels of mitochondrial NADH and ATP. Ca^{2+} influx into mitochondria enhances the proportion of pyruvate dehydrogenase complex in its active, dephosphorylated form. Ca^{2+} also increases the ability of isocitrate dehydrogenase and oxoglutarate dehydrogenase to bind their substrates. The result is regulation of mitochondrial TCA enzymes by cytosolic Ca^{2+} or "external regulation" as opposed to "internal regulation" that is caused by changes in matrix NAD/NADH and ADP/ATP ratio. In neurons, increases in cytosolic Ca^{2+} arise from depolarization of the plasma membrane with resultant increase in permeability to Ca^{2+} through glutamate receptors and voltage-gated Ca^{2+} channels. In addition, mitochondria receive Ca^{2+} from the ER which partners with mitochondria to respond to cytosolic Ca^{2+} loads. Elevated cytosolic Ca^{2+} enhances the release of Ca^{2+} from IP3 receptors directly into mitochondria within a tightly coupled space between the two organelles (Rizzuto et al. 2012). Oscillatory Ca^{2+} release from ER provided by IP3 receptors also contributes to influx of Ca^{2+} into the mitochondria and drives TCA cycle enzymes even more effectively than sustained elevations of mitochondrial Ca^{2+} (Hajnoczky et al. 2000), most likely because sustained Ca^{2+} allows re-equilibration of the signal in the matrix.

The interaction between cellular activity, Ca^{2+} influx into mitochondria, and energy generation was clarified in a study in drosophila neuromuscular junction (Chouhan et al. 2012). Using a complex set of imaging techniques including genetically encoded Ca^{2+}/pH indicators, it was shown that neuronal activity enhances Ca^{2+} uptake by mitochondria, increases mitochondrial NAD(P)H levels, and hyperpolarizes the mitochondrial inner membrane potential. These events are inhibited by pharmacological agents that block mitochondrial Ca^{2+} uptake. Levels of cytosolic Ca^{2+} remain similar in different neurons despite their very different firing rates, suggesting that a certain level of cytosolic Ca^{2+} is optimum for energy production during activity. An exciting implication of these novel findings is that during development or neural/synaptic plasticity, acute changes in cytosolic Ca^{2+} levels could theoretically produce plasticity in mitochondrial responses to adjust to changes in energy demands in neurons and to readjust cytosolic Ca^{2+} levels to optimize synaptic responses and neuronal excitability.

Mitochondrial management of Ca^{2+} in order to decrease cytosolic Ca^{2+} burden is observed in the large mammalian brain stem presynaptic terminals of the Calyx of Held. In this synapse, mitochondria remove Ca^{2+} during rises in cytosolic Ca^{2+} produced by Ca^{2+} influx through voltage-gated channels. Mitochondrial Ca^{2+} uptake rapidly dampens the overall rise in Ca^{2+} in the presynaptic terminal. The effect of this is to prevent synaptic depression by attenuating vesicle depletion and allowing for continued effective synaptic transmission during a train of action potentials (Billups and Forsythe 2002).

3.4 Mitochondrial Inner and Outer Membrane Channel Activity Regulates Ca^{2+} Rerelease from Mitochondria After Buffering

After Ca^{2+} buffering, Ca^{2+} rerelease from mitochondria occurs through exchangers and channels located on the mitochondrial inner membrane and via channels in the outer membrane. Ca^{2+}-sensitive ligand-gated mitochondrial channels are widely conserved and found in species from invertebrates to mammals. These channels open in response to elevated Ca^{2+} within the mitochondrial matrix. In the squid presynaptic terminal, opening of a Ca^{2+}-activated mitochondrial channel is correlated with enhanced neurotransmitter release (Fig. 3.1) (Jonas et al. 1999). Electrophysiological recordings (Jonas et al. 1997) demonstrate that within the resting presynaptic terminal, the conductance of mitochondrial membranes is low (Jonas et al. 1999). In contrast, during high-frequency electrical stimulation of the presynaptic nerve, a large increase in mitochondrial membrane ion channel activity takes place (Jonas et al. 1999). The delay in onset of the mitochondrial activity, the dependence of the mitochondrial channel activity on Ca^{2+} uptake across the plasma membrane, and the persistence of the mitochondrial activity after stimulation are in keeping with the role of a channel and/or exchanger in rereleasing Ca^{2+} from mitochondria during certain forms of short-term plasticity (Friel and Tsien 1994; Tang and Zucker 1997; Jonas et al. 1999). Furthermore, mitochondrial channel activity and short-term increases in Ca^{2+}-dependent synaptic transmitter release are both abrogated by applying the uncoupler FCCP (carbonyl cyanide *p*-trifluoromethoxyp henylhydrazone), which depolarizes mitochondria, preventing Ca^{2+} handling (Jonas et al. 1999).

Although exchangers as well as channels regulate the release of ions and metabolites across the inner membrane, the role of ion and metabolite release by the mitochondrial outer membrane is played in large part by the outer mitochondrial membrane channel known as the voltage-dependent anion channel (VDAC) which transfers Ca^{2+}, ATP, and other metabolites across the outer membrane into the cytosol (Vander Heiden et al. 2001; Gottlieb et al. 2002). VDAC also regulates the uptake of ADP and other metabolites during normal and pathological cell activities (Rostovtseva and Colombini 1997; Mannella and Kinnally 2008; Pang et al. 2010) and is regulated by cytoskeletal elements, in particular tubulin in its dimeric form (Rostovtseva and Bezrukov 2012).

3.5 Pathological Inner Membrane Depolarization: Mitochondrial Permeability Transition

An increase in mitochondrial outer membrane permeability (MOMP) may also be triggered by an acute inner membrane depolarization (Galluzzi et al. 2009), particularly after cytosolic and mitochondrial Ca^{2+} overload. Although, as described, Ca^{2+} uptake and rerelease from mitochondria is a normal physiological event in cells, accumulation of Ca^{2+} in the matrix can diminish energy production by the ATP synthase (Budd and Nicholls 1996) once the availability of other ions for exchange with Ca^{2+} has become exhausted. Ca^{2+} overload then produces an uncoupling process described historically as a rapid increase in permeability of the mitochondrial inner membrane to solutes and the halting of ATP production (Haworth and Hunter 1979; Hunter and Haworth 1979a, b). This phenomenon is termed permeability transition (PT).

PT can be reversible or irreversible (Haworth and Hunter 1979; Crompton 1999; Huser and Blatter 1999; Petronilli et al. 1999; Hausenloy et al. 2004; Wang et al. 2008; Korge et al. 2011). If not reversed by normalization of cellular conditions, a more extreme form of catastrophic PT takes place characterized by structural breakdown of the mitochondrial matrix accompanied by outer mitochondrial membrane rupture and cell death. The difference between this kind of mitochondrial cell death and apoptotic death produced by MOMP seems to be that pathological PT is associated with necrotic cell death such as is found in ischemia or injury, whereas MOMP occurring in the presence of sufficient amounts of ATP may have a more important role in developmental and genetically predetermined death (Baines 2011; Bonora and Pinton 2014). Intermembrane space pro-apoptotic factors such as cytochrome *c* and Smac/DIABLO are released during both forms of cell death. In MOMP, outer membrane permeabilization alone leads to release of these factors, whereas in prolonged PT, rupture of the outer membrane after inner membrane swelling releases pro-apoptotic factors into the cytosol (Galluzzi et al. 2009).

PT has been extensively studied for its role in ischemic injury in the brain, heart, and other organs as well as in neurodegenerative conditions (Bonora et al. 2014). In addition to Ca^{2+} influx into the matrix, PT is also induced by ROS, elevated inorganic phosphate, and intracellular acidification (Szabo et al. 1992; Giorgio et al. 2013b). In contrast, it is inhibited by ATP/ADP and Mg^{2+} (Kowaltowski et al. 1998; Crompton 1999). The pharmacological agent most efficient in inhibiting PT is cyclosporine A (CsA), an immunosuppressant drug which binds to cyclophilin D (CypD) and inhibits the channel activity associated with PT (see below). Inhibitors of ANT can either attenuate (bongkrekic acid) or enhance (atractyloside) PT opening (Hunter and Haworth 1979a; Szabo and Zoratti 1991; Giorgio et al. 2009). Recent reports have also confirmed increased activity of PT by polyphosphates, chains of 10s–100 s of repeating phosphates linked by ATP-like high energy bonds (Abramov et al. 2007; Seidlmayer et al. 2012; Holmstrom et al. 2013; Stotz et al. 2014). The actions of Ca^{2+} may also require polyhydroxybutyrate (PHB), which enters mitochondria and enhances the ability of Ca^{2+} to induce PT (Elustondo et al. 2013).

3.6 The Quest for an Inner Membrane Ca^{2+}-Sensitive Uncoupling Channel: The PT Pore

3.6.1 Electrophysiological Properties of the Mitochondrial PT Pore (mPTP)

Thus, PT is an important event that performs both physiologic and pathophysiologic functions. PT most likely begins as the opening of a Ca^{2+}-sensitive ion channel in the inner mitochondrial membrane similar to the ion channel activity initiated by physiological mitochondrial Ca^{2+} influx. The Ca^{2+} release channel is heavily regulated; therefore it is assumed that only after prolonged opening does pathological PT (with MOMP) occur (Bernardi 1999). The conversion of a physiological Ca^{2+} extrusion mechanism into a pathological channel opening is perhaps correlated with energy failure as a result of arrest of ATP-synthesizing activity and slowing of energy-dependent Ca^{2+} extrusion mechanisms. Although there are unknown factors that regulate the transition from physiological to pathophysiological events, reactive oxygen species, mitochondrial Ca^{2+} overload, changes in matrix pH, Pi, polyphosphate, and mitochondrial inner membrane depolarization are pathological circumstances contributing to PT. Nevertheless, identification of the molecular structure of the pore will provide the target for regulatory activities, allowing for a greater understanding of PT modulation during health and disease.

Description of the biophysical properties of the pore that opens in the inner membrane during PT (the mPTP) provided the earliest indication that PT was initiated by the opening of an ion channel. The first electrophysiological studies of mitochondrial inner membrane were published in 1987. This early report described a

~100 pS channel recorded by patch-clamping giant mouse liver mitochondria produced by cuprizone application (Sorgato et al. 1987). In the late 1980s, a putative mPTP was recorded by patch-clamping mitochondrial inner membrane or mitoplast preparations (Petronilli et al. 1989). The activity occurred at positive potentials of the patch pipette and was found either in whole organelle mode or in single-channel recordings in the organelle-attached configuration. The activity was slightly anion over cation selective with multiple subconductance states ranging from 30 pS to a peak single-channel conductance of 1.3 nS. Lower conductances were attributed to substates of the larger channel openings because of long periods lacking activity followed by periods of multi-conductance behavior (Petronilli et al. 1989). Conductances of 550 pS were frequently observed at positive potentials. Gating was less common at negative potentials, but this observation was consistent with the presence of prolonged openings and fewer subconductance states at negative patch potentials contrasted with increased flickering at positive potentials. The authors concluded that conductance levels were not sharply defined, consistent with the existence of many varied conductance levels of the channel.

In 1989, Kinnally et al. recorded a similar mitochondrial multi-conductance channel (MMC) in mouse liver mitoplasts (Kinnally et al. 1989). This channel changed over time, with low activity at the onset of the recording followed by progressively higher activity at later times during the recording. The channels were sometimes open more frequently at negative potentials, but at times channel activity was more frequent at positive potentials. Channel activity displayed multiple conductances ranging from 10 to 1,000 pS and was weakly cation selective. These early studies began to establish expected criteria for activity of mPTP.

Shortly after the first recordings of the putative mPTP were performed, similar inner membrane activity was found to be inhibited by CsA. In patch-clamp experiments performed in liver mitochondria, channel activity was rapidly inhibited by submicromolar concentrations of CsA in a manner consistent with the expression of the binding site on the matrix side of the inner membrane. Ca^{2+}-activated large conductance channel activity up to 1.3 nS was inhibitable, but a 107 pS inner membrane conductance similar to the first recorded inner mitochondrial membrane channel was also observed in the recordings. This smaller conductance was resistant to CsA, suggesting that this activity might be due to a separate ion channel (Szabo and Zoratti 1991). The large conductance channel was sensitive to Mg^{2+}, Mn^{2+}, Ba^{2+}, and Sr^{2+} in that order, which inhibited the activity in a competitive manner with Ca^{2+}, the main activator of the channel (Szabo et al. 1992).

3.6.2 *Characterization of a Molecular Complex Regulating the Pore*

The recent identification of the molecular structure matching the biophysical properties of mPTP was aided by seemingly unrelated sets of findings. One was that Bcl-x_L enhances metabolic efficiency (decreases uncoupling) by binding to the β-subunit of the ATP synthase. The second finding was that CypD, which had been known for many years to regulate PT, binds to the stator arm of ATP synthase, specifically on the OSCP subunit. The third finding suggested that closure of the mPTP is related to the level of CypD expression in a developmentally regulated manner as CypD falls at the onset of respiration in mammalian heart. The final project found that ATP synthase assembles into a very large complex with other proteins that may regulate the mPTP. These findings and their convergence that led to the definition of the molecular nature of the mPTP will now be discussed in greater detail.

3.6.3 Bcl-x_L Regulates Metabolic Efficiency by Binding to the β-Subunit of the ATP Synthase

Although Bcl-2 family proteins regulate mitochondrial outer membrane permeability to produce or inhibit cell death (Galonek and Hardwick 2006; Jonas 2009; Jonas et al. 2014; Park et al. 2014), Bcl-x_L also alters the metabolic properties of cells, even in the absence of cell death signals. Bcl-x_L increases the release of ATP through enhanced VDAC opening. Enhanced ATP release by mitochondria decreases the probability of MOMP in cancer cell lines by providing extra ATP to overcome cell death stimuli (Vander Heiden et al. 2001; Gottlieb et al. 2002). In the neuronal synapse, injection of either Bcl-x_L or ATP enhances synaptic transmitter release (Jonas et al. 2003), suggesting that Bcl-x_L helps increase ATP levels in the synapse during synaptic activity (Hickman et al. 2008).

Inefficiency of metabolism is correlated with cell death under conditions of neurodegeneration or acute cellular injury such as that occurring during PT (Brand 2005; Beal 2007; Dodson and Guo 2007). In contrast, a highly efficient state of metabolism requires maximally decreased uncoupling of the inner membrane (Hockenbery et al. 1990; Alavian et al. 2011; Chen et al. 2011). Therefore, Bcl-2 family proteins could form part of a large protein complex that regulates inner membrane coupling, and Bcl-x_L might regulate not only the release but also the production of ATP. In support of a role for Bcl-x_L in the manufacture of ATP, hippocampal neurons overexpressing Bcl-x_L show a large increase in cytoplasmic ATP levels. Surprisingly, this increase in ATP accompanies a decrease in neuronal oxygen uptake and glycolysis, consistent with an increase in mitochondrial bioenergetic efficiency (Alavian et al. 2011; Chen et al. 2011). Bcl-x_L depletion reverses these effects on metabolism (Li et al. 2008; Alavian et al. 2011). Furthermore, direct interaction of Bcl-x_L with the β-subunit of the ATP synthase maximizes the efficiency of ATP production (Alavian et al. 2011; Chen et al. 2011). These studies support a role for Bcl-x_L in closing a leak channel within the inner membrane upon binding to the β-subunit of the ATP synthase.

Relative closure of a leak within the inner membrane in the presence of Bcl-x_L aids actively firing neurons to increase neurotransmitter release (Li et al. 2008, 2013), consistent with a correlation between the increase in metabolic efficiency and the long-term higher efficacy of synaptic transmission found in Bcl-x_L-expressing neurons. In contrast, opening of the Bcl-x_L-regulated inner membrane leak decreases metabolic efficiency and predisposes neurons to death. Neurons lacking Bcl-x_L display a fluctuating mitochondrial inner membrane potential and a marked mitochondrial membrane depolarization in the presence of the ATP synthase inhibitor oligomycin (Chen et al. 2011). These data support the idea that Bcl-x_L regulates inner membrane coupling, prevention of energy failure, and cell death via direct effects on F_1F_O ATP synthase.

3.6.4 Cyclophilin D Binds to ATP Synthase and Regulates Permeability Transition

Another piece of the puzzle that helped determine the molecular components of mPTP was the discovery of the interaction between CypD and ATP synthase (Bernardi 2013). CypD is a chaperone protein and peptidyl-prolyl cis-trans isomerase that resides in the mitochondrial matrix. It regulates the mPTP by enhancing its sensitivity to Ca^{2+}. CsA inhibits PT by inhibiting cyclophilin D (CypD) (Crompton et al. 1988; Halestrap and Davidson 1990; McGuinness et al. 1990). CsA is still the gold standard of pharmacological tools to study the mPTP.

In 2005, experiments using CypD null mice demonstrated that CypD was not itself the pore of the mPTP but that it played an important regulatory role in the modulation of the mPTP by Ca^{2+} (Baines et al. 2005; Basso et al. 2005; Nakagawa et al. 2005; Schinzel et al. 2005). Four groups showed that deletion of CypD decreased sensitivity to ischemia-reperfusion injury in the heart and brain (Baines et al. 2005; Basso et al. 2005; Nakagawa et al. 2005), and the Molkentin group suggested that the physiologic function of CypD regulation of mPTP is to maintain "homeostatic mitochondrial Ca^{2+} levels to match metabolism with alterations in myocardial workload" (Elrod et al. 2010).

CypD expression varies widely among cell types. CypD is more highly expressed in aging hearts, and these changes in expression may regulate its association with a complex of proteins that increase mPTP opening during reperfusion (Zhu et al. 2013). Furthermore, CypD activity appears to be regulated by cell signaling and metabolic pathways (Hafner et al. 2010; Shulga and Pastorino 2010; Shulga et al. 2010; Di Lisa et al. 2011; Nguyen et al. 2011) and by developmental cues in differentiating myocytes.

Recent data support the idea that CypD regulates mPTP by binding to F_1F_O ATP synthase. The propensity toward PT is regulated by ATP hydrolysis and synthesis in as much as this regulates the membrane potential. Therefore, PT requires twice the Ca^{2+} load in mitochondria that are in the process of hydrolyzing ATP (making a membrane potential) versus synthesizing ATP (dissipating the membrane potential) (Hunter and Haworth 1979a, b; Giorgio et al. 2013b). CypD binds to proteins on the stator of ATP synthase (OSCP, b-, and d-subunits, (Giorgio et al. 2009)) as well as to the F_1F_O ATP synthase binding partners ANT (Crompton et al. 1998; Woodfield et al. 1998) and the phosphate carrier (PiC) (Leung and Halestrap 2008; Elrod et al. 2010; Gutierrez-Aguilar et al. 2014). Recent work on the binding of CypD to ATP synthase suggests that binding occurs exclusively between CypD and OSCP, and depletion of OSCP using siRNA also depletes CypD (Giorgio et al. 2013b) suggesting that the association regulates the expression of each protein. Interestingly, CsA decreases the propensity toward PT to the same degree (half) that decreases in OSCP expression increase the propensity toward PT (double), suggesting a privileged inhibitory role of OSCP in PT (Giorgio et al. 2013b). Third, purified ATP synthase dimers produce a current consistent with mPTP in artificial lipid membranes (Giorgio et al. 2013b). The single-channel activity of this current has a maximal conductance of 1–1.3 nS with subconductance states. The current is not stimulated by Ca^{2+} or inhibited by CsA presumably because CypD is not present in the purified dimer preparation although the enzymatic activity of the F_1F_O ATP synthase is preserved. The ATPase dimer current is also not sensitive to bongkrekic acid or atractyloside, agents that primarily affect ANT activity, making it unlikely that ANT forms part of the pore of the channel (Giorgio et al. 2013b). These data emphasize the idea that the regulation of the mPTP may occur via the interaction of CypD and other molecules with F_1F_O ATP synthase and its binding partners.

3.6.5 PT Activity Regulates Cardiac Development

A third line of evidence that helped unravel the identity of the mPTP was a series of studies of mitochondrial function during cardiac development. As described above, previous understanding held that opening of the mPTP was a devastating event that triggers cell death. However, over the last two decades, data have emerged suggesting that transient opening of the mPTP could serve a physiologic purpose. Some of these data in neurons have been discussed in this chapter, but additional data demonstrate that transient mPTP opening occurs in many other cell types (Crompton 1999; Huser and Blatter 1999; Petronilli et al. 1999; Hausenloy et al. 2004; Wang et al. 2008; Korge et al. 2011).

In the heart, physiologic variations in mPTP activity play a critical role in cardiac myocyte differentiation and cardiac development. Mitochondria in cardiac myocytes from adult hearts display transient depolarizations that may be associated with "superoxide flashes," and these depolarizations occur more often in neonatal myocytes (Wang et al. 2008). Interestingly, developmental studies demonstrate that the mPTP is open in myocytes in the early embryonic mouse heart, and this opening is not associated with any form of cell death. However, by the mid-embryonic stage, the mPTP is closed (Hom et al. 2011). This closure coincides with activation of complex I of the electron transport chain, assembly of electron transport chain supercomplexes called respirasomes, and activation of oxidative phosphorylation (Beutner et al. 2014). These changes cause a fall in mitochondrial-derived ROS that signals the myocyte to undergo further differentiation (Hom et al. 2011).

Furthermore, pharmacologically inhibiting mPTP or genetically inhibiting CypD enhances myocyte differentiation, while opening mPTP inhibits differentiation (Hom et al. 2011). These findings have been confirmed in cardiac stem cells (Fujiwara et al. 2011; Cho et al. 2014), and various reports have stressed the importance of the mPTP during cardiac development and myocyte differentiation (Drenckhahn 2011; Folmes et al. 2012).

3.6.6 Regulatory Molecules Do Not Form the Pore of mPTP

The F_1F_O ATP synthase interacts with a large number of proteins many of which have been candidates for mPTP. ANT was an early candidate to form the mPTP since atractyloside and bongkrekic acid, which inhibit ANT, affect the mPTP (Hunter and Haworth 1979b) and ANT was found to interact with CypD (Halestrap and Davidson 1990). VDAC was also an early candidate to form the mPTP due to its high conductance and its association with ANT in immunoprecipitation experiments (Crompton et al. 1998). In addition, it was shown that a complex of ANT, VDAC, hexokinase, and mitochondrial creatine kinase (mtCK) could form high-conductance pores when reconstituted into membranes (Beutner et al. 1996, 1998). Finally, the PiC is a more recent candidate to form the mPTP (Leung and Halestrap 2008).

However, genetic deletion of ANT1 and 2 and of the PiC demonstrated that these proteins were not essential to mPTP formation, but that they served regulatory roles (Kokoszka et al. 2004; Gutierrez-Aguilar et al. 2014; Kwong et al. 2014). Furthermore, deletion of VDAC did not affect pore formation (Baines et al. 2005). Additional data suggest that the conformation of ANT may be important for regulation of the mPTP (Gunter and Sheu 2009). Atractyloside induces mPTP opening and is known to stabilize the "c" conformation of ANT, such that the adenine nucleotide transport site faces the "*cytoplasmic*", or intermembrane space, side of the inner mitochondrial membrane (Gunter and Sheu 2009). In contrast, bongkrekic acid prevents mPTP opening and stabilizes ANT in its "m," or "*matrix*" facing conformation (Gunter and Sheu 2009). However, as both ATR and BKA inhibit ANT, it is unlikely that specific effects on ADP/ATP translocation regulate the mPTP, and a more likely scenario is that the conformation of ANT itself can regulate the mPTP.

The details of how these candidate molecules regulate the mPTP are not yet fully known, but evidence suggests that they participate as part of a large macromolecular structure with F_1F_O ATP synthase in the inner mitochondrial membrane. ANT and PiC form a complex with F_1F_O ATP synthase called the synthasome (Chen et al. 2004). In addition, the complex of ANT, VDAC, hexokinase, and mtCK is likely also involved in the regulation of ATP synthesis (Beutner et al. 1996, 1998). Therefore, each regulatory molecule may alter the structure and/or activity of F_1F_O ATP synthase, and, in so doing, modulate the opening of the mPTP.

Another mitochondrial inner membrane carrier, SGP7, a subunit of the mitochondrial *m*-AAA protease, has more recently been considered for its role in mPTP

(Shanmughapriya et al. 2015). m-AAA proteases preserve mitochondrial proteostasis, morphology, and OXPHOS activity. Mutations in the protease lead to neurodegeneration in spinocerebellar ataxia (SCA28) and hereditary spastic paraplegia (HSP7). Although depletion of SGP7 in cells was found to inhibit cell death, surprisingly, this protection was superior to the death protection of CsA, suggesting that another kind of cell death apart from mPTP could be interfering with the assay.

Regarding the role of SGP7 in permeability transition, it has been suggested previously that loss of the protease increases the propensity for PT (Maltecca et al. 2015). A recent report of an inducible expression of the dominant negative of AFG3L2 confirmed this (Konig et al. 2016), suggesting that the *m*-AAA proteases control the levels of partners of the mitochondrial Ca^{2+} uniporter (MCU). One of these partners is EMRE, a chaperone molecule that links MICU to MCU so that MICU may act as a gatekeeper to prevent mitochondrial Ca^{2+} overload. In the absence of functional AFG3L2/SGP7 heteromers, the EMRE/MCU complex goes into operation without MICU, leading to mitochondrial Ca^{2+} overload, mPT, and cell death. Therefore, depletion of the m-AAA proteases led to early PT, in contradistinction to the findings of Shanmughapriya et al.

3.7 The mPTP, a Molecular Definition

3.7.1 The c-Subunits of ATPases Form Pores

Many reports suggest that F_1F_O ATP synthase is a major factor in the formation of the mPTP. Recent evidence suggests that the F_O or membrane portion of F_1F_O ATP synthase in fact forms the pore (Bonora et al. 2013, 2014; Alavian et al. 2014; Azarashvili et al. 2014; Chinopoulos and Szabadkai 2014; Karch and Molkentin 2014) (Figs. 3.2 and 3.3). Mammalian F_1F_O ATP synthase is a ~600 kDa complex of 15 subunits. The membrane portion, or F_O, contains a ring of eight very hydrophobic c-subunits and subunits a, b, e, f, g, and A6L. A stalk composed of the δ, ε, and γ subunits connects the c-subunit ring to the catalytic F_1 component made of a hexamer of alternating α- and β-subunits, where ATP synthesis and hydrolysis occur. Finally, a stator containing the b, d, F6, and OSCP subunits connects the lateral portion of F_O to the top of the F_1. Movement of protons between the c-subunit and the a-subunit causes rotation of the c-subunit ring, the energy of which is transferred to F_1 to synthesize ATP (Pedersen 1994; Carbajo et al. 2005; Wittig and Schagger 2009; Jonckheere et al. 2012; Walker 2013).

Rat brain and heart mitochondrial F_1F_O ATP synthase complexes undergo age-dependent structural and functional alterations (Guerrieri et al. 1992). The Fo portion is present in the absence of an equivalent complement of F_1 at 3 months of age in heart mitochondria. Levels of F_1 then increase from 3 to 12 months as ATPase activity increases, and this accompanies a decrease in proton leak secondary to binding of F_1 to F_O. Decreases of F_1 content with respect to that observed for F_O are

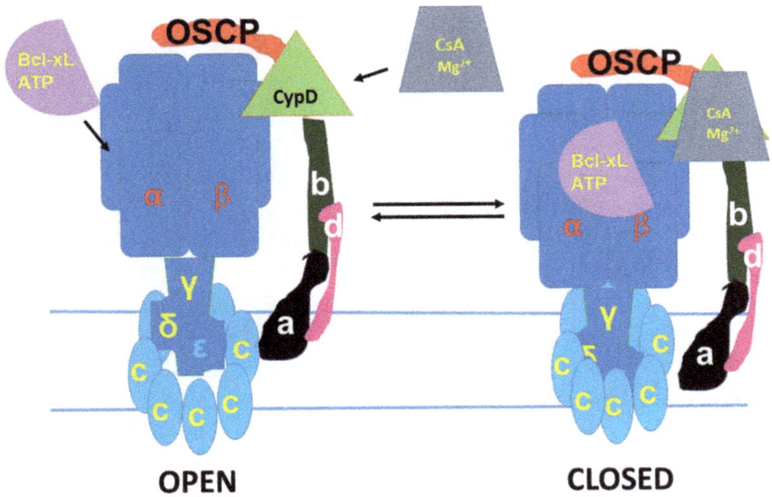

Fig. 3.2 Proposed model of enhanced efficiency of ATP production during synaptic plasticity. F_1 of the F_1F_O ATPase moves more closely toward the F_O of the ATP synthase in the presence of Bcl-x_L, ATP binding to the β-subunit or CsA binding to OSCP. This closes an inner membrane leak, increasing the efficiency of ATP production by the ATP synthase and enhancing synaptic transmission

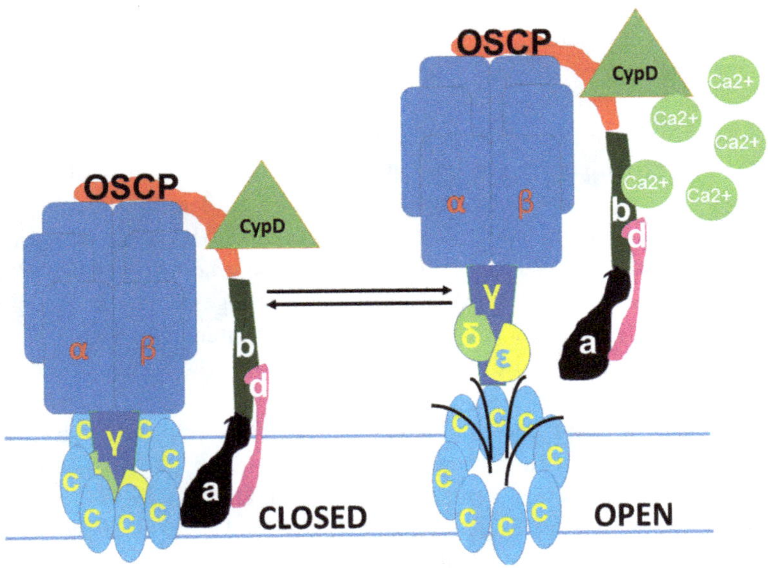

Fig. 3.3 Proposed model of the location of the mPTP pore and its gating. The F_1 or enzymatic portion of the F_1F_O ATP synthase serves as the gate of the channel, which resides in the F_O (within the c-subunit ring). Upon Ca^{2+} influx into the matrix, CypD assists the F_1 to lift off of the F_O, exposing the mouth of the pore which then conducts ions

detected in aging heart of 12–24-month-old animals, suggesting the presence of F_1-lacking F_O leak channels in the membranes of these aging mitochondria (Guerrieri et al. 1992).

Homologues of ATP synthase c-subunit are also present in the V_O subcomplex of various vacuolar H^+-ATPases, and V_OV_1-ATPase c-subunits line a water-accessible pore structure (Jones et al. 1995; Harrison et al. 2003) found both on the plasma membrane and on intracellular membranes (Mandel et al. 1988; Finbow and Harrison 1997). V_O c-subunits have also been shown to be involved in the formation of mega-channels in gap junctions between cells (Peters et al. 2001). Pore-forming ability, therefore, is an important feature of all homologous c-subunits which share similar amino acid sequence (Azarashvili et al. 2014).

The vacuolar ATPase's proton pumping function is responsible for the acidification of intracellular organelles including synaptic vesicles and lysosomal compartments. V-ATPase F_1 and F_O sectors are known to disassemble and reassemble readily in response to changes in nutrient state of the intracellular milieu. In its "naked" state, the V_O may serve as a channel (Couoh-Cardel et al. 2016). A recent report found that purified V_O monomers of VATPase form rings of outer and inner diameters of 8.6 and 3.5 nm, respectively (Couoh-Cardel et al. 2016). A density seems to fill the ring, either made of lipid or of an arm (N-terminal alpha helix) of the c" isomer that forms one of the c8c'c" partner molecules of the c-ring. When reconstituted into planar lipid membranes, this V_O c-ring forms voltage-dependent, multi-conductance rectified single-channel activity with an average single-channel conductance of 8.3 nS. The closings at negative potentials indicate that the positive or negative charges inherent in the c-subunit structure may move into the pore interior under polarized conditions. When the entire V_O is reconstituted into a lipid bilayer, the single-channel conductances are considerably smaller, ranging between 0.7 and 3.8 nS but averaging 1.8 nS. The d-subunit seemed to act as a gate in these recordings, particularly at negative potentials. At very positive potentials in holo V_O recordings, in some cases the d- and aNT subunits appeared to dissociate completely from c-subunit, yielding a very large single-channel conductance similar to that recorded with purified c-subunit alone. These findings indicate that the c-subunit of V_O forms a large pore that experiences voltage-dependent gating either by movement of its own charged moieties and/or by interacting with an arm and plug created by the aNT and d-subunits that move over the mouth of the pore to shut the channel.

In accordance with the findings for V_O, the membrane portion of the F_1F_O ATP synthase may comprise a large conductance channel that could produce PT under stress. To help determine this, cell death was measured after depletion of all three c-subunit isoforms or after overexpression of a tagged version of the wild-type c-subunit in HeLa cells (Bonora et al. 2013). Cell lines are able to utilize glycolysis for ATP production in normal glucose-containing medium; therefore depleting the c-subunit did not diminish ATP levels in the cells. In contrast, c-subunit depletion prevented CsA-sensitive PT measured by calcein-cobalt quench and mitochondrial morphological changes typical for PT. Furthermore, c-subunit depletion markedly attenuated H_2O_2-induced cell death and glutamate-induced excitotoxicity in neurons, suggesting that the c-subunit was required for PT.

Although the above study indicated that the F_1F_O ATP synthase c-subunit was an important component of the mPTP (Bonora et al. 2013), this work did not directly determine what portion of ATP synthase could form the pore of the mPTP. CypD and Bcl-x_L interact with the stator and β-subunit of ATPase, respectively, and these proteins are membrane attached but not embedded. The channel or pore-forming subunits, however, must span the mitochondrial inner membrane. Nine polypeptides form the F_O and the stator, but only three, a, b, and c, are required for proton translocation and are evolutionarily highly conserved, like PT. In addition, ρ0 cells that lack mitochondrial DNA do not contain an a-subunit but do undergo PT. Attention focused on the c-subunit, and not on the b-subunit, because the mitochondrial c-subunit had been shown previously to express ion channel activity (McGeoch and Guidotti 1997). The mammalian c-subunit undergoes conformational changes from an α-helix to a β-sheet when in contact with water, forming the walls of ion channels (McGeoch and McGeoch 2008) with a diameter of 2.3 nm allowing molecules up to 1.5 kDa to pass, similarly to PT.

3.7.2 *The c-Subunit of ATP Synthase Creates a High-Conductance mPTP Pore*

Recent experiments have directly tested the hypothesis that the main membrane embedded portion of mammalian F_1F_O ATP synthase, i.e., the c-subunit ring, forms the pore of the mPTP (Alavian et al. 2014; Azarashvili et al. 2014). Indeed, electrophysiological recordings of the purified mitochondrial c-subunit yield a multi-conductance, voltage-dependent channel with prominent subconductance states (Alavian et al. 2014). Patches contain a ~100-pS conductance in 150 mM KCl, which appears to be a subconductance state of a larger activity rather than a separate conductance. Recordings also reveal peak single-channel conductances of ~1.5–2 nS, similarly to activity described previously for the mitochondrial multiple conductance channel (MCC) (Kinnally et al. 1989). Also consistent with MCC, channel activity often but not always demonstrates negative rectification. At extreme positive patch pipette potentials of over 100 mV, single-channel conductances of ~1.5 nS and ~2 nS, are consistently observed. Single-channel events and gating are more likely to occur at positive potentials most likely because of the negative rectification, in keeping with previous reports for the mPTP (Petronilli et al. 1989).

Voltage dependence is an inherent property of the channel (Borjesson and Elinder 2008) and is not dependent on the mitochondrial inner membrane potential or on the solutions used to record the currents. When measured by varying the voltage across the membrane (the command or holding voltage), recent I–V plots on the purified c-subunit prove consistent with previously reported I–V plots of the mPTP currents (Kinnally et al. 1989; Petronilli et al. 1989; Alavian et al. 2014).

3.7.3 F_1 Regulates Biophysical Characteristics of the Purified c-Subunit

Purified c-subunit protein reconstituted into liposomes clearly lacks extrinsic regulatory moieties that are important for mPTP opening. The effects of Ca^{2+}, a critical regulatory molecule, demonstrate this property. Ca^{2+} binding sites have not been detected in mitochondrial F_O, perhaps because mammalian c-subunits lack the formyl Met at the N-terminus to which Ca^{2+} binds to *E.coli* or chloroplast F_1F_O ATP synthase c-subunits (Zakharov et al. 1996). In contrast, Ca^{2+} binds to specific, low-affinity, and moderate capacity sites on the β-subunit of F_1 ATP synthase (Hubbard and McHugh 1996). Other sites in other ATP synthase-interacting molecules may also be important for regulation by Ca^{2+} and other agents (Beutner et al. 1996, 1998). Therefore, although the new models of the mPTP pore must account for all inducers and inhibitors, these molecules may not interact directly with the c-subunit pore itself, but may instead bind to sites in the F_1 or to other molecules such as ANT and CypD that undergo structural rearrangements upon opening and closing of the pore.

In order to determine the location of the regulators, mitochondrial recordings were carried out using purified mitochondrial and F_1F_O ATP synthase preparations. In these studies, the absence of an effect of a modulator was taken as an indication that the ligand or binding site for that modulator had been removed by the purification process. For example, since in mitochondria or isolated inner membrane preparations Ca^{2+} activates the c-subunit leak channel while CsA and ATP/ADP inhibit it, the Ca^{2+} and CsA-sensitive sites must be present in these preparations. In contrast, removal of the F_1 and other peripheral membrane proteins by urea treatment of the inner membrane or removal of CypD by purification of ATP synthase monomers abrogates regulation of the c-subunit channel by CsA and Ca^{2+} and greatly diminishes sensitivity to ATP/ADP. These studies suggest that the CypD/calcium binding site is contained within or associated with the F_1 portion of the ATP synthase and that a second, low-affinity ATP binding site exists in the F_O (Alavian et al. 2014). These results are consistent with reports identifying the binding site of CypD and benzodiazepine 423, an ATP synthase-inhibitory and mPTP-sensitizing agent, on OSCP (Giorgio et al. 2009, 2013a), and suggest that the assembly of F_1F_O ATP synthase into monomers, dimers, and higher-order oligomers can regulate the formation of the mPTP.

Channel activity of the purified c-subunit is inhibited by the purified F_1, suggesting a structural rearrangement, whereby the stalk and F_1 of the ATP synthase inhibit opening of the c-subunit channel, aided by ADP/ATP/Bcl-xL binding to the β-subunit and opposed by CypD/Ca^{2+} interaction with OSCP (Fig. 3.2). Mitochondria treated with Ca^{2+} have a destabilized connection between the stalk and the c-subunit, disrupting protein/protein interaction between the c-subunit and F_1 (Fig. 3.3). A model incorporating these findings suggests that the channel of the mPTP forms within the c-subunit ring upon CypD and Ca^{2+}-dependent movement of the stalk away from the c-subunit (Alavian et al. 2014).

Loss of protein/protein interaction between F_1 and F_O requires fairly mild conditions such as 60 μM Ca^{2+} in the bath to initiate PT (Alavian et al. 2014). This concentration is well within the range of physiological Ca^{2+} concentrations found within the mitochondrial matrix (Rizzuto et al. 2000; Csordas et al. 2001; Rizzuto and Pozzan 2006) or concentrations measured adjacent to the mitochondria in Ca^{2+} microdomains at the plasma membrane or ER membrane (Schneggenburger and Neher 2005; Rizzuto et al. 2009). Destabilization of the F_1 and F_O connection is likely to be reversible such as has been shown upon chelation of Ca^{2+} in mitoplasts (Beutner et al. 1998), intact mitochondria (Roestenberg et al. 2012), intact neurons (Jonas et al. 1999), as well as in reconstituted dimers of F-ATP synthase (Gomez et al. 2007). Therefore the F_1 and the c-subunit may recombine to close the mPTP, reforming intact F_1F_O ATP synthase and reinitiating enzymatic function (Pedersen and Hullihen 1978). However, under certain conditions, this separation may become irreversible, forming pathophysiologic PT (with MOMP).

Another possible regulatory molecule is the ATPase inhibitory factor 1 (IF1), a small, nuclear-encoded endogenous polypeptide, which is involved in the regulation of the oligomeric state of the F_1F_O ATP synthase by facilitating enzyme dimerization of two F_1 domains (Garcia et al. 2006). IF1 is also known as an intrinsic inhibitor of the ATP synthase. It interacts with the catalytic β-subunit of ATP synthase and is responsible for inhibiting ATP hydrolysis but not synthesis. The binding of IF1 is pH dependent, and below neutrality its inhibitory activity increases (Cabezon et al. 2001). IF1 plays a protective role during hypoxic/ischemic conditions, when the electrochemical gradient across the membrane collapses and the enzyme switches from ATP synthesis to hydrolysis (Pullman and Monroy 1963; Jennings et al. 1991; Rouslin 1991). The absence of IF1 has been linked with the human pathological condition known as Luft's disease, characterized by densely packed mitochondrial cristae and significantly high ATPase rate (DiMauro et al. 1976; Yamada and Huzel 1992). IF1 expression is increased in many human carcinomas, supporting its anti-apoptotic and potentially tumorigenic role (Bravo et al. 2004; Sanchez-Cenizo et al. 2010). It protects cells from necrotic and apoptotic cell death by regulating mitochondrial cristae morphology, promoting the dimerization of ATP synthase, limiting release of Cyt c, and preventing the activation of Drp1 and oligomerization of pro-apoptotic Bax (Campanella et al. 2008; Faccenda et al. 2013a, b).

Additional regulation of the mPTP may be due to the association of other molecules with F_1F_O ATP synthase. As discussed above, F_1F_O ATP synthase may complex with ANT and PiC. In the study of F_1F_O ATP synthase dimers, bongkrekic acid, which inhibits ANT, fails to attenuate the mPTP channel activity (Giorgio et al. 2013b); this is recapitulated in studies of SMVs (Alavian et al. 2011). Therefore, the regulation of the mPTP by ANT and PiC may occur through their association with the peripheral membrane components of F_1F_O ATP synthase. Furthermore, the association of F_1F_O ATP synthase and ANT with PiC, mtCK, VDAC, and hexokinase may explain why mPTP regulation can occur via these molecules.

Some regulators of the mPTP also work directly on the F_1F_O ATP synthase itself. F_1 has binding sites that accommodate the effects of Ca^{2+}, Mg^{2+}, adenine nucleotides and P_i; and through CypD (un)binding those of H^+, CsA, and possibly of oxidants

(Kruse et al. 2008). Therefore, in summary, the new model of mPTP explains either direct or indirect interaction with all known inducers, inhibitors, and modulators of pore function (Figs. 3.2 and 3.3).

3.7.4 Structural Location of the Pore Within the c-Subunit Ring

The exact location of the ion-conducting pore of the c-subunit is becoming increasingly understood. Although it has been proposed that the pore sits between the two lateral stalks of a dimer of F_1F_O ATP synthases and not within the c-subunit ring (Giorgio et al. 2013b), there is currently no electrophysiological evidence for the formation of such a channel, and regulation of the mPTP by components of F_1 (Giorgio et al. 2013b; Alavian et al. 2014, 2015) argue against this.

Rather, it is likely that the leak is located either within the central portion of the c-subunit ring, between the individual c-subunit monomers, or between the c-subunit and the other F_O subunits, although the latter is less likely given the presence of PT in ρ0 cells that lack both mitochondrial DNA and the a-subunit (Bonora et al. 2013). In two separate experiments, it has been demonstrated that the c-subunit ring expands when it conducts ions, making it likely that the pore is formed by the c-subunit ring. The first experiment used fluorescent tetracysteine display. These studies showed that Ca^{2+} influx into cells causes an expansion of the diameter of the c-subunit ring, while CsA decreases ring diameter (Alavian et al. 2014). Mutagenesis to increase the diameter of the c-subunit ring also demonstrates that ring expansion is a means to increase conductance. (Norris et al. 1992; Alavian et al. 2014).

These findings support the hypothesis that the c-subunit is necessary and sufficient to produce the pore of mPTP. Although it has been suggested that phospholipids occupy the central cavity of the c-subunit ring in F_1F_O ATP synthases from different species (Meier et al. 2001; Oberfeld et al. 2006; Matthies et al. 2009), other evidence provides for formation of a proteolipid or proteophospholipid channel structure within the central lipid region (Pavlov et al. 2005; Abramov et al. 2007; McGeoch and McGeoch 2008; Elustondo et al. 2013; Azarashvili et al. 2014). Recent single-particle cryo-EM data also suggest that subunit e, DAPIT, or subunit 6.8 kDa may form a p-side density (PD) adjacent to the c-subunit pore within the structure of the bovine ATP synthase (Gerle 2016). The PD extends from the membrane adjacent to the c-ring toward the pore region of the c-ring. The part of the PD sitting in the membrane also contacts OSCP/b subunits in the peripheral stalk or stator part of the molecule, enhancing the possibility that PD movement could be regulated by conformational changes induced from the top of the $\alpha3\beta3$ hexamer through OSCP and b and eventually to the PD. These arrangements would fit with data suggesting that Ca^{2+}/Mg^{2+} and ATP binding to the F_1 affect gating of the mPTP. These data also support a model whereby strain is relieved on the ATP synthase by dimer and oligomer formation (Davies et al. 2012), making it more likely that monomeric ATP synthase would be predisposed to allow the PD to pull the lipids out of the central pore.

The data suggest a working model whereby the c-subunit pore forms within the proteolipid milieu upon activation of mPTP (e.g., by elevated matrix Ca^{2+}) whereupon the ring expands and F_1 shifts; the pore is closed by a decrease in diameter of the ring and inactivated by binding of the F_1 components to the ring (Figs. 3.2 and 3.3). The details of these changes and their regulation remain a work in progress.

3.8 mPTP Opening in Acute Neuronal Ischemia

3.8.1 Ischemia Triggers the Excitotoxic Pathway

Cerebral ischemia is a condition of insufficient blood and oxygen supply to the brain that causes neuronal energy deficits. This condition impairs ATP-dependent transporters that regulate ionic homeostasis in the neurons and allows uncontrolled ion influx, especially Ca^{2+} into the neuronal cytoplasm (Meyer 1989; Schwab et al. 2002; Bano et al. 2005). Ca^{2+} influx through glutamate receptors produces excitotoxicity in neurons (Choi 1987; Nishizawa 2001; Szydlowska and Tymianski 2010) resulting in structural and functional strain on intracellular organelles including mitochondria (Lo et al. 2003).

3.8.2 mPTP Mediates Neuronal Death in Ischemia

Ischemic insults impair electron transport which causes a deterioration in intracellular energy metabolism, production of reactive oxygen species (ROS) from mitochondrial proteins such as complex I and complex III (Massaad and Klann 2011; Holmstrom and Finkel 2014), and opening of mPTP (Baines 2009; Jonas 2009; Bernardi and Di Lisa 2015). Opening of mPTP disrupts mitochondrial membrane integrity, allows activation of glutathione (Reed and Savage 1995), and produces leakage of death-promoting factors (Du et al. 2000; Verhagen et al. 2000; Petronilli et al. 2001).

In contrast to the protective roles of Bcl-x_L in synaptic function, neuronal survival, and mPTP regulation, Bcl-x_L is also capable of initiating death signaling in ischemic brain. The formation of a caspase-dependent N-terminal cleavage product of Bcl-x_L (ΔN-Bcl-x_L) (Clem et al. 1998; Fujita et al. 1998; Seng et al. 2016) is enhanced during cerebral ischemia in rodent models, and strategies to prevent ΔN-Bcl-x_L formation protect the brain against ischemia-induced injury (Miyawaki et al. 2008; Ofengeim et al. 2012). ΔN-Bcl-x_L is missing a BH4 domain that exerts antiapoptotic function in full length Bcl-x_L (Hirotani et al. 1999; Sugioka et al. 2003). Therefore, ΔN-Bcl-x_L exhibits pro-death characteristics. Injection of

ΔN-Bcl-x_L induces large multi-conductance channels in the mitochondrial membrane but not the plasma membrane and causes synaptic depression (Jonas et al. 2004; Hickman et al. 2008). In addition, application of a caspase inhibitor prevents hypoxia-induced mitochondrial channel activity and improves synaptic responses (Jonas et al. 2005) indicating that ΔN-Bcl-x_L works in part to cause loss of mitochondrial membrane potential, suggesting loss of energy and attenuated Ca^{2+} buffering capacity. Since full length Bcl-x_L binds with ATP synthase (Alavian et al. 2011; Veas-Perez de Tudela et al. 2015) and ATP synthase undergoes conformational changes that favor closing mPTP (Alavian et al. 2011, 2014a), by analogy ischemia-induced ΔN-Bcl-x_L formation may interfere with mPTP closure by sequestering functional Bcl-x_L, preventing full length Bcl-x_L interaction with ATP synthase.

Ischemia also changes the dynamics of other Bcl2 family proteins that may regulate mPTP. Ischemic stimuli activate death receptors that release initiator caspases (Matsushita et al. 2000; Broughton et al. 2009; Park et al. 2015) and activate pro-death BH3-only molecules and pro-apoptotic Bax (Li et al. 1998; Luo et al. 1998; Lindsten et al. 2000). Pro-apoptotic Bid cleavage appears in both an in vitro oxygen glucose deprivation model and in an in vivo middle cerebral artery occlusion model. Bid and Bax interact with possible mPTP regulators such as ANT and VDAC (Narita et al. 1998; Shimizu et al. 1999; Brenner et al. 2000; Zamzami et al. 2000; Cao et al. 2001) further supporting a regulatory role of Bcl2 family proteins in mPTP during ischemia.

3.8.3 Strategies to Block mPTP in Ischemic Models

Due to the significance of mPTP in ischemia-induced neuronal death, blockade of mPTP has been suggested to protect the brain from ischemic insults. CypD-depleted mitochondria are more resistant to Ca^{2+}-induced mitochondrial swelling in vitro, and CypD-deficient mice show significantly decreased infarct size after middle cerebral artery occlusion-induced brain injury in vivo (Schinzel et al. 2005). CsA protects the brain from ischemia-induced mPTP opening (Matsumoto et al. 1999). However, another series of studies have reported that depletion of CypD does not completely eliminate mPTP (Basso et al. 2005; Nakagawa et al. 2005; Baines et al. 2007).

Since PT occurs acutely during excitotoxicity or high ROS during brain or cardiac ischemia (Baines 2009), this predicts that genetic depletion of the c-subunit may prevent PT. Indeed, excitotoxic and ROS-induced cell death is greatly attenuated upon depletion of the c-subunit by shRNA in neurons and other cells (Bonora et al. 2013; Alavian et al. 2014); cell death protection by c-subunit depletion is not further attenuated by CsA, suggesting that the c-subunit forms the inner mitochondrial membrane target of the CsA-sensitive complex (Alavian et al. 2014). In contrast to c-subunit depletion, overexpression of the wild-type c-subunit (Bonora et al. 2013) or mutation of the c-subunit to form a high-conductance leaky c-subunit (Alavian et al. 2014) predisposes to enhanced cell death upon excitotoxic or ROS stimulation. Death under these conditions is not sensitive to CsA, presumably because the leaky pore prevents normal regulation by components of F_1. Despite the powerful neuroprotective property of ATP synthase c-subunit depletion against an ischemia-like environment such as excitotoxic and oxidative stress in vitro (Bonora et al. 2013; Alavian et al. 2014), knocking down the c-subunit of ATP synthase has not yet been fully explored in an in vivo ischemic model. Therefore, further investigation is required to address mPTP as a therapeutic target of ischemic brain disease.

3.9 mPTP in Neurodegenerative Disease

3.9.1 Alzheimer's Disease and the mPTP

Unremitting metabolic demand on neurons throughout life can lead to high neuronal stress, which contributes to the development of neurodegenerative diseases. Mitochondrial dysfunction, which is often accompanied by protein misfolding and the release of proapoptotic factors, can be used as a marker for the development of neurodegenerative diseases (Schon and Manfredi 2003).

The hallmark feature of Alzheimer's disease (AD) in brain pathology is the extracellular plaque containing Aβ peptide, the result of exuberant APP processing. Aβ has the ability to penetrate both the outer and inner mitochondrial membranes through the translocase of the outer and inner membrane (TOM and TIM) complexes (Reddy 2009; Cadonic et al. 2015). Aβ binds to heme groups in the first four complexes of the electron transport chain (ETC), preventing these groups from performing their redox functions (Atamna and Frey 2004). Aβ also binds to amyloid beta binding dehydrogenase (ABAD), impairing complex IV's enzymatic activity (Rao et al. 2014) and causing increased production of reactive oxygen species (ROS). In addition to the calcium overload of the cytosolic compartment common in affected neurons in AD, this leads to further mitochondrial deterioration (Supnet and Bezprozvanny 2010). Aβ also depolarizes the IMM directly (Qiao et al. 2005; Cha et al. 2012), leading to a functional uncoupling. These abnormalities of mitochondria caused by Aβ are associated with early opening of the mPTP (Du et al. 2010).

Reports have suggested that Aβ's interaction with CypD is a trigger for mPTP formation (Du et al. 2008). CypD expression increases in areas of the brain that are most adversely affected by AD (Du et al. 2010, 2014; Du and Yan 2010; Guo et al. 2013). Comparing CypD expression in wild-type and Aβ-containing cortical mitochondria from humans and the mAPP mouse model of AD, it is clear that in both the temporal pole and hippocampus, Aβ-containing mitochondria express significantly higher levels of CypD than wild-type mitochondria. CypD expression levels in mAPP mouse brains increase over time in comparison to those of non-transgenic mice, consistent with progression of neurodegeneration (Du and Yan 2010).

Through in vitro binding assays, it has been shown that oligomeric Aβ40 and Aβ42 bind to recombinant human CypD protein with high affinity (Du et al. 2010), prevented by antibodies targeting Aβ or CypD. CypD and Aβ form a complex in mAPP mouse brain and human brain tissue affected by AD (Du et al. 2008) but not in age-matched wild-type mouse brains and control human brain tissue (Du et al. 2008.

CypD-induced mPTP may also impair other mitochondrial functions in AD brain including calcium handling, membrane potential, and ROS regulation. From 3 to 24 months, the calcium buffering capacity of mAPP, in comparison to non-transgenic mice, decreases at a much faster rate. Interestingly, mAPP mice lacking CypD (mAPP/Ppif$^{-/-}$) surpass even the age-matched non-transgenic mice in calcium uptake (Du and Yan 2010), which suggests that loss of CypD may protect against both AD and the aging phenotype. Additionally, mAPP mice treated with CsA exhibit a markedly improved increase in calcium uptake ability and improvement on learning and memory tasks (Du and Yan 2010). CypD-induced mPTP opening also alters heme metabolism, increases ROS decreases ATP production (Reynolds 1999; Blanchard et al. 2002; Atamna and Frey 2004).

One mechanism behind the CypD-Aβ-induced mPTP opening may be related to the Bcl-2 family of proteins (Chen et al. 2015; Veas-Perez de Tudela et al. 2015). The mitotic protein cyclin B1 accumulates aberrantly in damaged mitochondria in degenerating neurons, activating cyclin-dependent kinase-1 (Cdk1) which phosphorylates Bcl-x_L, leading to dissociation of Bcl-x_L from the β-subunit of ATP synthase, inhibition of ATP synthase enzymatic activity, mitochondrial membrane depolarization, and neuronal death (Veas-Perez de Tudela et al. 2015). In contrast, Bcl-x_L protects axons from degeneration in neurons undergoing Aβ-induced toxicity (Alobuia et al. 2013). Aβ may also bind to the ATP synthase directly, disrupting interaction of ATP synthase with other enzyme systems that normally enhance enzymatic rate, thereby impairing ATP synthesis (Cha et al. 2015).

3.9.2 *Relationship of Synaptic Mitochondrial Channel Activity to Long-Term Depression of Synaptic Responses During Neurodegeneration*

Long-term synaptic depression (LTD) caused by low-frequency stimulation or by cell signaling is a normal mechanism of synaptic plasticity opposite in some ways to long-term potentiation (LTP), the latter of which is brought on by high-frequency stimulation (Malenka and Bear 2004). Despite its role in normal synaptic plasticity, however, long-term depression can also serve as a marker for a pre-degenerative synaptic state. In hippocampal CA3 to CA1 synapse, low synaptic activity leads to a long-lasting decline in synaptic efficacy, brought about in part by removal of postsynaptic receptors (Malinow and Malenka 2002; Malenka and Bear 2004); this state can be quite stable and may never lead to somatic demise. Mitochondria are important for a form of LTD associated with normal synaptic plasticity in hippocampal CA1 neurons. In this form of physiological LTD in the CA1 dendrite, low-frequency activity causes Bcl-x_L-sensitive mitochondrially mediated release of cytochrome c followed by low-level activation of caspase 3, which leads to the removal of postsynaptic glutamate receptors from the plasma membrane (Li et al. 2010).

Despite the role of LTD in normal synaptic plasticity, age or developmentally related degenerative changes may be set in motion by prolonged synaptic depression. It is now well accepted that release of toxic Aβ from axonal endings decreases synaptic dendritic spine number in partner neurons, contributing to a decline in synaptic efficacy and loss of LTP (Wei et al. 2010). In synapses from Bax −/− mice treated with the toxic Aβ protein, prevention of LTP by Aβ was attenuated, implying that actions of the pro-apoptotic protein Bax at mitochondria may be necessary for the relative loss of synapses during Aβ-induced toxicity (Olsen and Sheng 2012; Erturk et al. 2014). In a model of developmental axonal targeting in spinal neurons, both mitochondrial Bax and caspase 6 activation were found to contribute to axonal loss in response to nerve growth factor withdrawal (Nikolaev et al. 2009). In this scenario, the N-terminus of amyloid precursor protein (APP) binds to death receptor 6 (DR6) to initiate an intracellular cascade resulting in mitochondrial-dependent axonal demise.

The role of mPTP in neuritic and synaptic loss is inferred in a recent study of neurite growth arrest in cultured neurons depleted of Bcl-x_L by siRNA. The work shows that declining Bcl-x_L levels prevents normal outgrowth and branching of neuronal processes over 4 weeks in culture before any somatic death occurs (Park et al. 2014), associated with upregulation of death receptor 6 (DR6) and Bax. In contrast to the slowly occurring growth arrest found upon Bcl-x_L depletion, loss of neurites takes place with a much more rapid timescale after a hypoxic stimulus. Hypoxia-induced loss is greatly attenuated in neurons by Bcl-x_L and by depletion of DR6, suggesting reversal of detrimental metabolic changes and amelioration of relative opening of mPTP in dendritic and axonal mitochondria.

3.9.3 mPTP and Parkinson's Disease

Parkinson's disease (PD) is the second most common neurodegenerative disorder after Alzheimer's disease, affecting about 1% of the population. PD is characterized by the selective loss of dopaminergic neurons in the substantia nigra pars compacta (SNpc), deficiency of dopamine in the striatum, and the presence of excess α -synuclein protein in presynaptic neural cells (Banerjee et al. 2009). So far, several familial PD genes have been identified, notably α-synuclein (α-syn), Parkin, PTEN-induced putative kinase 1 (PINK1), leucine-rich repeat kinase 2 (LRRK2), and DJ-1 (Banerjee et al. 2009).

In the past several decades, mitochondrial dysfunction has been implicated in the pathogenesis of PD. Most studies have focused on the mitochondrial respiratory chain, particularly complex1 dysfunction, but recent studies have highlighted the role of PINK1 and Parkin as key activators of mitophagy.

The mitochondrial permeability transition pore (mPTP) has received attention as a contributor to PD. It is reported, for example, that toxic MPP+, 6-hydroxydopamine, or dopamine strongly stimulates Ca^{2+} release from mitochondria and hydrolysis of intra-mitochondrial pyridine nucleotides. This is accompanied by cytochrome c release and mitochondrial membrane depolarization. Interestingly, these events are prevented by CsA. These findings comprise the first set of evidence linking Ca^{2+}-induced mPTP with PD-related cell death (Frei and Richter 1986; Cassarino et al. 1999). Mitochondria isolated from CypD knockout mouse brain are less sensitive to MPP+-induced membrane depolarization, and these mitochondria are also relatively protected from free radical generation compared to wild-type mice (Thomas et al. 2012). The ventral midbrain mitochondria from MPP+-treated CypD KO mice also exhibit less damage than those isolated from wild-type mice when judged by respiratory chain complex I activity, state 3 respiration rate, and respiratory control index (Thomas et al. 2012).

Recently, the Thy1-A53T hαSyn tg C57BL/6 mouse model was established (Martin et al. 2014). These transgenic mice develop a severe, age-related, fatal PD-like movement disorder and robust brainstem neurodegeneration during development (Martin et al. 2014). In this model, CypD level is modestly increased in the brainstem, striatum, and cortex of early and late symptomatic mice, whereas ANT and VDAC levels remain unchanged. Mitochondria appear aggregated and swollen in the SNc neurons. The disease onset of this Thy1-A53T tg mouse is significantly delayed when reducing the levels of CypD by genetic ablation, and the mouse lifespan is extended. Given that CypD expression is positively correlated with mPTP opening, the model establishes a direct cause-effect relationship between the mPTP and PD disease mechanisms (Martin et al. 2014).

The above discoveries have also been confirmed in several other familial PD mouse models. It was reported that the mitochondrial transmembrane potential ($\Delta\Psi m$) is reduced in PINK1−/− MEFs and neurons, and the reduction of $\Delta\Psi m$ in PINK1−/− cells is associated with increased opening of mPTP. It is also described that inhibition of mPTP reverses the depolarization of the mitochondrial inner membrane and respiration defects seen in PINK1−/− cells (Gautier et al. 2012). Mitochondria from PINK1−/− mouse brain have altered Ca^{2+} storage capacity and increased mPTP opening (Rasola and Bernardi 2011).

Fibroblasts harboring PARK2 mutations from juvenile Parkinson's disease (JPD) patients contain impaired mitochondria; ATP levels and $\Delta\Psi m$ are reduced (Zanellati et al. 2015). In a twist on the story of Parkin targeting to mitochondria during mitophagy, in SH-SY5Y neuroblastoma cells, the Parkin protein instead departs the mitochondria and locates itself to other intracellular structures when treated with inhibitors of electron transport chain such as rotenone, mitochondrial uncouplers, and cell cycle blockers (Kuroda et al. 2006). This unusual scenario is brought on by opening of the mPTP under toxic conditions, and the study suggests that the removal of Parkin from the mitochondria signals for transcription of mitochondrial proteins including those of ATP synthase (mitochondrial biogenesis).

DJ1 mutations cause a rare form of familial disease that accounts for only 1–2% of early onset PD. Several studies demonstrate that the ATP concentration is decreased and mitochondrial transmembrane potential is reduced in the DJ-1 −/− MEFs, whereas mPTP opening is increased, measured by the calcein-cobalt assay (Giaime et al. 2012). Despite these tantalizing findings, so far, there is no direct evidence that connects selective nigrostriatal neuron death with mPTP opening. Nevertheless, considering the high metabolic demands placed on nigrostriatal neurons, it is easy to understand why this type of neuron would be among the most susceptible cells to mPTP opening.

How mitochondrial failure triggers PD during slow degeneration is still unclear. Based on the current knowledge of mPTP in PD, therapeutic strategies by targeting mPTP might halt or slow the progression of PD. Some research groups have already started identifying novel mPTP inhibitors in small molecule libraries through unbiased high-throughput screening (HTS) (Rasheed et al. 2016).

3.10 Conclusion

For many years investigators have sought to identify the molecular structure underlying acute alterations in mitochondrial morphology and increases in inner membrane conductance associated with acute cell death known collectively as mitochondrial PT. Early evidence asserted that PT was caused by opening of an inner membrane ion channel. More recent data have shown that the c-subunit of the F_1F_O ATP synthase forms a channel with similar biophysical characteristics to mPTP but whose Ca^{2+}, CsA, CypD, and Bcl-xL regulatory sites are contained in the F_1 including the stator and the catalytic portions. Depletion of c-subunit isoforms in cells blocks CsA-dependent PT and subsequent cell death. Inhibitors and activators may also work through peripheral regulatory moieties such as ANT, PiC, and VDAC that exist in a large complex of proteins with the F_1F_O ATP synthase. Lipids and polyphosphates also may play an important role in pore gating or formation.

Activators of the mPTP appear to open the pore by a gating mechanism in which F_1 moves away from the mouth of the c-subunit ring while the ring expands (Fig. 3.3). The process is reversible, perhaps due to binding of F_1 components to the ring or by the reassociation of the entire F_1 onto the ring (Fig. 3.2). Although a wealth of information regarding the molecular structure and regulation of mPTP has been unearthed recently, there is still much to be learned about this fascinating complex.

The study of mPTP in enhanced inner membrane coupling during development of oxidative phosphorylation, in aging and in supercomplex formation, comprises a rapidly changing field. In neurons, it is now becoming clear that the metabolic efficiency of mitochondria regulates neuronal survival, neurotransmitter vesicle recycling, and synaptic development. During neurodegeneration, metabolic inefficiency within neurons or neuronal processes may lead to depression of synaptic responses, proapoptotic mitochondrial conductance changes, metabolic compromise, and eventual loss of neurite outgrowth accompanied by increases in the probability of mPTP opening. Ongoing studies will illuminate the molecular structural changes associated with mitochondrial channel activity during cell development, plasticity, and during stressful or degenerative events.

Acknowledgements The authors would like to thank Dr. Leonard Kaczmarek for the artwork in Fig. 3.1.

References

Abramov AY, Fraley C, Diao CT, Winkfein R, Colicos MA, Duchen MR, French RJ, Pavlov E (2007) Targeted polyphosphatase expression alters mitochondrial metabolism and inhibits calcium-dependent cell death. Proc Natl Acad Sci U S A 104:18091–18096

Alavian KN, Li H, Collis L, Bonanni L, Zeng L, Sacchetti S, Lazrove E, Nabili P, Flaherty B, Graham M, Chen Y, Messerli SM, Mariggio MA, Rahner C, McNay E, Shore GC, Smith PJ, Hardwick JM, Jonas EA (2011) Bcl-xL regulates metabolic efficiency of neurons through interaction with the mitochondrial F1FO ATP synthase. Nat Cell Biol 13:1224–1233

Alavian KN, Beutner G, Lazrove E, Sacchetti S, Park HA, Licznerski P, Li H, Nabili P, Hockensmith K, Graham M, Porter GA Jr, Jonas EA (2014) An uncoupling channel within the c-subunit ring of the F1FO ATP synthase is the mitochondrial permeability transition pore. Proc Natl Acad Sci U S A 111:10580–10585

Alavian KN, Dworetzky SI, Bonanni L, Zhang P, Sacchetti S, Li H, Signore AP, Smith PJ, Gribkoff VK, Jonas EA (2015) The mitochondrial complex v-associated large-conductance inner membrane current is regulated by cyclosporine and dexpramipexole. Mol Pharmacol 87:1–8

Alobuia WM, Xia W, Vohra BP (2013) Axon degeneration is key component of neuronal death in amyloid-beta toxicity. Neurochem Int 63:782–789

Andrews ZB, Diano S, Horvath TL (2005) Mitochondrial uncoupling proteins in the CNS: in support of function and survival. Nat Rev Neurosci 6:829–840

Atamna H, Frey WH 2nd (2004) A role for heme in Alzheimer's disease: heme binds amyloid beta and has altered metabolism. Proc Natl Acad Sci U S A 101:11153–11158

Azarashvili T, Odinokova I, Bakunts A, Ternovsky V, Krestinina O, Tyynela J, Saris NE (2014) Potential role of subunit c of F0F1-ATPase and subunit c of storage body in the mitochondrial permeability transition. Effect of the phosphorylation status of subunit c on pore opening. Cell Calcium 55:69–77

Baines CP (2009) The mitochondrial permeability transition pore and ischemia-reperfusion injury. Basic Res Cardiol 104:181–188

Baines CP (2011) The mitochondrial permeability transition pore and the cardiac necrotic program. Pediatr Cardiol 32:258–262

Baines CP, Kaiser RA, Purcell NH, Blair NS, Osinska H, Hambleton MA, Brunskill EW, Sayen MR, Gottlieb RA, Dorn GW, Robbins J, Molkentin JD (2005) Loss of cyclophilin D reveals a critical role for mitochondrial permeability transition in cell death.[see comment]. Nature 434:658–662

Baines CP, Kaiser RA, Sheiko T, Craigen WJ, Molkentin JD (2007) Voltage-dependent anion channels are dispensable for mitochondrial-dependent cell death. Nat Cell Biol 9:550–555

Banerjee R, Starkov AA, Beal MF, Thomas B (2009) Mitochondrial dysfunction in the limelight of Parkinson's disease pathogenesis. Biochim Biophys Acta 1792:651–663

Bano D, Young KW, Guerin CJ, Lefeuvre R, Rothwell NJ, Naldini L, Rizzuto R, Carafoli E, Nicotera P (2005) Cleavage of the plasma membrane Na+/Ca2+ exchanger in excitotoxicity. Cell 120:275–285

Basso E, Fante L, Fowlkes J, Petronilli V, Forte MA, Bernardi P (2005) Properties of the permeability transition pore in mitochondria devoid of cyclophilin D. J Biol Chem 280:18558–18561

Beal MF (2007) Mitochondria and neurodegeneration. Novartis Found Symp 287:183–192. discussion 192–186

Bernardi P (1999) Mitochondrial transport of cations: channels, exchangers, and permeability transition. Physiol Rev 79:1127–1155

Bernardi P (2013) The mitochondrial permeability transition pore: a mystery solved? Front Physiol 4:95

Bernardi P, Di Lisa F (2015) The mitochondrial permeability transition pore: molecular nature and role as a target in cardioprotection. J Mol Cell Cardiol 78:100–106

Beutner G, Ruck A, Riede B, Welte W, Brdiczka D (1996) Complexes between kinases, mitochondrial porin and adenylate translocator in rat brain resemble the permeability transition pore. FEBS Lett 396:189–195

Beutner G, Ruck A, Riede B, Brdiczka D (1998) Complexes between porin, hexokinase, mitochondrial creatine kinase and adenylate translocator display properties of the permeability transition pore. Implication for regulation of permeability transition by the kinases. Biochim Biophys Acta 1368:7–18

Beutner G, Eliseev RA, Porter GA Jr (2014) Initiation of electron transport chain activity in the embryonic heart coincides with the activation of mitochondrial complex 1 and the formation of supercomplexes. PLoS One 9:e113330

Billups B, Forsythe ID (2002) Presynaptic mitochondrial calcium sequestration influences transmission at mammalian central synapses. J Neurosci 22:5840–5847

Blanchard BJ, Thomas VL, Ingram VM (2002) Mechanism of membrane depolarization caused by the Alzheimer Abeta1-42 peptide. Biochem Biophys Res Commun 293:1197–1203

Bonora M, Pinton P (2014) Shedding light on molecular mechanisms and identity of mPTP. Mitochondrion 21:11

Bonora M, Bononi A, De Marchi E, Giorgi C, Lebiedzinska M, Marchi S, Patergnani S, Rimessi A, Suski JM, Wojtala A, Wieckowski MR, Kroemer G, Galluzzi L, Pinton P (2013) Role of the c subunit of the FO ATP synthase in mitochondrial permeability transition. Cell Cycle 12:674–683

Bonora M, Wieckowski MR, Chinopoulos C, Kepp O, Kroemer G, Galluzzi L, Pinton P (2014) Molecular mechanisms of cell death: central implication of ATP synthase in mitochondrial permeability transition. Oncogene 34:1475. 0

Borjesson SI, Elinder F (2008) Structure, function, and modification of the voltage sensor in voltage-gated ion channels. Cell Biochem Biophys 52:149–174

Brand MD (2005) The efficiency and plasticity of mitochondrial energy transduction. Biochem Soc Trans 33:897–904

Bravo C, Minauro-Sanmiguel F, Morales-Rios E, Rodriguez-Zavala JS, Garcia JJ (2004) Overexpression of the inhibitor protein IF(1) in AS-30D hepatoma produces a higher association with mitochondrial F(1)F(0) ATP synthase compared to normal rat liver: functional and cross-linking studies. J Bioenerg Biomembr 36:257–264

Brenner C, Cadiou H, Vieira HL, Zamzami N, Marzo I, Xie Z, Leber B, Andrews D, Duclohier H, Reed JC, Kroemer G (2000) Bcl-2 and Bax regulate the channel activity of the mitochondrial adenine nucleotide translocator. Oncogene 19:329–336

Broughton BR, Reutens DC, Sobey CG (2009) Apoptotic mechanisms after cerebral ischemia. Stroke J Cereb Circ 40:e331–e339

Budd SL, Nicholls DG (1996) Mitochondria, calcium regulation, and acute glutamate excitotoxicity in cultured cerebellar granule cells. J Neurochem 67:2282–2291

Cabezon E, Runswick MJ, Leslie AG, Walker JE (2001) The structure of bovine IF(1), the regulatory subunit of mitochondrial F-ATPase. EMBO J 20:6990–6996

Cadonic C, Sabbir MG, Albensi BC (2015) Mechanisms of mitochondrial dysfunction in Alzheimer's disease. Mol Neurobiol 53:6078

Campanella M, Casswell E, Chong S, Farah Z, Wieckowski MR, Abramov AY, Tinker A, Duchen MR (2008) Regulation of mitochondrial structure and function by the F1Fo-ATPase inhibitor protein, IF1. Cell Metab 8:13–25

Cao G, Minami M, Pei W, Yan C, Chen D, O'Horo C, Graham SH, Chen J (2001) Intracellular Bax translocation after transient cerebral ischemia: implications for a role of the mitochondrial apoptotic signaling pathway in ischemic neuronal death. J Cereb Blood Flow Metab Off J Intern Soc Cereb Blood Flow Metab 21:321–333

Carbajo RJ, Kellas FA, Runswick MJ, Montgomery MG, Walker JE, Neuhaus D (2005) Structure of the F1-binding domain of the stator of bovine F1Fo-ATPase and how it binds an alpha-subunit. J Mol Biol 351:824–838

Cassarino DS, Parks JK, Parker WD Jr, Bennett JP Jr (1999) The parkinsonian neurotoxin MPP+ opens the mitochondrial permeability transition pore and releases cytochrome c in isolated mitochondria via an oxidative mechanism. Biochim Biophys Acta 1453:49–62

Caviston TL, Ketchum CJ, Sorgen PL, Nakamoto RK, Cain BD (1998) Identification of an uncoupling mutation affecting the b subunit of F1F0 ATP synthase in *Escherichia coli*. FEBS Lett 429:201–206

Cha MY, Han SH, Son SM, Hong HS, Choi YJ, Byun J, Mook-Jung I (2012) Mitochondria-specific accumulation of amyloid beta induces mitochondrial dysfunction leading to apoptotic cell death. PLoS One 7:e34929

Cha MY, Cho HJ, Kim C, Jung YO, Kang MJ, Murray ME, Hong HS, Choi YJ, Choi H, Kim DK, Choi H, Kim J, Dickson DW, Song HK, Cho JW, Yi EC, Kim J, Jin SM, Mook-Jung I (2015) Mitochondrial ATP synthase activity is impaired by suppressed O-GlcNAcylation in Alzheimer's disease. Hum Mol Genet 24:6492–6504

Chen C, Ko Y, Delannoy M, Ludtke SJ, Chiu W, Pedersen PL (2004) Mitochondrial ATP synthasome: three-dimensional structure by electron microscopy of the ATP synthase in complex formation with carriers for Pi and ADP/ATP. J Biol Chem 279:31761–31768

Chen YB, Aon MA, Hsu YT, Soane L, Teng X, McCaffery JM, Cheng WC, Qi B, Li H, Alavian KN, Dayhoff-Brannigan M, Zou S, Pineda FJ, O'Rourke B, Ko YH, Pedersen PL, Kaczmarek LK, Jonas EA, Hardwick JM (2011) Bcl-xL regulates mitochondrial energetics by stabilizing the inner membrane potential. J Cell Biol 195:263–276

Chen Q, Xu H, Xu A, Ross T, Bowler E, Hu Y, Lesnefsky EJ (2015) Inhibition of Bcl-2 sensitizes mitochondrial permeability transition pore (MPTP) opening in ischemia-damaged mitochondria. PLoS One 10:e0118834

Chinopoulos C, Szabadkai G (2014) What makes you can also break you, part III: mitochondrial permeability transition pore formation by an uncoupling channel within the C-subunit ring of the F1FO ATP synthase? Front Oncol 4:235

Cho SW, Park JS, Heo HJ, Park SW, Song S, Kim I, Han YM, Yamashita JK, Youm JB, Han J, Koh GY (2014) Dual modulation of the mitochondrial permeability transition pore and redox signaling synergistically promotes cardiomyocyte differentiation from pluripotent stem cells. J Am Heart Assoc 3:e000693

Choi DW (1987) Ionic dependence of glutamate neurotoxicity. J Neurosci Off J Soc Neurosci 7:369–379

Chouhan AK, Ivannikov MV, Lu Z, Sugimori M, Llinas RR, Macleod GT (2012) Cytosolic calcium coordinates mitochondrial energy metabolism with presynaptic activity. J Neurosci 32:1233–1243

Clem RJ, Cheng EH, Karp CL, Kirsch DG, Ueno K, Takahashi A, Kastan MB, Griffin DE, Earnshaw WC, Veliuona MA, Hardwick JM (1998) Modulation of cell death by Bcl-XL through caspase interaction. Proc Natl Acad Sci U S A 95:554–559

Couoh-Cardel S, Hsueh YC, Wilkens S, Movileanu L (2016) Yeast V-ATPase proteolipid ring acts as a large-conductance transmembrane protein pore. Sci Rep 6:24774

Crompton M (1999) The mitochondrial permeability transition pore and its role in cell death. Biochem J 341:233–249

Crompton M, Ellinger H, Costi A (1988) Inhibition by cyclosporin A of a Ca2+-dependent pore in heart mitochondria activated by inorganic phosphate and oxidative stress. Biochem J 255:357–360

Crompton M, Virji S, Ward JM (1998) Cyclophilin-D binds strongly to complexes of the voltage-dependent anion channel and the adenine nucleotide translocase to form the permeability transition pore. Eur J Biochem 258:729–735

Csordas G, Thomas AP, Hajnoczky G (2001) Calcium signal transmission between ryanodine receptors and mitochondria in cardiac muscle. Trends Cardiovasc Med 11:269–275

D'Alessandro M, Turina P, Melandri BA (2008) Intrinsic uncoupling in the ATP synthase of *Escherichia coli*. Biochim Biophys Acta 1777:1518–1527

Davies KM, Anselmi C, Wittig I, Faraldo-Gomez JD, Kuhlbrandt W (2012) Structure of the yeast F1Fo-ATP synthase dimer and its role in shaping the mitochondrial cristae. Proc Natl Acad Sci U S A 109:13602–13607

De Stefani D, Rizzuto R (2014) Molecular control of mitochondrial calcium uptake. Biochem Biophys Res Commun 449:373–376

Denton RM (2009) Regulation of mitochondrial dehydrogenases by calcium ions. Biochim Biophys Acta 1787:1309–1316

Di Lisa F, Carpi A, Giorgio V, Bernardi P (2011) The mitochondrial permeability transition pore and cyclophilin D in cardioprotection. Biochim Biophys Acta 1813:1316–1322

DiMauro S, Bonilla E, Lee CP, Schotland DL, Scarpa A, Conn H Jr, Chance B (1976) Luft's disease. Further biochemical and ultrastructural studies of skeletal muscle in the second case. J Neurol Sci 27:217–232

Divakaruni AS, Brand MD (2011) The regulation and physiology of mitochondrial proton leak. Physiology 26:192–205

Dodson MW, Guo M (2007) Pink1, Parkin, DJ-1 and mitochondrial dysfunction in Parkinson's disease. Curr Opin Neurobiol 17:331–337

Drenckhahn JD (2011) Heart development: mitochondria in command of cardiomyocyte differentiation. Dev Cell 21:392–393

Du H, Yan SS (2010) Mitochondrial permeability transition pore in Alzheimer's disease: cyclophilin D and amyloid beta. Biochim Biophys Acta 1802:198–204

Du C, Fang M, Li Y, Li L, Wang X (2000) Smac, a mitochondrial protein that promotes cytochrome c-dependent caspase activation by eliminating IAP inhibition. Cell 102:33–42

Du H, Guo L, Fang F, Chen D, Sosunov AA, McKhann GM, Yan Y, Wang C, Zhang H, Molkentin JD, Gunn-Moore FJ, Vonsattel JP, Arancio O, Chen JX, Yan SD (2008) Cyclophilin D deficiency attenuates mitochondrial and neuronal perturbation and ameliorates learning and memory in Alzheimer's disease. Nat Med 14:1097–1105

Du H, Guo L, Yan S, Sosunov AA, McKhann GM, Yan SS (2010) Early deficits in synaptic mitochondria in an Alzheimer's disease mouse model. Proc Natl Acad Sci U S A 107:18670–18675

Du H, Guo L, Wu X, Sosunov AA, McKhann GM, Chen JX, Yan SS (2014) Cyclophilin D deficiency rescues Abeta-impaired PKA/CREB signaling and alleviates synaptic degeneration. Biochim Biophys Acta 1842:2517–2527

Elrod JW, Wong R, Mishra S, Vagnozzi RJ, Sakthievel B, Goonasekera SA, Karch J, Gabel S, Farber J, Force T, Brown JH, Murphy E, Molkentin JD (2010) Cyclophilin D controls mitochondrial pore-dependent Ca(2+) exchange, metabolic flexibility, and propensity for heart failure in mice. J Clin Invest 120:3680–3687

Elustondo PA, Angelova PR, Kawalec M, Michalak M, Kurcok P, Abramov AY, Pavlov EV (2013) Polyhydroxybutyrate targets mammalian mitochondria and increases permeability of plasmalemmal and mitochondrial membranes. PLoS One 8:e75812

Erturk A, Wang Y, Sheng M (2014) Local pruning of dendrites and spines by caspase-3-dependent and proteasome-limited mechanisms. J Neurosci 34:1672–1688

Faccenda D, Tan CH, Duchen MR, Campanella M (2013a) Mitochondrial IF(1) preserves cristae structure to limit apoptotic cell death signaling. Cell Cycle 12:2530–2532

Faccenda D, Tan CH, Seraphim A, Duchen MR, Campanella M (2013b) IF1 limits the apoptotic-signalling cascade by preventing mitochondrial remodelling. Cell Death Differ 20:686–697

Fedorenko A, Lishko PV, Kirichok Y (2012) Mechanism of fatty-acid-dependent UCP1 uncoupling in brown fat mitochondria. Cell 151:400–413

Finbow ME, Harrison MA (1997) The vacuolar H+-ATPase: a universal proton pump of eukaryotes. Biochem J 324(Pt 3):697–712

Folmes CD, Dzeja PP, Nelson TJ, Terzic A (2012) Mitochondria in control of cell fate. Circ Res 110:526–529

Frei B, Richter C (1986) N-methyl-4-phenylpyridine (MMP+) together with 6-hydroxydopamine or dopamine stimulates Ca2+ release from mitochondria. FEBS Lett 198:99–102

Friel DD, Tsien RW (1994) An FCCP-sensitive Ca2+ store in bullfrog sympathetic neurons and its participation in stimulus-evoked changes in [Ca2+]i. J Neurosci 14:4007–4024

Fujita N, Nagahashi A, Nagashima K, Rokudai S, Tsuruo T (1998) Acceleration of apoptotic cell death after the cleavage of Bcl-XL protein by caspase-3-like proteases. Oncogene 17:1295–1304

Fujiwara M, Yan P, Otsuji TG, Narazaki G, Uosaki H, Fukushima H, Kuwahara K, Harada M, Matsuda H, Matsuoka S, Okita K, Takahashi K, Nakagawa M, Ikeda T, Sakata R, Mummery CL, Nakatsuji N, Yamanaka S, Nakao K, Yamashita JK (2011) Induction and enhancement of cardiac cell differentiation from mouse and human induced pluripotent stem cells with cyclosporin-A. PLoS One 6:e16734

Galluzzi L, Blomgren K, Kroemer G (2009) Mitochondrial membrane permeabilization in neuronal injury. Nat Rev Neurosci 10:481–494

Galonek HL, Hardwick JM (2006) Upgrading the BCL-2 network.[comment]. Nat Cell Biol 8:1317–1319

Garcia JJ, Morales-Rios E, Cortes-Hernandez P, Rodriguez-Zavala JS (2006) The inhibitor protein (IF1) promotes dimerization of the mitochondrial F1F0-ATP synthase. Biochemistry 45:12695–12703

Gautier CA, Giaime E, Caballero E, Nunez L, Song Z, Chan D, Villalobos C, Shen J (2012) Regulation of mitochondrial permeability transition pore by PINK1. Mol Neurodegener 7:22

Gellerich FN, Gizatullina Z, Trumbeckaite S, Nguyen HP, Pallas T, Arandarcikaite O, Vielhaber S, Seppet E, Striggow F (2010) The regulation of OXPHOS by extramitochondrial calcium. Biochim Biophys Acta 1797:1018–1027

Gerle C (2016) On the structural possibility of pore-forming mitochondrial FoF1 ATP synthase. Biochim Biophys Acta 1857:1191–1196

Giaime E, Yamaguchi H, Gautier CA, Kitada T, Shen J (2012) Loss of DJ-1 does not affect mitochondrial respiration but increases ROS production and mitochondrial permeability transition pore opening. PLoS One 7:e40501

Giorgio V, Bisetto E, Soriano ME, Dabbeni-Sala F, Basso E, Petronilli V, Forte MA, Bernardi P, Lippe G (2009) Cyclophilin D modulates mitochondrial F0F1-ATP synthase by interacting with the lateral stalk of the complex. J Biol Chem 284:33982–33988

Giorgio V, von Stockum S, Antoniel M, Fabbro A, Fogolari F, Forte M, Glick GD, Petronilli V, Zoratti M, Szabo I, Lippe G, Bernardi P (2013a) Dimers of mitochondrial ATP synthase form the permeability transition pore. PNAS 110:5887–5892

Giorgio V, von Stockum S, Antoniel M, Fabbro A, Fogolari F, Forte M, Glick GD, Petronilli V, Zoratti M, Szabo I, Lippe G, Bernardi P (2013b) Dimers of mitochondrial ATP synthase form the permeability transition pore. Proc Natl Acad Sci U S A 110:5887–5892

Gomez L, Thibault H, Gharib A, Dumont JM, Vuagniaux G, Scalfaro P, Derumeaux G, Ovize M (2007) Inhibition of mitochondrial permeability transition improves functional recovery and reduces mortality following acute myocardial infarction in mice. Am J Physiol Heart Circ Physiol 293:H1654–H1661

Gottlieb E, Armour SM, Thompson CB (2002) Mitochondrial respiratory control is lost during growth factor deprivation. Proc Natl Acad Sci U S A 99:12801–12806

Guerrieri F, Capozza G, Kalous M, Papa S (1992) Age-related changes of mitochondrial F0F1 ATP synthase. Ann N Y Acad Sci 671:395–402

Gunter TE, Sheu SS (2009) Characteristics and possible functions of mitochondrial Ca(2+) transport mechanisms. Biochim Biophys Acta 1787:1291–1308

Guo L, Du H, Yan S, Wu X, McKhann GM, Chen JX, Yan SS (2013) Cyclophilin D deficiency rescues axonal mitochondrial transport in Alzheimer's neurons. PLoS One 8:e54914

Gutierrez-Aguilar M, Douglas DL, Gibson AK, Domeier TL, Molkentin JD, Baines CP (2014) Genetic manipulation of the cardiac mitochondrial phosphate carrier does not affect permeability transition. J Mol Cell Cardiol 72:316–325

Hafner AV, Dai J, Gomes AP, Xiao CY, Palmeira CM, Rosenzweig A, Sinclair DA (2010) Regulation of the mPTP by SIRT3-mediated deacetylation of CypD at lysine 166 suppresses age-related cardiac hypertrophy. Aging 2:914–923

Hajnoczky G, Csordas G, Krishnamurthy R, Szalai G (2000) Mitochondrial calcium signaling driven by the IP3 receptor. J Bioenerg Biomembr 32:15–25

Halestrap AP, Davidson AM (1990) Inhibition of Ca2(+)-induced large-amplitude swelling of liver and heart mitochondria by cyclosporin is probably caused by the inhibitor binding to mitochondrial-matrix peptidyl-prolyl cis-trans isomerase and preventing it interacting with the adenine nucleotide translocase. Biochem J 268:153–160

Harrison M, Durose L, Song CF, Barratt E, Trinick J, Jones R, Findlay JB (2003) Structure and function of the vacuolar H+-ATPase: moving from low-resolution models to high-resolution structures. J Bioenerg Biomembr 35:337–345

Hausenloy D, Wynne A, Duchen M, Yellon D (2004) Transient mitochondrial permeability transition pore opening mediates preconditioning-induced protection. Circulation 109:1714–1717

Haworth RA, Hunter DR (1979) The Ca2+-induced membrane transition in mitochondria. II. Nature of the Ca2+ trigger site. Arch Biochem Biophys 195:460–467

Hickman JA, Hardwick JM, Kaczmarek LK, Jonas EA (2008) Bcl-xL inhibitor ABT-737 reveals a dual role for Bcl-xL in synaptic transmission. J Neurophysiol 99:1515–1522

Hirotani M, Zhang Y, Fujita N, Naito M, Tsuruo T (1999) NH2-terminal BH4 domain of Bcl-2 is functional for heterodimerization with Bax and inhibition of apoptosis. J Biol Chem 274:20415–20420

Hockenbery D, Nunez G, Milliman C, Schreiber RD, Korsmeyer SJ (1990) Bcl-2 is an inner mitochondrial membrane protein that blocks programmed cell death. Nature 348:334–336

Holmstrom KM, Finkel T (2014) Cellular mechanisms and physiological consequences of redox-dependent signalling. Nat Rev Mol Cell Biol 15:411–421

Holmstrom KM, Marina N, Baev AY, Wood NW, Gourine AV, Abramov AY (2013) Signalling properties of inorganic polyphosphate in the mammalian brain. Nat Commun 4:1362

Hom JR, Quintanilla RA, Hoffman DL, de Mesy Bentley KL, Molkentin JD, Sheu SS, Porter GA Jr (2011) The permeability transition pore controls cardiac mitochondrial maturation and myocyte differentiation. Dev Cell 21:469–478

Hubbard MJ, McHugh NJ (1996) Mitochondrial ATP synthase F1-beta-subunit is a calcium-binding protein. FEBS Lett 391:323–329

Hunter DR, Haworth RA (1979a) The Ca2+-induced membrane transition in mitochondria. III. Transitional Ca2+ release. Arch Biochem Biophys 195:468–477

Hunter DR, Haworth RA (1979b) The Ca2+−induced membrane transition in mitochondria. I. The protective mechanisms. Arch Biochem Biophys 195:453–459

Huser J, Blatter LA (1999) Fluctuations in mitochondrial membrane potential caused by repetitive gating of the permeability transition pore. Biochem J 343(Pt 2):311–317

Jennings RB, Reimer KA, Steenbergen C (1991) Effect of inhibition of the mitochondrial ATPase on net myocardial ATP in total ischemia. J Mol Cell Cardiol 23:1383–1395

Jonas E (2006) BCL-xL regulates synaptic plasticity. Mol Interv 6:208–222

Jonas EA (2009) Molecular participants in mitochondrial cell death channel formation during neuronal ischemia. Exp Neurol 218:203–212

Jonas EA, Knox RJ, Kaczmarek LK (1997) Giga-ohm seals on intracellular membranes: a technique for studying intracellular ion channels in intact cells. Neuron 19:7–13

Jonas EA, Buchanan J, Kaczmarek LK (1999) Prolonged activation of mitochondrial conductances during synaptic transmission. Science 286:1347–1350

Jonas EA, Hoit D, Hickman JA, Brandt TA, Polster BM, Fannjiang Y, McCarthy E, Montanez MK, Hardwick JM, Kaczmarek LK (2003) Modulation of synaptic transmission by the BCL-2 family protein BCL-xL. J Neurosci 23:8423–8431

Jonas EA, Hickman JA, Chachar M, Polster BM, Brandt TA, Fannjiang Y, Ivanovska I, Basanez G, Kinnally KW, Zimmerberg J, Hardwick JM, Kaczmarek LK (2004) Proapoptotic N-truncated BCL-xL protein activates endogenous mitochondrial channels in living synaptic terminals. Proc Natl Acad Sci USA 101:13590–13595

Jonas EA, Hickman JA, Hardwick JM, Kaczmarek LK (2005) Exposure to hypoxia rapidly induces mitochondrial channel activity within a living synapse. J Biol Chem 280:4491–4497

Jonas EA, Porter GA, Alavian KN (2014) Bcl-xL in neuroprotection and plasticity. Front Physiol 5:355

Jonckheere AI, Smeitink JA, Rodenburg RJ (2012) Mitochondrial ATP synthase: architecture, function and pathology. J Inherit Metab Dis 35:211–225

Jones PC, Harrison MA, Kim YI, Finbow ME, Findlay JB (1995) The first putative transmembrane helix of the 16 kDa proteolipid lines a pore in the Vo sector of the vacuolar H(+)-ATPase. Biochem J 312(Pt 3):739–747

Kang JS, Tian JH, Pan PY, Zald P, Li C, Deng C, Sheng ZH (2008) Docking of axonal mitochondria by syntaphilin controls their mobility and affects short-term facilitation. Cell 132:137–148

Karch J, Molkentin JD (2014) Identifying the components of the elusive mitochondrial permeability transition pore. Proc Natl Acad Sci USA 111:10396–10397

Kinnally KW, Campo ML, Tedeschi H (1989) Mitochondrial channel activity studied by patch-clamping mitoplasts. J Bioenerg Biomembr 21:497–506

Kokoszka JE, Waymire KG, Levy SE, Sligh JE, Cai J, Jones DP, MacGregor GR, Wallace DC (2004) The ADP/ATP translocator is not essential for the mitochondrial permeability transition pore. [see comment] Nature 427:461–465

Konig T, Troder SE, Bakka K, Korwitz A, Richter-Dennerlein R, Lampe PA, Patron M, Muhlmeister M, Guerrero-Castillo S, Brandt U, Decker T, Lauria I, Paggio A, Rizzuto R, Rugarli EI, De Stefani D, Langer T (2016) The m-AAA protease associated with neurodegeneration limits MCU activity in mitochondria. Mol Cell 64:148–162

Korge P, Yang L, Yang JH, Wang Y, Qu Z, Weiss JN (2011) Protective role of transient pore openings in calcium handling by cardiac mitochondria. J Biol Chem 286:34851–34857

Kowaltowski AJ, Naia-da-Silva ES, Castilho RF, Vercesi AE (1998) Ca2+-stimulated mitochondrial reactive oxygen species generation and permeability transition are inhibited by dibucaine or Mg2+. Arch Biochem Biophys 359:77–81

Kruse SE, Watt WC, Marcinek DJ, Kapur RP, Schenkman KA, Palmiter RD (2008) Mice with mitochondrial complex I deficiency develop a fatal encephalomyopathy. Cell Metab 7:312–320

Kuroda Y, Mitsui T, Kunishige M, Shono M, Akaike M, Azuma H, Matsumoto T (2006) Parkin enhances mitochondrial biogenesis in proliferating cells. Hum Mol Genet 15:883–895

Kwong JQ, Davis J, Baines CP, Sargent MA, Karch J, Wang X, Huang T, Molkentin JD (2014) Genetic deletion of the mitochondrial phosphate carrier desensitizes the mitochondrial permeability transition pore and causes cardiomyopathy. Cell Death Differ 21:1209–1217

Leung AW, Halestrap AP (2008) Recent progress in elucidating the molecular mechanism of the mitochondrial permeability transition pore. Biochim Biophys Acta 1777:946–952

Li H, Zhu H, Xu CJ, Yuan J (1998) Cleavage of BID by caspase 8 mediates the mitochondrial damage in the Fas pathway of apoptosis. Cell 94:491–501

Li H, Chen Y, Jones AF, Sanger RH, Collis LP, Flannery R, McNay EC, Yu T, Schwarzenbacher R, Bossy B, Bossy-Wetzel E, Bennett MV, Pypaert M, Hickman JA, Smith PJ, Hardwick JM, Jonas EA (2008) Bcl-xL induces Drp1-dependent synapse formation in cultured hippocampal neurons. Proc Natl Acad Sci U S A 105:2169–2174

Li Z, Jo J, Jia JM, Lo SC, Whitcomb DJ, Jiao S, Cho K, Sheng M (2010) Caspase-3 activation via mitochondria is required for long-term depression and AMPA receptor internalization. Cell 141:859–871

Li H, Alavian KN, Lazrove E, Mehta N, Jones A, Zhang P, Licznerski P, Graham M, Uo T, Guo J, Rahner C, Duman RS, Morrison RS, Jonas EA (2013) A Bcl-xL-Drp1 complex regulates synaptic vesicle membrane dynamics during endocytosis. Nat Cell Biol 15:773–785

Lindsten T et al (2000) The combined functions of proapoptotic Bcl-2 family members bak and bax are essential for normal development of multiple tissues. Mol Cell 6:1389–1399

Lo EH, Dalkara T, Moskowitz MA (2003) Mechanisms, challenges and opportunities in stroke. Nat Rev Neurosci 4:399–415

Lopreiato R, Giacomello M, Carafoli E (2014) The plasma membrane calcium pump: new ways to look at an old enzyme. J Biol Chem 289:10261–10268

Luo X, Budihardjo I, Zou H, Slaughter C, Wang X (1998) Bid, a Bcl2 interacting protein, mediates cytochrome c release from mitochondria in response to activation of cell surface death receptors. Cell 94:481–490

Malenka RC, Bear MF (2004) LTP and LTD: an embarrassment of riches. Neuron 44:5–21

Malinow R, Malenka RC (2002) AMPA receptor trafficking and synaptic plasticity. Annu Rev Neurosci 25:103–126

Maltecca F, Baseggio E, Consolato F, Mazza D, Podini P, Young SM Jr, Drago I, Bahr BA, Puliti A, Codazzi F, Quattrini A, Casari G (2015) Purkinje neuron Ca2+ influx reduction rescues ataxia in SCA28 model. J Clin Invest 125:263–274

Mandel M, Moriyama Y, Hulmes JD, Pan YC, Nelson H, Nelson N (1988) cDNA sequence encoding the 16-kDa proteolipid of chromaffin granules implies gene duplication in the evolution of H+-ATPases. Proc Natl Acad Sci USA 85:5521–5524

Mannella CA, Kinnally KW (2008) Reflections on VDAC as a voltage-gated channel and a mitochondrial regulator. J Bioenerg Biomembr 40:149–155

Martin LJ, Semenkow S, Hanaford A, Wong M (2014) Mitochondrial permeability transition pore regulates Parkinson's disease development in mutant alpha-synuclein transgenic mice. Neurobiol Aging 35:1132–1152

Massaad CA, Klann E (2011) Reactive oxygen species in the regulation of synaptic plasticity and memory. Antioxid Redox Signal 14:2013–2054

Matsumoto S, Friberg H, Ferrand-Drake M, Wieloch T (1999) Blockade of the mitochondrial permeability transition pore diminishes infarct size in the rat after transient middle cerebral artery occlusion. J Cereb Blood Flow Metab Off J Intern Soc Cereb Blood Flow Metab 19:736–741

Matsushita K, Wu Y, Qiu J, Lang-Lazdunski L, Hirt L, Waeber C, Hyman BT, Yuan J, Moskowitz MA (2000) Fas receptor and neuronal cell death after spinal cord ischemia. J Neurosci Off J Soc Neurosci 20:6879–6887

Matthies D, Preiss L, Klyszejko AL, Muller DJ, Cook GM, Vonck J, Meier T (2009) The c13 ring from a thermoalkaliphilic ATP synthase reveals an extended diameter due to a special structural region. J Mol Biol 388:611–618

McGeoch JE, Guidotti G (1997) A 0.1-700 Hz current through a voltage-clamped pore: candidate protein for initiator of neural oscillations. Brain Res 766:188–194

McGeoch JE, McGeoch MW (2008) Entrapment of water by subunit c of ATP synthase. J R Soc Interface R Soc 5:311–318

McGuinness O, Yafei N, Costi A, Crompton M (1990) The presence of two classes of high-affinity cyclosporin A binding sites in mitochondria. Evidence that the minor component is involved in the opening of an inner-membrane Ca(2+)-dependent pore. Eur J Biochem FEBS 194:671–679

Meier T, Matthey U, Henzen F, Dimroth P, Muller DJ (2001) The central plug in the reconstituted undecameric c cylinder of a bacterial ATP synthase consists of phospholipids. FEBS Lett 505:353–356

Meyer FB (1989) Calcium, neuronal hyperexcitability and ischemic injury. Brain Res Brain Res Rev 14:227–243

Mitchell P (1961) Coupling of phosphorylation to electron and hydrogen transfer by a chemiosmotic type of mechanism. Nature 191:144–148

Miyawaki T, Mashiko T, Ofengeim D, Flannery RJ, Noh KM, Fujisawa S, Bonanni L, Bennett MV, Zukin RS, Jonas EA (2008) Ischemic preconditioning blocks BAD translocation, Bcl-xL cleavage, and large channel activity in mitochondria of postischemic hippocampal neurons. Proc Natl Acad Sci U S A 105:4892–4897

Nakagawa T, Shimizu S, Watanabe T, Yamaguchi O, Otsu K, Yamagata H, Inohara H, Kubo T, Tsujimoto Y (2005) Cyclophilin D-dependent mitochondrial permeability transition regulates some necrotic but not apoptotic cell death. Nature 434:652–658

Narita M, Shimizu S, Ito T, Chittenden T, Lutz RJ, Matsuda H, Tsujimoto Y (1998) Bax interacts with the permeability transition pore to induce permeability transition and cytochrome c release in isolated mitochondria. Proc Natl Acad Sci U S A 95:14681–14686

Neher E, Sakaba T (2008) Multiple roles of calcium ions in the regulation of neurotransmitter release. Neuron 59:861–872

Nguyen TT, Stevens MV, Kohr M, Steenbergen C, Sack MN, Murphy E (2011) Cysteine 203 of cyclophilin D is critical for cyclophilin D activation of the mitochondrial permeability transition pore. J Biol Chem 286:40184–40192

Nicholls DG, Rial E (1999) A history of the first uncoupling protein, UCP1. J Bioenerg Biomembr 31:399–406

Nikolaev A, McLaughlin T, O'Leary DD, Tessier-Lavigne M (2009) APP binds DR6 to trigger axon pruning and neuron death via distinct caspases. Nature 457:981–989

Nishizawa Y (2001) Glutamate release and neuronal damage in ischemia. Life Sci 69:369–381

Norris U, Karp PE, Fimmel AL (1992) Mutational analysis of the glycine-rich region of the c subunit of the *Escherichia coli* F0F1 ATPase. J Bacteriol 174:4496–4499

Oberfeld B, Brunner J, Dimroth P (2006) Phospholipids occupy the internal lumen of the c ring of the ATP synthase of *Escherichia coli*. Biochemistry 45:1841–1851

Ofengeim D, Chen YB, Miyawaki T, Li H, Sacchetti S, Flannery RJ, Alavian KN, Pontarelli F, Roelofs BA, Hickman JA, Hardwick JM, Zukin RS, Jonas EA (2012) N-terminally cleaved Bcl-xL mediates ischemia-induced neuronal death. Nat Neurosci 15:574–580

Olsen KM, Sheng M (2012) NMDA receptors and BAX are essential for Abeta impairment of LTP. Sci Rep 2:225

Pang ZP, Cao P, Xu W, Sudhof TC (2010) Calmodulin controls synaptic strength via presynaptic activation of calmodulin kinase II. J Neurosci 30:4132–4142

Park HA, Licznerski P, Alavian KN, Shanabrough M, Jonas EA (2014) Bcl-xL is necessary for neurite outgrowth in hippocampal neurons. Antioxid Redox Signal 22:93

Park HA, Licznerski P, Alavian KN, Shanabrough M, Jonas EA (2015) Bcl-xL is necessary for neurite outgrowth in hippocampal neurons. Antioxid Redox Signal 22:93–108

Pavlov E, Zakharian E, Bladen C, Diao CT, Grimbly C, Reusch RN, French RJ (2005) A large, voltage-dependent channel, isolated from mitochondria by water-free chloroform extraction. Biophys J 88:2614–2625

Pedersen PL (1994) ATP synthase. The machine that makes ATP. Curr Biol CB 4:1138–1141

Pedersen PL, Hullihen J (1978) Adenosine triphosphatase of rat liver mitochondria. Capacity of the homogeneous F1 component of the enzyme to restore ATP synthesis in urea-treated membranes. J Biol Chem 253:2176–2183

Peters C, Bayer MJ, Buhler S, Andersen JS, Mann M, Mayer A (2001) Trans-complex formation by proteolipid channels in the terminal phase of membrane fusion. Nature 409:581–588

Petronilli V, Szabo I, Zoratti M (1989) The inner mitochondrial membrane contains ion-conducting channels similar to those found in bacteria. FEBS Lett 259:137–143

Petronilli V, Miotto G, Canton M, Brini M, Colonna R, Bernardi P, Di Lisa F (1999) Transient and long-lasting openings of the mitochondrial permeability transition pore can be monitored directly in intact cells by changes in mitochondrial calcein fluorescence. Biophys J 76:725–734

Petronilli V, Penzo D, Scorrano L, Bernardi P, Di Lisa F (2001) The mitochondrial permeability transition, release of cytochrome c and cell death. Correlation with the duration of pore openings in situ. J Biol Chem 276:12030–12034

Pullman ME, Monroy GC (1963) A naturally occurring inhibitor of mitochondrial adenosine triphosphatase. J Biol Chem 238:3762–3769

Qiao H, Koya RC, Nakagawa K, Tanaka H, Fujita H, Takimoto M, Kuzumaki N (2005) Inhibition of Alzheimer's amyloid-beta peptide-induced reduction of mitochondrial membrane potential and neurotoxicity by gelsolin. Neurobiol Aging 26:849–855

Raffaello A, De Stefani D, Rizzuto R (2012) The mitochondrial Ca(2+) uniporter. Cell Calcium 52:16–21

Rao VK, Carlson EA, Yan SS (2014) Mitochondrial permeability transition pore is a potential drug target for neurodegeneration. Biochim Biophys Acta 1842:1267–1272

Rasheed MZ, Tabassum H, Parvez S (2016) Mitochondrial permeability transition pore: a promising target for the treatment of Parkinson's disease. Protoplasma 254:33

Rasola A, Bernardi P (2011) Mitochondrial permeability transition in Ca(2+)-dependent apoptosis and necrosis. Cell Calcium 50:222–233

Reddy PH (2009) Amyloid beta, mitochondrial structural and functional dynamics in Alzheimer's disease. Exp Neurol 218:286–292

Reed DJ, Savage MK (1995) Influence of metabolic inhibitors on mitochondrial permeability transition and glutathione status. Biochim Biophys Acta 1271:43–50

Reynolds IJ (1999) Mitochondrial membrane potential and the permeability transition in excitotoxicity. Ann NY Acad Sci 893:33–41

Rizzuto R, Pozzan T (2006) Microdomains of intracellular Ca2+: molecular determinants and functional consequences. Physiol Rev 86:369–408

Rizzuto R, Bernardi P, Pozzan T (2000) Mitochondria as all-round players of the calcium game. J Physiol 529(Pt 1):37–47

Rizzuto R, Marchi S, Bonora M, Aguiari P, Bononi A, De Stefani D, Giorgi C, Leo S, Rimessi A, Siviero R, Zecchini E, Pinton P (2009) Ca(2+) transfer from the ER to mitochondria: when, how and why. Biochim Biophys Acta 1787:1342–1351

Rizzuto R, De Stefani D, Raffaello A, Mammucari C (2012) Mitochondria as sensors and regulators of calcium signalling. Nat Rev Mol Cell Biol 13:566–578

Roestenberg P, Manjeri GR, Valsecchi F, Smeitink JA, Willems PH, Koopman WJ (2012) Pharmacological targeting of mitochondrial complex I deficiency: the cellular level and beyond. Mitochondrion 12:57–65

Rostovtseva TK, Bezrukov SM (2012) VDAC inhibition by tubulin and its physiological implications. Biochim Biophys Acta 1818:1526–1535

Rostovtseva T, Colombini M (1997) VDAC channels mediate and gate the flow of ATP: implications for the regulation of mitochondrial function. Biophys J 72:1954–1962

Rouslin W (1991) Regulation of the mitochondrial ATPase in situ in cardiac muscle: role of the inhibitor subunit. J Bioenerg Biomembr 23:873–888

Sanchez-Cenizo L, Formentini L, Aldea M, Ortega AD, Garcia-Huerta P, Sanchez-Arago M, Cuezva JM (2010) Up-regulation of the ATPase inhibitory factor 1 (IF1) of the mitochondrial H+–ATP synthase in human tumors mediates the metabolic shift of cancer cells to a Warburg phenotype. J Biol Chem 285:25308–25313

Schinzel AC, Takeuchi O, Huang Z, Fisher JK, Zhou Z, Rubens J, Hetz C, Danial NN, Moskowitz MA, Korsmeyer SJ (2005) Cyclophilin D is a component of mitochondrial permeability transition and mediates neuronal cell death after focal cerebral ischemia. Proc Natl Acad Sci U S A 102:12005–12010

Schneggenburger R, Neher E (2005) Presynaptic calcium and control of vesicle fusion. Curr Opin Neurobiol 15:266–274

Schon EA, Manfredi G (2003) Neuronal degeneration and mitochondrial dysfunction. J Clin Invest 111:303–312

Schwab BL, Guerini D, Didszun C, Bano D, Ferrando-May E, Fava E, Tam J, Xu D, Xanthoudakis S, Nicholson DW, Carafoli E, Nicotera P (2002) Cleavage of plasma membrane calcium pumps by caspases: a link between apoptosis and necrosis. Cell Death Differ 9:818–831

Seidlmayer LK, Blatter LA, Pavlov E, Dedkova EN (2012) Inorganic polyphosphate – an unusual suspect of the mitochondrial permeability transition mystery. Channels 6:463–467

Seng NS, Megyesi J, Tarcsafalvi A, Price PM (2016) Mimicking Cdk2 phosphorylation of Bcl-xL at Ser73 results in caspase activation and Bcl-xL cleavage. Cell Death Discov 2:16001–16006

Shanmughapriya S et al (2015) SPG7 is an essential and conserved component of the mitochondrial permeability transition pore. Mol Cell 60:47–62
Shimizu S, Narita M, Tsujimoto Y (1999) Bcl-2 family proteins regulate the release of apoptogenic cytochrome c by the mitochondrial channel VDAC. Nature 399:483–487
Shulga N, Pastorino JG (2010) Ethanol sensitizes mitochondria to the permeability transition by inhibiting deacetylation of cyclophilin-D mediated by sirtuin-3. J Cell Sci 123:4117–4127
Shulga N, Wilson-Smith R, Pastorino JG (2010) Sirtuin-3 deacetylation of cyclophilin D induces dissociation of hexokinase II from the mitochondria. J Cell Sci 123:894–902
Sorgato MC, Keller BU, Stuhmer W (1987) Patch-clamping of the inner mitochondrial membrane reveals a voltage-dependent ion channel. Nature 330:498–500
Stotz SC, Scott LO, Drummond-Main C, Avchalumov Y, Girotto F, Davidsen J, Gomez-Garcia MR, Rho JM, Pavlov EV, Colicos MA (2014) Inorganic polyphosphate regulates neuronal excitability through modulation of voltage-gated channels. Mol Brain 7:42
Sugioka R, Shimizu S, Funatsu T, Tamagawa H, Sawa Y, Kawakami T, Tsujimoto Y (2003) BH4-domain peptide from Bcl-xL exerts anti-apoptotic activity in vivo. Oncogene 22:8432–8440
Supnet C, Bezprozvanny I (2010) The dysregulation of intracellular calcium in Alzheimer disease. Cell Calcium 47:183–189
Szabo I, Zoratti M (1991) The giant channel of the inner mitochondrial membrane is inhibited by cyclosporin A. J Biol Chem 266:3376–3379
Szabo I, Bernardi P, Zoratti M (1992) Modulation of the mitochondrial megachannel by divalent cations and protons. J Biol Chem 267:2940–2946
Szydlowska K, Tymianski M (2010) Calcium, ischemia and excitotoxicity. Cell Calcium 47:122–129
Tang Y, Zucker RS (1997) Mitochondrial involvement in post-tetanic potentiation of synaptic transmission. Neuron 18:483–491
Thomas B, Banerjee R, Starkova NN, Zhang SF, Calingasan NY, Yang L, Wille E, Lorenzo BJ, Ho DJ, Beal MF, Starkov A (2012) Mitochondrial permeability transition pore component cyclophilin D distinguishes nigrostriatal dopaminergic death paradigms in the MPTP mouse model of Parkinson's disease. Antioxid Redox Signal 16:855–868
Vander Heiden MG, Li XX, Gottleib E, Hill RB, Thompson CB, Colombini M (2001) Bcl-xL promotes the open configuration of the voltage-dependent anion channel and metabolite passage through the outer mitochondrial membrane. J Biol Chem 276:19414–19419
Veas-Perez de Tudela M, Delgado-Esteban M, Maestre C, Bobo-Jimenez V, Jimenez-Blasco D, Vecino R, Bolanos JP, Almeida A (2015) Regulation of Bcl-xL-ATP synthase interaction by mitochondrial cyclin B1-cyclin-dependent kinase-1 determines neuronal survival. J Neurosci 35:9287–9301
Verhagen AM, Ekert PG, Pakusch M, Silke J, Connolly LM, Reid GE, Moritz RL, Simpson RJ, Vaux DL (2000) Identification of DIABLO, a mammalian protein that promotes apoptosis by binding to and antagonizing IAP proteins. Cell 102:43–53
Walker JE (2013) The ATP synthase: the understood, the uncertain and the unknown. Biochem Soc Trans 41:1–16
Wang W, Fang H, Groom L, Cheng A, Zhang W, Liu J, Wang X, Li K, Han P, Zheng M, Yin J, Mattson MP, Kao JP, Lakatta EG, Sheu SS, Ouyang K, Chen J, Dirksen RT, Cheng H (2008) Superoxide flashes in single mitochondria. Cell 134:279–290
Watt IN, Montgomery MG, Runswick MJ, Leslie AG, Walker JE (2010) Bioenergetic cost of making an adenosine triphosphate molecule in animal mitochondria. Proc Natl Acad Sci U S A 107:16823–16827
Wei W, Nguyen LN, Kessels HW, Hagiwara H, Sisodia S, Malinow R (2010) Amyloid beta from axons and dendrites reduces local spine number and plasticity. Nat Neurosci 13:190–196
Wittig I, Schagger H (2009) Supramolecular organization of ATP synthase and respiratory chain in mitochondrial membranes. Biochim Biophys Acta 1787:672–680
Woodfield K, Ruck A, Brdiczka D, Halestrap AP (1998) Direct demonstration of a specific interaction between cyclophilin-D and the adenine nucleotide translocase confirms their role in the mitochondrial permeability transition. Biochem J 336(Pt 2):287–290

Yamada EW, Huzel NJ (1992) Distribution of the ATPase inhibitor proteins of mitochondria in mammalian tissues including fibroblasts from a patient with Luft's disease. Biochim Biophys Acta 1139:143–147

Zakharov SD, Li X, Red'ko TP, Dilley RA (1996) Calcium binding to the subunit c of *E. coli* ATP-synthase and possible functional implications in energy coupling. J Bioenerg Biomembr 28:483–494

Zamzami N, El Hamel C, Maisse C, Brenner C, Munoz-Pinedo C, Belzacq AS, Costantini P, Vieira H, Loeffler M, Molle G, Kroemer G (2000) Bid acts on the permeability transition pore complex to induce apoptosis. Oncogene 19:6342–6350

Zanellati MC, Monti V, Barzaghi C, Reale C, Nardocci N, Albanese A, Valente EM, Ghezzi D, Garavaglia B (2015) Mitochondrial dysfunction in Parkinson disease: evidence in mutant PARK2 fibroblasts. Front Genet 6:78

Zhu J, Rebecchi MJ, Glass PS, Brink PR, Liu L (2013) Interactions of GSK-3beta with mitochondrial permeability transition pore modulators during preconditioning: age-associated differences. J Gerontol A Biol Sci Med Sci 68:395–403

Zucker RS, Regehr WG (2002) Short-term synaptic plasticity. Annu Rev Physiol 64:355–405

Chapter 4
Mitochondrial Calcium Uptake in Activation of the Permeability Transition Pore and Cell Death

Maria E. Solesio and Evgeny V. Pavlov

4.1 Introduction

Calcium signaling is central for regulation of the multiple physiological processes in all living organisms. Cells are able to maintain significant gradients of calcium between extracellular media, cytosol, and various intracellular organelles. The presence of such gradients allows to rapidly change calcium concentration by opening of the specific calcium-selective channels and generate calcium signal event. Well-described examples of such calcium signaling events in mammalian organisms include excitation-contraction coupling in the muscle and neuronal synaptic signal transmission in the brain. Thus, disruption of normal calcium homeostasis can have profound negative impact on organism function. Further, uncontrolled increase in cytosolic calcium concentration not only prevents normal calcium signaling but can be direct cause of the cell death and result in tissue necrosis. Here we will discuss one specific mechanism of cell death induced by high calcium concentrations, which involves damaging effects of high calcium on the mitochondrial function. Specifically, we will review details of the cell death mechanisms linked to the calcium-induced activation of the mitochondrial permeability transition pore (mPTP). It is generally established that during calcium overload in such conditions as ischemia, mitochondria can accumulate abnormally high amounts of calcium. Eventually, such calcium overload leads to the increased permeability of the mitochondrial inner membrane – permeability transition. Permeability transition is caused by the opening of the nonselective pore – permeability transition pore or mPTP. Further, we will review what is currently known about possible molecular

M.E. Solesio • E.V. Pavlov (✉)
Department of Basic Sciences, New York University, College of Dentistry,
345 East 24th Street, Schwartz Building, Room 1030, New York, NY 10010, USA
e-mail: ep37@nyu.edu

T.K. Rostovtseva (ed.), *Molecular Basis for Mitochondrial Signaling*,
Biological and Medical Physics, Biomedical Engineering,
DOI 10.1007/978-3-319-55539-3_4

mechanisms responsible for two key events leading to the mPTP development: mitochondrial calcium uptake and activation of mPTP by calcium increase inside the mitochondria.

4.2 Calcium-Activated mPTP in Isolated Mitochondria and Its Relationship to In Vivo Processes

Investigation of the function of the isolated mitochondria was one of the first methods used by the field of mitochondrial physiology (Kennedy and Lehninger 1949). In this approach, first intact mitochondria are isolated from the whole animal tissues or cultured cells. Such isolated mitochondria maintain all their key activities including respiration, ATP synthesis, and ion transport. The main advantage of this approach is that it allows to study mitochondrial function directly and independently of other cellular processes. Very early experiments with isolated mitochondria established that in the presence of orthophosphate, they are capable to accumulate significant amounts of calcium (Greenawalt et al. 1964). It was also noted that following certain threshold of calcium accumulation, mitochondrial membrane became highly permeable (Haworth and Hunter 1979; Hunter and Haworth 1979a, b). This resulted in calcium release and mitochondrial swelling. Later size exclusion experiments confirmed that excessive calcium accumulation makes mitochondrial inner membrane to undergo permeability transition that allows free flux of the molecules up to 1,500 Da in size (Haworth and Hunter 1979). At the time of this discovery, it was not clear whether such calcium-induced increase in membrane permeability is an effect that only occurs in isolated mitochondria or has some pathological or physiological significance. This changed when it was demonstrated that permeability transition can be significantly delayed by the immunosuppressant drug cyclosporine A (CSA) (Broekemeier et al. 1989). The use of CSA not only allowed to establish that this process can be regulated pharmacologically at the level of isolated mitochondria but also allowed to suggest that PTP might play an important role in cell pathology. Indeed, CSA not only inhibited PTP in isolated mitochondria but also prevented mitochondrial depolarization in the living cells (Halestrap et al. 1997). Further, the use of CSA has shown protection against ischemia-reperfusion injury in human patients, suggesting that PTP plays important role in vivo and can also be a promising target for treatment against tissue damage caused by acute stress.

4.3 Mechanisms of Calcium Activation of mPTP

Since the discovery of mPTP, calcium activation remains the defining feature of this phenomenon. Despite many years of mPTP studies, the molecular mechanism of calcium activation of mPTP is not entirely understood. mPTP activation by calcium doesn't follow mechanisms characteristic to other calcium-induced physiological processes. Typically, calcium activation occurs through direct binding of calcium ion to the selective calcium binding site of the target molecule. In order for this process to occur, concentration of free calcium has to significantly increase. In case of mitochondria, however, the process of calcium activation seems to be different. mPTP activation requires accumulation of the large amounts of this ion in the mitochondrial matrix; however during this process, the concentration of free (bioavailable) calcium remains remarkably steady (Chalmers and Nicholls 2003). Thus it can be concluded that hypothetical mPTP calcium-sensing site does not directly respond to calcium increase. Most likely activation of mPTP by calcium is not direct but involves calcium-phosphate interactions. During increase in total calcium accumulation inside mitochondria, it becomes buffered by orthophosphate. Thus, while bioavailable calcium remains constant, there is a significant increase in calcium-phosphate precipitates that appear as electron dense granules and can be detected by electron microscopy (Kristian et al. 2007). This suggests that calcium activation of mPTP does not involve direct binding of calcium to the putative calcium binding site of the mPTP complex but rather occurs through increased formation of calcium-phosphate aggregates. Interestingly calcium phosphate precipitates formation in vitro normally is irreversible, and it is not expected that after mitochondrial calcium accumulation and precipitates formation they can be easily dissolved (Nicholls and Chalmers 2004). This prediction is in contrast with the experimental observations that show rapid release of free calcium from the mitochondria upon the mPTP activation. It has been suggested that reversibility of the calcium-phosphate aggregation is determined by the fact that in addition of the orthophosphate; phosphate might present in variety of forms, including ATP and inorganic polyphosphates (polyP) (Nicholls and Chalmers 2004; Solesio et al. 2016a). Such a notion is supported by the observation that calcium-to-phosphate ratio found in calcium-phosphate aggregates is not constant and depends on the mitochondrial calcium load. Specifically, when in isolated mitochondria, mPTP activation is delayed by the presence of CSA; the calcium/phosphate ratio found in precipitates is significantly increased (Kristian et al. 2007).

The mechanistic link between the increased amount of calcium-phosphate precipitates and mPTP is not entirely clear. One of possibilities is that calcium activation of mPTP involves its interactions with polyP (Solesio et al. 2016b). Calcium-polyP induction of ion channel formation has been first described in bacteria. In this case incubation of mitochondria with calcium leads to the assembly of the complex of calcium, polyP, and polyhydroxybutyrate (PHB) (Castuma et al. 1995). Such macromolecular complex forms ion channels and was proposed to play role in development of competence. In case of mitochondria, the polyP/Ca^{2+}/PHB channel has all the characteristics of mPTP (Pavlov et al. 2005). Mitochondrial analog of the bacterial channel forms a large, weakly selective pores. The levels of such channel are non-detectable in the intact mitochondria but increase dramatically in the mitochondria in which mPTP was activated by calcium. Complex formation is inhibited in the presence of mPTP blocker CSA. Interestingly mitochondrial channel-forming complex appears to be closely associated with the

C-subunit of the mitochondrial ATP synthase (C-subunit) (Elustondo et al. 2016). In normal conditions, C-subunit resides in the mitochondrial inner membrane in the form of oligomers and does not form high conductance channels. However in pathological conditions C-subunit has been directly linked to the mPTP channel formation (Alavian et al. 2014), it can be suggested that active mPTP requires the presence of all three components: polyP, PHB, and C-subunit, which assembly into active channel is triggered by interaction between calcium and polyP. This scenario can explain the molecular mechanism of mPTP activation by calcium which does not involve increase in free calcium concentration. Involvement of the polyP in calcium-induced mPTP is further supported by the observation that in isolated mitochondrial high conductance mPTP is inhibited by spermine – polyamine with high affinity to polyP (Elustondo et al. 2015). Further experiments will help to clarify the details of such model.

4.4 Electrophysiological Properties of mPTP

Experiments with isolated mitochondria that led to discovery of permeability transition phenomenon could not answer the question on the molecular nature of this permeability increase. Generally, mechanisms of membrane permeation might occur through the number of pathways. These could involve transporter mechanism, nonspecific loss of lipid bilayer integrity, or formation of the large water-filled pore in the membrane. The identity of mPTP as a large pore was established by the experiments that involved patch clamp of the mitochondrial inner membrane (Szabo and Zoratti 1992; Kinnally et al. 1996). It should be noted that due to the nature of the technique, patch-clamp experiments cannot be performed in the mitochondria of intact cells or functional isolated mitochondria. Indeed, to make inner membrane accessible to the patch pipette, mitochondria need to be isolated from the cells, and their outer membrane needs to be ruptured either by swelling in low osmotic solution or by resuspending mitochondria in high osmotic solution and pressing preparation through French press. Thus, although it is generally agreed that mPTP is a large channel, the evidence for this which relies exclusively on patch-clamp data is correlative in nature. More direct evidence for this would require establishing the molecular identity of mPTP pore and demonstration genetic and pharmacological regulation of the mPTP channel.

To date the following electrophysiological properties of the mitochondrial channel have been assigned to mPTP: it is a large voltage-sensitive, weakly selective channel of the maximal conductance size of 1.5 nS and multiple sub-conductance states. Some aspects of the regulation of this channel resemble the regulation of permeability transition in the isolated mitochondria. Perhaps most importantly, this channel is not readily detected in the patch clamp when intact mitochondria were swollen in the absence of calcium. However, mPTP was detected in 95% of patches of the mitochondrial inner membrane derived from the mitochondria in which mPTP was activated by addition of calcium to the intact organelle prior to the swelling (Kinnally et al. 1991). This is important since as discussed above in the intact mitochondria, mPTP cannot be activated by the increase of free calcium concentration, and thus it is not expected that calcium activation phenomenon can be reproduced in the patch-clamp experiment of the native membrane that hasn't developed mPTP. Indeed, in order to induce mPTP-like channel activity, millimolar concentrations of calcium were needed to be added to the solution (Szabo et al. 1992). This is three orders of magnitude higher than free calcium concentrations experienced by the matrix of intact mitochondria prior to mPTP activation. Interestingly, it has been demonstrated in experiments with whole-mitoplast configuration that, even at millimolar concentrations of calcium, mPTP is not activated (Kirichok et al. 2004). This is an important finding since, unlike excised patch configuration which might or might not contain mPTP, whole mitoplast includes all the membrane and, thus, presumably contained mPTP channels but did not demonstrate mPTP even in high calcium. This further stresses the notion that, consistently with experiments with intact isolated mitochondria, mPTP is unlikely activated by rise in free calcium concentration. Another important parameter of mPTP channel is its regulation by CSA. Unfortunately, the use of CSA, which specifically blocks mPTP at the level of isolated mitochondria, was not conclusive for mPTP identification in case of single-channel electrophysiological studies. This is likely due to the fact that CSA does not act directly on the channel part of mPTP but provides inhibitory effect through interaction with peripheral protein cyclophilin D (CypD). If CypD is lost during the patch-lamp experiment than mPTP, channel loses its sensitivity to CSA blocker.

4.5 Molecular Structure of the mPTP

Although complete understanding of the molecular organization of the mPTP is still lacking, it is established that mPTP is formed by a large multiprotein complex. Several proteins are closely involved in mPTP activity in the inner or outer mitochondrial membranes. These include ANT, VDAC, CypD, phosphate carrier, ATP synthase, and as well a recently identified novel component of mPTP protein SPG7 (Shanmughapriya et al. 2015). Knockout experiments confirmed various important roles of these proteins in mPTP but also demonstrated that ANT, VDAC, and CypD appear to be nonessential for mPTP function. The most important question that is currently remains unresolved is what forms the core "pore" part of mPTP. Electrophysiological data suggest that pore structure is expected to be similar to other known large channels, like mitochondrial outer membrane channel VDAC and bacterial porins (Szabo et al. 1993). Over the past two decades, several proteins have been purified from the mitochondria and demonstrated activity compatible with the activity of mPTP as seen in patch-clamp experiments on native mitoplasts. This activity has been shown for mitochondrial VDAC and ANT

(Brustovetsky et al. 2002). At various time points, these data along with experiments using isolated mitochondria lead to conclusions that pore part of mPTP is composed of these proteins. Furthermore, in the model system when inner mitochondrial membrane is reconstituted into giant liposomes by dehydration-rehydration procedure, channel of similar size and selectivity can be formed by the translocator of the inner membrane (TIM) protein (Muro et al. 2003). In fact, even SPG7 has been considered as a core component of mPTP (Shanmughapriya et al. 2015), although this view has been disputed (Bernardi and Forte 2015). Taking into account that a large number of mitochondrial proteins can be converted into the pore, it is interesting to consider the possibility that depending on conditions mPTP might occur through a number of different proteins formed pores. This could potentially be the case of knockout experiments when, for example, in ANT knockout cells mPTP activation might occur by other pore-forming mechanism.

4.6 ATP Synthase as a Candidate for mPTP Pore

Despite the possibility that various mechanisms of mPTP might be present, currently most likely mechanism is that this pore is formed with participation of ATP synthase (Bonora et al. 2013; Giorgio et al. 2013; Bernardi et al. 2015; Mnatsakanyan et al. 2016). More specifically, the pore part is formed by the oligomers of C-subunit (Alavian et al. 2014). In normal mitochondria, ATP synthase is a multiprotein complex with oligomers of C-subunit located in the mitochondria inner membrane and in the absence of mPTP is not permeable for ions. It has been proposed that during mPTP activation, C-subunit is transformed into the mPTP pore. This hypothesis is supported by the observation that purified C-subunit forms ion channels in the artificial lipid membranes. Further, genetic deletion of C-subunit prevents mPTP activation (Bonora et al. 2013). It is very unlikely though that C-subunit alone forms mPTP. In order to allow ion and small molecule passage, mPTP pore should be hydrophilic inside. Such an arrangement for C-subunit is very unfavorable taking into account that this is highly hydrophobic protein. Indeed, based on these considerations, it is believed that to form stable structure, C-subunit oligomer should contain lipids in their central core (Walker 2013). The presence of lipids would make such structure stable but on the other hand would not allow formation of the high conductance pore. Recent studies that show possible association between C-subunit and PHB can help to resolve such a controversy. Indeed, PHB is an amphipathic molecule which can be present in the highly hydrophobic environment and, at the same time, allows ion passage (Seebach and Fritz 1999). PHB can mediate a slow diffusion of ions across bilayer as well as form large ion-conducting pores. In mitochondria PHB is an endogenous ionophore that in normal conditions contributes to the mitochondrial calcium uptake (Smithen et al. 2013). As discussed above, calcium- and polyP-induced association between PHB and C-subunit might contribute to the formation of high conductance pore.

4.7 Possible Physiological Significance of mPTP Activation

While the main focus of the chapter is on review of the details of calcium activation of mPTP and induction of pathological membrane conductance, we should note that regulation of this channel is a very complicated process that involves many different mechanisms and molecules. Thus in this section we briefly summarize possible roles of mPTP opening which have been proposed. These include not only pathological consequences of mPTP activation but possible physiological significance of this event.

Traditionally, it has been postulated that mPTP opening uncouples the respiratory chain, collapses the mitochondrial membrane potential, and arrests the mitochondrial ATP synthesis. All these processes together will irretrievably lead to cell death (Halestrap 2009). However, recent studies showed that opening the mPTP is a double-edged sword for cell survival: on one hand it is an important structure for mitochondrial physiology, essential for the proper functioning of the organelle. On the other, it is a deleterious process in many pathological processes, such as different cardiac diseases (Kwong and Molkentin 2015) and neurodegenerative processes (Green and Kroemer 2004), or in nervous and muscular dystrophies (Fiskum 2000; Bernardi and Bonaldo 2008).

The actual hypothesis is that while prolonged opening of the mPTP induces mitochondrial dysfunction and necrotic cell death, the brief opening of the channel is needed for the proper mitochondrial calcium homeostasis and trafficking. In fact, the pore is not always closed in resting cells but quickly changing between open and closed states of low conductance. When specific inputs reach mitochondria, the opening state lasts longer, inducing bioenergetics impairments, mitochondrial dysfunction, and cell death (Petronilli et al. 1999). This dual action of the pore makes this structure to be essential both under physiological and pathological conditions.

The fact that mPTP can be modulated by direct molecular effects, as Ca^{+2}, fatty acids, Bax, Bid, p53, or indirect pathophysiological effects, as hypoxia, exercise, and aging (Marzo et al. 1998; Schonfeld et al. 2000; Zamzami et al. 2000; Marcil et al. 2006; Martel et al. 2012; Vaseva et al. 2012), strongly supports the important role of mPTP in mitochondrial physiology. In fact, even if classically the pore has been postulated to be anchored to the inner mitochondrial membrane, mPTP could also be localized in the contact sites between the inner and the outer mitochondrial membranes. In this case, mPTP would be involved in energy transfer processes and in the apoptotic cell death pathway, by facilitating specific protein-protein interactions, between proteins located in the inner and the outer mitochondrial membranes, as well as by inducing protein conformational changes (Brdiczka 1991; Vyssokikh and Brdiczka 2003). All these process are needed for the proper mitochondrial physiology.

Additionally, mPTP has also been described as a protective mechanism against the accumulation of old, defective mitochondria, which drives cells to malfunctioning and death in many pathologies, as well as in regular aging (Carew and Huang 2002; Chen and Chan 2009; Cui et al. 2012). Mitochondrial dynamics and mitophagy are the physiological tools to prevent this dysfunction. However, the exact mechanisms driving these processes are still unclear. It has been described that mPTP may be involved in removing aging mitochondria from cells, by selective autophagy (Rodriguez-Enriquez et al. 2004). The authors described that aging mitochondria are clearly more exposed to oxidative stress, becoming more susceptible to mPTP opening. When this stress will exceed a certain threshold, the increased amount of ROS will be enough to induce the opening of the mPTP, even at resting Ca^{+2} concentration. This fact may be understood by the autophagy machinery as a signal to activate the process and to delete these aging mitochondria, allowing the rest of the organelles and the cell to survive.

Interestingly, under physiological conditions, mPTP is also able to exist in a harmless, low conductance state, which may not actually induce mitochondrial swelling (Ichas and Mazat 1998) and which is extremely influenced by the mitochondrial matrix pH (Ichas et al. 1997). The low conductance pore allows the pass of molecules and ions sized 300 kDa, such as H^+, K^+, and, importantly, Ca^{+2}, between the mitochondrial matrix and the other mitochondrial spaces. In fact, Ca^{+2} could use this mechanism to leave mitochondria, in order to keep an adequate cell Ca^{+2} homeostasis and trafficking (Hunter and Haworth 1979a, b; Altschuld et al. 1992). Thus, this version of the pore promotes mitochondrial depolarization spikes and Ca^{+2} exchange between mitochondria, which in turns generates Ca^{+2} waves, amplifying Ca^{+2} signals produced by the endoplasmic reticulum (Ichas et al. 1997; Ichas and Mazat 1998). While the opening of the low conductance pore which is present under physiological conditions is induced by Ca^{+2} and blocked by CsA, the unregulated form of the pore, formed under stress and dysfunctional conditions, is independent of Ca^{+2} and insensitive to CsA (He and Lemasters 2002). Importantly, it has been demonstrated that the high conductance and the long-lasting mPTP, that is, the one present under pathological conditions, is the preferred version of the pore (Brenner and Moulin 2012).

Under pathological conditions, mPTP can be involved in both apoptotic and necrotic cell death. It has been described that in the presence of ATP, apoptosis is the preferred cell death pathway, while in the absence of this energetic molecule, as well as when there is mitochondrial Ca^{+2} overload and excessive ROS generation, it is necrosis (Leist et al. 1997; Nicotera and Orrenius 1998; Kim et al. 2003). However, both cell death pathways should not be seen as independent processes, as recent evidence showed a clear interconnection between them, at different levels, (Christofferson and Yuan 2010; Vandenabeele et al. 2010; Galluzzi et al. 2012).

It has also been described that the pore is extremely harmful for cell survival when excessive opening and closing cycles that are lasting longer occurred. In these cycles, different posttranslational modifications of proteins have been postulated to act as metabolic regulators of mPTP. This is the case of sirtuin-3, which is a mitochondrial deacetylase, (Giralt and Villarroya 2012). Unbalanced ROS/antioxidant ratio could also modify the equilibrium between high and low conductance pore and induce long-lasting pore opening, mitochondrial dysfunction, and cell death (Zamzami et al. 1998; Curtis et al. 2012).

The understanding of mPTP functioning, especially under physiological conditions, is still incomplete. Extended studies of this problem should be conducted, in order to elucidate the real role of this phenomenon, which seems to be needed both for cell survival and death. Modulation of mPTP represents a feasible and promising therapeutic tool against many different diseases and pathologies, ranging from ischemia-reperfusion or heart failure to cancer or neurodegeneration.

Acknowledgments Authors would like to acknowledge the support from the American Heart Association and from the National Institute of Health.

References

Alavian KN, Beutner G, Lazrove E, Sacchetti S, Park HA, Licznerski P, Li H, Nabili P, Hockensmith K, Graham M, Porter GA Jr, Jonas EA (2014) An uncoupling channel within the c-subunit ring of the F1FO ATP synthase is the mitochondrial permeability transition pore. Proc Natl Acad Sci U S A 111(29):10580–10585

Altschuld RA, Hohl CM, Castillo LC, Garleb AA, Starling RC, Brierley GP (1992) Cyclosporin inhibits mitochondrial calcium efflux in isolated adult rat ventricular cardiomyocytes. Am J Phys 262(6 Pt 2):H1699–H1704

Bernardi P, Bonaldo P (2008) Dysfunction of mitochondria and sarcoplasmic reticulum in the pathogenesis of collagen VI muscular dystrophies. Ann N Y Acad Sci 1147:303–311

Bernardi P, Forte M (2015) Commentary: SPG7 is an essential and conserved component of the mitochondrial permeability transition pore. Front Physiol 6:320

Bernardi P, Rasola A, Forte M, Lippe G (2015) The mitochondrial permeability transition pore: channel formation by F-ATP synthase, integration in signal transduction, and role in pathophysiology. Physiol Rev 95(4):1111–1155

Bonora M, Bononi A, De Marchi E, Giorgi C, Lebiedzinska M, Marchi S, Patergnani S, Rimessi A, Suski JM, Wojtala A, Wieckowski MR, Kroemer G, Galluzzi L, Pinton P (2013) Role of the c subunit of the FO ATP synthase in mitochondrial permeability transition. Cell Cycle 12(4): 674–683

Brdiczka D (1991) Contact sites between mitochondrial envelope membranes. Structure and function in energy- and protein-transfer. Biochim Biophys Acta 1071(3):291–312

Brenner C, Moulin M (2012) Physiological roles of the permeability transition pore. Circ Res 111(9):1237–1247

Broekemeier KM, Dempsey ME, Pfeiffer DR (1989) Cyclosporin A is a potent inhibitor of the inner membrane permeability transition in liver mitochondria. J Biol Chem 264(14): 7826–7830

Brustovetsky N, Tropschug M, Heimpel S, Heidkamper D, Klingenberg M (2002) A large Ca2+-dependent channel formed by recombinant ADP/ATP carrier from Neurospora crassa resembles the mitochondrial permeability transition pore. Biochemistry 41(39):11804–11811

Carew JS, Huang P (2002) Mitochondrial defects in cancer. Mol Cancer 1:9

Castuma CE, Huang R, Kornberg A, Reusch RN (1995) Inorganic polyphosphates in the acquisition of competence in *Escherichia coli*. J Biol Chem 270(22):12980–12983

Chalmers S, Nicholls DG (2003) The relationship between free and total calcium concentrations in the matrix of liver and brain mitochondria. J Biol Chem 278(21):19062–19070

Chen H, Chan DC (2009) Mitochondrial dynamics – fusion, fission, movement, and mitophagy – in neurodegenerative diseases. Hum Mol Genet 18(R2):R169–R176

Christofferson DE, Yuan J (2010) Necroptosis as an alternative form of programmed cell death. Curr Opin Cell Biol 22(2):263–268

Cui H, Kong Y, Zhang H (2012) Oxidative stress, mitochondrial dysfunction, and aging. J Signal Transduct 2012:646354

Curtis JM, Hahn WS, Long EK, Burrill JS, Arriaga EA, Bernlohr DA (2012) Protein carbonylation and metabolic control systems. Trends Endocrinol Metab 23(8):399–406

Elustondo PA, Negoda A, Kane CL, Kane DA, Pavlov EV (2015) Spermine selectively inhibits high-conductance, but not low-conductance calcium-induced permeability transition pore. Biochim Biophys Acta 1847(2):231–240

Elustondo PA, Nichols M, Negoda A, Thirumaran A, Zakharian E, Robertson GS, Pavlov EV (2016) Mitochondrial permeability transition pore induction is linked to formation of the complex of ATPase C-subunit, polyhydroxybutyrate and inorganic polyphosphate. Cell Death Discov 2:16070

Fiskum G (2000) Mitochondrial participation in ischemic and traumatic neural cell death. J Neurotrauma 17(10):843–855

Galluzzi L, Vitale I, Abrams JM, Alnemri ES, Baehrecke EH, Blagosklonny MV, Dawson TM, Dawson VL, El-Deiry WS, Fulda S, Gottlieb E, Green DR, Hengartner MO, Kepp O, Knight RA, Kumar S, Lipton SA, Lu X, Madeo F, Malorni W, Mehlen P, Nunez G, Peter ME, Piacentini M, Rubinsztein DC, Shi Y, Simon HU, Vandenabeele P, White E, Yuan J, Zhivotovsky B, Melino G, Kroemer G (2012) Molecular definitions of cell death subroutines: recommendations of the Nomenclature Committee on Cell Death 2012. Cell Death Differ 19(1):107–120

Giorgio V, von Stockum S, Antoniel M, Fabbro A, Fogolari F, Forte M, Glick GD, Petronilli V, Zoratti M, Szabo I, Lippe G, Bernardi P (2013) Dimers of mitochondrial ATP synthase form the permeability transition pore. Proc Natl Acad Sci U S A 110(15):5887–5892

Giralt A, Villarroya F (2012) SIRT3, a pivotal actor in mitochondrial functions: metabolism, cell death and aging. Biochem J 444(1):1–10

Green DR, Kroemer G (2004) The pathophysiology of mitochondrial cell death. Science 305(5684):626–629

Greenawalt JW, Rossi CS, Lehninger AL (1964) Effect of active accumulation of calcium and phosphate ions on the structure of rat liver mitochondria. J Cell Biol 23:21–38

Halestrap AP (2009) What is the mitochondrial permeability transition pore? J Mol Cell Cardiol 46(6):821–831

Halestrap AP, Connern CP, Griffiths EJ, Kerr PM (1997) Cyclosporin A binding to mitochondrial cyclophilin inhibits the permeability transition pore and protects hearts from ischaemia/reperfusion injury. Mol Cell Biochem 174(1–2):167–172

Haworth RA, Hunter DR (1979) The Ca2+-induced membrane transition in mitochondria. II. Nature of the Ca2+ trigger site. Arch Biochem Biophys 195(2):460–467

He L, Lemasters JJ (2002) Regulated and unregulated mitochondrial permeability transition pores: a new paradigm of pore structure and function? FEBS Lett 512(1–3):1–7

Hunter DR, Haworth RA (1979a) The Ca2+-induced membrane transition in mitochondria. I. The protective mechanisms. Arch Biochem Biophys 195(2):453–459

Hunter DR, Haworth RA (1979b) The Ca2+-induced membrane transition in mitochondria. III. Transitional Ca2+ release. Arch Biochem Biophys 195(2):468–477

Ichas F, Mazat JP (1998) From calcium signaling to cell death: two conformations for the mitochondrial permeability transition pore. Switching from low- to high-conductance state. Biochim Biophys Acta 1366(1–2):33–50

Ichas F, Jouaville LS, Mazat JP (1997) Mitochondria are excitable organelles capable of generating and conveying electrical and calcium signals. Cell 89(7):1145–1153

Kennedy EP, Lehninger AL (1949) Oxidation of fatty acids and tricarboxylic acid cycle intermediates by isolated rat liver mitochondria. J Biol Chem 179(2):957–972

Kim JS, He L, Lemasters JJ (2003) Mitochondrial permeability transition: a common pathway to necrosis and apoptosis. Biochem Biophys Res Commun 304(3):463–470

Kinnally KW, Zorov D, Antonenko Y, Perini S (1991) Calcium modulation of mitochondrial inner membrane channel activity. Biochem Biophys Res Commun 176(3):1183–1188

Kinnally KW, Lohret TA, Campo ML, Mannella CA (1996) Perspectives on the mitochondrial multiple conductance channel. J Bioenerg Biomembr 28(2):115–123

Kirichok Y, Krapivinsky G, Clapham DE (2004) The mitochondrial calcium uniporter is a highly selective ion channel. Nature 427(6972):360–364

Kristian T, Pivovarova NB, Fiskum G, Andrews SB (2007) Calcium-induced precipitate formation in brain mitochondria: composition, calcium capacity, and retention. J Neurochem 102(4):1346–1356

Kwong JQ, Molkentin JD (2015) Physiological and pathological roles of the mitochondrial permeability transition pore in the heart. Cell Metab 21(2):206–214

Leist M, Single B, Castoldi AF, Kuhnle S, Nicotera P (1997) Intracellular adenosine triphosphate (ATP) concentration: a switch in the decision between apoptosis and necrosis. J Exp Med 185(8):1481–1486

Marcil M, Bourduas K, Ascah A, Burelle Y (2006) Exercise training induces respiratory substrate-specific decrease in Ca2+-induced permeability transition pore opening in heart mitochondria. Am J Physiol Heart Circ Physiol 290(4):H1549–H1557

Martel C, Huynh le H, Garnier A, Ventura-Clapier R, Brenner C (2012) Inhibition of the mitochondrial permeability transition for cytoprotection: direct versus indirect mechanisms. Biochem Res Int 2012:213403

Marzo I, Brenner C, Zamzami N, Susin SA, Beutner G, Brdiczka D, Remy R, Xie ZH, Reed JC, Kroemer G (1998) The permeability transition pore complex: a target for apoptosis regulation by caspases and bcl-2-related proteins. J Exp Med 187(8):1261–1271

Mnatsakanyan N, Beutner G, Porter GA, Alavian KN, Jonas EA (2016) Physiological roles of the mitochondrial permeability transition pore. J Bioenerg Biomembr 49:13–25.

Muro C, Grigoriev SM, Pietkiewicz D, Kinnally KW, Campo ML (2003) Comparison of the TIM and TOM channel activities of the mitochondrial protein import complexes. Biophys J 84(5):2981–2989

Nicholls DG, Chalmers S (2004) The integration of mitochondrial calcium transport and storage. J Bioenerg Biomembr 36(4):277–281

Nicotera P, Orrenius S (1998) The role of calcium in apoptosis. Cell Calcium 23(2–3):173–180

Pavlov E, Zakharian E, Bladen C, Diao CT, Grimbly C, Reusch RN, French RJ (2005) A large, voltage-dependent channel, isolated from mitochondria by water-free chloroform extraction. Biophys J 88(4):2614–2625

Petronilli V, Miotto G, Canton M, Brini M, Colonna R, Bernardi P, Di Lisa F (1999) Transient and long-lasting openings of the mitochondrial permeability transition pore can be monitored directly in intact cells by changes in mitochondrial calcein fluorescence. Biophys J 76(2): 725–734

Rodriguez-Enriquez S, He L, Lemasters JJ (2004) Role of mitochondrial permeability transition pores in mitochondrial autophagy. Int J Biochem Cell Biol 36(12):2463–2472

Schonfeld P, Wieckowski MR, Wojtczak L (2000) Long-chain fatty acid-promoted swelling of mitochondria: further evidence for the protonophoric effect of fatty acids in the inner mitochondrial membrane. FEBS Lett 471(1):108–112

Seebach D, Fritz MG (1999) Detection, synthesis, structure, and function of oligo(3-hydroxyalkanoates): contributions by synthetic organic chemists. Int J Biol Macromol 25(1–3): 217–236

Shanmughapriya S, Rajan S, Hoffman NE, Higgins AM, Tomar D, Nemani N, Hines KJ, Smith DJ, Eguchi A, Vallem S, Shaikh F, Cheung M, Leonard NJ, Stolakis RS, Wolfers MP, Ibetti J, Chuprun JK, Jog NR, Houser SR, Koch WJ, Elrod JW, Madesh M (2015) SPG7 is an essential and conserved component of the mitochondrial permeability transition pore. Mol Cell 60(1):47–62

Smithen M, Elustondo PA, Winkfein R, Zakharian E, Abramov AY, Pavlov E (2013) Role of polyhydroxybutyrate in mitochondrial calcium uptake. Cell Calcium 54(2):86–94

Solesio ME, Demirkhanyan L, Zakharian E, Pavlov EV (2016a) Contribution of inorganic polyphosphate towards regulation of mitochondrial free calcium. Biochim Biophys Acta 1860(6):1317–1325

Solesio ME, Elustondo PA, Zakharian E, Pavlov EV (2016b) Inorganic polyphosphate (polyP) as an activator and structural component of the mitochondrial permeability transition pore. Biochem Soc Trans 44(1):7–12
Szabo I, Zoratti M (1992) The mitochondrial megachannel is the permeability transition pore. J Bioenerg Biomembr 24(1):111–117
Szabo I, Bernardi P, Zoratti M (1992) Modulation of the mitochondrial megachannel by divalent cations and protons. J Biol Chem 267(5):2940–2946
Szabo I, De Pinto V, Zoratti M (1993) The mitochondrial permeability transition pore may comprise VDAC molecules. II. The electrophysiological properties of VDAC are compatible with those of the mitochondrial megachannel. FEBS Lett 330(2):206–210
Vandenabeele P, Galluzzi L, Vanden Berghe T, Kroemer G (2010) Molecular mechanisms of necroptosis: an ordered cellular explosion. Nat Rev Mol Cell Biol 11(10):700–714
Vaseva AV, Marchenko ND, Ji K, Tsirka SE, Holzmann S, Moll UM (2012) p53 opens the mitochondrial permeability transition pore to trigger necrosis. Cell 149(7):1536–1548
Vyssokikh MY, Brdiczka D (2003) The function of complexes between the outer mitochondrial membrane pore (VDAC) and the adenine nucleotide translocase in regulation of energy metabolism and apoptosis. Acta Biochim Pol 50(2):389–404
Walker JE (2013) The ATP synthase: the understood, the uncertain and the unknown. Biochem Soc Trans 41(1):1–16
Zamzami N, Marzo I, Susin SA, Brenner C, Larochette N, Marchetti P, Reed J, Kofler R, Kroemer G (1998) The thiol crosslinking agent diamide overcomes the apoptosis-inhibitory effect of Bcl-2 by enforcing mitochondrial permeability transition. Oncogene 16(8):1055–1063
Zamzami N, El Hamel C, Maisse C, Brenner C, Munoz-Pinedo C, Belzacq AS, Costantini P, Vieira H, Loeffler M, Molle G, Kroemer G (2000) Bid acts on the permeability transition pore complex to induce apoptosis. Oncogene 19(54):6342–6350

Part III
Mitochondrial Outer Membrane Transport Systems: Structure, Function, and Physiological Implications

Chapter 5
Voltage-Dependent Anion Channels and Tubulin: Bioenergetic Controllers in Cancer Cells

Eduardo N. Maldonado, David N. DeHart, and John J. Lemasters

5.1 Introduction

5.1.1 Warburg Phenotype and Cell Proliferation

The interdependence between bioenergetics, catabolism, and anabolism differs in cancer and other proliferating cells compared to differentiated cells. A metabolic phenotype characterized by enhanced glycolysis and suppression of mitochondrial

E.N. Maldonado
Department of Drug Discovery & Biomedical Sciences, Medical University of South Carolina, DD506 Drug Discovery Building, 70 President Street, MSC 139, Charleston, SC 29425, USA

Hollings Cancer Center, Medical University of South Carolina, Charleston, SC 29425, USA

Center for Cell Death, Injury & Regeneration, Medical University of South Carolina, Charleston, SC 29425, USA
e-mail: maldona@musc.edu

D.N. DeHart
Departments of Drug Discovery and Biomedical Sciences and the Hollings Cancer Center, Medical University of South Carolina, Charleston, SC 29425, USA

J.J. Lemasters
Department of Biochemistry & Molecular Biology, Medical University of South Carolina, Charleston, SC 29425, USA

Institute of Theoretical and Experimental Biophysics, Russian Academy of Sciences, Pushchino, Russian Federation

Department of Drug Discovery & Biomedical Sciences, Medical University of South Carolina, DD506 Drug Discovery Building, 70 President Street, MSC 139, Charleston, SC 29425, USA

Hollings Cancer Center, Medical University of South Carolina, Charleston, SC 29425, USA

Center for Cell Death, Injury & Regeneration, Medical University of South Carolina, Charleston, SC 29425, USA

T.K. Rostovtseva (ed.), *Molecular Basis for Mitochondrial Signaling*, Biological and Medical Physics, Biomedical Engineering,
DOI 10.1007/978-3-319-55539-3_5

metabolism even in the presence of physiological levels of oxygen was first described by Otto Warburg in the early twentieth century (Warburg et al. 1927; Warburg 1956). Warburg also postulated that irreversible but not completely damaged respiration led to cancer. According to Warburg, cells compensate for lower energy production associated with damaged respiration by increasing the conversion of glucose to lactic acid (fermentation). Cells capable of increasing fermentation through successive divisions to compensate for defective respiration eventually become neoplastic (Warburg 1956). The lack of function of mitochondria in tumor tissues was challenged by Weinhouse and others demonstrating both high glycolysis and oxidative metabolism in cancer tissues (Weinhouse 1956). Since the early work of Warburg, several investigations showed active mitochondrial metabolism in cancer cells and their isolated mitochondria as determined by measurements of ATP generation, NADH production, and mitochondrial membrane potential ($\Delta\Psi$) among other functional parameters (Lim et al. 2011; Maldonado et al. 2010; Mathupala et al. 2010; Moreno-Sanchez et al. 2014; Nakashima et al. 1984; Pedersen 1978; Singleterry et al. 2014).

Although functional, the contribution of mitochondria to ATP generation in cancer cells through oxidative phosphorylation (OXPHOS) is lower compared to differentiated cells. Differentiated cells produce about 95% of total ATP by OXPHOS and the remaining 5% through aerobic glycolysis. By contrast in cancer and other proliferating cells, 20–90% of total ATP production derives from glycolysis with the remainder coming from mitochondrial oxidation of pyruvate, fatty acids, and glutamine (6, 11). Accordingly, tumor cells have increased uptake of glucose compared to differentiated cells. This glucose avidity of tumors can be used to diagnose primary tumors, recurrences, and metastases by positron emission tomography (PET) of the glucose analog ^{18}fluorodeoxyglucose (Zhu et al. 2011). Enhanced glycolysis in cancer cells is associated with a high rate of cell proliferation (Griguer et al. 2005; Guppy et al. 2002; Moreno-Sanchez et al. 2007; Scott et al. 2011). Nonetheless, bioenergetic profiles can be different among tumor types and even in cells from the same type of tumor. Subsets of cells with either high glycolysis or high levels of OXPHOS have been identified in gliomas and large B cell lymphomas (Beckner et al. 2005; Bouzier et al. 1998; Caro et al. 2012).

Incomplete breakdown of glucose through glycolysis generates only 2 moles of ATP per mole of glucose, whereas mitochondrial oxidation of the 2 moles of pyruvate generated from glucose to CO_2 and H_2O generates about an additional 31 moles of ATP taking into account currently accepted proton stoichiometries for respiration, ATP synthesis, ATP/ADP•Pi exchange, and the malate/aspartate shuttle, although actual ATP yields will be less due to proton leak and possible molecular "slippage" of the respiratory complexes (Brand 2005; Rich 2003; Rich and Marechal 2010; Walker 2013; Wikstrom et al. 2015). In cancer cells, lower efficiency of ATP generation by aerobic glycolysis appears to be offset by greater glycolytic rates (Locasale and Cantley 2010). It is also proposed that the ATP necessary for biosynthesis of macromolecules is lower than the energy requirements of basal cellular processes making unlikely that ATP generation is rate limiting in proliferating cells (Kilburn et al. 1969).

The metabolic requirements of cell division are not simply limited to energy generation. A dividing cell must double its biomass (lipids, proteins, and nucleic acids) before mitosis. This biosynthetic demand requires carbon backbones for the synthesis of new macromolecules. Full oxidation of glucose, glutamine, and fatty acids in mitochondria generates maximum ATP but not residual carbon backbones. By contrast, incomplete breakdown of glucose to lactate and possibly decreased mitochondrial degradation of glutamine and fatty acids provides precursors for biomass formation (Cairns 2015; DeBerardinis et al. 2008; Keibler et al. 2016; Liberti and Locasale 2016; Lunt and Vander Heiden 2011). Specifically, the by-products of glucose catabolism, glucose-6-phosphate, glyceraldehyde-3-phosphate, and 3-phosphoglycerate contribute to the synthesis of nucleotides, lipids, and amino acids, respectively. High glycolytic flux also increases NADPH production by the pentose phosphate pathway for reductive biosynthesis. Glutamine and other fuels also generate biosynthetic precursors in the Krebs cycle, including citrate for lipid biosynthesis and oxaloacetate and α-ketoglutarate for synthesis of nonessential amino acids (Fig. 5.1) (DeBerardinis and Cheng 2010). In addition, one-carbon metabolism, a set of reactions that transfer one carbon units from serine and glycine, plays an important role for de novo synthesis of purines and thymidylate during rapid tumor growth (Meiser and Vazquez 2016). In summary, the Warburg metabolic phenotype is a complex network of interrelated processes involving glycolysis and mitochondrial metabolism.

5.1.2 *Cytosolic ATP/ADP Ratio: A Key to Sustain Glycolysis*

Maximal mitochondrial oxidation of respiratory substrates, including pyruvate, fatty acyl-CoA, glutamine, and amino acids, by OXPHOS generates a maximum yield of ATP per mole of respiratory substrates and minimal residual carbon backbones. Newly synthesized ATP in the mitochondrial matrix is transported to the cytosol by the electrogenic adenine nucleotide translocator (ANT) because of the coupling to $\Delta\Psi$ of mitochondrial ATP^{-4} release for ADP^{-3} uptake. In differentiated cells with predominantly oxidative metabolism, cytosolic ATP/ADP ratios can be 50–100 times higher than in the mitochondrial matrix (Schwenke et al. 1981). A high cytosolic ATP/ADP ratio suppresses glycolysis through inhibition of phosphofructokinase-1 (PFK-1) although other mechanisms may be involved. ATP is a strong allosteric inhibitor, and ADP and AMP are activators of PFK-1 (Mor et al. 2011; Moreno-Sanchez et al. 2007). In cancer cells, suppression of mitochondrial metabolism contributes to a low cytosolic ATP/ADP ratio, which releases this brake on glycolysis. Recently, we demonstrated that closing of the voltage-dependent anion channels (VDAC) promoted by free tubulin limits ingress of respiratory substrates into mitochondria and limits ATP production, whereas replacement of electrogenic ATP/ADP exchange by ANT with a non-electrogenic exchange mechanism decreases cytosolic ATP/ADP ratios. These two independent mechanisms contribute to suppress mitochondrial metabolism and to maintain a low cytosolic ATP/ADP ratio favoring aerobic glycolysis in cancer cells (Maldonado et al. 2013, 2016; Maldonado and Lemasters 2014).

5.2 VDAC Modulation of Cancer Bioenergetics

5.2.1 VDAC and the Warburg Phenotype

The bioenergetics of cancer cells depends on chemical reactions occurring in two functional, interconnected, and interdependent cellular compartments separated by the mitochondrial outer membrane (MOM) (Fig. 5.1). VDAC, the most abundant protein in the MOM, is the gateway through which most respiratory substrates, ADP, and Pi enter mitochondria and ATP exits. The subcellular localization of VDAC determines that the closing or the opening of the channels regulates the flux of metabolites that enter or leave mitochondria. Thus, VDAC is positioned to be a global controller or governator of mitochondrial metabolism and whole cellular bioenergetics (Lemasters and Holmuhamedov 2006; Maldonado et al. 2013; Maldonado and Lemasters 2012, 2014).

The influx of polar metabolites through VDAC is determined mostly by their charge and size (Colombini 1980, 2004). Metabolites that reach the intermembrane space are further transported to the matrix by numerous different transporters located in the mitochondrial inner membrane (MIM). Respiratory substrates in the matrix are catabolized in the Krebs cycle generating NADH and $FADH_2$ that enters the respiratory chain. The transfer of electrons from NADH and $FADH_2$ to the final acceptor O_2 produces proton translocation across MIM by Complexes I, III, and IV to generate a negative transmembrane $\Delta\Psi$ and positive ΔpH, the components of the proton motive force (Δp). Δp then drives ATP synthesis from ADP and Pi by Complex V (F_1F_O-ATP synthase) (Fig. 5.1).

Based on its role in metabolite exchange between mitochondria and the cytosol, VDAC is proposed to be a regulated governor or "governator" that limits global mitochondrial metabolism (Lemasters and Holmuhamedov 2006). Interactions with tubulin and possibly other proteins, such as hexokinase (Pastorino and Hoek 2003; Wolf et al. 2011), modulate the open/closed sate of VDAC. Single and double knockdown of the three different VDAC isoforms support this concept that VDAC serves as a master regulator of mitochondrial metabolism in cancer cells (Maldonado et al. 2013). Thus, VDAC regulation by free tubulin emerges as a mechanism to block or promote OXPHOS and indirectly regulate glycolysis through the cytosolic ATP/ADP ratio. Ultimately, disruption of VDAC-tubulin interactions may be a pharmacological target to increase mitochondrial metabolism in cancer cells and to revert Warburg metabolism.

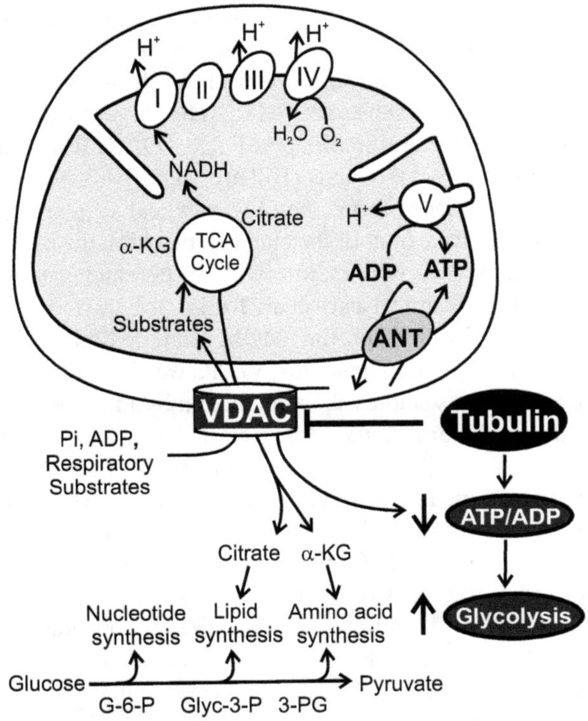

Fig. 5.1 VDAC in Warburg metabolism. Metabolites cross mitochondrial outer membranes through VDAC. Oxidation of respiratory substrates in the tricarboxylic acid cycle generates NADH and $FADH_2$, which feed into the respiratory chain (Complexes I–IV). Proton translocation by the respiratory chain across MIM generates $\Delta\Psi$. ATP is synthesized from ADP and Pi by the F_1F_O-ATP synthase (Complex V) driven by protons moving back across MIM into the matrix. Glucose-6-phosphate (G-6-P), glyceraldehyde 3-phosphate (Glyc-3-P), and 3-phosphoglycerate (3-PG) originating from the catabolism of glucose and intermediates of the Krebs cycle are used for synthesis of nucleotides, lipids, and amino acids. In cancer cells, high free tubulin blocks VDAC conductance, suppresses mitochondrial metabolism, and decreases cytosolic ATP/ADP to favor glycolysis. *α-KG* α-ketoglutarate; *MIM* Mitochondrial inner membrane

5.2.2 *VDAC Structure and Regulation of Mitochondrial Metabolism*

The three isoforms of VDAC present in all eukaryotic cells, VDAC1, VDAC2, and VDAC3, are encoded by separate genes. VDAC1 and VDAC2 are the main isoforms in most differentiated mammalian cells. The minor isoform VDAC3 is abundant only in testis (Sampson et al. 1997, 2001). In cancer cells VDAC1 and VDAC2 are also the major isoforms accounting for 90% of the total. The least abundant isoform, VDAC3, comprises the remaining 10% (De Pinto et al. 2010; Huang et al. 2014; Maldonado et al. 2013). Gating and selectivity of VDAC1 and VDAC2 are highly conserved among mammals (Blachly-Dyson and Forte 2001).

VDAC in humans and mice is a ~30 kDa protein enclosing an aqueous channel of ~3-nm internal diameter that allows the passage of molecules up to ~5 kDa (Colombini 1980, 2012; Song and Colombini 1996). In the closed state, the flux through VDAC of respiratory substrates, ATP, ADP, Pi, and other mostly anionic metabolites is blocked. Structural studies reveal that VDAC1 has a barrel configuration with staves formed by 19 β-strands (Hiller et al. 2010; Ujwal et al. 2008). An additional N-terminal sequence forms the only α-helical segment. The N-terminal helix appears to move to the center of the channel, blocking the passage of metabolites. Recently, a similar β barrel structure with 19 β-strands has been shown for VDAC2 from zebra fish (Schredelseker et al. 2014).

Because of its localization in the MOM and central role in mediating mitochondria-cytosol fluxes of metabolites, VDAC was initially considered constitutively open, but numerous studies show regulation by multiple factors, including hexokinase (Al Jamal 2005; Azoulay-Zohar et al. 2004; Nakashima et al. 1988), Bcl2 family members (Tsujimoto and Shimizu 2000), glutamate (Gincel et al. 2000), ethanol (Holmuhamedov and Lemasters 2009; Lemasters and Holmuhamedov 2006), and NADH (Zizi et al. 1994). VDAC phosphorylation by protein kinases, including glycogen synthase 3β (GSK3β), protein kinase A (PKA), and protein kinase C epsilon (PKCε), blocks or inhibits association of VDAC with other proteins, such as Bax and *t*Bid, and also regulates VDAC opening (Azoulay-Zohar et al. 2004; Baines et al. 2003; Das et al. 2008; Lee et al. 1994; Rostovtseva et al. 2004; Vander Heiden et al. 2000, 2001). PKA-dependent VDAC phosphorylation decreases VDAC conductance (Bera et al. 1995), whereas GSK3β-mediated VDAC2 phosphorylation induces VDAC opening (Das et al. 2008). Here, we will focus on the inhibitory effect of free tubulin on VDAC in cancer cells as a regulatory mechanism of VDAC opening (Maldonado et al. 2010, 2013; Palmieri and Pierri 2010).

5.3 VDAC-Tubulin Interaction

5.3.1 VDAC Inhibition by Free Tubulin

Mitochondrial ΔΨ in cancer cells can be generated both by the respiratory chain and from hydrolysis of glycolytic ATP by the mitochondrial F_1F_O-ATPase working in reverse. Pharmacological interventions to destabilize microtubules with nocodazole and colchicine or stabilize microtubules with paclitaxel increase and decrease, respectively, cytosolic free tubulin. Such high and low cytosolic free tubulin promotes low and high mitochondrial ΔΨ, respectively (Maldonado et al. 2010). In nonproliferating cells like cultured rat hepatocytes, free tubulin is much lower compared to hepatoma cells, since nonproliferating hepatocytes do not need a reservoir of tubulin for spindle formation at mitosis. Thus, microtubule stabilization with paclitaxel does not increase ΔΨ in hepatocytes, because free tubulin is already very low, whereas microtubule destabilization still increases tubulin and, in turn, decreases $\Delta\Psi$. These findings imply that VDAC is indeed constitutively open in nonproliferating hepatocytes under normal incubation. By contrast, since paclitaxel increases and nocodazole/colchicine decreases $\Delta\Psi$ in tumor cells, the conclusion can be made that VDAC is partially closed in tumor cells under the regulation of endogenous free tubulin (Maldonado et al. 2010). Negative modulation of $\Delta\Psi$ by tubulin through VDAC closure is a mechanism that explains, at least in part, the suppression of mitochondrial metabolism in the Warburg phenotype. Our studies performed in intact cancer cells are in agreement with earlier work showing that heterodimeric αβ-tubulin closes VDAC inserted into lipid bilayers and decreases respiration in isolated brain mitochondria and permeabilized synaptosomes (Rostovtseva et al. 2008; Timohhina et al. 2009).

Knockdown studies of VDAC1, VDAC2, and VDAC3 in HepG2 cells further characterized the role of VDAC in mitochondrial metabolism in cancer cells. Single knockdown of each of the three VDAC isoforms, especially the minor isoform VDAC3, decreased mitochondrial $\Delta\Psi$, indicating that all VDAC isoforms contribute to $\Delta\Psi$ formation. Knockdown of VDAC3 not only caused the greatest drop in $\Delta\Psi$ but also decreased cellular ATP and ADP and the NAD(P)H/NAD(P)$^+$ ratio, suggesting that the VDAC3 contributed most to MOM permeability despite being the least abundant isoform (Maldonado et al. 2013). Double knockdown of VDAC isoforms in all possible combinations allowed determination of the response of each individual isoform to tubulin inhibition. All single and double knockdowns partially blocked suppression of $\Delta\Psi$ induced by increased free tubulin (Maldonado et al. 2013). Further studies showed an almost identical voltage gating and response to dimeric αβ-tubulin of constitutive VDAC isolated from wild-type HepG2 cells compared to VDAC from heart and liver mitochondria. VDAC1 and VDAC2 isolated from double knockdown HepG2 cells inserted in lipid bilayers were almost equally sensitive to tubulin inhibition, whereas VDAC3 was insensitive even at tubulin concentrations fivefold higher than those used to inhibit VDAC1 and VDAC2 (Maldonado et al. 2013). The knockdown studies supported the conclusion that VDAC3, at least in HepG2 cells, is constitutively open, whereas VDAC1 and VDAC2 are totally or partially closed by free tubulin.

5.3.2 VDAC-Tubulin Influence on Warburg Metabolism During Cell Cycle

During the cell cycle, biosynthetic processes to generate a new cell occur during G1, S, and G2. Presumably, Warburg metabolism is maximal during these phases, and mitochondrial metabolism is suppressed. VDAC closing by a pool of constitutive free tubulin appears to contribute to mitochondrial suppression during these growth stages. Most of the cell cycle of cancer cells is composed by G1, S, and G2 phases. The actual cell division occurs during the M or mitotic phase lasting only about 30 min of a cell cycle lasting 30 h or more (Hahn et al. 2009). During mitosis, energy demand increases sharply to support chromosome separation and cytokinesis. At this point, a Warburg metabolic phenotype may not be beneficial since all the new macromolecules have been already synthesized. Moreover, mitochondrial activation and full oxidation of respiratory substrates may be required to meet the ATP demands of cell division. A possible scenario is that as the spindle forms during prophase, the free tubulin pool decreases abruptly, releasing tubulin inhibition of VDAC. VDAC opening then promotes increased mitochondrial metabolism reverting the Warburg phenotype precisely when the energy demand is maximal. After mitosis, the pool of free tubulin increases again, and cells return to a high glycolytic, pro-proliferative phenotype during the non-mitotic stages of the cell cycle (Maldonado and Lemasters 2012).

5.3.3 Mitochondrial Contribution to Metabolic Heterogeneity in Tumors

The extent to which cancer cell metabolism is glycolytic or oxidative is not a permanent feature and is under epigenetic control. Tumor cells are metabolically flexible, and the relative contribution of OXPHOS can vary substantially over time depending on multiple factors, including availability to different fuels, proximity to newly formed vs. mature blood vessels, and the release of soluble factors such as lactate from neighboring cells, both cancerous and noncancerous. Hypoxia can decrease the OXPHOS flux depending on time of hypoxic exposure, cell type, and environmental conditions. In MCF-7 and HeLa cells that predominantly depend on OXPHOS for ATP supply, prolonged hypoxia increases glycolysis only in MCF-7 (Rodriguez-Enriquez et al. 2010). The respiratory chain of tumor cells can be fully functional at oxygen levels as low as 0.5%, which is biologically relevant because in solid tumors with heterogeneous perfusion, tumor cells exposed to 2% or less of oxygen can still produce ATP by OXPHOS.

Inadequate blood perfusion in rapidly growing tumors not only exposes cells to hypoxia but to a less frequently considered lower supply of nutrients such as glucose. The importance of nutrient availability on the bioenergetic profile of cancer cells is illustrated by the switch from aerobic glycolysis to OXPHOS in breast cancer cell lines and lymphoma cells cultured in glucose-free media (Robinson et al. 2012; Smolkova et al. 2010). Tumor cells also adapt to oxidize other substrates when glucose or glutamine are limited, including lactate, methionine, asparagine, leucine, arginine, cysteine, acetate, and even proteins and lipids from the environment (Chung et al. 2005; Clavell et al. 1986; Comerford et al. 2014; Commisso et al. 2013; Keenan and Chi 2015; Kennedy et al. 2013; Kreis et al. 1980; Mashimo et al. 2014; Scott et al. 2000; Sheen et al. 2011; Sonveaux et al. 2008). While glucose deprivation promotes a switch to oxidative metabolism, inhibition of Complex III by antimycin and Complex I by piericidin A triggers a compensatory increase in the uptake and consumption of glucose in myoblasts. Total cellular ATP production before and after OXPHOS inhibition was similar indicating that the loss of ATP generation by OXPHOS was fully compensated by increased glycolytic ATP generation (Liemburg-Apers et al. 2015). This metabolic flexibility of tumors and the potential to switch from a predominantly glycolytic to an oxidative metabolism and vice versa underscore the importance of mechanisms like VDAC regulation that underlie these adaptive changes.

Most research efforts to target tumor metabolism have been directed toward inhibition of glycolysis (Doherty and Cleveland 2013; Pelicano et al. 2006). Only recently has mitochondrial metabolism emerged as a chemotherapeutic target (Bhat et al. 2015; Weinberg and Chandel 2015). Most approaches attempt to inhibit mitochondrial metabolism in cancer cells. The observation that the antidiabetic drug metformin decreased the prevalence of certain types of cancer triggered an interest in the role of mitochondrial inhibition as a mechanism to suppress abnormal cell proliferation (Giovannucci et al. 2010; Libby et al. 2009). Although metformin decreases OXPHOS by inhibiting Complex I of the respiratory chain, metformin also inhibits the mammalian target of rapamycin (mTOR), interferes with folate metabolism, and activates AMP kinase (AMPK) (Jara and Lopez-Munoz 2015). Other approaches to inhibit mitochondrial metabolism in various cancer cell models include etomoxir to inhibit carnitine *O*-palmitoyltransferase 1 and consequent mitochondrial fatty acid oxidation (leukemia), tigecycline to inhibit mitochondrial protein translation (leukemia), glutaminase inhibitors (breast cancer, lymphoma), and the compound VLX600 to inhibit OXPHOS (colon cancer) (Samudio et al. 2010; Skrtic et al. 2011; Wang et al. 2010; Zhang et al. 2014). By contrast, other anticancer, antiproliferative strategies attempt to promote mitochondrial metabolism. For example, the pyruvate analog dichloroacetate activates pyruvate dehydrogenase to increase mitochondrial metabolism, which promotes cell killing in several cancer cell lines and in some in vivo models (Sutendra and Michelakis 2013).

5.3.4 VDAC Opening: A Metabolic Switch

The relative closure of VDAC by free tubulin in cancer cells and the broad metabolic consequences of VDAC opening make VDAC-tubulin interaction a novel pharmacological target to revert the Warburg phenotype. Antagonizing the constitutive inhibition of VDAC by free tubulin would be expected to increase mitochondrial metabolism and to have an anti-Warburg effect. Our group reported the first antagonist of the inhibitory effect of free tubulin on VDAC, the small molecule erastin (Maldonado et al. 2013). Erastin selectively induces non-apoptotic cell death in human cells engineered to harbor small T oncoprotein and the oncogenic allele of HRAS, v-Ha-ras Harvey rat sarcoma viral oncogene homologue RAS^{v12} (Dolma et al. 2003). Erastin non-apoptotic-induced cell death is blocked by antioxidants, such as α-tocopherol, butylated hydroxytoluene, and desferal, but not by pan-caspase inhibitors (Dolma et al. 2003). Other cell lines harboring the v-Ki-ras2 Kirsten rat sarcoma viral oncogene homologue (KRAS) and an activating V600E mutation in v-raf-murine sarcoma viral oncogene homologue B1 (BRAF) are moderately sensitive to erastin. Erastin is proposed to bind to VDAC2 and VDAC3, leading to oxidative stress and cell death in cells with activated RAS-RAF-MEK signaling (Yagoda et al. 2007).

Erastin in wild-type HepG2 cells and other cell lines promotes mitochondrial hyperpolarization and prevents depolarization induced by microtubule destabilizers. In addition, erastin added after microtubule destabilizers restores mitochondrial $\Delta\Psi$, indicating that erastin prevents and reverts the inhibitory effect of free tubulin on VDAC (Maldonado et al. 2013). Erastin also completely blocks the inhibitory effect of free tubulin on VDAC conductance of wild-type VDAC from HepG2 cells inserted into planar lipid bilayers. Erastin alone did not modify the voltage dependence of VDAC closure, indicating that the effect of erastin was specific for tubulin-dependent inhibition of VDAC (Maldonado et al. 2013). Following the identification of erastin as a VDAC-tubulin antagonist, we identified a group of "erastin-like" compounds using a high-throughput cell-based screening. These erastin-like compounds were selected based on their capability of hyperpolarizing mitochondria in the presence of microtubule destabilizers (DeHart et al. 2015).

5.3.5 VDAC Opening-Related Effects in Cancer Cells

VDAC opening leads to three main biological effects: increased mitochondrial metabolism, decreased glycolysis, and increased formation of reactive oxygen species (ROS). After VDAC opening, flux of pyruvate, fatty acids, and other metabolic substrates into mitochondria fuels the tricarboxylic acid cycle to produce NADH that enters the electron transport chain. Increased mitochondrial $\Delta\Psi$ and increased reduction of respiratory chain components lead to superoxide anion ($O_2^{\bullet-}$) generation (Chance et al. 1979; Suski et al. 2012). Quantitatively mitochondria are the most important source of ROS, with Complex III (Site III_{Qo}), Complex I (Site I_Q), and Complex II (Site II_F) being the main ROS-producing sites out of seven major mitochondrial sites (Chen et al. 2003; Quinlan et al. 2012; Tribble et al. 1988), (Skulachev 1996). $O_2^{\bullet-}$ formed at Complexes I and II is released to the matrix, whereas $O_2^{\bullet-}$ generated at Complex III is released in large part to the intermembrane space and hence to the cytosol through VDAC (Brand 2010; Han et al. 2003; Muller et al. 2004). $O_2^{\bullet-}$ is rapidly converted to H_2O_2 by superoxide dismutases located in the mitochondrial matrix (manganese-containing enzyme MnSOD or SOD2) and the cytosol (copper-and-zinc-containing enzyme Cu, ZnSOD, or SOD1) (Fridovich 1997). H_2O_2, the least reactive of ROS, diffuses across membranes and is a cell signaling molecule that does not necessarily disrupt redox homeostasis (Morgan et al. 2011; Veal et al. 2007). For example, H_2O_2 modulates the pro-survival HIF-1 and MAP/ERK, PI3K/akt/mTOR pathways that favor tumorigenesis and metastasis (Clerkin et al. 2008; Giles 2006; Ushio-Fukai and Nakamura 2008). Alternatively, H_2O_2 can accept an electron from free and loosely bound Fe^{2+} to form the highly reactive hydroxyl radical (OH•) by the Fenton reaction. O_2•⁻ and especially the highly reactive OH• are damaging for cells.

VDAC opening promotes mitochondrial ROS formation by increasing mitochondrial $\Delta\Psi$ and the reduction of the respiratory chain. Continued ROS production eventually overcomes the antioxidant capacity of cancer cells leading to cytotoxicity. Opening of VDAC by antagonism of the inhibitory effect of free tubulin on VDAC selectively affects cancer and other proliferating cells, since free tubulin is low and does not inhibit VDAC in differentiated cells (Maldonado et al. 2010; Maldonado and Lemasters 2012). In cancer cells, ROS can be cytostatic, favor tumor growth, or be cytotoxic (Marengo et al. 2016; Panieri and Santoro 2016; Sullivan and Chandel 2014). Although basal levels of ROS are higher in cancer cells compared to differentiated cells, these higher ROS levels are compensated by the higher content of scavenging enzymes and antioxidants, including glutathione-linked enzymes that reduce protein disulfide bonds, catalase that converts H_2O_2 to H_2O and O_2, and SODs (Liou and Storz 2010; Panieri and Santoro 2016; Sullivan and Chandel 2014; Venditti et al. 2013). Oxidative stress is reported to induce cancer cell cycle arrest, senescence, apoptosis, or necrosis (Liou and Storz 2010). Chemotherapeutic agents including cisplatin, adriamycin, the anthracyclines doxorubicin, epirubicin, and daunorubicin among others promote oxidative stress and depletion of the antioxidant capacity of tumor cells leading to a tumoricidal effect (Conklin 2004; Faber et al. 1995; Ladner et al. 1989; Weijl et al. 1998).

The effects of mitochondrially generated ROS on cellular structures depend on the specific ROS. The lifetimes of H_2O_2 and O_2•⁻ allow them to react both with mitochondria and extramitochondrial structures. By contrast, OH• is so reactive that its effects are almost completely restricted to mitochondria. Both O_2•⁻ and OH• inactivate mitochondrial proteins, including ATP synthase, NADH oxidase, and NADH dehydrogenase (Zhang et al. 1990). Beyond proteins, ROS damage mitochondrial DNA and lipids in the MIM. Cardiolipin, a MIM phospholipid rich in polyunsaturated fatty, is peroxidized by ROS, and peroxidized cardiolipin is considered an early event in apoptosis (Schenkel and Bakovic 2014). Cytosolic ROS, in turn, activate members of the MAPK family of serine/threonine kinases, especially c-Jun N-terminal kinase (JNK), the extracellular signal-regulated kinase (ERK 1/ERK 2), and p38 whose signaling can cause mitochondrial dysfunction (Kamata et al. 2005; Son et al. 2011).

5.3.6 A Metabolic Double Hit: Anti-Warburg Effect and Oxidative Stress

Heterogeneity of metabolism among cells within a tumor is a complicating factor for the success of cancer chemotherapy (Dang 2012; Eason and Sadanandam 2016; Gerlinger et al. 2012; Yun et al. 2012). However, nearly all cancer cells display some level of enhanced glycolysis, suggesting some degree of contribution of VDAC closure to suppression of mitochondrial metabolism (Griguer et al. 2005;

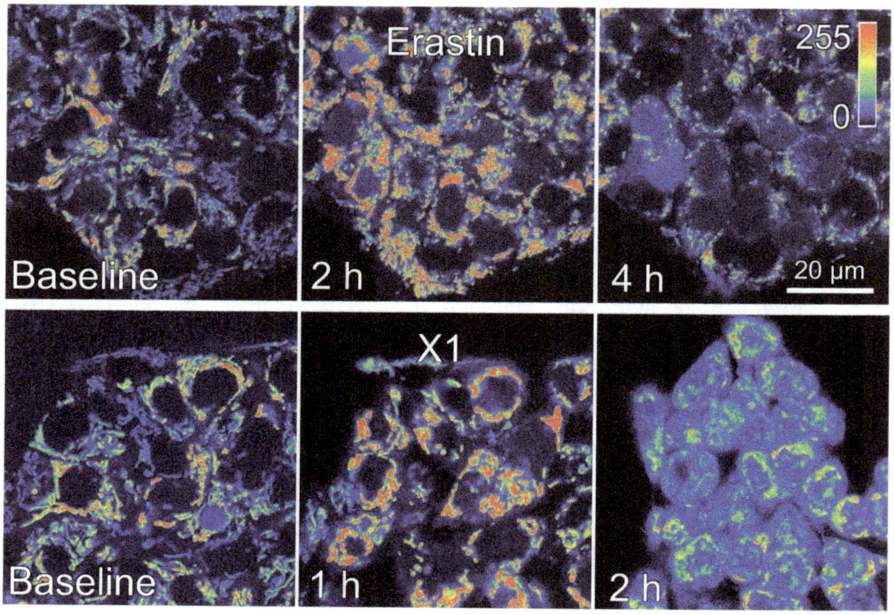

Fig. 5.2 Erastin and X1-dependent mitochondrial dysfunction. Initial mitochondrial hyperpolarization induced by erastin (*center upper panel*) and X1 (*center lower panel*) was followed by mitochondrial depolarization indicative of mitochondrial dysfunction (*right upper and lower panels*)

Guppy et al. 2002; Moreno-Sanchez et al. 2007; Scott et al. 2011). Antagonism of the inhibitory effect of tubulin on VDAC triggers two distinct and nearly simultaneous effects: (1) activation of OXPHOS with consequent decrease of glycolysis (anti-Warburg effect) and (2) an increase in ROS formation leading to oxidative stress. The antiproliferative effect of derepression of mitochondrial function (anti-Warburg effect) may be quantitatively more important in highly glycolytic tumors, whereas oxidative stress may cause tumoristatic and tumoricidal effects on a more broad population of cells.

The VDAC-tubulin antagonist erastin and erastin-like compounds cause mitochondrial hyperpolarization followed by mitochondrial depolarization indicative of mitochondrial dysfunction in human hepatocarcinoma cells (Fig. 5.2). The initial increase in $\Delta\Psi$ is just in advance of the increase in ROS generation, whereas subsequent JNK activation precedes mitochondrial dysfunction. A lead erastin-like compound identified by small molecule screening also decreases glycolysis as evidenced by a decrease in lactate release (DeHart 2015). The combination of reversal of Warburg metabolism and oxidative stress by the lead compound causes cell death to human hepatocarcinoma cell lines in culture and to xenografted Huh7 hepatocarcinoma cells (DeHart et al. 2015). Thus, erastin and lead erastin-like compound by causing "two hits" of anti-Warburg metabolism and promotion of oxidative stress represent a potential new class of cancer chemotherapeutic agents (Fig. 5.3).

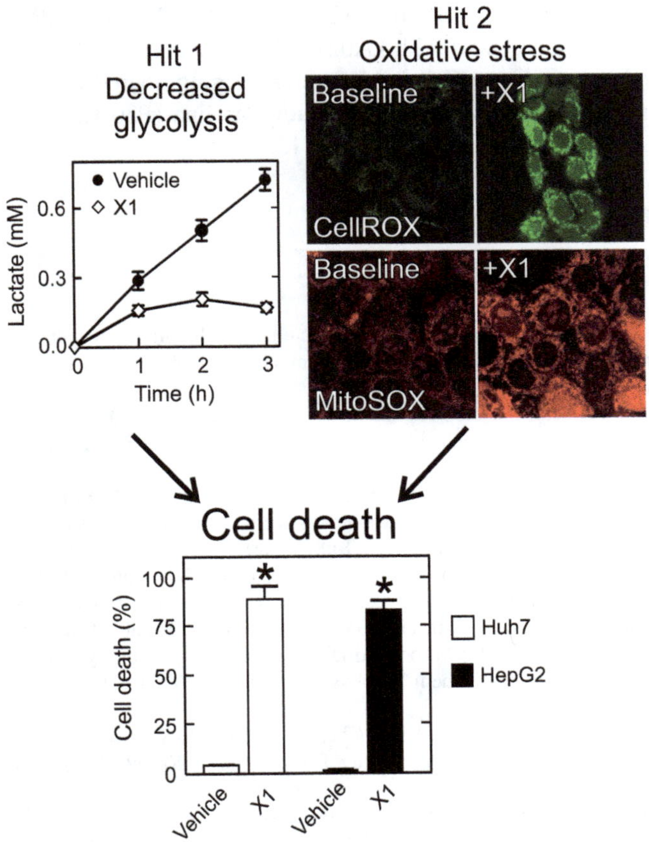

Fig. 5.3 Mechanisms to promote cell death after VDAC opening. X1 decreased lactate release by 80% in Huh7 cells (anti-Warburg effect, Hit 1). Lead compound X1 also increased fluorescence of the cellular ROS indicator CellROX *green* and the mitochondrial superoxide anion indicator MitoSOX *red* in Huh7 cells (oxidative stress, Hit 2). The two-hit mechanism led to over 90% cell death in Huh7 cells and over 80% in HepG2 cells

5.4 Concluding Remarks

VDAC-tubulin interaction in cancer cells is a global bioenergetic controller. Drug-induced VDAC opening increases mitochondrial metabolism and decreases glycolysis. Opening of the VDAC switch triggers two "hits" – an anti-Warburg effect that promotes a nonproliferative metabolic phenotype and an increase in ROS formation leading to mitochondrial dysfunction and cell death. ROS may be lethal for some cells and sublethal for others, whereas the anti-Warburg effect will decrease or stop cell proliferation. In summary, VDAC-tubulin is a new pharmacological target to turn a pro-proliferative into a nonproliferative phenotype and to induce oxidative death to cancer cells.

Funding R01CA184456, GM103542 and ACS 13-041-01-IRG to ENM, T32DK083262 to DND, and R21 AA021191, AA022815 and DK073336 to JJL.

Supported in part by the Cell & Molecular Imaging Shared Resource, Hollings Cancer Center, Medical University of South Carolina (P30 CA138313) and the Shared Instrumentation Grant S10 OD018113.

References

Al Jamal JA (2005) Involvement of porin N,N-dicyclohexylcarbodiimide-reactive domain in hexokinase binding to the outer mitochondrial membrane. Protein J 24(1):1–8. Available from: PM:15756812

Azoulay-Zohar H, Israelson A, Abu-Hamad S, Shoshan-Barmatz V (2004) In self-defence: hexokinase promotes voltage-dependent anion channel closure and prevents mitochondria-mediated apoptotic cell death. Biochem J 377(Pt 2):347–355

Baines CP, Song CX, Zheng YT, Wang GW, Zhang J, Wang OL, Guo Y, Bolli R, Cardwell EM, Ping P (2003) Protein kinase Cepsilon interacts with and inhibits the permeability transition pore in cardiac mitochondria. Circ Res 92(8):873–880

Beckner ME, Gobbel GT, Abounader R, Burovic F, Agostino NR, Laterra J, Pollack IF (2005) Glycolytic glioma cells with active glycogen synthase are sensitive to PTEN and inhibitors of PI3K and gluconeogenesis. Lab Investig 85(12):1457–1470. Available from: PM:16170333

Bera AK, Ghosh S, Das S (1995) Mitochondrial VDAC can be phosphorylated by cyclic AMP-dependent protein kinase. Biochem Biophys Res Commun 209(1):213–217. Available from: PM:7537039

Bhat TA, Kumar S, Chaudhary AK, Yadav N, Chandra D (2015) Restoration of mitochondria function as a target for cancer therapy. Drug Discov Today 20(5):635–643. Available from: PM:25766095

Blachly-Dyson E, Forte M (2001) VDAC channels. IUBMB Life 52(3–5):113–118

Bouzier AK, Voisin P, Goodwin R, Canioni P, Merle M (1998) Glucose and lactate metabolism in C6 glioma cells: evidence for the preferential utilization of lactate for cell oxidative metabolism. Dev Neurosci 20(4–5):331–338. Available from: PM:9778569

Brand MD (2005) The efficiency and plasticity of mitochondrial energy transduction. Biochem Soc Trans 33(Pt 5):897–904. Available from: PM:16246006

Brand MD (2010) The sites and topology of mitochondrial superoxide production. Exp Gerontol 45(7–8):466–472. Available from: PM:20064600

Cairns RA (2015) Drivers of the Warburg phenotype. Cancer J 21(2):56–61. Available from: PM:25815844

Caro P, Kishan AU, Norberg E, Stanley IA, Chapuy B, Ficarro SB, Polak K, Tondera D, Gounarides J, Yin H, Zhou F, Green MR, Chen L, Monti S, Marto JA, Shipp MA, Danial NN (2012) Metabolic signatures uncover distinct targets in molecular subsets of diffuse large B cell lymphoma. Cancer Cell 22(4):547–560. Available from: PM:23079663

Chance B, Sies H, Boveris A (1979) Hydroperoxide metabolism in mammalian organs. Physiol Rev 59:527–605

Chen Q, Vazquez EJ, Moghaddas S, Hoppel CL, Lesnefsky EJ (2003) Production of reactive oxygen species by mitochondria: central role of complex III. J Biol Chem 278(38):36027–36031

Chung WJ, Lyons SA, Nelson GM, Hamza H, Gladson CL, Gillespie GY, Sontheimer H (2005) Inhibition of cystine uptake disrupts the growth of primary brain tumors. J Neurosci 25(31):7101–7110. Available from: PM:16079392

Clavell LA, Gelber RD, Cohen HJ, Hitchcock-Bryan S, Cassady JR, Tarbell NJ, Blattner SR, Tantravahi R, Leavitt P, Sallan SE (1986) Four-agent induction and intensive asparaginase therapy for treatment of childhood acute lymphoblastic leukemia. N Engl J Med 315(11):657–663. Available from: PM:2943992

Clerkin JS, Naughton R, Quiney C, Cotter TG (2008) Mechanisms of ROS modulated cell survival during carcinogenesis. Cancer Lett 266(1):30–36. Available from: PM:18372105
Colombini M (1980) Structure and mode of action of a voltage dependent anion-selective channel (VDAC) located in the outer mitochondrial membrane. Ann NY Acad Sci 341:552–563. Available from: PM:6249159
Colombini M (2004) VDAC: the channel at the interface between mitochondria and the cytosol. Mol Cell Biochem 256–257(1–2):107–115
Colombini M (2012) VDAC structure, selectivity, and dynamics. Biochim Biophys Acta 1818(6):1457–1465. Available from: PM:22240010
Comerford SA, Huang Z, Du X, Wang Y, Cai L, Witkiewicz AK, Walters H, Tantawy MN, Fu A, Manning HC, Horton JD, Hammer RE, McKnight SL, Tu BP (2014) Acetate dependence of tumors. Cell 159(7):1591–1602. Available from: PM:25525877
Commisso C, Davidson SM, Soydaner-Azeloglu RG, Parker SJ, Kamphorst JJ, Hackett S, Grabocka E, Nofal M, Drebin JA, Thompson CB, Rabinowitz JD, Metallo CM, Vander Heiden MG, Bar-Sagi D (2013) Macropinocytosis of protein is an amino acid supply route in Ras-transformed cells. Nature 497(7451):633–637. Available from: PM:23665962
Conklin KA (2004) Chemotherapy-associated oxidative stress: impact on chemotherapeutic effectiveness. Integr Cancer Ther 3(4):294–300. Available from: PM:15523100
Dang CV (2012) Links between metabolism and cancer. Genes Dev 26(9):877–890. Available from: PM:22549953
Das S, Wong R, Rajapakse N, Murphy E, Steenbergen C (2008) Glycogen synthase kinase 3 inhibition slows mitochondrial adenine nucleotide transport and regulates voltage-dependent anion channel phosphorylation. Circ Res 103(9):983–991. Available from: PM:18802025
De Pinto V, Guarino F, Guarnera A, Messina A, Reina S, Tomasello FM, Palermo V, Mazzoni C (2010) Characterization of human VDAC isoforms: a peculiar function for VDAC3? Biochim Biophys Acta 1797(6–7):1268–1275. Available from: PM:20138821
DeBerardinis RJ, Cheng T (2010) Q's next: the diverse functions of glutamine in metabolism, cell biology and cancer. Oncogene 29(3):313–324. Available from: PM:19881548
DeBerardinis RJ, Sayed N, Ditsworth D, Thompson CB (2008) Brick by brick: metabolism and tumor cell growth. Curr Opin Genet Dev 18(1):54–61. Available from: PM:18387799
DeHart DN, Gooz M, Lemasters JJ, Maldonado EN (2015) Small anti-Warburg molecules kill hepatocarcinoma cells by opening voltage dependent anion channels and promoting mitochondrial oxidative stress. The Liver Meeting 2015, San Francisco. Ref Type: Abstract
Doherty JR, Cleveland JL (2013) Targeting lactate metabolism for cancer therapeutics. J Clin Invest 123(9):3685–3692. Available from: PM:23999443
Dolma S, Lessnick SL, Hahn WC, Stockwell BR (2003) Identification of genotype-selective antitumor agents using synthetic lethal chemical screening in engineered human tumor cells. Cancer Cell 3(3):285–296. Available from: PM:12676586
Eason K, Sadanandam A (2016) Molecular or metabolic reprograming: what triggers tumor subtypes? Cancer Res 76(18):5195–5200. Available from: PM:27635042
Faber M, Coudray C, Hida H, Mousseau M, Favier A (1995) Lipid peroxidation products, and vitamin and trace element status in patients with cancer before and after chemotherapy, including adriamycin. A preliminary study. Biol Trace Elem Res 47(1–3):117–123. Available from: PM:7779537
Fridovich I (1997) Superoxide anion radical (O2-.), superoxide dismutases, and related matters. J Biol Chem 272(30):18515–18517. Available from: PM:9228011
Gerlinger M, Rowan AJ, Horswell S, Larkin J, Endesfelder D, Gronroos E, Martinez P, Matthews N, Stewart A, Tarpey P, Varela I, Phillimore B, Begum S, McDonald NQ, Butler A, Jones D, Raine K, Latimer C, Santos CR, Nohadani M, Eklund AC, Spencer-Dene B, Clark G, Pickering L, Stamp G, Gore M, Szallasi Z, Downward J, Futreal PA, Swanton C (2012) Intratumor heterogeneity and branched evolution revealed by multiregion sequencing. N Engl J Med 366(10):883–892. Available from: PM:22397650

Giles GI (2006) The redox regulation of thiol dependent signaling pathways in cancer. Curr Pharm Des 12(34):4427–4443. Available from: PM:17168752

Gincel D, Silberberg SD, Shoshan-Barmatz V (2000) Modulation of the voltage-dependent anion channel (VDAC) by glutamate. J Bioenerg Biomembr 32(6):571–583. Available from: PM:15254371

Giovannucci E, Harlan DM, Archer MC, Bergenstal RM, Gapstur SM, Habel LA, Pollak M, Regensteiner JG, Yee D (2010) Diabetes and cancer: a consensus report. CA Cancer J Clin 60(4):207–221. Available from: PM:20554718.

Griguer CE, Oliva CR, Gillespie GY (2005) Glucose metabolism heterogeneity in human and mouse malignant glioma cell lines. J Neurooncol 74(2):123–133. Available from: PM:16193382

Guppy M, Leedman P, Zu X, Russell V (2002) Contribution by different fuels and metabolic pathways to the total ATP turnover of proliferating MCF-7 breast cancer cells. Biochem J 364(Pt 1):309–315. Available from: PM:11988105

Hahn AT, Jones JT, Meyer T (2009) Quantitative analysis of cell cycle phase durations and PC12 differentiation using fluorescent biosensors. Cell Cycle 8(7):1044–1052. Available from: PM:19270522

Han D, Antunes F, Canali R, Rettori D, Cadenas E (2003) Voltage-dependent anion channels control the release of the superoxide anion from mitochondria to cytosol. J Biol Chem 278(8):5557–5563. Available from: PM:12482755

Hiller S, Abramson J, Mannella C, Wagner G, Zeth K (2010) The 3D structures of VDAC represent a native conformation. Trends Biochem Sci 35(9):514–521. Available from: PM:20708406

Holmuhamedov E, Lemasters JJ (2009) Ethanol exposure decreases mitochondrial outer membrane permeability in cultured rat hepatocytes. Arch Biochem Biophys 481(2):226–233. Available from: PM:19014900

Huang H, Shah K, Bradbury NA, Li C, White C (2014) Mcl-1 promotes lung cancer cell migration by directly interacting with VDAC to increase mitochondrial Ca2+ uptake and reactive oxygen species generation. Cell Death Dis 5:e1482. Available from: PM:25341036

Jara JA, Lopez-Munoz R (2015) Metformin and cancer: between the bioenergetic disturbances and the antifolate activity. Pharmacol Res 101:102–108. Available from: PM:26277279

Kamata H, Honda S, Maeda S, Chang L, Hirata H, Karin M (2005) Reactive oxygen species promote TNFalpha-induced death and sustained JNK activation by inhibiting MAP kinase phosphatases. Cell 120(5):649–661

Keenan MM, Chi JT (2015) Alternative fuels for cancer cells. Cancer J 21(2):49–55. Available from: PM:25815843

Keibler MA, Wasylenko TM, Kelleher JK, Iliopoulos O, Vander Heiden MG, Stephanopoulos G (2016) Metabolic requirements for cancer cell proliferation. Cancer Metab 4:16. Available from: PM:27540483

Kennedy KM, Scarbrough PM, Ribeiro A, Richardson R, Yuan H, Sonveaux P, Landon CD, Chi JT, Pizzo S, Schroeder T, Dewhirst MW (2013) Catabolism of exogenous lactate reveals it as a legitimate metabolic substrate in breast cancer. PLoS One 8(9):e75154. Available from: PM:24069390

Kilburn DG, Lilly MD, Webb FC (1969) The energetics of mammalian cell growth. J Cell Sci 4(3):645–654. Available from: PM:5817088

Kreis W, Baker A, Ryan V, Bertasso A (1980) Effect of nutritional and enzymatic methionine deprivation upon human normal and malignant cells in tissue culture. Cancer Res 40(3):634–641. Available from: PM:6937240

Ladner C, Ehninger G, Gey KF, Clemens MR (1989) Effect of etoposide (VP16-213) on lipid peroxidation and antioxidant status in a high-dose radiochemotherapy regimen. Cancer Chemother Pharmacol 25(3):210–212. Available from: PM:2513140

Lee AC, Zizi M, Colombini M (1994) Beta-NADH decreases the permeability of the mitochondrial outer membrane to ADP by a factor of 6. J Biol Chem 269(49):30974–30980

Lemasters JJ, Holmuhamedov E (2006) Voltage-dependent anion channel (VDAC) as mitochondrial governator – thinking outside the box. Biochim Biophys Acta 1762(2):181–190. Available from: PM:16307870

Libby G, Donnelly LA, Donnan PT, Alessi DR, Morris AD, Evans JM (2009) New users of metformin are at low risk of incident cancer: a cohort study among people with type 2 diabetes. Diabetes Care 32(9):1620–1625. Available from: PM:19564453

Liberti MV, Locasale JW (2016) The Warburg effect: how does it benefit cancer cells? Trends Biochem Sci 41(3):211–218. Available from: PM:26778478

Liemburg-Apers DC, Schirris TJ, Russel FG, Willems PH, Koopman WJ (2015) Mitoenergetic dysfunction triggers a rapid compensatory increase in steady-state glucose flux. Biophys J 109(7):1372–1386. Available from: PM:26445438

Lim HY, Ho QS, Low J, Choolani M, Wong KP (2011) Respiratory competent mitochondria in human ovarian and peritoneal cancer. Mitochondrion 11(3):437–443. Available from: PM:21211574

Liou GY, Storz P (2010) Reactive oxygen species in cancer. Free Radic Res 44(5):479–496. Available from: PM:20370557

Locasale JW, Cantley LC (2010) Altered metabolism in cancer. BMC Biol 8:88. Available from: PM:20598111

Lunt SY, Vander Heiden MG (2011) Aerobic glycolysis: meeting the metabolic requirements of cell proliferation. Annu Rev Cell Dev Biol 27:441–464. Available from: PM:21985671

Maldonado EN, Lemasters JJ (2012) Warburg revisited: regulation of mitochondrial metabolism by voltage-dependent anion channels in cancer cells. J Pharmacol Exp Ther 342(3):637–641. Available from: PM:22700429

Maldonado EN, Lemasters JJ (2014) ATP/ADP ratio, the missed connection between mitochondria and the Warburg effect. Mitochondrion 19(Pt A):78–84. Available from: PM:25229666

Maldonado EN, Patnaik J, Mullins MR, Lemasters JJ (2010) Free tubulin modulates mitochondrial membrane potential in cancer cells. Cancer Res 70(24):10192–10201. Available from: PM:21159641

Maldonado EN, Sheldon KL, DeHart DN, Patnaik J, Manevich Y, Townsend DM, Bezrukov SM, Rostovtseva TK, Lemasters JJ (2013) Voltage-dependent anion channels modulate mitochondrial metabolism in cancer cells: regulation by free tubulin and erastin. J Biol Chem 288(17):11920–11929. Available from: PM:23471966

Maldonado EN, DeHart DN, Patnaik J, Klatt SC, Beck GM, Lemasters JJ (2016) ATP/ADP turnover and import of glycolytic ATP into mitochondria in cancer cells is independent of the adenine nucleotide translocator. J Biol Chem 291:19642. Available from: PM:27458020

Marengo B, Nitti M, Furfaro AL, Colla R, Ciucis CD, Marinari UM, Pronzato MA, Traverso N, Domenicotti C (2016) Redox homeostasis and cellular antioxidant systems: crucial players in cancer growth and therapy. Oxid Med Cell Longev 2016:6235641. Available from: PM:27418953

Mashimo T, Pichumani K, Vemireddy V, Hatanpaa KJ, Singh DK, Sirasanagandla S, Nannepaga S, Piccirillo SG, Kovacs Z, Foong C, Huang Z, Barnett S, Mickey BE, DeBerardinis RJ, Tu BP, Maher EA, Bachoo RM (2014) Acetate is a bioenergetic substrate for human glioblastoma and brain metastases. Cell 159(7):1603–1614. Available from: PM:25525878

Mathupala SP, Ko YH, Pedersen PL (2010) The pivotal roles of mitochondria in cancer: Warburg and beyond and encouraging prospects for effective therapies. Biochim Biophys Acta 1797(6–7):1225–1230. Available from: PM:20381449

Meiser J, Vazquez A (2016) Give it or take it: the flux of one-carbon in cancer cells. FEBS J 283:3695. Available from: PM:27042806

Mor I, Cheung EC, Vousden KH (2011) Control of glycolysis through regulation of PFK1: old friends and recent additions. Cold Spring Harb Symp Quant Biol 76:211–216. Available from: PM:22096029

Moreno-Sanchez R, Rodriguez-Enriquez S, Marin-Hernandez A, Saavedra E (2007) Energy metabolism in tumor cells. FEBS J 274(6):1393–1418. Available from: PM:17302740

Moreno-Sanchez R, Marin-Hernandez A, Saavedra E, Pardo JP, Ralph SJ, Rodriguez-Enriquez S (2014) Who controls the ATP supply in cancer cells? Biochemistry lessons to understand cancer energy metabolism. Int J Biochem Cell Biol 50:10–23. Available from: PM:24513530

Morgan B, Sobotta MC, Dick TP (2011) Measuring E(GSH) and H2O2 with roGFP2-based redox probes. Free Radic Biol Med 51(11):1943–1951. Available from: PM:21964034

Muller FL, Liu Y, Van RH (2004) Complex III releases superoxide to both sides of the inner mitochondrial membrane. J Biol Chem. 279(47):49064–49073. Available from: PM:15317809

Nakashima RA, Paggi MG, Pedersen PL (1984) Contributions of glycolysis and oxidative phosphorylation to adenosine 5′-triphosphate production in AS-30D hepatoma cells. Cancer Res 44(12 Pt 1):5702–5706

Nakashima RA, Paggi MG, Scott LJ, Pedersen PL (1988) Purification and characterization of a bindable form of mitochondrial bound hexokinase from the highly glycolytic AS-30D rat hepatoma cell line. Cancer Res 48(4):913–919. Available from: PM:3338084

Palmieri F, Pierri CL (2010) Mitochondrial metabolite transport. Essays Biochem 47:37–52. Available from: PM:20533899

Panieri E, Santoro MM (2016) ROS homeostasis and metabolism: a dangerous liason in cancer cells. Cell Death Dis 7(6):e2253. Available from: PM:27277675

Pastorino JG, Hoek JB (2003) Hexokinase II: the integration of energy metabolism and control of apoptosis. Curr Med Chem 10(16):1535–1551

Pedersen PL (1978) Tumor mitochondria and the bioenergetics of cancer cells. Prog Exp Tumor Res 22:190–274. Available from: PM:149996

Pelicano H, Martin DS, Xu RH, Huang P (2006) Glycolysis inhibition for anticancer treatment. Oncogene 25(34):4633–4646. Available from: PM:16892078

Quinlan CL, Orr AL, Perevoshchikova IV, Treberg JR, Ackrell BA, Brand MD (2012) Mitochondrial complex II can generate reactive oxygen species at high rates in both the forward and reverse reactions. J Biol Chem 287(32):27255–27264. Available from: PM:22689576

Rich PR (2003) The molecular machinery of Keilin's respiratory chain. Biochem Soc Trans 31(Pt 6):1095–1105. Available from: PM:14641005

Rich PR, Marechal A (2010) The mitochondrial respiratory chain. Essays Biochem 47:1–23. Available from: PM:20533897

Robinson GL, Dinsdale D, MacFarlane M, Cain K (2012) Switching from aerobic glycolysis to oxidative phosphorylation modulates the sensitivity of mantle cell lymphoma cells to TRAIL. Oncogene 31(48):4996–5006. Available from: PM:22310286

Rodriguez-Enriquez S, Carreno-Fuentes L, Gallardo-Perez JC, Saavedra E, Quezada H, Vega A, Marin-Hernandez A, Olin-Sandoval V, Torres-Marquez ME, Moreno-Sanchez R (2010) Oxidative phosphorylation is impaired by prolonged hypoxia in breast and possibly in cervix carcinoma. Int J Biochem Cell Biol 42(10):1744–1751. Available from: PM:20654728

Rostovtseva TK, Antonsson B, Suzuki M, Youle RJ, Colombini M, Bezrukov SM (2004) Bid, but not Bax, regulates VDAC channels. J Biol Chem 279(14):13575–13583

Rostovtseva TK, Sheldon KL, Hassanzadeh E, Monge C, Saks V, Bezrukov SM, Sackett DL (2008) Tubulin binding blocks mitochondrial voltage-dependent anion channel and regulates respiration. Proc Natl Acad Sci USA 105(48):18746–18751. Available from: PM:19033201

Sampson MJ, Lovell RS, Craigen WJ (1997) The murine voltage-dependent anion channel gene family. Conserved structure and function. J Biol Chem 272(30):18966–18973. Available from: PM:9228078

Sampson MJ, Decker WK, Beaudet AL, Ruitenbeek W, Armstrong D, Hicks MJ, Craigen WJ (2001) Immotile sperm and infertility in mice lacking mitochondrial voltage-dependent anion channel type 3. J Biol Chem 276(42):39206–39212. Available from: PM:11507092

Samudio I, Harmancey R, Fiegl M, Kantarjian H, Konopleva M, Korchin B, Kaluarachchi K, Bornmann W, Duvvuri S, Taegtmeyer H, Andreeff M (2010) Pharmacologic inhibition of fatty acid oxidation sensitizes human leukemia cells to apoptosis induction. J Clin Invest 120(1):142–156. Available from: PM:20038799

Schenkel LC, Bakovic M (2014) Formation and regulation of mitochondrial membranes. Int J Cell Biol 2014:709828. Available from: PM:24578708

Schredelseker J, Paz A, Lopez CJ, Altenbach C, Leung CS, Drexler MK, Chen JN, Hubbell WL, Abramson J (2014) High resolution structure and double electron-electron resonance of the zebrafish voltage-dependent anion channel 2 reveal an oligomeric population. J Biol Chem 289(18):12566–12577. Available from: PM:24627492

Schwenke WD, Soboll S, Seitz HJ, Sies H (1981) Mitochondrial and cytosolic ATP/ADP ratios in rat liver in vivo. Biochem J 200(2):405–408. Available from: PM:7340839

Scott L, Lamb J, Smith S, Wheatley DN (2000) Single amino acid (arginine) deprivation: rapid and selective death of cultured transformed and malignant cells. Br J Cancer 83(6):800–810. Available from: PM:10952786

Scott DA, Richardson AD, Filipp FV, Knutzen CA, Chiang GG, Ronai ZA, Osterman AL, Smith JW (2011) Comparative metabolic flux profiling of melanoma cell lines: beyond the Warburg effect. J Biol Chem 286(49):42626–42634. Available from: PM:21998308

Sheen JH, Zoncu R, Kim D, Sabatini DM (2011) Defective regulation of autophagy upon leucine deprivation reveals a targetable liability of human melanoma cells in vitro and in vivo. Cancer Cell 19(5):613–628. Available from: PM:21575862

Singleterry J, Sreedhar A, Zhao Y (2014) Components of cancer metabolism and therapeutic interventions. Mitochondrion 17C:50–55. Available from: PM:24910195

Skrtic M, Sriskanthadevan S, Jhas B, Gebbia M, Wang X, Wang Z, Hurren R, Jitkova Y, Gronda M, Maclean N, Lai CK, Eberhard Y, Bartoszko J, Spagnuolo P, Rutledge AC, Datti A, Ketela T, Moffat J, Robinson BH, Cameron JH, Wrana J, Eaves CJ, Minden MD, Wang JC, Dick JE, Humphries K, Nislow C, Giaever G, Schimmer AD (2011) Inhibition of mitochondrial translation as a therapeutic strategy for human acute myeloid leukemia. Cancer Cell 20(5):674–688. Available from: PM:22094260

Skulachev VP (1996) Role of uncoupled and non-coupled oxidations in maintenance of safely low levels of oxygen and its one-electron reductants. Q Rev Biophys 29(2):169–202. Available from: PM:8870073

Smolkova K, Bellance N, Scandurra F, Genot E, Gnaiger E, Plecita-Hlavata L, Jezek P, Rossignol R (2010) Mitochondrial bioenergetic adaptations of breast cancer cells to aglycemia and hypoxia. J Bioenerg Biomembr 42(1):55–67. Available from: PM:20084539

Son Y, Cheong YK, Kim NH, Chung HT, Kang DG, Pae HO (2011) Mitogen-activated protein kinases and reactive oxygen species: how can ROS activate MAPK pathways? J Signal Transduct 2011:792639. Available from: PM:21637379

Song J, Colombini M (1996) Indications of a common folding pattern for VDAC channels from all sources. J Bioenerg Biomembr 28(2):153–161. Available from: PM:9132414

Sonveaux P, Vegran F, Schroeder T, Wergin MC, Verrax J, Rabbani ZN, De Saedeleer CJ, Kennedy KM, Diepart C, Jordan BF, Kelley MJ, Gallez B, Wahl ML, Feron O, Dewhirst MW (2008) Targeting lactate-fueled respiration selectively kills hypoxic tumor cells in mice. J Clin Invest 118(12):3930–3942. Available from: PM:19033663

Sullivan LB, Chandel NS (2014) Mitochondrial reactive oxygen species and cancer. Cancer Metab 2:17. Available from: PM:25671107

Suski JM, Lebiedzinska M, Bonora M, Pinton P, Duszynski J, Wieckowski MR (2012) Relation between mitochondrial membrane potential and ROS formation. Methods Mol Biol 810:183–205. Available from: PM:22057568

Sutendra G, Michelakis ED (2013) Pyruvate dehydrogenase kinase as a novel therapeutic target in oncology. Front Oncol 3:38. Available from: PM:23471124

Timohhina N, Guzun R, Tepp K, Monge C, Varikmaa M, Vija H, Sikk P, Kaambre T, Sackett D, Saks V (2009) Direct measurement of energy fluxes from mitochondria into cytoplasm in permeabilized cardiac cells in situ: some evidence for mitochondrial interactosome. J Bioenerg Biomembr 41(3):259–275. Available from: PM:19597977

Tribble DL, Jones DP, Edmondson DE (1988) Effect of hypoxia on tert-butylhydroperoxide-induced oxidative injury in hepatocytes. Molec Pharmacol 34:413–420

Tsujimoto Y, Shimizu S (2000) VDAC regulation by the Bcl-2 family of proteins. Cell Death Differ 7(12):1174–1181. Available from: PM:11175254

Ujwal R, Cascio D, Colletier JP, Faham S, Zhang J, Toro L, Ping P, Abramson J (2008) The crystal structure of mouse VDAC1 at 2.3 A resolution reveals mechanistic insights into metabolite gating. Proc Natl Acad Sci USA 105(46):17742–17747. Available from: PM:18988731

Ushio-Fukai M, Nakamura Y (2008) Reactive oxygen species and angiogenesis: NADPH oxidase as target for cancer therapy. Cancer Lett 266(1):37–52. Available from: PM:18406051

Vander Heiden MG, Chandel NS, Li XX, Schumacker PT, Colombini M, Thompson CB (2000) Outer mitochondrial membrane permeability can regulate coupled respiration and cell survival. Proc Natl Acad Sci USA 97(9):4666–4671

Vander Heiden MG, Li XX, Gottleib E, Hill RB, Thompson CB, Colombini M (2001) Bcl-xL promotes the open configuration of the voltage-dependent anion channel and metabolite passage through the outer mitochondrial membrane. J Biol Chem 276(22):19414–19419

Veal EA, Day AM, Morgan BA (2007) Hydrogen peroxide sensing and signaling. Mol Cell 26(1):1–14. Available from: PM:17434122

Venditti P, Di SL, Di MS (2013) Mitochondrial metabolism of reactive oxygen species. Mitochondrion 13(2):71–82. Available from: PM:23376030

Walker JE (2013) The ATP synthase: the understood, the uncertain and the unknown. Biochem Soc Trans 41(1):1–16. Available from: PM:23356252

Wang JB, Erickson JW, Fuji R, Ramachandran S, Gao P, Dinavahi R, Wilson KF, Ambrosio AL, Dias SM, Dang CV, Cerione RA (2010) Targeting mitochondrial glutaminase activity inhibits oncogenic transformation. Cancer Cell 18(3):207–219. Available from: PM:20832749

Warburg O (1956) On the origin of cancer cells. Science 123(3191):309–314. Available from: PM:13298683

Warburg O, Wind F, Negelein E (1927) The metabolism of tumors in the body. J Gen Physiol 8(6):519–530. Available from: PM:19872213

Weijl NI, Hopman GD, Wipkink-Bakker A, Lentjes EG, Berger HM, Cleton FJ, Osanto S (1998) Cisplatin combination chemotherapy induces a fall in plasma antioxidants of cancer patients. Ann Oncol 9(12):1331–1337. Available from: PM:9932164

Weinberg SE, Chandel NS (2015) Targeting mitochondria metabolism for cancer therapy. Nat Chem Biol 11(1):9–15. Available from: PM:25517383

Weinhouse S (1956) On respiratory impairment in cancer cells. Science 124(3215):267–269. Available from: PM:13351638

Wikstrom M, Sharma V, Kaila VR, Hosler JP, Hummer G (2015) New perspectives on proton pumping in cellular respiration. Chem Rev 115(5):2196–2221. Available from: PM:25694135

Wolf A, Agnihotri S, Micallef J, Mukherjee J, Sabha N, Cairns R, Hawkins C, Guha A (2011) Hexokinase 2 is a key mediator of aerobic glycolysis and promotes tumor growth in human glioblastoma multiforme. J Exp Med 208(2):313–326. Available from: PM:21242296

Yagoda N, von Rechenberg RM, Zaganjor E, Bauer AJ, Yang WS, Fridman DJ, Wolpaw AJ, Smukste I, Peltier JM, Boniface JJ, Smith R, Lessnick SL, Sahasrabudhe S, Stockwell BR (2007) RAS-RAF-MEK-dependent oxidative cell death involving voltage-dependent anion channels. Nature 447(7146):864–868. Available from: PM:17568748

Yun J, Johnson JL, Hanigan CL, Locasale JW (2012) Interactions between epigenetics and metabolism in cancers. Front Oncol 2:163. Available from: PM:23162793

Zhang Y, Marcillat O, Giulivi C, Ernster L, Davies KJ (1990) The oxidative inactivation of mitochondrial electron transport chain components and ATPase. J Biol Chem 265(27):16330–16336. Available from: PM:2168888

Zhang X, Fryknas M, Hernlund E, Fayad W, De MA, Olofsson MH, Gogvadze V, Dang L, Pahlman S, Schughart LA, Rickardson L, D'Arcy P, Gullbo J, Nygren P, Larsson R, Linder S (2014) Induction of mitochondrial dysfunction as a strategy for targeting tumour cells in metabolically compromised microenvironments. Nat Commun 5:3295. Available from: PM:24548894

Zhu A, Lee D, Shim H (2011) Metabolic positron emission tomography imaging in cancer detection and therapy response. Semin Oncol 38(1):55–69. Available from: PM:21362516

Zizi M, Forte M, Blachly-Dyson E, Colombini M (1994) NADH regulates the gating of VDAC, the mitochondrial outer membrane channel. J Biol Chem 269(3):1614–1616. Available from: PM:7507479

Chapter 6
An Assessment of How VDAC Structures Have Impacted Our Understanding of Their Function

Lucie Bergdoll, Michael Grabe, and Jeff Abramson

6.1 Introduction

Mitochondria are the control centers for respiration in all eukaryotic cells producing the vast majority of the universal cellular energy currency, ATP. Efficient exchange of anions, cations, and metabolites between the cytoplasm and the intermembrane space (IMS) of the mitochondria is essential for cellular homeostasis, and this exchange is mediated by the most abundant protein in the outer mitochondrial membrane – the voltage-dependent anion channel (VDAC) (Schein et al. 1976; Rostovtseva and Colombini 1997). Given its role as the primary conduit between the cytoplasm and the mitochondria, VDAC represents an essential cog in the mitochondrial machine's capabilities of modulating mitochondrial activity.

In mammals, there are three VDAC isoforms – VDAC1, VDAC2, and VDAC3 – which have a high degree of sequence similarity (~80%) and wide and overlapping tissue distribution (Fig. 6.1). All isoforms have the ability to transport metabolites and ions (Craigen and Graham 2008); however, they have distinct physiological roles.

L. Bergdoll
Department of Physiology, David Geffen School of Medicine, University of California, Los Angeles, CA, USA

M. Grabe
Cardiovascular Research Institute, Department of Pharmaceutical Chemistry, University of California, San Francisco, CA, USA

J. Abramson (✉)
Department of Physiology, David Geffen School of Medicine, University of California, Los Angeles, CA, USA

Institute for Stem Cell Biology and Regenerative Medicine (inStem), National Centre for Biological Sciences–Tata Institute of Fundamental Research, Bellary Road, Bangalore 560065, Karnataka, India
e-mail: JAbramson@mednet.ucla.edu

T.K. Rostovtseva (ed.), *Molecular Basis for Mitochondrial Signaling*, Biological and Medical Physics, Biomedical Engineering, DOI 10.1007/978-3-319-55539-3_6

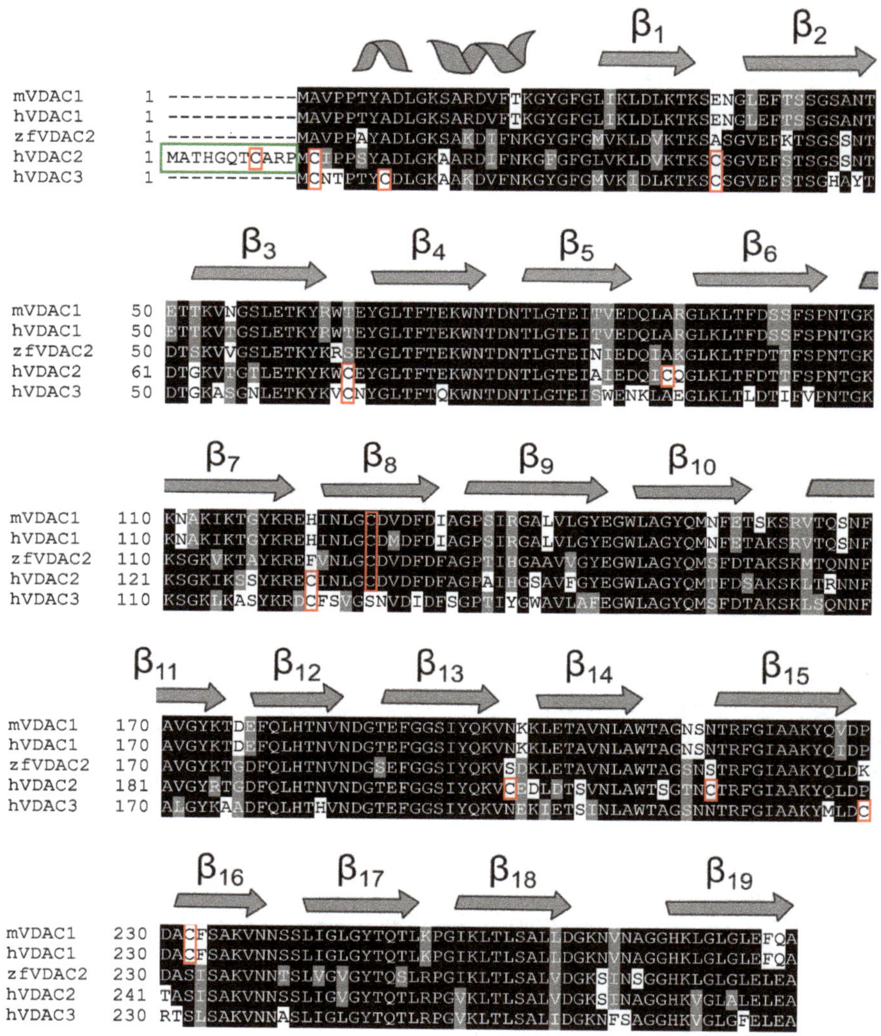

Fig. 6.1 Comparison of VDAC isoforms. Sequence alignment of several VDAC isoforms: mVDAC1 (Q60932), hVDAC1 (P217796), zfVDAC2 (Q8AWD0), hVDAC2 (P45880), and hVDAC3 (Q9Y277). The secondary structure elements are represented on top of the sequence for reference. The cysteine residues are shown in *red boxes*, and the N-terminal extension of 11 amino acids of hVDAC2 is shown in *green*

VDAC1 is the prototypical isoform common to all Eukaryotes and by far the most well characterized (Yamamoto et al. 2006). It is responsible for most of the metabolite transport across the OMM. The human VDAC2 isoform has a unique N-terminal extension of 11 residues and contains 7 additional cysteines compared to hVDAC1 (shown in red in Fig. 6.1). The functional significances of these alterations remain unknown as they have very similar ion and metabolite transport activity. From a physiological standpoint, VDAC2 has a number of unique properties including favorable calcium transport (Shimizu et al. 2015) and forming complexes with proteins of the Bcl2 family (Roy et al. 2009). VDAC3 is the least abundant isoform, and very little is known about its function. It is noteworthy that while VDAC1 is the most widely expressed isoform, knockout mice lacking VDAC1 display only a mild phenotype. On the contrary, VDAC2 knockout mice are embryonically lethal (Cheng et al. 2003) and knockout mice of VDAC3 lead to male sterility (Sampson et al. 2001).

In 2008, after three decades of biophysical and biochemical characterization, the structure of VDAC1 was resolved by three independent groups using NMR and X-ray crystallography (Bayrhuber et al. 2008; Hiller et al. 2008; Ujwal et al. 2008). Protein structure is a valuable tool for interpreting functional data and driving new experiments to test the proteins' functional properties. Since publishing the structure of murine VDAC1 (Ujwal et al. 2008), its atomic blueprint has been prodded by a large number of cellular, biochemical, and biophysical studies (in excess of 300). In this chapter, we will provide an overview of our current structural knowledge of VDACs and discuss the progress in the field since the structural models have been released. We will also highlight the remaining structural challenges and questions that have yet to be answered.

6.2 The Structure

In 2008, the first structures of VDAC1 were solved by three independent groups: two structures of hVDAC1 – one by NMR (Hiller et al. 2008) and the other using a combination of NMR and X-ray crystallography (Bayrhuber et al. 2008) – and a high-resolution crystal structure of murine VDAC1 at 2.3 Å (mVDAC1) (Ujwal et al. 2008). Murine and human VDAC1 have nearly identical sequences differing by only two conservative amino acid substitutions. The overall structure forms a β-barrel composed of 19 β-strands forming a large pore of 27 Å, with an overall height of 40 Å (Fig. 6.2). As anticipated, the strands are arranged in an antiparallel fashion with the exception of strands 1 and 19, which associate in a parallel manner. The mVDAC1 structure established the position of an N-terminal distorted α-helical segment located approximately halfway down the pore held by strong hydrophobic interactions with β-strand 8 through 18 of the barrel wall. The secondary structure coincides nicely with early circular dichroism studies that predicted a protein composed of both α-helix and β-sheet segments (Shanmugavadivu et al. 2007). These VDAC1 structures, with an odd number of strands, represented a new membrane protein fold since all other known β-barrel membrane proteins are composed of even number of strands (Zeth 2010). In addition, VDAC is one of the two mammalian β-barrel membrane protein structures solved to date with the lymphocyte perforin protein from *Mus musculus* (Law et al. 2010) being the other. By far, the vast majority of β-barrel membrane protein structures are from prokaryotic origin (http://blanco.biomol.uci.edu/mpstruc/).

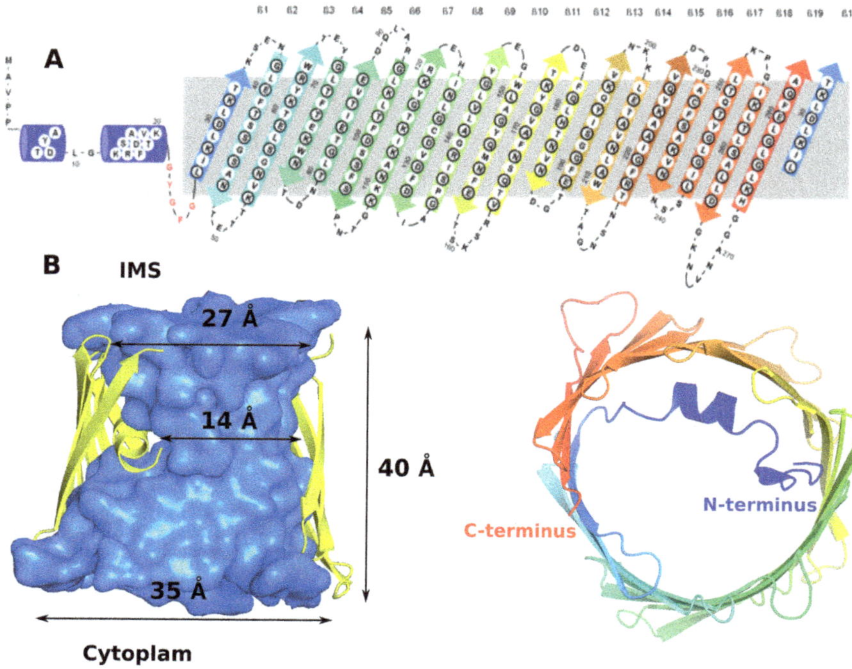

Fig. 6.2 Structural overview of mVDAC1. (**a**) Topology of the primary sequence. (**b**) Cartoon representation of mVDAC1 based on PDB file 3EMN. The left panel is the view from the membrane with β-Strands 3–7 removed for clarity, and the right panel is the top down view from the inter-membrane space (Adapted from Ujwal et al. (2008))

In 2014, a structure of the second isoform from zebrafish (zfVDAC2) was solved at 2.8 Å resolution by X-ray crystallography (Schredelseker et al. 2014). A structural alignment between zfVDAC2 (PDB accession code 4BUM) and mVDAC1 (PDB accession code 3EMN) reveals a small Cα root mean square deviation of 1 Å (between all residues), underscoring their high degree of structural similarity and explaining their similar functional properties (Fig. 6.3). However, it is noteworthy that some of the most striking sequence differences between VDAC1 and VDAC2 discussed previously – namely, the 11 amino acid N-terminal extension and 7 additional cysteines present in mammalian VDAC2 – are not present in zfVDAC2. In this regard, zfVDAC2 has a sequence more similar to VDAC1 than many other VDAC2 subtypes. Nevertheless, a rescue study using zfVDAC2 to rescue the phenotype of VDAC2$^{-/-}$ mouse embryonic fibroblast unambiguously demonstrated that zfVDAC2 had the same functional properties as mammalian VDAC2 (Naghdi et al. 2015). Currently, there are no representative structures from isoform 3 or mammalian isoform 2. The structural difficulties with these later two subtypes are mainly due to difficulties in refolding and obtaining sufficient quantities for structural studies.

The structures of both VDAC1 (mVDAC1 and hVDAC1) and zfVDAC2 isoforms have a glutamate at position 73 (E73), which is facing toward the lipidic milieu (Fig. 6.4). Charged side chains are not frequently observed in such hydrophobic

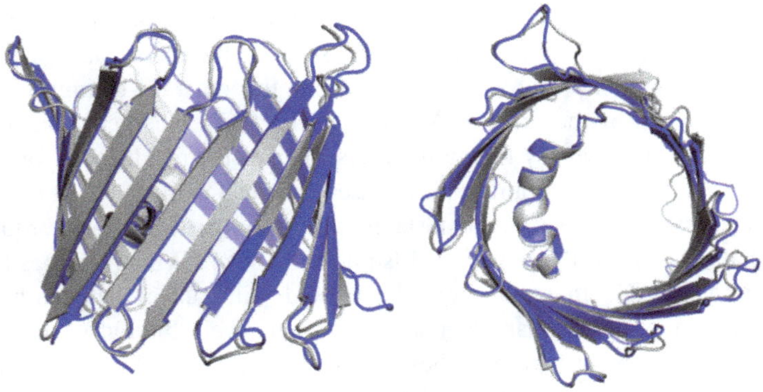

Fig. 6.3 Superimposition of the crystal structures of mVDAC1 (PDB code 3EMN, colored in *gray*) and zfVDAC2 (PDB code 4BUM, colored in *blue*) present a Cα root mean square deviation of 0.98 Å. *Left panel*: view from the membrane; *right panel*: view from the cytosol

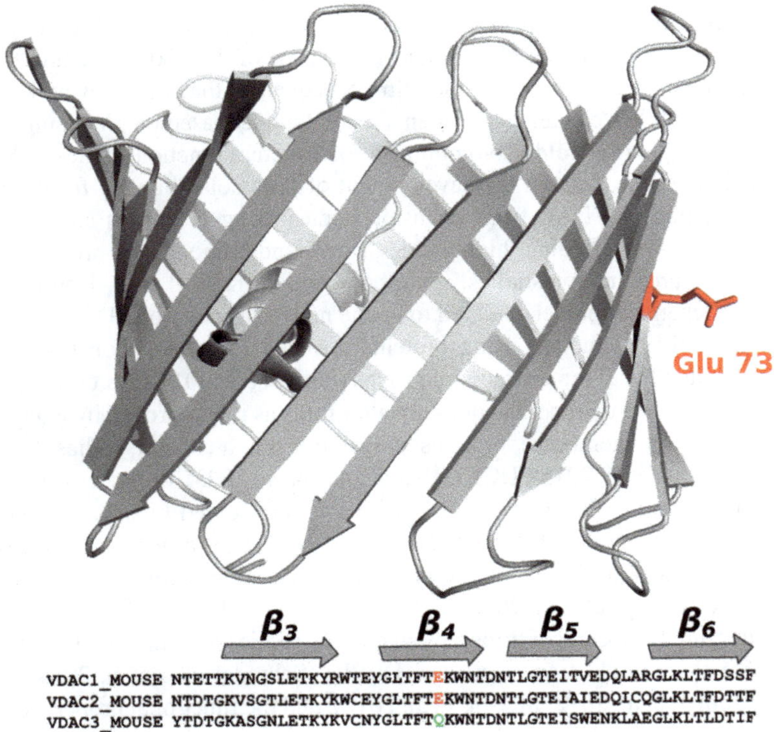

Fig. 6.4 Glutamate 73 is a charged side chain residue pointing toward the membrane. *Bottom*: Sequence alignment of the three mouse VDAC isoforms centered on glutamate 73. VDAC1 and VDAC2 possess a glutamate (*E*) in position 73, whereas VDAC3 possesses a glutamine (*Q*)

environments, but when present they often have a specific functional role. In the case of the transporter LeuT, the basic residue K288 faces the membrane environment, potentially affecting the rate of transition from the outward-facing state to the inward-facing state of the protein (Mondal et al. 2013). Charged residues in the membrane can also serve key functional roles, such as the residue D61 on the rotor of the F_1-F_O ATPase, which binds and releases protons (Valiyaveetil et al. 2002) as they move across the inner mitochondrial membrane imparting rotary torque on the gamma shaft to drive ATP production (Elston et al. 1998)[1]. What role does E73 play in regulating VDAC? Interestingly, cells treated with dicyclohexylcarbodiimide (DCCD), a compound that has been shown to exclusively bind at position E73 in VDAC1, prevent the formation of hexokinase-VDAC1 complex (De Pinto et al. 1993). E73 is also implicated in calcium binding (Ge et al. 2016; Israelson et al. 2007) and calcium-mediated oligomerization (Keinan et al. 2013). The precise functional roles of E73 are not fully understood; however, the third isoform – VDAC3 – has a glutamine at this position (Fig. 6.4) likely leading to isoform-specific function of VDAC3.

Like most high-resolution structures, the structures of VDACs answered many questions regarding their function, but many more questions arose. The most striking one revolved around the physiological relevance of the 2008 structures. It was originally hypothesized, based on the primary sequence, that VDAC would adopt a 19 β-strand fold (Forte et al. 1987); then a second model based on biochemical data suggested a 13 β-strand fold (Colombini 2009). Structure function studies on VDAC were made possible through the development of protocols allowing for the large-scale production of VDAC from inclusion bodies using *E. coli* as an expression host. After in vitro refolding, high yields of pure and homogenous protein, necessary for structure-function studies, could be obtained (Koppel et al. 1998). The 13 β-strand biochemical model, claimed to be the native structure of VDAC in opposition to the solved structure using proteins refolded from inclusion bodies, was strongly refuted by all three structural groups (Hiller et al. 2010). Refolding protein is a commonly used method that does not alter folds as it was proven by many additional studies on membrane proteins prepared from inclusion bodies (such as TOM40 (Kuszak et al. 2015), UCP (Jaburek and Garlid 2003), or Bacteriorhodopsin (Popot et al. 1987)). Furthermore, the structure were solved by solution-NMR in detergent micelles (Hiller et al. 2008; Bayrhuber et al. 2008) as well as by detergent- (Bayrhuber et al. 2008) and lipidic-based crystallization techniques (Ujwal and Abramson 2012; Ujwal et al. 2008) yielding the same 19 β-strand fold. The validity of the structure has been confirmed by numerous studies and methodology including double electron-electron resonance (DEER) (Schredelseker et al. 2014), cross-linking (Teijido et al. 2012; Mertins et al. 2012), and the functional properties of the structure have been confirmed by molecular dynamics simulations and continuum electrostatics calculations – both of which agree well with the known physiological

[1] For a comprehensive list of all residues, on all multipass membrane proteins predicted to be electrostatically destabilizing in the membrane, please see Ref. Marcoline et al. (2015).

ion selectivity values, single-channel conductance values, and ATP permeation rates (Choudhary et al. 2014; Choudhary et al. 2010). To the best of our knowledge, there has not been a single study demonstrating that the 19 β-strand fold is incorrect.

6.3 Conduction Properties and Topology of VDAC

VDAC facilitates the flow of ions and metabolites through the mitochondria. Given its large pore size as revealed in the structures, it is not surprising that it exhibits a high conductance of 0.45–0.58 nS in 0.1 M KCl (Colombini 1989). VDAC's functional activity is routinely monitored electrophysiologically using planar bilayer systems. At low voltage, VDAC adopts an open state allowing ion permeation, with a slight anion selectivity of 1.7–1.9 chloride to potassium ions in a 1.0–0.1 asymmetric gradient (Colombini 1989). A network of charged residues in the pore creating an electric field determines the ion selectivity of VDAC; at high salt concentration the charged residues get shielded and the selectivity decreases (Krammer et al. 2011), the channel is therefore more anion selective at low salt concentration. VDAC is also responsible for metabolite passage, at a flow of 2×10^6 ATP/s under saturating ATP conditions and an estimated 10,000 ATP/s under physiological values (at 1 mM ATP) (Rostovtseva and Colombini 1997). When the voltage is increased, either in the positive or negative direction, the conductance drops to ~0.5 of the open state, corresponding to a closing of the channel, resulting in a bell-shaped current voltage dependence (Fig. 6.5). The closed state displays a negligible metabolite flux but is still permeable to ions and becomes cation selective. There appears to be a single open state; however, the closed configuration is not unique and displays a number of low-conductance states.

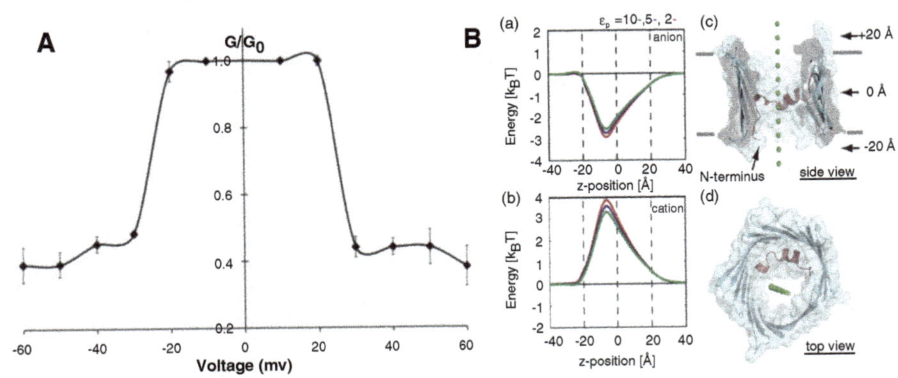

Fig. 6.5 Conduction properties of VDAC. (**A**) Conductance profile of mVDAC1 shows a *bell-shaped curve* with high conductance near 0 mV but reduced conductance at values below −30 mV and above +30 mV (Ujwal et al. 2008). (**B**) The electrostatic energy profile for anions (*a*) and cations (*b*) to move through the channel. The ion pathways traversed in panels *a*, *b* are shown from the side view from the membrane (*c*) and from the cytoplasm (*d*). The negative energy in panel *a* indicates that anions are favored in the pore (Reprinted from Choudhary et al. (2010) with permission from Elsevier)

6.3.1 *Computational Studies*

The availability of the 3D structure provided a new platform to study VDAC's function. The computational community was essential in validating the structures by relating the physiological properties observed from electrophysiological and biochemical experiments to the 3D architecture. Here we highlight a few of these studies, but a more extensive review on computational technics used to study VDAC can be found in Ref. Noskov et al. (2016).

6.3.1.1 Electrostatic Calculations

The structure revealed a large pore 27 Å in diameter, which is compatible with the high ion flux measured for the open state. This hypothesis was quantitatively probed using a combination of Poisson-Boltzmann (PB) electrostatic calculations and Poisson-Nernst Planck (PNP) theory (Choudhary et al. 2010). The PNP calculations revealed a large single-channel conductance incompatible with the closed state and 1.8–2.3 times higher than the experimental value, which is a typical overestimation for this level of theory when applied to large pores (Im and Roux 2002). However, dramatic improvements of calculated conductance were made possible by the use of long molecular dynamic (MD) simulations (Choudhary et al. 2014), with a calculated value of 0.96 nS in 142 mM NaCl, close to the experimental range of 0.64 to 0.83 nS under these conditions. Interestingly, the presence of ATP in the pore decreases VDAC conductance by 42%, perfectly corroborating the experimental result of 43% obtained on single channels in saturating ATP conditions (Rostovtseva and Bezrukov 1998). Furthermore, the calculations showed a slight preference for anions of 1.75 (Choudhary et al. 2010) in agreement with the experimental anion-to-cation selectivity values of 1.7–1.9 in 1 M KCl (Fig. 6.5). In summary, the electrostatics and MD simulations exploring ion conduction are all *quantitatively* consistent with a high-conductance open state that is selective for anions (Choudhary et al. 2010; Rui et al. 2011; Krammer et al. 2011).

6.3.1.2 Metabolite Permeation

In a recent publication (Choudhary et al. 2014), the permeation pathways of ATP through mVDAC1's pore were deduced using an innovative all-atom Markov state model approach – a first for membrane systems. The structure of VDAC in complex with ATP was also solved (Fig. 6.6) by soaking VDAC crystals in a high concentration of ATP (50 mM). The ATP presented a weak electron density, compatible with a low-affinity site, suggesting a high mobility in the pore. This structure (PDB code 4C69) along with the apo structure (PDB code 3EMN) with ATP placed randomly in solution were used to seed molecular dynamics simulations to investigate ATP permeation through the VDAC pore.

Initially, several multi-microsecond-long simulations carried out using the Anton special-purpose supercomputer at the Pittsburgh Supercomputing Center failed to

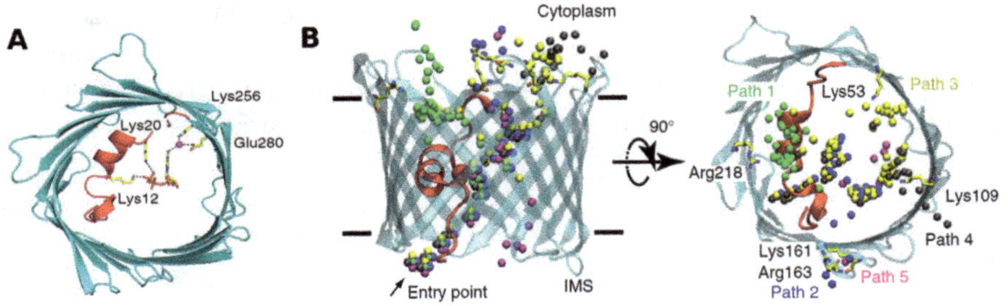

Fig. 6.6 ATP permeation through VDAC. (**A**) Cartoon representation of mVDAC1 in complex with ATP (PDB code 4C69). (**B**) Representation of the five primary pathways of ATP permeation, *dots* represent the γ−phosphate of ATP. ATP permeates the pore via several distinct pathways each lined with basic residues that interact electrostatically with the phosphate group. The *arrow* indicates the common entry point (Reprinted from Choudhary et al. (2014))

reveal ATP permeation. Rather, ATP entered the channel pore from solution where it interacted with the N-terminal helix. As an alternative approach, a Markov state model of ATP permeation was constructed by combining hundreds of short simulations (100–200 ns in length) together to determine how ATP moves through the pore. The simulations showed that ATP moves from one side of the channel to the other in many distinct pathways (top 5 shown in Fig. 6.6B) utilizing the complex network of basic residues facing the pore to "hop" from one position in the pore to the next. The predicted mean first passage time is 32 μs, and the average rate based on this MFPT is ~49,000 ATP/s – a result in excellent agreement with the experimental value of 50,000 ATP/s recorded from *Neurospora crassa* VDAC channels recorded in high ATP concentrations and extrapolated down to 5 mM ATP (Rostovtseva and Colombini 1996). Thus, the ATP flux computed from the MSM again confirms that the crystallographic mVDAC1 structure represents the native open conformation.

6.3.2 *VDAC's Orientation in the Membrane*

The high-resolution structures did not resolve the debate regarding VDAC's orientation in the membrane, largely due to conflicting biochemical data. This information is essential, since VDAC is a known binding partner for OMM proteins as well as cytosolic and inner membrane space proteins; thus, the correct orientation of the channel is necessary to identify the binding motifs of those various partners. The structure revealed that both N- and C-termini face the same side of the membrane (Ujwal et al. 2008), which was clouded by earlier contradictory studies. Using a combination of antibodies against the N- and C-termini moieties, one study concluded that both extremities where facing the IMS (Stanley et al. 1995). Contrary to this initial claim, a second study found that the N-terminus faces the cytosol and that the C-terminus was buried in the membrane (De Pinto et al. 1991). However, in

2013, this issue was ultimately resolved through studies on VDAC1 in intact cells using a combination of two C-terminal tags on VDAC1 separated by a caspase cleavage site (Tomasello et al. 2013). These researchers conclusively demonstrated that both the N- and C-termini face the IMS of the mitochondria.

6.4 Gating

Since the biochemical and computational studies all indicate that the current VDAC structures represent the open state, the next structural hurdle is to define the closed state(s) and determine how the channel transitions between states. Importantly, it is likely that VDAC channels adopt multiple closed states. There are several known factors that can modulate gating – the transition from open to closed state – including pH (Teijido et al. 2014), voltage (Colombini 1989), lipid composition (Rostovtseva et al. 2006), and salt concentration (Colombini 1989). Although the conformation of the closed state is unknown, the transition from open to closed presumably involves large conformational rearrangements that both hinder the passage of metabolites and alter ion selectivity.

6.4.1 The N-Terminal Helix and Its Influence on Gating

Initial gating hypotheses, developed from visual inspection of the structure, were centered on movements of the N-terminal α-helix. The helix lines the center of the pore causing the diameter to taper from 27 Å at the ends to 14 Å at the center (Fig. 6.2), yet this restriction does not hinder metabolite transport (Choudhary et al. 2014). It was originally postulated that a displacement of the helix away from the wall toward the center of the pore would narrow the cavity even further preventing metabolite permeation (Ujwal et al. 2008; Tornroth-Horsefield and Neutze 2008). Moreover, the helix is rich in charged amino acids and may act as the voltage sensor to facilitate voltage gating, as previously described (Colombini 1989).

This hypothesis was first questioned through the use of continuum electrostatic calculations in which two distinct gating scenarios were evaluated (Choudhary et al. 2010). First, the helix from the mVDAC1 structure was rigidly displaced toward the center of the pore to a position that would dramatically reduce the space available for metabolite permeation (a), and second, the helix was completely removed from the pore (b) leaving only the 19-stranded barrel with a wider pore domain (Fig. 6.7D).

Fig. 6.7 (continued) (**D**) Voltage dependence of gating for mVDAC1 X-ray structure compared to two hypothetical gating motions. In panel (*a*) the N-terminal helix is moved from the wall to the center of the pore. In panel (*b*) the helix is moved out of the pore. The motion in (*a*) produces no voltage-dependent energy change in the system (*blue line* in panel *c*), while the motion in (*b*) produces a modest voltage dependence corresponding to 1.5 gating charges (*red line* in panel *c*) (Reprinted from Choudhary et al. (2010) with permission from Elsevier)

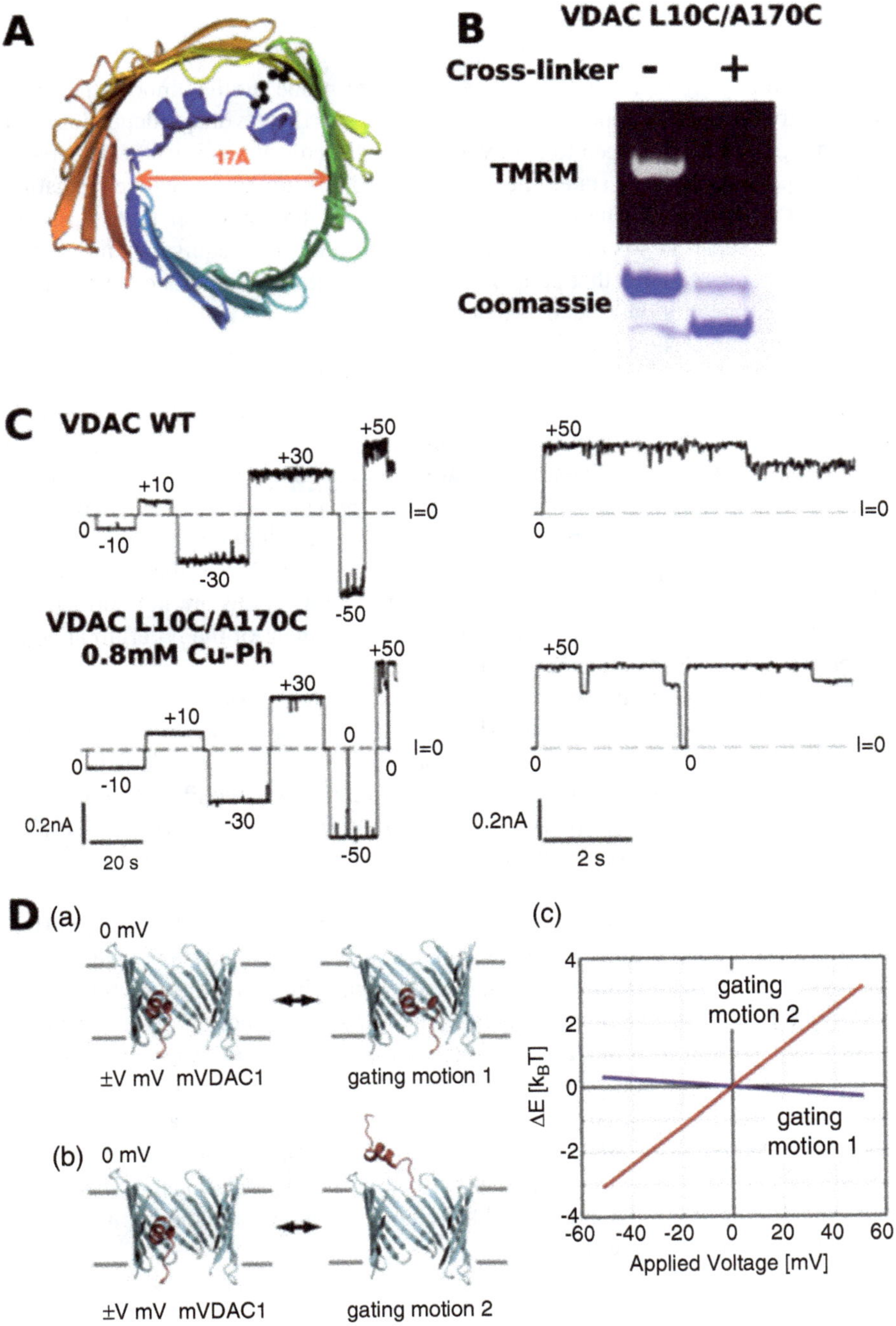

Fig. 6.7 Investigating VDAC's gating mechanism. (**A**) Cartoon representation of mVDAC1 (3EMN) and mutated residues L10C and A170C are displayed in ball and stick form. (**B**) On gel migration of mVDAC 1 L10C/A170C in the absence and presence of cross-linker (CuPh). Cross-linked samples (VDAC L10C/A170C + Cu-Ph) displayed no fluorescence because Cys residues are cross-linked and were not available to react with TMRM. (**C**) Representative records of ion currents through two channels formed by VDAC WT and cross-linked VDAC (Teijido et al. 2012).

Two electrophysiological signatures were used to probe the likelihood that these altered structures had properties of the closed state: (i) the voltage dependence of the gating motion from open to the hypothetical closed state and (ii) the selectivity of the hypothetical closed state. The first quantity indicates how sensitive the gating motion is to changes in the membrane voltage, and the parameter that is used to quantify this sensitivity is the gating charge. This value is the equivalent number of fundamental charge units that pass all the way through the membrane electric field during the gating motion. Channels that are very sensitive to membrane voltage, like voltage-gated potassium channels, exhibit gating charge values between 12 and 14 (Schoppa et al. 1992), and VDAC has a much weaker dependence measured between 2.5 and 4.5 (Colombini 1989; Hiller et al. 2008). As discussed previously, the second physiological property of the closed channel is a slight preference for cations over anions. Analysis of both gating scenarios revealed they failed on both accounts. Both hypothetical closed state models remained anion selective, and the helix hinge motion in the first scenario produce no gating charge at all, while the second helix remove motion only produces a charge of 1.5 – both well below the experimental range of 2.5–4.5 charge units. While these specific transitions are not correct, the calculations could not rule out a more complex movement of the N-terminal helix in combination with other channel conformational changes.

Next, to experimentally test the N-terminal gating model, we employed a combination of electrophysiology and cross-linking experiments (Teijido et al. 2012). The N-terminal α-helix was affixed to the wall of the barrel by creating a covalent bond between Leu-10 and Ala-170 (Fig. 6.7A). Cross-linking was monitored using tetramethylrhodamine (TMRM) (Chaptal et al. 2010) and band-shift on SDS-PAGE (Fig. 6.7B). The effects of the fixation of the helix on gating were monitored in planar lipid bilayers, where the N-terminal cross-linked VDAC formed functional channels of conductance and gating behavior similar to the WT (Fig. 6.7C). This study answered several important questions. First, it was an additional validation of the crystal structure showing that cross-links designed based on the structure could readily be formed in functional studies in the bilayer. Second, the results demonstrated that the gating motion does not involve independent movements of the N-terminal helix with respect to strand 11 on the wall. Surprisingly, another study using similar methodology revealed "asymmetric" gating behavior and preferential selection to a specific closed state (Mertins et al. 2012). Additional experimental studies using a wide array of methodologies coupled with validation through gating charge and selectivity calculations will be required to understand the implication of the N-terminal helix in gating.

6.4.2 A Major Rearrangement of the Barrel

An alternative model of gating is emerging in which the channel elongates in the direction of the principal axis of the N-terminal helix (Teijido et al. 2012; Zachariae et al. 2012). This motion involves VDAC morphing from a cylindrical shape to an ovular configuration that significantly constricts the pore potentially preventing metabolite passage through steric effects. Consistent with this barrel-constriction hypothesis, MD simulations revealed that VDAC is a flexible barrel that undergoes breathing motions (Villinger et al. 2010). This flexibility is also reflected in the crystallographic b-factors (Ujwal et al. 2008) and the ensemble of structures produced by NMR measurements in solution (Hiller et al. 2008).

Investigating this gating mechanism, Zachariae et al. (2012) carried out MD simulations under applied lateral pressure using a model of mVDAC1 that lacked the N-terminal helix and observed that the channel entered a semi-collapse state inducing an elliptical barrel shape. While this study represents an important step toward understanding channel gating, there are several concerns and unanswered questions: (i) in the presence of the N-terminal helix, this elliptic semi-collapsed state could not be reached, probably because the helix stabilizes the barrel structure, (ii) the elliptic states do not match the cation selectivity observed in the closed state of VDAC, and (iii) the gating charge of this motion was not determined.

From a structural standpoint, investigating VDAC's closed state and gating mechanism is a challenging task due to the difficulty of inducing the closed state in vitro. Indeed, voltage, used to induce gating in lipidic bilayers, cannot be used in solution nor in classical structural biochemistry methods such as NMR or X-ray crystallography. However, other parameters shown to improve gating, such as low pH (Teijido et al. 2014), pressure (Rostovtseva et al. 2006), or salt concentration (Colombini 1989), can be valuable tools to investigate gating.

6.5 Oligomerization

Several lines of evidence suggest that VDAC can adopt various oligomeric states in many different environments ranging from the native mitochondrial membrane to detergent micelles. Clearly this is an area for further investigations as VDAC oligomerization is suggested to play a large role in apoptosis (Keinan et al. 2010; Zalk et al. 2005).

In 2007, prior to the publication of the first high-resolution structures, atomic force microscopy (AFM) studies on the outer mitochondrial membrane of potato tubers (Hoogenboom et al. 2007) and *Saccharomyces cerevisiae* (Goncalves et al. 2007) revealed the organization of VDAC in its native lipidic environment. These studies revealed pores 38 Å wide, which nicely match the diameter values observed in the high-resolution structures. The AFM images showed the presence of monomers, dimers, tetramers, hexamers, and even higher-order oligomers, suggesting that VDAC certainly self-interacts, but the most frequent pore-to-pore distance observed was 53 Å, corresponding to two neighboring pores (Goncalves et al. 2007).

Analysis of crystal packing gave hints to the dimeric organization of VDAC. mVDAC1 crystallizes as an antiparallel dimer (Ujwal et al. 2009), and thus, this interaction is very unlikely to have a physiological relevance, since it is hard to imagine that the channel inserts into the OMM in more than one topological arrangement. However, both hVDAC1 (Bayrhuber et al. 2008) and zfVDAC2 (Schredelseker et al. 2014) interact as parallel dimers in solution and in the crystal, respectively, with inter-protein interactions occurring across strands, β1, β2, β17, β18, and β19. For both channel isoforms, cross-linking experiments in detergent micelles validated the dimeric association.

6.6 Protein Interactions and Metabolite Binding

6.6.1 Metabolite Binding

VDAC is responsible for metabolite passage between the cytosol and the mitochondria. The most characterized metabolite transported by VDAC is ATP, which can flow through the open state at a rate of two million molecules per second in vitro under saturating ATP concentrations and up to 10,000 molecules/s under physiological conditions (Rostovtseva and Colombini 1996, 1997). With such a high flux, it is not surprising that all reported metabolites binding to VDAC have low affinities, which makes identifying potential binding sites in the channel difficult. A structure of mVDAC1 in complex with ATP was obtained by soaking crystals in a 50 mM ATP solution, revealing a weak binding site for ATP inside the pore, where it associates with basic residues on the N-terminal helix (Fig. 6.6A) (Choudhary et al. 2014). This structure confirms predictions from earlier experiments suggesting that ATP weakly interacts with VDAC's pore (Rostovtseva and Bezrukov 1998; Rostovtseva et al. 2002; Yehezkel et al. 2006). Mass spectrometry was used to reveal two ATP binding site regions – one in the N-terminal helix and another in the region between positions 110–120 of the barrel (numbering in mVDAC1 and hVDAC1) (Yehezkel et al. 2006). The findings from mass spectrometry were later corroborated by the co-crystal structure and molecular dynamic simulations of ATP permeation through the pore, which showed that the phosphate tail of ATP interacts with residues Lys-113 and Lys-115 on β-barrel wall as well as basic residues Lys-12, Arg-15, and Lys-20 on the helix (Choudhary et al. 2014; Noskov et al. 2013).

The NMR structure of hVDAC1 revealed an additional interaction site for the small-molecule NADH on β-strands 17 and 18 implicating 6 residues (Gly-242, Leu-243, Gly-244, Ala-261, Leu-263, Asp-264) (Hiller et al. 2008). However, all those amino acid side chains are facing out of the pore, indicating a binding of the molecule that would be more a regulatory role than for transport. The crystal structure of mVDAC1 clearly indicates that there is not sufficient space to bind a NADH molecule at this specific position. Nonetheless, the described NADH-binding site is located in the vicinity of the flexible region of the N-terminal segment where the helix unwinds and enters β-strand 1, suggesting that binding of NADH in this area could possibly trigger a structural change in this connecting region between the two structural domains of the helix and barrel. In addition to ATP and ADP, a number of glycolysis intermediates like pyruvate and malate must make their way through the OMM into the IMS through VDAC. In total, there is still very little data on metabolite binding to VDAC, how molecules move through the channel, and how small molecule binding might bias the conformation of the channel. An important future goal is to design experiments to uncover more of these interaction sites, understand how binding impacts the channel structure, and conversely understand how the channel structure aids small-molecule passage into and out of the mitochondria.

6.6.2 VDAC's Interaction with Protein Partners

In addition to being the primary shuttle for ATP and ADP movement across the outer membrane, VDAC is also an essential component of mitochondrial regulation, which it orchestrates through its binding with various protein partners. There are many reports of VDAC binding to other proteins in the OMM, such as the translocator protein (TSPO) (Fan and Papadopoulos 2013), the proteins forming the mitochondrial permeability transition pore (mPTP) (Shoshan-Barmatz et al. 2008), proteins of the inner mitochondrial membrane (IMM) like ANT (Vyssokikh and Brdiczka 2003), as well as cytosolic proteins, such as hexokinase (De Pinto et al. 1993; Shoshan-Barmatz et al. 2015) and tubulin. Tubulin binding to VDAC results in an interaction that impedes metabolite passage (Rostovtseva and Bezrukov 2008), suggesting a powerful way in which cytoplasmic signals could influence energy production by the mitochondria. With the availability of both VDAC and tubulin structures, a hypothetical docking model was generated (Noskov et al. 2013) in which the negatively charged C-terminal tail of tubulin slides into the positively charged pore of VDAC causing occlusion of the pore that would sterically block ATP passage.

VDAC is also regulated by proteins in the Bcl2 family of proteins (Bak, Bax, and tBid) that control cellular apoptosis. During the early stages of apoptosis, caspase activation induces truncation of Bid to form tBid. tBid then relocates to the OMM triggering oligomerization of Bak into homo-oligomers and hetero-oligomers containing Bax and Bak (Wei et al. 2001; Korsmeyer et al. 2000). This oligomerization process leads to the permeabilization of the OMM. VDAC2 is specifically required for Bak import to the outer mitochondrial membrane and tBid-induced apoptosis (Cheng et al. 2003; Roy et al. 2009). In 2015, an elegant study used the structural information from mVDAC1 and zfVDAC2 to design protein chimeras between VDAC1 and VDAC2 to perform a rescue assay in VDAC2$^{-/-}$ fibroblasts (Naghdi et al. 2015). These chimeras identified the structural motif of VDAC2 necessary for Bak and tBid recruitment and induction of apoptosis. This study not only provided further validation that the VDAC1 and VDAC2 structures are biologically relevant, since the chimeras based on them were successful, but also pinpointed two critical residues in VDAC required for Bak recruitment, Thr-168 and Asp-170. Both of these residues are located on the cytoplasmic side of β-strand 10 of VDAC2 positioned in an ideal place to interact with a soluble protein in the cytoplasm. Superposition of the mVDAC1 structure onto zfVDAC2 revealed a binding pocket in VDAC2, positioned on the cytoplasmic side of the membrane (Naghdi et al. 2015).

This site could explain the specificity of VDAC2 for tBid and Bak binding (Naghdi et al. 2015). Interestingly, even though zfVDAC2 differs from mammalian VDAC2 due to its lack of the N-terminal extension and the absence of several cysteines (Fig. 6.1), it is able to recruit Bak and cause apoptosis in mouse embryonic fibroblasts, indicating that zfVDAC2 is a suitable model for mammalian VDAC2.

While valuable information is available both for tubulin and Bak interaction with VDAC, verification of the binding configuration via structural techniques is missing, yet essential for furthering our understanding of the regulation of outer membrane transport and mitochondria-mediated apoptosis.

6.7 Conclusion and Perspectives

During the last 8 years, impressive progress has been made toward understanding VDAC's function. Most of the new data acquired over this time would not have been possible without the availability of a high-resolution 3D structure, yet there are still many challenges ahead.

Going forward, the field must continue to probe the specific physiological roles of all three VDAC isoforms. From a structural and biophysical approach, gating remains a fundamental target: understanding the nature of the closed state and the gating mechanism of VDAC will provide a breakthrough in the understanding of VDAC function and mitochondrial regulation. What does the closed state look like, and how does VDAC transition from an open to a closed conformation? It appears that the closed conformation may not be a stable state, and as such, the use of conventional structural techniques may fail, and we may have to turn to alternative methods that capture dynamic processes. Finally, from a structural standpoint, VDAC complexes are clearly an obtainable goal but require the formation of stable complexes between VDAC and binding partners. Taken together, the regulation of ion and metabolite permeation through VDAC depends on both channel gating and the channel's interaction with protein-binding partners. These various mechanisms likely give rise to a rich set of channel biophysical properties that cannot be described by a simple two-state gating model. As we uncover this new information, we will gain a deeper understanding of VDAC's ability to regulate the mitochondria under a diverse set of physiological conditions.

Acknowledgment This work was supported by the National Institutes of Health Grant R01 GM 089740 (Grabe) and R01GM078844 (Abramson).

References

Bayrhuber M, Meins T, Habeck M, Becker S, Giller K, Villinger S, Vonrhein C, Griesinger C, Zweckstetter M, Zeth K (2008) Structure of the human voltage-dependent anion channel. Proc Natl Acad Sci U S A 105:15370–15375

Chaptal V, Ujwal R, Nie Y, Watanabe A, Kwon S, Abramson J (2010) Fluorescence detection of heavy atom labeling (FD-HAL): a rapid method for identifying covalently modified cysteine residues by phasing atoms. J Struct Biol 171:82–87

Cheng EH, Sheiko TV, Fisher JK, Craigen WJ, Korsmeyer SJ (2003) VDAC2 inhibits BAK activation and mitochondrial apoptosis. Science 301:513–517

Choudhary OP, Ujwal R, Kowallis W, Coalson R, Abramson J, Grabe M (2010) The electrostatics of VDAC: implications for selectivity and gating. J Mol Biol 396:580–592

Choudhary OP, Paz A, Adelman JL, Colletier JP, Abramson J, Grabe M (2014) Structure-guided simulations illuminate the mechanism of ATP transport through VDAC1. Nat Struct Mol Biol 21:626–632

Colombini M (1989) Voltage gating in the mitochondrial channel, VDAC. J Membr Biol 111:103–111

Colombini M (2009) The published 3D structure of the VDAC channel: native or not? Trends Biochem Sci 34:382–389

Craigen WJ, Graham BH (2008) Genetic strategies for dissecting mammalian and Drosophila voltage-dependent anion channel functions. J Bioenerg Biomembr 40:207–212

De Pinto V, Prezioso G, Thinnes F, Link TA, Palmieri F (1991) Peptide-specific antibodies and proteases as probes of the transmembrane topology of the bovine heart mitochondrial porin. Biochemistry 30:10191–10200

De Pinto V, Al Jamal JA, Palmieri F (1993) Location of the dicyclohexylcarbodiimide-reactive glutamate residue in the bovine heart mitochondrial porin. J Biol Chem 268:12977–12982

Elston T, Wang H, Oster G (1998) Energy transduction in ATP synthase. Nature 391:510–513

Fan J, Papadopoulos V (2013) Evolutionary origin of the mitochondrial cholesterol transport machinery reveals a universal mechanism of steroid hormone biosynthesis in animals. PLoS One 8:e76701

Forte M, Guy HR, Mannella CA (1987) Molecular genetics of the VDAC ion channel: structural model and sequence analysis. J Bioenerg Biomembr 19:341–350

Ge L, Villinger S, Mari SA, Giller K, Griesinger C, Becker S, Muller DJ, Zweckstetter M (2016) Molecular plasticity of the human voltage-dependent Anion Channel embedded into a membrane. Structure 24:585–594

Goncalves RP, Buzhynskyy N, Prima V, Sturgis JN, Scheuring S (2007) Supramolecular assembly of VDAC in native mitochondrial outer membranes. J Mol Biol 369:413–418

Hiller S, Garces RG, Malia TJ, Orekhov VY, Colombini M, Wagner G (2008) Solution structure of the integral human membrane protein VDAC-1 in detergent micelles. Science 321:1206–1210

Hiller S, Abramson J, Mannella C, Wagner G, Zeth K (2010) The 3D structures of VDAC represent a native conformation. Trends Biochem Sci 35:514–521

Hoogenboom BW, Suda K, Engel A, Fotiadis D (2007) The supramolecular assemblies of voltage-dependent anion channels in the native membrane. J Mol Biol 370:246–255

Im W, Roux B (2002) Ion permeation and selectivity of OmpF porin: a theoretical study based on molecular dynamics, Brownian dynamics, and continuum electrodiffusion theory. J Mol Biol 322:851–869

Israelson A, Abu-Hamad S, Zaid H, Nahon E, Shoshan-Barmatz V (2007) Localization of the voltage-dependent anion channel-1 Ca2+-binding sites. Cell Calcium 41:235–244

Jaburek M, Garlid KD (2003) Reconstitution of recombinant uncoupling proteins: UCP1, -2, and -3 have similar affinities for ATP and are unaffected by coenzyme Q10. J Biol Chem 278:25825–25831

Keinan N, Tyomkin D, Shoshan-Barmatz V (2010) Oligomerization of the mitochondrial protein voltage-dependent anion channel is coupled to the induction of apoptosis. Mol Cell Biol 30:5698–5709

Keinan N, Pahima H, Ben-Hail D, Shoshan-Barmatz V (2013) The role of calcium in VDAC1 oligomerization and mitochondria-mediated apoptosis. Biochim Biophys Acta 1833: 1745–1754

Koppel DA, Kinnally KW, Masters P, Forte M, Blachly-Dyson E, Mannella CA (1998) Bacterial expression and characterization of the mitochondrial outer membrane channel. Effects of n-terminal modifications. J Biol Chem 273:13794–13800

Korsmeyer SJ, Wei MC, Saito M, Weiler S, Oh KJ, Schlesinger PH (2000) Pro-apoptotic cascade activates BID, which oligomerizes BAK or BAX into pores that result in the release of cytochrome c. Cell Death Differ 7:1166–1173

Krammer EM, Homble F, Prevost M (2011) Concentration dependent ion selectivity in VDAC: a molecular dynamics simulation study. PLoS One 6:e27994

Kuszak AJ, Jacobs D, Gurnev PA, Shiota T, Louis JM, Lithgow T, Bezrukov SM, Rostovtseva TK, Buchanan SK (2015) Evidence of distinct channel conformations and substrate binding affinities for the mitochondrial outer membrane protein translocase pore Tom40. J Biol Chem 290:26204–26217

Law RH, Lukoyanova N, Voskoboinik I, Caradoc-Davies TT, Baran K, Dunstone MA, D'Angelo ME, Orlova EV, Coulibaly F, Verschoor S, Browne KA, Ciccone A, Kuiper MJ, Bird PI, Trapani JA, Saibil HR, Whisstock JC (2010) The structural basis for membrane binding and pore formation by lymphocyte perforin. Nature 468:447–451

Marcoline FV, Bethel N, Guerriero CJ, Brodsky JL, Grabe M (2015) Membrane protein properties revealed through data-rich electrostatics calculations. Structure 23:1526–1537

Mertins B, Psakis G, Grosse W, Back KC, Salisowski A, Reiss P, Koert U, Essen LO (2012) Flexibility of the N-terminal mVDAC1 segment controls the channel's gating behavior. PLoS One 7:e47938

Mondal S, Khelashvili G, Shi L, Weinstein H (2013) The cost of living in the membrane: a case study of hydrophobic mismatch for the multi-segment protein LeuT. Chem Phys Lipids 169:27–38

Naghdi S, Varnai P, Hajnoczky G (2015) Motifs of VDAC2 required for mitochondrial Bak import and tBid-induced apoptosis. Proc Natl Acad Sci U S A 112:E5590–E5599

Noskov SY, Rostovtseva TK, Bezrukov SM (2013) ATP transport through VDAC and the VDAC-tubulin complex probed by equilibrium and nonequilibrium MD simulations. Biochemistry 52:9246–9256

Noskov SY, Rostovtseva TK, Chamberlin AC, Teijido O, Jiang W, Bezrukov SM (2016) Current state of theoretical and experimental studies of the voltage-dependent anion channel (VDAC). Biochim Biophys Acta 1858:1778–1790

Popot JL, Gerchman SE, Engelman DM (1987) Refolding of bacteriorhodopsin in lipid bilayers. A thermodynamically controlled two-stage process. J Mol Biol 198:655–676

Rostovtseva TK, Bezrukov SM (1998) ATP transport through a single mitochondrial channel, VDAC, studied by current fluctuation analysis. Biophys J 74:2365–2373

Rostovtseva TK, Bezrukov SM (2008) VDAC regulation: role of cytosolic proteins and mitochondrial lipids. J Bioenerg Biomembr 40:163–170

Rostovtseva T, Colombini M (1996) ATP flux is controlled by a voltage-gated channel from the mitochondrial outer membrane. J Biol Chem 271:28006–28008

Rostovtseva T, Colombini M (1997) VDAC channels mediate and gate the flow of ATP: implications for the regulation of mitochondrial function. Biophys J 72:1954–1962

Rostovtseva TK, Komarov A, Bezrukov SM, Colombini M (2002) Dynamics of nucleotides in VDAC channels: structure-specific noise generation. Biophys J 82:193–205

Rostovtseva TK, Kazemi N, Weinrich M, Bezrukov SM (2006) Voltage gating of VDAC is regulated by nonlamellar lipids of mitochondrial membranes. J Biol Chem 281:37496–37506

Roy SS, Ehrlich AM, Craigen WJ, Hajnoczky G (2009) VDAC2 is required for truncated BID-induced mitochondrial apoptosis by recruiting BAK to the mitochondria. EMBO Rep 10:1341–1347

Rui H, Lee KI, Pastor RW, Im W (2011) Molecular dynamics studies of ion permeation in VDAC. Biophys J 100:602–610

Sampson MJ, Decker WK, Beaudet AL, Ruitenbeek W, Armstrong D, Hicks MJ, Craigen WJ (2001) Immotile sperm and infertility in mice lacking mitochondrial voltage-dependent anion channel type 3. J Biol Chem 276:39206–39212

Schein SJ, Colombini M, Finkelstein A (1976) Reconstitution in planar lipid bilayers of a voltage-dependent anion-selective channel obtained from paramecium mitochondria. J Membr Biol 30:99–120

Schoppa NE, Mccormack K, Tanouye MA, Sigworth FJ (1992) The size of gating charge in wild-type and mutant Shaker potassium channels. Science 255:1712–1715

Schredelseker J, Paz A, Lopez CJ, Altenbach C, Leung CS, Drexler MK, Chen JN, Hubbell WL, Abramson J (2014) High resolution structure and double electron-electron resonance of the zebrafish voltage-dependent anion channel 2 reveal an oligomeric population. J Biol Chem 289:12566–12577

Shanmugavadivu B, Apell HJ, Meins T, Zeth K, Kleinschmidt JH (2007) Correct folding of the beta-barrel of the human membrane protein VDAC requires a lipid bilayer. J Mol Biol 368:66–78

Shimizu H, Schredelseker J, Huang J, Lu K, Naghdi S, Lu F, Franklin S, Fiji HD, Wang K, Zhu H, Tian C, Lin B, Nakano H, Ehrlich A, Nakai J, Stieg AZ, Gimzewski JK, Nakano A, Goldhaber JI, Vondriska TM, Hajnoczky G, Kwon O, Chen JN (2015) Mitochondrial Ca(2+) uptake by the voltage-dependent anion channel 2 regulates cardiac rhythmicity. Elife 4:e04801

Shoshan-Barmatz V, Keinan N, Zaid H (2008) Uncovering the role of VDAC in the regulation of cell life and death. J Bioenerg Biomembr 40:183–191

Shoshan-Barmatz V, Ben-Hail D, Admoni L, Krelin Y, Tripathi SS (2015) The mitochondrial voltage-dependent anion channel 1 in tumor cells. Biochim Biophys Acta 1848:2547–2575

Stanley S, Dias JA, D'Arcangelis D, Mannella CA (1995) Peptide-specific antibodies as probes of the topography of the voltage-gated channel in the mitochondrial outer membrane of Neurospora crassa. J Biol Chem 270:16694–16700

Teijido O, Ujwal R, Hillerdal CO, Kullman L, Rostovtseva TK, Abramson J (2012) Affixing N-terminal alpha-helix to the wall of the voltage-dependent anion channel does not prevent its voltage gating. J Biol Chem 287:11437–11445

Teijido O, Rappaport SM, Chamberlin A, Noskov SY, Aguilella VM, Rostovtseva TK, Bezrukov SM (2014) Acidification asymmetrically affects voltage-dependent anion channel implicating the involvement of salt bridges. J Biol Chem 289:23670–23682

Tomasello MF, Guarino F, Reina s, Messina A, De Pinto V (2013) The voltage-dependent anion selective channel 1 (VDAC1) topography in the mitochondrial outer membrane as detected in intact cell. PLoS One 8:e81522

Tornroth-Horsefield S, Neutze R (2008) Opening and closing the metabolite gate. Proc Natl Acad Sci U S A 105:19565–19566

Ujwal R, Abramson, J (2012) High-throughput crystallization of membrane proteins using the lipidic bicelle method. J Vis Exp 59:e3383

Ujwal R, Cascio D, Colletier JP, Faham S, Zhang J, Toro L, Ping P, Abramson J (2008) The crystal structure of mouse VDAC1 at 2.3 A resolution reveals mechanistic insights into metabolite gating. Proc Natl Acad Sci U S A 105:17742–17747

Ujwal R, Cascio D, Chaptal V, Ping P, Abramson J (2009) Crystal packing analysis of murine VDAC1 crystals in a lipidic environment reveals novel insights on oligomerization and orientation. Channels (Austin) 3:167–170

Valiyaveetil F, Hermolin J, Fillingame RH (2002) pH dependent inactivation of solubilized F1F0 ATP synthase by dicyclohexylcarbodiimide: pK(a) of detergent unmasked aspartyl-61 in *Escherichia coli* subunit c. Biochim Biophys Acta 1553:296–301

Villinger s, Briones R, Giller K, Zachariae U, Lange A, De Groot BL, Griesinger C, Becker S, Zweckstetter M (2010) Functional dynamics in the voltage-dependent anion channel. Proc Natl Acad Sci U S A 107:22546–22551

Vyssokikh MY, Brdiczka D (2003) The function of complexes between the outer mitochondrial membrane pore (VDAC) and the adenine nucleotide translocase in regulation of energy metabolism and apoptosis. Acta Biochim Pol 50:389–404

Wei MC, Zong WX, Cheng EH, Lindsten T, Panoutsakopoulou V, Ross AJ, Roth KA, Macgregor GR, Thompson CB, Korsmeyer SJ (2001) Proapoptotic BAX and BAK: a requisite gateway to mitochondrial dysfunction and death. Science 292:727–730

Yamamoto T, Yamada A, Watanabe M, Yoshimura Y, Yamazaki N, Yoshimura Y, Yamauchi T, Kataoka M, Nagata T, Terada H, Shinohara Y (2006) VDAC1, having a shorter N-terminus than

VDAC2 but showing the same migration in an SDS-polyacrylamide gel, is the predominant form expressed in mitochondria of various tissues. J Proteome Res 5:3336–3344

Yehezkel G, Hadad N, Zaid H, Sivan S, Shoshan-Barmatz V (2006) Nucleotide-binding sites in the voltage-dependent anion channel: characterization and localization. J Biol Chem 281:5938–5946

Zachariae U, Schneider R, Briones R, Gattin Z, Demers JP, Giller K, Maier E, Zweckstetter M, Griesinger C, Becker S, Benz R, De Groot BL, LANGE A (2012) Beta-barrel mobility underlies closure of the voltage-dependent anion channel. Structure 20:1540–1549

Zalk R, Israelson A, Garty ES, Azoulay-Zohar H, Shoshan-Barmatz V (2005) Oligomeric states of the voltage-dependent anion channel and cytochrome c release from mitochondria. Biochem J 386:73–83

Zeth K (2010) Structure and evolution of mitochondrial outer membrane proteins of beta-barrel topology. Biochim Biophys Acta 1797:1292–1299

Chapter 7
Plant VDAC Permeability: Molecular Basis and Role in Oxidative Stress

Fabrice Homblé, Hana Kmita, Hayet Saidani, and Marc Léonetti

Abbreviations

BiFC	Bimolecular fluorescence complementation
HK and HXK	Hexokinase
Trx	Thioredoxin
VDAC	Voltage-dependent anion channel

7.1 Introduction

The plant mitochondria are the place of a variety of processes including respiratory substrate oxidation, ATP synthesis, calcium and redox signaling, and regulated cell death. Their biochemical activity requires exchange of ions and molecules through the mitochondrial outer membrane (MOM) and the mitochondrial inner membrane (MIM).

F. Homblé (✉) • H. Saidani
Structure et Fonction des Membranes Biologiques, Université Libre de Bruxelles (ULB), Boulevard du Triomphe CP 206/2, 1050 Brussels, Belgium
e-mail: fhomble@ulb.ac.be; hayettesaiidany@gmail.com

H. Kmita
Laboratory of Bioenergetics, Institute of Molecular Biology and Biotechnology, Faculty of Biology, Adam Mickiewicz University, Umultowska 89, 61-614 Poznan, Poland
e-mail: kmita@amu.edu.pl

M. Léonetti
I.R.P.H.E, Aix-Marseille Université, CNRS, Technopôle de Château-Gombert, 13384 Marseille Cedex 13, France
e-mail: leonetti@irphe.univ-mrs.fr

T.K. Rostovtseva (ed.), *Molecular Basis for Mitochondrial Signaling*, Biological and Medical Physics, Biomedical Engineering,
DOI 10.1007/978-3-319-55539-3_7

Due to the coordination of the metabolic pathways between different cellular compartments, they also play a role in other plant cellular processes such as photosynthesis, photorespiration, and nitrogen metabolism (Sweetlove et al. 2002; Dutilleul et al. 2003; Foyer et al. 2011; Bauwe et al. 2012; Araújo et al. 2014).

The voltage-dependent anion-selective channel (VDAC) is the major protein in plant MOM (Mannella and Bonner 1975). It is the main transport pathway for ions (Krammer et al. 2014), metabolites (Alcántar-Aguirre et al. 2013; Krammer et al. 2015), and nucleic acids (Koulintchenko et al. 2003; Salinas et al. 2006, 2008). By using molecules of increasing size, it has been shown that VDAC of mung bean has a size exclusion limit of about 5 kDa (Zalman et al. 1980). Like VDAC of other organisms, VDAC isolated from different plants is voltage-dependent after reconstitution in a planar lipid bilayer. This denotes that at voltage $|V| \geq 20$ mV, the channel exhibits transition between the open state and lower conductance states. This change in conductance might be of prime importance for the physiological role of plant VDAC because it has been shown that VDAC from *Neurospora crassa* is permeable to ATP in the open state but not in the subconductance states (Rostovtseva and Colombini 1996, 1997). This change in permeability to metabolites suggests that VDAC may not only facilitate but also control the metabolic flux through MOM.

Phylogenetic analysis points to an early evolutionary divergence of VDAC. VDAC from algae, plant, fungi, and animals belong to different clades. This suggests that they all derive from a common ancestor (Young et al. 2007; Homblé et al. 2012). Despite differences in their amino acid sequence, at least one VDAC isoform (named "canonical VDAC") in each clade bears extremely conserved structural and functional features. For instance, the most abundant mammalian isoform VDAC1 shares similar basic electrophysiological properties, secondary structure, and fold with the most abundant VDAC isoform of plants and fungi (e.g., Colombini 1989; Homblé et al. 2012; Mertins et al. 2014).

Plant VDAC proteins were purified and characterized from maize (Smack and Colombini 1985; Aljamal et al. 1993), potato tuber (Heins et al. 1994), pea seedlings and roots (Schmid et al. 1992; Fischer et al. 1994; Clausen et al. 2004), wheat seedlings (Blumenthal et al. 1993), and runner bean seeds (Abrecht et al. 2000b). The conductance, ion selectivity, and voltage dependence of these VDACs are very similar and correspond to isoforms highly abundant in the tissue used for purification. Two major mitochondrial VDAC isoforms were purified from bean and lentil seed cotyledons (Abrecht et al. 2000b) and potato tubers (Heins et al. 1994). However, proteomics studies indicate that more VDAC protein isoforms exists in plant tissues. For instance, there is a perfect matching between the transcriptomic and proteomic data for the model plant *Arabidopsis thaliana*, indicating the existence of four VDAC protein isoforms (Marmagne et al. 2004; Heazlewood et al. 2004; Tateda et al. 2011; Rao et al. 2016). The "green plants" (Viridiplantae) VDACs are encoded in the nuclear genome and belong to a small multigene family which is larger than that found in mammals (three) and fungi (two but only one in *Neurospora crassa*). Most green plants are sessile, and this large diversity of VDAC genes might be the source of more physiological plasticity to cope with environmental stresses. In plant, abiotic and biotic stresses are known to impact mitochondrial function and to be correlated with ROS imbalance known to alter VDAC permeability in mammalian and yeast mitochondria (Rostovtseva et al. 2006; Owsianowski et al. 2008; Homblé et al. 2012; McCommis and Baines 2012; Takahashi and Tateda 2013).

The 3D structure of the mouse and human VDAC isoform1 (VDAC1) and that of the zebrafish VDAC isoform2 (VDAC2) was resolved at atomic resolution (Hiller et al. 2008; Bayrhuber et al. 2008; Ujwal et al. 2008; Schneider et al. 2010; Schredelseker et al. 2014). They all display a common fold, which consists in a 19 β-strands β-barrel and 1 N-terminal helical segment lying inside the pore parallel to the membrane plane at about mid-height of the channel. This structure corresponds most probably to the open state of VDAC. There are clues suggesting that this structure might be conserved at least in the canonical VDACs of the different phylogenetic clades because (a) amino acid sequence analysis has put forward the presence of conserved motifs, long stretches of alternating hydrophobic and hydrophilic amino acid residues typical of β-barrel structure and a region featuring properties of α-helix at N-terminus (Young et al. 2007; Kutik et al. 2008; Imai et al. 2011; Homblé et al. 2012; Jores et al. 2016); (b) the barrel fold is also confirmed by circular dichroism (CD) and attenuated total reflection Fourier transform infrared (ATR-FTIR) spectroscopy (Shao et al. 1996; Abrecht et al. 2000a; Shanmugavadivu et al. 2007); and (c) the electron and atomic force microscopy images corroborate the 3D structure (Hiller et al. 2010).

In its open state, VDAC possesses a slight preference for inorganic monatomic anions over cations (Smack and Colombini 1985; Schmid et al. 1992; Blumenthal et al. 1993; Heins et al. 1994; Abrecht et al. 2000b). Remarkably, most inorganic ions, metabolites, and tRNA transported through VDAC are negatively charged. This raises the question of the mechanism(s) used to transport such diverse compounds.

In this review, we address two key functional issues about plant VDAC: (1) How inorganic ions and metabolites are transported through the pore? (2) What is the role of VDAC in the response of plant cells to oxidative stress? For this purpose, we compare the data available for plants and other organisms to highlight the similarities and differences between orthologues.

7.2 Transport and Selectivity

The main physiological role of VDAC is the transport of chemical compounds of varying size, composition, structure, and charge, e.g., K^+, Cl^-, Ca^{2+}, HPO_4^{2-}, succinate^{2-}, ATP^{4-}, or highly charged tRNA through MOM. Since the discovery of VDAC from *Paramecium aurelia* mitochondria (Schein et al. 1976), the transport function of VDAC has been investigated by electrophysiology following its reconstitution in a planar lipid bilayer. For the sake of clarity, we describe separately the transport of inorganic monatomic ions (consisting of a single atom) and that of inorganic polyatomic ions (consisting of several atoms) and metabolites.

7.2.1 Inorganic Monatomic Ions

The functional properties of plant VDAC were mainly inferred from transport studies of inorganic monatomic ions (K^+, Na^+, Cl^-). The obtained data indicate that the open state of plant VDAC can sustain a flow of about 10^7ions/s in the presence of 0.1MKCl and 10 mV and its selectivity for chloride ions over potassium ions is weak (~2Cl^-:1 K^+) (Smack and Colombini 1985; Aljamal et al. 1993; Blumenthal et al. 1993; Heins et al. 1994; Abrecht et al. 2000b). These values are akin to those found for VDACs of other organisms (Colombini 1989).

The experimental parameter measured to get information on the channel selectivity is the reversal potential, which is a voltage set up in the presence of a salt concentration gradient across the membrane when there is no net flow of ions (zero current) through the channel. Knowing the value of the reversal potential (mV), an electrodiffusion model is required to quantify the selectivity. Two models are commonly used: the Planck model and the Goldman-Hodgkin-Katz model. Solving the Nernst-Planck equation under the assumption of either electroneutrality (Planck) or constant electric field (Goldman-Hodgkin-Katz) in a channel gives two simple equations to quantify the selectivity (Krammer et al. 2014). The Goldman-Hodgkin-Katz equation, commonly known as GHK equation, is the most popular one and was applied to quantify the selectivity of plant VDAC using the value of reversal potential measured at a single high salt concentration gradient (e.g., 1 M/0.1 M KCl) across the membrane. Surprisingly, in a recent work on VDAC purified from *Phaseolus coccineus* (PcVDAC) (Krammer et al. 2014), it has been shown that PcVDAC selectivity value calculated by GHK equation changes with KCl concentration gradient, which disagrees with the assumption of a constant permeability ratio in the GHK model. Moreover, the GHK equation did not fit properly to the data showing the change in reversal potential versus the concentration gradient (Krammer et al. 2014). The inappropriateness of the GHK and Planck models for the description of the selectivity was further demonstrated in an experiment showing the change in reversal potential as a function of salt concentration, at a constant KCl concentration gradient across the membrane (Fig. 7.1). Under the condition of the constant gradient, the two equations predict that the reversal potential is constant whatever the salt concentration is and are in contradiction with the experimental results showing that the selectivity toward chloride ion increases at low KCl concentration. The dependence of the reversal potential on the salt concentration indicates the involvement of positive-charged amino acid residues at low salt concentration and their screening at high concentration.

The PcVDAC selectivity was correctly predicted taking into account the net fixed charge in the pore of the channel. A one-compartment macroscopic electrodiffusion model including both ion diffusion and an effective fixed charge in the pore was sufficient to describe properly the selectivity of the channel (Fig. 7.1). Thus, in agreement with published data on yeast *Saccharomyces cerevisiae* (Blachly-Dyson et al. 1990; Peng et al. 1992; Zambrowicz and Colombini 1993), the fixed charge in the pore governs the selectivity of plant VDAC.

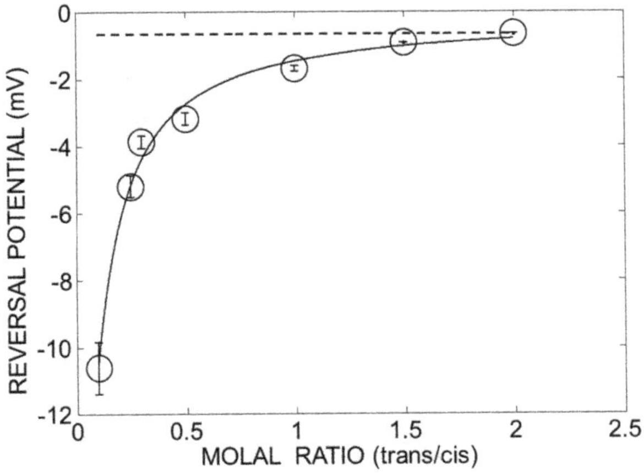

Fig. 7.1 Effect of ionic strength on the reversal potential of PcVDAC at a fixed concentration ratio trans/cis = 2. A nonlinear least square regression was used to fit the data to a fixed charge model (*continuous line*) (DCl/DK = 1.02, effective fixed charge concentration X = 0.1 M, goodness of the fit $R^2 = 0.994$). The reversal potential calculated using the voltage GHK equation (with the PCl/PK = 1.10 at concentration ratio 2/1 M KCl (trans/cis)) is shown as a *dashed line* (Reprinted with permission from Mitochondrion (Krammer et al. 2014) with slight modification)

Several plant genomes have been sequenced. The sequences coding for putative VDAC proteins indicate that in plant VDACs, the number of positively charged amino acids is equal or higher than that of negatively charged amino acids (Fig. 7.2), which is consistent with a selectivity favoring chloride ions relative to potassium ions. However, an accurate assessment of the selectivity would require knowledge of the charge distribution within the protein structure.

The electrostatic screening effect is also clearly evidenced when the VDAC conductance normalized in respect to the bulk concentration (G/[KCl] ratio) is plotted versus the bulk concentration (Fig. 7.3).

At low KCl concentration (0.1 M), the screening of the fixed charges of the channel is weak, which favors the accumulation of chloride ions inside the channel because of the excess of positive fixed charges. Consequently, both G/[KCl] ratio and selectivity toward chloride increase. At high KCl concentration (1 M), there is a strong screening of the fixed charges, and VDAC behaves like a wide neutral pore. Thus, the ion concentration inside the channel will be close to that in solution, and thus, the G/[KCl] ratio reaches its lowest value. At 1 M KCl, the selectivity toward chloride ions is weak because the diffusion coefficient of chloride and potassium ions has a similar value. As shown in Figs. 7.1 and 7.3, the fixed charge has a major impact on the G/[KCl] ratio and selectivity at concentrations that prevails in vivo. Therefore, the measurement of the reversal potential at high concentrations is of limited interest to assess plant VDAC selectivity because of the electrostatic screening effect.

Molecular modeling has been used to get insight into the microscopic mechanism for the monatomic ions transport through VDAC. Different simulation

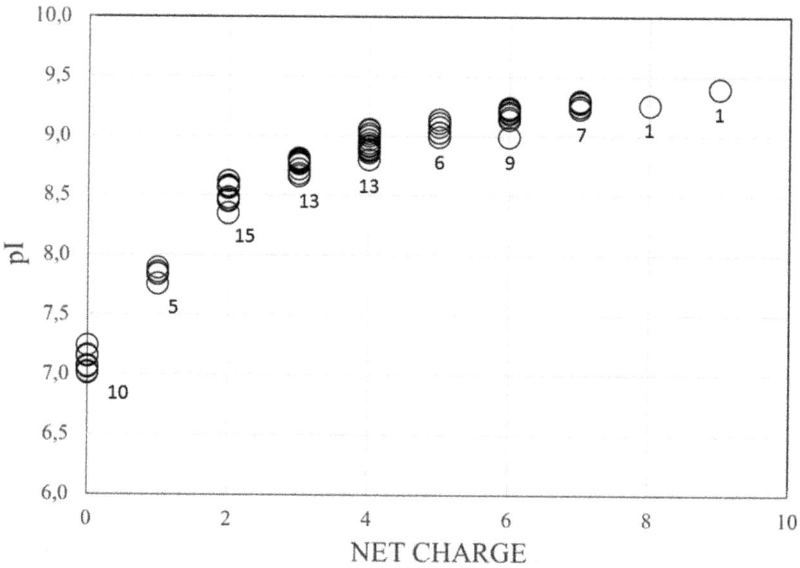

Fig. 7.2 Net charge of plant VDAC proteins. Data were obtained from a set of 80 gene sequences encoding putative VDAC proteins retrieved from full-sequenced genomes of representatives of 16 different plant genera (https://phytozome.jgi). These VDAC proteins contain from 274 to 280 amino acids and have theoretical isoelectric point (pI) consistent with the data found for VDAC in 2D-PAGE. The number attached to the symbol indicates the number of sequences with the specified net charge with an average value of 3.5 ± 2.3 (mean ± S.D.)

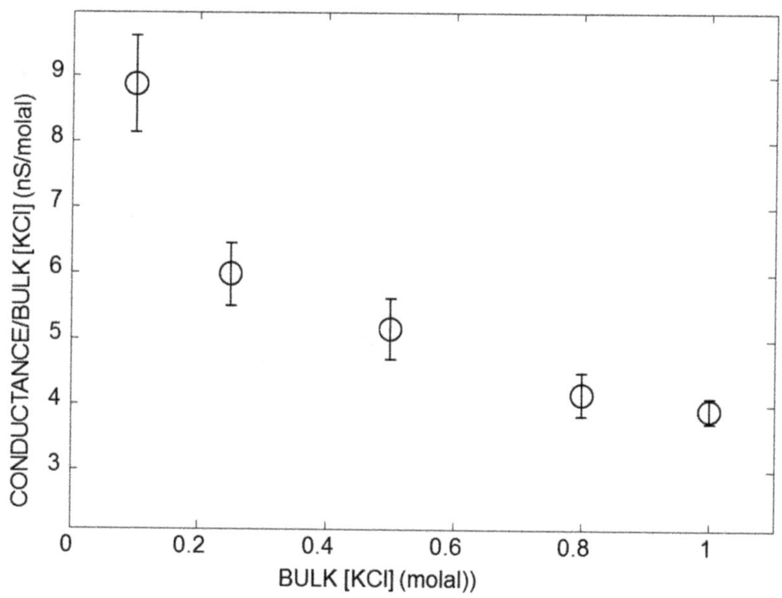

Fig. 7.3 VDAC conductance normalized in respect to the bulk concentration (G/[KCl] ratio) as a function of plotted versus the bulk concentration

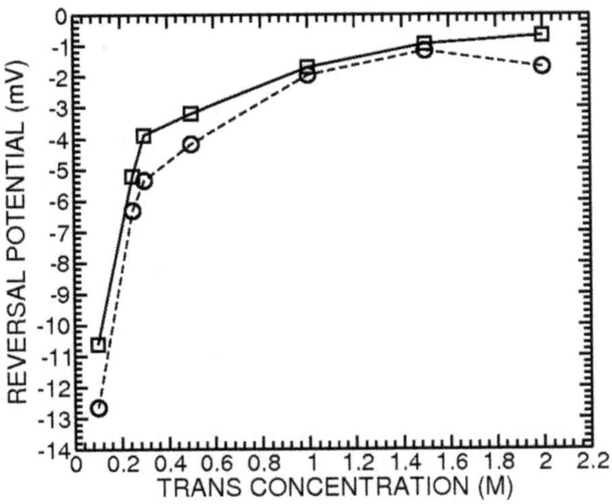

Fig. 7.4 Effect of ionic strength on the reversal potential of PcVDAC at a fixed concentration ratio trans/cis = 2. The reversal potential computed from BD trajectories of mVDAC1 (*circle, dashed line*) and experimentally determined on PcVDAC (*square, solid line*) (Reprinted with permission from Mitochondrion (Krammer et al. 2014))

techniques have been used to investigate the structure-function relationship of VDAC (Noskov et al. 2013). They require the knowledge of the protein structure at an atomic resolution, which does not exist for plant VDAC. The PcVDAC and the mouse VDAC1 (mVDAC1) share similar electrophysiological and secondary structure features suggesting a conserved mechanism of ion transport and a similar folding. Thus, a 3D model of PcVDAC was constructed using homology modeling and the 3D structure of mVDAC1 determined at a resolution of 2.3 Å as template (Homblé et al. 2012). It features an electrostatic energy ion profile, a β-strand dynamics, and a distribution of the hydrophobic and charged residues comparable to those of mVDAC1 (Homblé et al. 2012). These similarities do not prove that the model is the right structure, but they provide a rational basis to use it as a working hypothesis to interpret the data collected on plant VDACs. The conductance and selectivity values of the PcVDAC 3D model calculated using Brownian dynamics (BD) simulations are consistent with the experimental data, albeit some discrepancies are observed. In addition, BD and molecular dynamics (MD) simulations performed for mVDAC1 structure reproduce properly the concentration dependence of the PcVDAC selectivity (Fig. 7.4) (Krammer et al. 2011, 2013, 2014). The free energy profile of the ion translocation along the pore of mVDAC1 shows a concentration dependence (Fig. 7.5). The free energy profile of chloride ions is lower than that of potassium ions favoring the transport of chloride ions. The difference between the free energy profiles of these ions decreases at high salt concentration in agreement with the weakest selectivity observed for PcVDAC at 1 M KCl. The maximum of the free energy profile of ion permeation coincides with the channel region made up of the N-terminal helix and the surrounding beta-barrel strip, which might be the rate-limiting step for ion transport through VDAC. These conclusions are consistent with results obtained by BD and MD simulations on mVDAC1 (Krammer et al. 2011, 2013). MD also shows that ion transport through

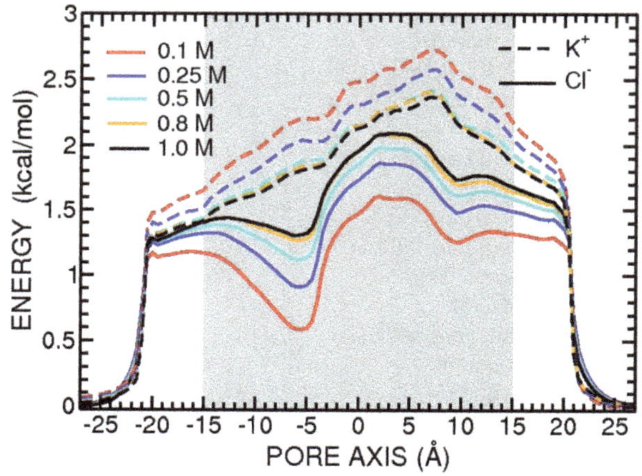

Fig. 7.5 Ion concentration-dependent energy profile of mVDAC1. Chloride (*solid lines*) and potassium (*dashed lines*) concentration-dependent energy profiles computed from BD simulations performed at 0.1 M (*red*), 0.25 M (*blue*), 0.5 M (*cyan*), 0.8 M (*orange*), and 1.0 M (*black*) KCl (Reprinted with permission from Mitochondrion (Krammer et al. (2014)))

mVDAC1 is driven by the global electrostatics due to the distribution of charges in the pore and not by specific pathways or interactions with amino acid residues (Krammer et al. 2011).

In regards to the transport process of inorganic monatomic ions through VDAC, the molecular simulations emphasize the major role of the nonuniform charge distribution in VDAC pore. Despite the simplistic assumptions underpinning the macroscopic model, it is remarkable that it fits pretty well to the experimental data, and thereby, it provides an easy practical way to quantify the selectivity measured experimentally. The preference for anions relative to cations serves the major physiological role of VDAC, i.e., the transport of metabolites, most of which are anions. This major VDAC function might have undergone pressure of the natural selection, which might explain the conserved transport process of metabolites between the different phylogenetic clades. In agreement with this suggestion, the weak VDAC selectivity of all species studied for monatomic inorganic ions appears as a consequence of the electrostatic properties of the pore (Colombini 2016).

7.2.2 Inorganic Polyatomic Ions and Metabolites

The major function of mitochondria is the transformation of energy for cellular needs. This requires the exchange of inorganic polyatomic ions (e.g., $H_2PO_4^-$, HPO_4^{2-}) and metabolites (e.g., ADP^{2-}, ATP^{4-}, succinate^{2-}) between the mitochondrial matrix and the cytosol. The experiment with purified VDAC from *Neurospora crassa* (NcVDAC) has provided indirect and direct evidences for the permeation of several compounds through VDAC. The flux through VDAC estimated from selectivity measurement indicated that VDAC can discriminate between different compounds according to the sequence $H_2PO_4^- > HPO_4^{2-}$ = succinate^{2-}>citrate^{3-} and that the flux of these compounds is strongly reduced when VDAC switched from the open to a subconductance state (Hodge and Colombini 1997). The specificity of nucleotide-VDAC binding was shown for mammalian VDAC by the channel labeling with [α-^{32}P] BzATP in the presence of various nucleotides (Yehezkel et al. 2006). Combining electrophysiology and luciferin/luciferase assay, Rostovtseva and Colombini (1996, 1997) have achieved a direct evidence demonstrating that ATP^{4-} permeates VDAC in the open state but not in subconductance states. Assuming ATP concentration of 1–10 mM and a voltage of 10 mV, one can estimate that ATP can permeate VDAC at a rate of about 10^4–10^5 ATP/s, i.e., at least two order of magnitude lower than that of monatomic inorganic ions. The dependence of ATP flux through VDAC channel on ATP concentration is linear up to 100 mM ATP suggesting a low-affinity binding site, which is required for an efficient translocation rate.

The VDAC occupancy by ATP interferes with the permeation of inorganic monatomic ions, which results in a decrease of VDAC conductance (Rostovtseva and Bezrukov 1998). Thus, the change in conductance is an indirect way to probe the occupancy of VDAC channel by metabolites. Mononucleotides (AMP^{2-}, ADP^{3-}, UTP^{4-}) and dinucleotides (NAD^{-}, $NADH^{2-}$, and $NADPH^{4-}$) were found to decrease the NcVDAC conductance (Rostovtseva et al. 2002). Accordingly, the conductance of mouse VDAC2 and PcVDAC decreases in the presence of ATP (Komarov et al. 2005; Krammer et al. 2015). In the case of PcVDAC, the decrease in channel conductance is larger in the presence of ATP than AMP. The magnitude of the conductance change is attenuated at high salt concentration (1 M) (Krammer et al. 2015) probably due to the electrostatic screening of the metabolites and channel fixed charge.

Altogether, these experimental results point to a common mechanism of metabolite transport shared by VDACs of all organisms. They emphasize firstly that VDACs are channels for the transport of metabolites and organic polyatomic ions and secondly that they may regulate the flow of these compounds. The regulatory property is related to a change in VDAC conductance. Though it is well established that this change in conductance can be triggered by a change in the membrane potential, there is numerous evidences concerning mammalian cells that it could be also induced by interaction with proteins, peptides, or small compounds (Rostovtseva and Bezrukov 2008, 2012; Shoshan-Barmatz et al. 2010). For example, NADH was found to regulate the gating of potato tuber VDAC (Lee et al. 1994; Zizi et al. 1994).

MD simulations of $H_2PO_4^-$ diffusion through mVDAC1 revealed a permeation process very similar to that of chloride ions. Thus, both free energy profiles of permeation displayed the same features, and no specific pathway was reported for the monovalent anion $H_2PO_4^-$, though a few long-lived interactions with some amino acid residues were noticed (Krammer et al. 2015). A completely different scenario was found for the permeation of the divalent anion HPO_4^{2-}, which followed a specific pathway and long-lasting interactions with a limited number of positively charged amino acid residues: K12, R15, K20, K96, K119, and R218 (Krammer et al. 2015).

The molecular simulation of metabolites permeation through VDAC is all but trivial. According to the experimental flux measurements (Rostovtseva and Colombini 1996), the transport of one ATP through the VDAC lasts about 10–100 μs,

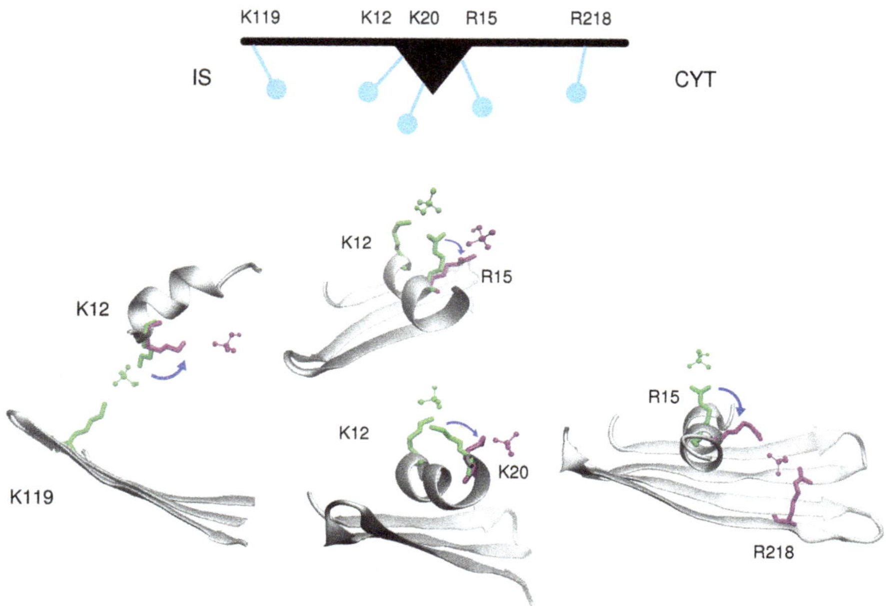

Fig. 7.6 The charged brush mechanism of VDAC. Schematic representation of the positively charged key residues acting as a charged brush to facilitate permeation of the divalent form of Pi, AMP, and ATP along the VDAC pore. Below are represented successive MD conformations (in *green* and *purple*) showing these positively charged side chains undergoing a sweeping motion assisting the translocation of Pi along the pore. Pi are shown as *balls* and *sticks* and the basic residues as *thick sticks*. CYT and IS stand for the cytosolic and the intermembrane side, respectively (Reprinted with permission from PLoS ONE Krammer et al. (2015))

and this is a major bottleneck regarding MD simulations. Several approaches were used to unravel the mechanism of ATP permeation by the use of molecular simulation (Noskov et al. 2013). They all indicated that the phosphate moiety of ATP interacts with positively charged amino acid residues of the pore (R15, K20, K96) and point the N-terminus α-helix as a probable weak binding site for ATP (Noskov et al. 2013; Choudhary et al. 2014; Krammer et al. 2015) that corresponds to experimental data (Yehezkel et al. 2006, 2007; Villinger et al. 2014). As ATP, AMP interacts with the binding site but more weakly. Though there is a general agreement about the identification of the binding site, the question on how ATP arrives to and leaves from the binding site is a matter of debate. Choudhary et al. (2014) identified five different pathways for ATP permeation, while Krammer et al. (2015) pointed to a preferential pathway recruiting K12, R15, K20, K96, K119, and R218. Their key position in the structure and their long and flexible side chains might favor the permeation of the phosphate ions and metabolites through VDAC via a brush-like mechanism (Yehezkel et al. 2006; Villinger et al. 2014; Krammer et al. 2015) (Fig. 7.6). Even though the occurrence of long-lasting interaction between $H_2PO_4^-$ and some amino acid residues is relatively less frequent, it occurs notably with R15 and K20 of the binding site. Thus, according to the minimum in the free energy profile of the studied compounds, the probability of interaction with the binding site decreases according to the sequence ATP^{4-}>HPO_4^{2-}>AMP^{2-}>$H_2PO_4^-$.

As regard to the mechanism of permeation through plant VDAC, conserved features have been noticed (Krammer et al. 2015). Namely, the N-terminal helical segment and the cluster of positively charged amino acid residues forming the binding site, K12, R15, and K20 are conserved in PcVDAC. When the effect of ATP and AMP on the PcVDAC conductance is considered, it is observed that the decrease in conductance is higher in the presence of ATP because of its long dwelling time inside the pore, which restricts the diffusion of inorganic monatomic ions. At high NaCl concentration (1 M), change in the conductance is weaker than at 0.1 M NaCl because the electrostatic screening of the charges inside the pore shallows the free energy profile for ATP and AMP, that decreases their probability of interaction with the binding site and thereof reduces dwelling time inside the VDAC pore.

7.3 VDAC Involvement in Oxidative Stress and Regulated Cell Death

7.3.1 The Role of Mitochondria in Oxidative Stress

In eukaryotic cells, oxidative stress can be defined as a process resulting from an imbalance between the production of reactive oxygen species (ROS) and capacity of cellular antioxidant defenses. The defense mechanisms include upregulation of various endogenous molecules, such as glutathione, ROS catalytic removal by enzymes like superoxide dismutase, catalase, and peroxidase, and repair of ROS-modified molecules (Barber et al. 2006; Radak et al. 2011; Mailloux et al. 2013). Among ROS, there are the free radicals of oxygen (e.g., singlet oxygen ($^1O_2$), superoxide anion ($O_2^{\bullet-}$), and hydroxyl radical (•OH)), which are atoms or molecules with an unpaired electron that can be rapidly donated making them highly reactive molecular species. Hydrogen peroxide (H_2O_2) is not a free radical, but it belongs to ROS, and it is capable of producing oxidative stress. The main site of ROS production in cells are mitochondria (e.g., Ott et al. 2007; Murphy 2009) and chloroplasts in photosynthetic tissues (Woodson and Chory 2008; Van Aken and Van Breusegem 2015). The details of ROS production in chloroplasts are beyond the scope of the chapter as VDAC is highly abundant in mitochondria. In mitochondria, they are mainly generated at Complex I and III of the respiratory chain as a result of respiratory chain reactions. The one-electron reduction of molecular oxygen produces a relatively stable intermediate, i.e., $O_2^{\bullet-}$, which serves as the precursor of most ROS but the detail mechanism of their generation is beyond the scope of the review (for review, see, e.g., Murphy 2009).

$O_2^{\bullet-}$ produced by mitochondria Complex I is released into the mitochondrial matrix whereas $O_2^{\bullet-}$ produced by Complex III appears on both sides of the inner membrane. As summarized by Ott et al. (2007), within the mitochondrial matrix,

MnSOD (manganese containing superoxide dismutase) converts $O_2^{\bullet-}$ to H_2O_2, which can be further metabolized by glutathione peroxidase 1 (Gpx1) and peroxiredoxin III (PrxIII), or diffuse from the mitochondria into the cytosol. $O_2^{\bullet-}$ located in the intermembrane space can be released into the cytosol and/or converted into H_2O_2 by CuZnSOD (copper and zinc containing superoxide dismutase). Accordingly, mitochondria contribute 20–30% of the cytosolic steady-state concentration of H_2O_2 (Han et al. 2003). The conversion of $O_2^{\bullet-}$ into H_2O_2 proceeds also in the cytosol due to the presence of cytosolic CuZnSOD, and the H_2O_2 may be then decomposed to water and oxygen by catalase located mainly in peroxisomes (Del Río et al. 2008). Importantly, about 1% of ROS escapes elimination (Ott et al. 2007). Likewise excessive $O_2^{\bullet-}$ generation by the mitochondrial respiratory chain and consequently other ROS production can lead to oxidation of macromolecules including DNA, proteins, and membrane lipids that may influence cell viability and trigger cell death[1] (e.g., Shoshan-Barmatz et al. 2015). Moreover, it is proposed that the steady-state concentration of $O_2^{\bullet-}$ in the mitochondrial matrix is about five- to tenfold higher than that in the cytosol or nucleus making mitochondria a possible primary target for the damaging effects of ROS (Ott et al. 2007). This denotes that mitochondria may generate and/or propagate oxidative stress. However, the mechanisms of $O_2^{\bullet-}$ transport via the outer membrane of mitochondria and the control of this process are still not clear.

7.3.2 VDAC and the Regulated Cell Death

In mammals, the $O_2^{\bullet-}$ release from rat heart mitochondria is inhibited by known VDAC inhibitor (Han et al. 2003). It has been also reported that closure of VDAC1 may impede the efflux of $O_2^{\bullet-}$ from the intermembrane space to the cytosol that results in internal oxidative stress and promotion of mitochondrial membrane permeability changes related to regulated cell death (Tikunov et al. 2010). In addition, silencing of VDAC1 expression results in inhibition of oxidative stress-induced regulated cell death whereas overexpression of the protein enhances the regulated cell death in mammalian cells (e.g., Liu et al. 2006; Simamura et al. 2006; Tomasello et al. 2009; Chen et al. 2014).

In yeasts, it has been shown that isogenic wild-type and *Δpor1 S. cerevisiae* mitochondria depleted of YVDAC1 differ distinctly in the level of $O_2^{\bullet-}$ release (Budzińska et al. 2009). The expression level of CuZnSOD and MnSOD proteins is influenced by the absence of either YVDAC isoforms, and both YVDAC1 and YVDAC2 are required for the homeostasis of the mitochondrial redox state (Galganska et al. 2010).

There are strong evidences indicating that regulated cell death occurs in plants and it is coupled to the mtROS production (Rhoads et al. 2006; Love et al. 2008;

---

[1] Within the chapter, we will follow the proposition of the Nomenclature Committee on Cell Death (Galluzzi et al. 2015).

Zancani et al. 2015; Van Aken and Van Breusegem 2015; Huang et al. 2016). The synthesis of mtROS occurs during respiratory activity of plant cells. Under normal conditions, mtROS detoxification by dismutation and redox reactions prevents their deleterious effects on proteins, lipids, and DNA molecules. In addition, the electron transport chain of plant mitochondria contains an alternative oxidase (AOX) that catalyzes the oxidation of ubiquinol and reduces oxygen to water, which minimizes the formation of mtROS (Møller 2001; Vanlerberghe 2013). There is a growing body of evidence pointing at enhanced mtROS production and consecutive redox signaling during the plant response to biotic and abiotic stresses (Love et al. 2008; Van Aken and Van Breusegem 2015).

When mtROS formation exceeds the normal ROS level despite the existence of an endogenous protective capacity, it may induced mitochondrial retrograde regulation as well as regulated cell death (Garcia-Brugger et al. 2006; Rhoads et al. 2006; Woodson and Chory 2008; Huang et al. 2016). It has been suggested that VDAC is involved in plant stress responses through its involvement in regulated cell death (Vianello et al. 2012; Takahashi and Tateda 2013). Several experimental data support this hypothesis. Either the overexpression of the rice OsVDAC4 or the human HVDAC1 in mammalian cell lines (Yurkat T-cell) induces apoptosis (Godbole et al. 2003). This indicates that OsVDAC4 can substitute for HVDAC1to trigger PCD in mammalian cells. Regulated cell death was also induced by the overexpression of OsVDAC4 in tobacco BY2 cells and leaves (Godbole et al. 2013) or after the overexpression of *Pennisetum glaucum* VDAC in rice (Desai et al. 2006). In addition, regulated cell death induced by biotic or abiotic stress is correlated to an upregulation of VDAC transcript and protein level (Lacomme and Roby 1999; Kawai-Yamada et al. 2001; Al Bitar et al. 2003; Swidzinski et al. 2004; Desai et al. 2006; Tateda et al. 2009; Kusano et al. 2009). The mammalian anti-apoptotic factors Bcl-xL and its functional homologue Ced-9 in *Caenorhabditis* are known to suppress regulated cell death in non-plant cells. Though there is no homologue protein to the BCL-2 family in plant genome, the heterologous expression of Bcl-2, Bcl-xL, and Ced-9 cDNA in tobacco plants under the control of a strong promoter inhibits plant regulated death induced by different abiotic stresses or by the non-virulent pathogen tobacco mosaic virus (Mitsuhara et al. 1999; Qiao et al. 2002). The overexpression of mammalian Bax gene in tobacco plants causes regulated cell death (Lacomme and Santa Cruz 1999; Tateda et al. 2009) in agreement with the pro-apoptotic function of Bax proteins.

Altogether, these data highlight the involvement of plant VDAC in the regulated cell death.

7.3.3 VDAC-Protein Interaction

The function of a protein is often regulated through interaction with other proteins. The occurrence of protein-protein interactions is ubiquitous and is essential for proper functioning of cells. Cytosolic and mitochondrial proteins have been found to interact directly with the mammalian VDAC1 to mediate signaling between cytosol and mitochondria (Rostovtseva and Bezrukov 2008, 2012; Shoshan-Barmatz et al. 2015). Here, we will focus on hexokinase and thioredoxin, which have been studied in plants.

7.3.4 VDAC-Hexokinase Interaction

Interaction of mammalian VDAC1 with hexokinase (HK) isoforms is important for cell protection against oxidative stress and consecutive cell death (e.g., Shoshan-Barmatz et al. 2015). It has been observed for HEK cells that HK I and HK II binding to VDAC1 decreases ROS release from mitochondria (Sun et al. 2008). HK interaction with VDAC1 affects its channel properties, but the question of whether the binding to VDAC leads to its closure or to its opening is still not answered although the available data allow for suggestion that binding of HK II to VDAC1 protects it from closing (Robey and Hay 2006; Sun et al. 2008) whereas binding of HK I induces it closure (Azoulay-Zohar et al. 2004). Such alteration of the channel conductance may have detrimental effect for cell because it has been shown that excessive VDAC closure induces mitochondrial swelling and regulated cell death in mammalian cells (Majewski et al. 2004; Rostovtseva et al. 2005).

Plant hexokinases, named HXKs, are the only enzymes that can phosphorylate glucose[2] (Granot 2007). The HXKs that interact with mitochondria harbor an N-terminal putative membrane anchor sequence and are classified in "type B HXK" (HXK1) (Granot et al. 2014). Hexokinase1 (HXK1) has a dual function: it catalyzes glucose phosphorylation and acts as a glucose sensor (Jang and Sheen 1994; Granot et al. 2014).

In plants, hexokinase has been shown to play a role in pathogen-induced and H_2O_2-induced cell death (Kim et al. 2006; Camacho-Pereira et al. 2009; Kusano et al. 2009). Kim's seminal work (Kim et al. 2006) has shown that *Nicotiana benthamiana* HXK1was associated with the mitochondria. Using virus-induced gene silencing, they have shown that the downregulation of *HXK1* activated regulated cell death whereas the overexpression of the mitochondria-associated *Arabidopsis* hexokinases, *HXK1* and *HXK2*, enhanced the resistance to regulated cell death of cells subjected to (experiencing undergoing) an oxidative stress. In addition, HXK1 prevented the H_2O_2-induced cytochrome c release if it harbors its N-terminal membrane anchor.

Another study has shown that transiently expressed OsVDAC4 and NtKxK3 *Nicotiana benthamiana* hexokinase into tobacco BY2 cells are localize to mitochondria. The co-expression of the two proteins in plant leaves were more efficient to prevent the induction of regulated cell death than NtKxK3 alone (Godbole et al. 2013). Finally, a detergent-solubilized membrane proteins from beetroot

[2] The affinity for glucose is one to three orders of magnitude higher than that of other sugars.

mitochondria was shown to contain a complex HXK-VDAC suggesting that VDAC and HXK might interact in the MOM (Alcántar-Aguirre et al. 2013).

These results indicate that mitochondrial-associated hexokinase plays a role in the regulated cell death and suggest that the interaction between HK and VDAC might be the functional unit.

These data support the conclusion that plant VDAC has a role in the regulated cell death through its interaction with soluble proteins. It has long been thought that VDAC was a constituent of a protein complex named the permeability transition pore (PTP) that increases the permeability of MIM to solutes. However, the existence of the PTP is now questioned because the permeability transition occurs in the absence of the major components of the PTP. For instance, the peripheral benzodiazepine receptor (TSPO), the adenosine nucleotide translocase (ANT), and the VDAC are dispensable for the permeability transition (Kokoszka et al. 2004; Krauskopf et al. 2006; Baines et al. 2007; Šileikytė et al. 2014; Gutiérrez-Aguilar et al. 2014). Recent data point to the role of the F-ATP synthase dimer in the formation of the PTP in MIM (Zancani et al. 2015).

7.3.5 VDAC-Thioredoxin Interaction

Thioredoxins are regulatory disulfide proteins that mediate the cellular redox homeostasis allowing disulfide bridge reduction of the target protein. Trx are recruited by Trx-dependent peroxidases (Prxs) to scavenge hydrogen peroxide. In addition, through the control of the redox state of proteins, they may also regulate the ROS signaling (Sevilla et al. 2015).

Four VDAC isoforms have been characterized in *Arabidopsis* (Yan et al. 2009; Tateda et al. 2011; Bauwe et al. 2012; Robert et al. 2012; Tateda et al. 2012; Li et al. 2013). Recently, Zhang et al. (2015) have performed a yeast two-hybrid screen to identify the interaction between the *Arabidopsis thaliana* VDAC isoform 3 (AtVDAC3) and the chloroplast protein thioredoxin m2 (AtTrx m2) in the regulation of the ROS signaling occurring in response to abiotic stress. This interaction is specific for AtVDAC3 since AtVDAC1, AtVDAC2, and AtVDAC4 failed to interact with AtTrx m2 in the Y2H assay. The co-expression of AtTrx m2-GFP^N and AtVDAC3-GFP^C in BY2 tobacco cells localize the two proteins in the mitochondria. The localization of AtTrx m2 at the mitochondrial is not surprising since it has been shown that it is targeted to the cytosol when interacting with cytosolic proteins (Meyer et al. 2011; Hölscher et al. 2014).

The authors present evidence that AtVDAC3 and AtTrx m2 participate to the regulated cell death. They have shown that hat *AtVDAC3* and *AtTrx m2* genes were upregulated when wild-type *Arabidopsis* plants were treated with H_2O_2, mannitol, or NaCl.

Zhang et al. (2015) have also analyzed transgenic Arabidopsis plants that overexpress either *AtVDAC3*^{OE} or *AtTrx m2*^{OE}. Both wildtype and transgenic plants were treated with either NaCl (salt stress) or H_2O_2. Interestingly, *AtVDAC3*^{OE} and

AtTrx m2^{OE} transgenic lines displayed opposite effects in the presence of the abiotic stress. In regards to the wildtype, the seedling growth was lower and the ROS production was enhanced in *AtVDAC3*^{OE} cell line whereas the seedling growth was higher and the ROS production reduced in *AtTrx m2*^{OE} cell line. Altogether, these data suggest that AtVDAC3 and AtTrx m2 might be involved in the regulated cell death. How their interaction might affect the regulated cell death deserves further investigations.

7.4 Conclusion

Despite early evolutionary divergence of VDAC proteins and their weak amino acid sequence similarity between different organisms, both the experimental and theoretical data indicate at conservation of the permeation process. The deciphering of the mechanisms underlying inorganic ion and metabolite permeation through VDAC is important, as the most of the chemical species entering or leaving mitochondria are anions significant for energy transformation. The change in VDAC permeability may be imposed by oxidative stress. It has been observed for plant cells that oxidative stress may result in VDAC interaction with other proteins such as cytosolic hexokinase and thioredoxin although the way of their binding to plant VDAC and their impact on its functional properties remain to be elucidated. This in turn should allow for explanation of VDAC contribution to plant cell regulated death.

Acknowledgments FH is a research director at the F.R.S.-FNRS (Belgium).

References

Abrecht H, Goormaghtigh E, Ruysschaert JM, Homble F (2000a) Structure and orientation of two voltage-dependent anion-selective channel isoforms – an attenuated total reflection Fourier-transform infrared spectroscopy study. J Biol Chem 275:40992–40999. doi:10.1074/jbc.M006437200

Abrecht H, Wattiez R, Ruysschaert JM, Homble F (2000b) Purification and characterization of two voltage-dependent anion channel isoforms from plant seeds. Plant Physiol 124:1181–1190. doi:10.1104/pp.124.3.1181

Al Bitar F, Roosens N, Smeyers M, Vauterin M, Van Boxtel J, Jacobs M, Homblé F (2003) Sequence analysis, transcriptional and posttranscriptional regulation of the rice vdac family. Biochim Biophys Acta Gene Struct Expr 1625:43–51. doi:10.1016/S0167-4781(02)00590-0

Alcántar-Aguirre FC, Chagolla A, Tiessen A, Délano JP, de la Vara LE (2013) ATP produced by oxidative phosphorylation is channeled toward hexokinase bound to mitochondrial porin (VDAC) in beetroots (*Beta vulgaris*). Planta 237:1571–1583. doi:10.1007/s00425-013-1866-4

Aljamal JA, Genchi G, De Pinto V, Stefanizzi L, De Santis A, Benz R, Palmieri F (1993) Purification and characterization of porin from corn (*Zea mays* L.) mitochondria. Plant Physiol 102:615–621

Araújo WL, Nunes-Nesi A, Fernie AR (2014) On the role of plant mitochondrial metabolism and its impact on photosynthesis in both optimal and sub-optimal growth conditions. Photosynth Res 119:141–156. doi:10.1007/s11120-013-9807-4

Azoulay-Zohar H, Israelson A, Abu-Hamad S, Shoshan-Barmatz V (2004) In self-defence: hexokinase promotes voltage-dependent anion channel closure and prevents mitochondria-mediated apoptotic cell death. Biochem J 377:347–355. doi:10.1042/bj20031465

Baines CP, Kaiser RA, Sheiko T, Craigen WJ, Molkentin JD (2007) Voltage-dependent anion channels are dispensable for mitochondrial-dependent cell death. Nat Cell Biol 9:550–555. doi:10.1038/ncb1575

Barber SC, Mead RJ, Shaw PJ (2006) Oxidative stress in ALS: a mechanism of neurodegeneration and a therapeutic target. Biochim Biophys Acta Mol basis Dis 1762:1051–1067. doi:10.1016/j.bbadis.2006.03.008

Bauwe H, Hagemann M, Kern R, Timm S (2012) Photorespiration has a dual origin and manifold links to central metabolism. Curr Opin Plant Biol 15:269–275. doi: http://dx.doi.org/10.1016/j.pbi.2012.01.008

Bayrhuber M, Meins T, Habeck M, Becker S, Giller K, Villinger S, Vonrhein C, Griesinger C, Zweckstetter M, Zeth K (2008) Structure of the human voltage-dependent anion channel. ProcNatlAcadSciUSA 105:15370–15375. doi:10.1073/pnas.0808115105

Blachly-Dyson E, Peng S, Colombini M, Forte M (1990) Selectivity changes in site-directed mutants of the VDAC ion channel: structural implications. Science (80-) 247:1233–1236. doi:10.1126/science.1690454

Blumenthal A, Kahn K, Beja O, Galun E, Colombini M, Breiman A (1993) Purification and characterization of the voltage-dependent anion-selective channel protein from wheat mitochondrial membranes. Plant Physiol 101:579–587. doi:10.1104/pp.101.2.579

Budzińska M, Gałgańska H, Karachitos A, Wojtkowska M, Kmita H (2009) The TOM complex is involved in the release of superoxide anion from mitochondria. J Bioenerg Biomembr 41:361–367. doi:10.1007/s10863-009-9231-9

Camacho-Pereira J, Meyer LE, Machado LB, Oliveira MF, Galina A (2009) Reactive oxygen species production by potato tuber mitochondria is modulated by mitochondrially bound hexokinase activity. Plant Physiol 149:1099–1110. doi:10.1104/pp.108.129247

Chen H, Gao W, Yang Y, Guo S, Wang H, Wang W, Zhang S, Zhou Q, Xu H, Yao J, Tian Z, Li B, Cao W, Zhang Z, Tian Y (2014) Inhibition of VDAC1 prevents Ca2+-mediated oxidative stress and apoptosis induced by 5-aminolevulinic acid mediated sonodynamic therapy in THP-1 macrophages. Apoptosis 19:1712–1726. doi:10.1007/s10495-014-1045-5

Choudhary OP, Paz A, Adelman JL, Colletier J-P, Abramson J, Grabe M (2014) Structure-guided simulations illuminate the mechanism of ATP transport through VDAC1. Nat Struct Mol Biol 21:626–632. doi:10.1038/nsmb.2841

Clausen C, Ilkavets I, Thomson R, Philippar K, Vojta A, Mohlmann T, Neuhaus E, Fulgosi H, Soll J (2004) Intracellular localization of VDAC proteins in plants. Planta 220:30–37

Colombini M (1989) Voltage gating in the mitochondrial channel, VDAC. J Membr Biol 111:103–111. doi:10.1007/BF01871775

Colombini M (2016) The VDAC channel: molecular basis for selectivity. Biochim Biophys Acta, Mol Cell Res 1863:2498–2502. doi:10.1016/j.bbamcr.2016.01.019

Del Río LA, Corpas FJ, Sandalio LM, Palma JM, Barroso JB (2008) Plant peroxisomes, reactive oxygen metabolism and nitric oxide. IUBMB Life 55:71–81. doi:10.1002/tbmb.718540875

Desai MK, Mishra RN, Verma D, Nair S, Sopory SK, Reddy MK (2006) Structural and functional analysis of a salt stress inducible gene encoding voltage dependent anion channel (VDAC) from pearl millet (*Pennisetum glaucum*). Plant Physiol Biochem 44:483–493. doi:10.1016/j.plaphy.2006.08.008

Dutilleul C, Driscoll S, Cornic G, De Paepe R, Foyer CH, Noctor G (2003) Functional mitochondrial complex I is required by tobacco leaves for optimal photosynthetic performance in photorespiratory conditions and during transients. Plant Physiol 131:264–275. doi:10.1104/pp.011155

Fischer K, Weber A, Brink S, Arbinger B, Schunemann D, Borchert S, Heldt HW, Popp B, Benz R, Link TA, Eckerskorn C, Flugge UI (1994) Porins from plants. Molecular cloning and functional characterization of two new members of the porin family. J Biol Chem 269: 25754–25760

Foyer CH, Noctor G, Hodges M (2011) Respiration and nitrogen assimilation: targeting mitochondria-associated metabolism as a means to enhance nitrogen use efficiency. J Exp Bot 62:1467–1482. doi:10.1093/jxb/erq453

Galganska H, Karachitos A, Wojtkowska M, Stobienia O, Budzinska M, Kmita H (2010) Communication between mitochondria and nucleus: putative role for VDAC in reduction/oxidation mechanism. Biochim Biophys Acta Bioenerg 1797:1276–1280. doi:10.1016/j.bbabio.2010.02.004

Galluzzi L, Bravo-San Pedro JM, Vitale I, Aaronson SA, Abrams JM, Adam D, Alnemri ES, Altucci L, Andrews D, Annicchiarico-Petruzzelli M, Baehrecke EH, Bazan NG, Bertrand MJ, Bianchi K, Blagosklonny MV, Blomgren K, Borner C, Bredesen DE, Brenner C, Campanella M, Candi E, Cecconi F, Chan FK, Chandel NS, Cheng EH, Chipuk JE, Cidlowski JA, Ciechanover A, Dawson TM, Dawson VL, De Laurenzi V, De Maria R, Debatin K-M, Di Daniele N, Dixit VM, Dynlacht BD, El-Deiry WS, Fimia GM, Flavell RA, Fulda S, Garrido C, Gougeon M-L, Green DR, Gronemeyer H, Hajnoczky G, Hardwick JM, Hengartner MO, Ichijo H, Joseph B, Jost PJ, Kaufmann T, Kepp O, Klionsky DJ, Knight RA, Kumar S, Lemasters JJ, Levine B, Linkermann A, Lipton SA, Lockshin RA, López-Otín C, Lugli E, Madeo F, Malorni W, Marine J-C, Martin SJ, Martinou J-C, Medema JP, Meier P, Melino S, Mizushima N, Moll U, Muñoz-Pinedo C, Nuñez G, Oberst A, Panaretakis T, Penninger JM, Peter ME, Piacentini M, Pinton P, Prehn JH, Puthalakath H, Rabinovich GA, Ravichandran KS, Rizzuto R, Rodrigues CM, Rubinsztein DC, Rudel T, Shi Y, Simon H-U, Stockwell BR, Szabadkai G, Tait SW, Tang HL, Tavernarakis N, Tsujimoto Y, Vanden Berghe T, Vandenabeele P, Villunger A, Wagner EF, Walczak H, White E, Wood WG, Yuan J, Zakeri Z, Zhivotovsky B, Melino G, Kroemer G (2015) Essential versus accessory aspects of cell death: recommendations of the NCCD 2015. Cell Death Differ 22:58–73. doi:10.1038/cdd.2014.137

Garcia-Brugger A, Lamotte O, Vandelle E, Bourque S, Lecourieux D, Poinssot B, Wendehenne D, Pugin A (2006) Early signaling events induced by elicitors of plant defenses. Mol Plant-Microbe Interact 19:711–724. doi:10.1094/MPMI-19-0711

Godbole A, Varghese J, Sarin A, Mathew MK (2003) VDAC is a conserved element of death pathways in plant and animal systems. Biochim Biophys Acta-Mol Cell Res 1642:87–96

Godbole A, Dubey AK, Reddy PS, Udayakumar M, Mathew MK (2013) Mitochondrial VDAC and hexokinase together modulate plant programmed cell death. Protoplasma 250:875–884. doi:10.1007/s00709-012-0470-y

Granot D (2007) Role of tomato hexose kinases. Funct Plant Biol 34:564–570

Granot D, Kelly G, Stein O, David-Schwartz R (2014) Substantial roles of hexokinase and fructokinase in the effects of sugars on plant physiology and development. J Exp Bot 65:809–819. doi:10.1093/jxb/ert400

Gutiérrez-Aguilar M, Douglas DL, Gibson AK, Domeier TL, Molkentin JD, Baines CP (2014) Genetic manipulation of the cardiac mitochondrial phosphate carrier does not affect permeability transition. J Mol Cell Cardiol 72:316–325. doi:10.1016/j.yjmcc.2014.04.008

Han D, Antunes F, Canali R, Rettori D, Cadenas E (2003) Voltage-dependent anion channels control the release of the superoxide anion from mitochondria to cytosol. J Biol Chem 278:5557–5563. doi:10.1074/jbc.M210269200

Heazlewood JL, Tonti-Filippini JS, Gout AM, Day DA, Whelan J, Millar AH (2004) Experimental analysis of the Arabidopsis mitochondrial proteome highlights signaling and regulatory components, provides assessment of targeting prediction programs, and indicates plant-specific mitochondrial proteins. Plant Cell 16:241–256. doi:10.1105/tpc.016055

Heins L, Mentzel H, Schmid A, Benz R, Schmitz UK (1994) Biochemical, molecular, and functional characterization of porin isoforms from potato mitochondria. J Biol Chem 269: 26402–26410

Hiller S, Garces RG, Malia TJ, Orekhov VY, Colombini M, Wagner G (2008) Solution structure of the integral human membrane protein VDAC-1 in detergent micelles. Science (80–) 321:1206–1210. doi:10.1126/science.1161302

Hiller S, Abramson J, Mannella C, Wagner G, Zeth K (2010) The 3D structures of VDAC represent a native conformation. Trends Biochem 35:514–521

Hodge T, Colombini M (1997) Regulation of metabolite flux through voltage-gating of VDAC channels. J Membr Biol 157:271–279. doi:10.1007/s002329900235

Hölscher C, Meyer T, von Schaewen A (2014) Dual-targeting of arabidopsis 6-phosphogluconolactonase 3 (PGL3) to chloroplasts and peroxisomes involves interaction with Trx m2 in the cytosol. Mol Plant 7:252–255. doi:10.1093/mp/sst126

Homblé F, Krammer E-M, Prévost M (2012) Plant VDAC: facts and speculations. Biochim Biophys Acta Biomembr 1818:1486–1501. doi:10.1016/j.bbamem.2011.11.028

Huang S, Van Aken O, Schwarzländer M, Belt K, Millar AH (2016) Roles of mitochondrial reactive oxygen species in cellular signalling and stress response in plants. Plant Physiol 171:1551–1559. doi:10.1104/pp.16.00166

Imai K, Fujita N, Gromiha MM, Horton P (2011) Eukaryote-wide sequence analysis of mitochondrial beta-barrel outer membrane proteins. BMC Genom 12:79. doi:10.1186/1471-2164-12-79

Jang JC, Sheen J (1994) Sugar sensing in higher plants. Plant Cell 6:1665–1679. doi:10.1105/tpc.6.11.1665

Jores T, Klinger A, Groß LE, Kawano S, Flinner N, Duchardt-Ferner E, Wöhnert J, Kalbacher H, Endo T, Schleiff E, Rapaport D (2016) Characterization of the targeting signal in mitochondrial β-barrel proteins. Nat Commun 7:12036. doi:10.1038/ncomms12036

Kawai-Yamada M, Jin L, Yoshinaga K, Hirata A, Uchimiya H (2001) Mammalian Bax-induced plant cell death can be down-regulated by overexpression of Arabidopsis Bax Inhibitor-1 (AtBI-1). Proc Natl Acad Sci 98:12295–12300. doi:10.1073/pnas.211423998

Kim M, Lim JH, Ahn CS, Park K, Kim GT, Kim WT, Pai HS (2006) Mitochondria-associated hexokinases play a role in the control of programmed cell death in Nicotiana benthamiana. Plant Cell 18:2341–2355

Kokoszka JE, Waymire KG, Levy SE, Sligh JE, Cai J, Jones DP, MacGregor GR, Wallace DC (2004) The ADP/ATP translocator is not essential for the mitochondrial permeability transition pore. Nature 427:461–465. doi:10.1038/nature02229

Komarov AG, Deng DF, Craigen WJ, Colombini M (2005) New insights into the mechanism of permeation through large channels. Biophys J 89:3950–3959

Koulintchenko M, Konstantinov Y, Dietrich A (2003) Plant mitochondria actively import DNA via the permeability transition pore complex. EMBO J 22:1245–1254

Krammer E-M, Homblé F, Prévost M (2011) Concentration dependent ion selectivity in VDAC: a molecular dynamics simulation study. PLoS One 6:e27994. doi:10.1371/journal.pone.0027994

Krammer E-M, Homblé F, Prévost M (2013) Molecular origin of VDAC selectivity towards inorganic ions: a combined molecular and Brownian dynamics study. Biochim Biophys Acta Biomembr 1828:1284–1292. doi:10.1016/j.bbamem.2012.12.018

Krammer E-M, Saidani H, Prévost M, Homblé F (2014) Origin of ion selectivity in *Phaseolus coccineus* mitochondrial VDAC. Mitochondrion 19:206–213. doi:10.1016/j.mito.2014.04.003

Krammer E-M, Vu GT, Homblé F, Prévost M (2015) Dual mechanism of ion permeation through VDAC revealed with inorganic phosphate ions and phosphate metabolites. PLoS One 10:e0121746. doi:10.1371/journal.pone.0121746

Krauskopf A, Eriksson O, Craigen WJ, Forte MA, Bernardi P (2006) Properties of the permeability transition in VDAC1−/− mitochondria. Biochim Biophys Acta Bioenerg 1757:590–595. doi:10.1016/j.bbabio.2006.02.007

Kusano T, Tateda C, Berberich T, Takahashi Y (2009) Voltage-dependent anion channels: their roles in plant defense and cell death. Plant Cell Rep 28:1301–1308. doi:10.1007/s00299-009-0741-z

Kutik S, Stojanovski D, Becker L, Becker T, Meinecke M, Kruger V, Prinz C, Meisinger C, Guiard B, Wagner R, Pfanner N, Wiedemann N (2008) Dissecting membrane insertion of mitochondrial beta-barrel proteins. Cell 132:1011–1024. doi:10.1016/j.cell.2008.01.028

Lacomme C, Roby D (1999) Identification of new early markers of the hypersensitive response in *Arabidopsis thaliana*. FEBS Lett 459:149–153. doi:10.1016/S0014-5793(99)01233-8

Lacomme C, Santa Cruz S (1999) Bax-induced cell death in tobacco is similar to the hypersensitive response. Proc Natl Acad Sci 96:7956–7961. doi:10.1073/pnas.96.14.7956

Lee AC, Zizi M, Colombini M (1994) Beta-NADH decreases the permeability of the mitochondrial outer membrane to ADP by a factor of 6. J Biol Chem 269:30974–30980

Li Z-Y, Xu Z-S, He G-Y, Yang G-X, Chen M, Li L-C, Ma Y (2013) The voltage-dependent Anion Channel 1 (AtVDAC1) negatively regulates plant cold responses during germination and seedling development in Arabidopsis and interacts with calcium sensor CBL1. Int J Mol Sci 14:701–713. doi:10.3390/ijms14010701

Liu S, Ishikawa H, Tsuyama N, Li F-J, Abroun S, Otsuyama K, Zheng X, Ma Z, Maki Y, Iqbal MS, Obata M, Kawano MM (2006) Increased susceptibility to apoptosis in CD45+ myeloma cells accompanied by the increased expression of VDAC1. Oncogene 25:419–429. doi:10.1038/sj.onc.1208982

Love AJ, Milner JJ, Sadanandom A (2008) Timing is everything: regulatory overlap in plant cell death. Trends Plant Sci 13:589–595

Mailloux RJ, McBride SL, Harper M-E (2013) Unearthing the secrets of mitochondrial ROS and glutathione in bioenergetics. Trends Biochem Sci 38:592–602. doi:10.1016/j.tibs.2013.09.001

Majewski N, Nogueira V, Bhaskar P, Coy PE, Skeen JE, Gottlob K, Chandel NS, Thompson CB, Robey RB, Hay N (2004) Hexokinase-mitochondria interaction mediated by Akt is required to inhibit apoptosis in the presence or absence of Bax and Bak. Mol Cell 16:819–830. doi:10.1016/j.molcel.2004.11.014

Mannella CA, Bonner WD (1975) X-ray diffraction from oriented outer mitochondrial membranes. Biochim Biophys Acta Biomembr 413:226–233. doi:10.1016/0005-2736(75)90106-6

Marmagne A, Rouet MA, Ferro M, Rolland N, Alcon C, Joyard J, Garin J, Barbier-Brygoo H, Ephritikhine G (2004) Identification of new intrinsic proteins in Arabidopsis plasma membrane proteome. Mol Cell Proteomics 3:675–691

McCommis KS, Baines CP (2012) The role of VDAC in cell death: friend or foe? Biochim Biophys Acta Biomembr 1818:1444–1450. doi:10.1016/j.bbamem.2011.10.025

Mertins B, Psakis G, Essen L-O (2014) Voltage-dependent anion channels: the wizard of the mitochondrial outer membrane. Biol Chem 395:1435–1442. doi:10.1515/hsz-2014-0203

Meyer T, Hölscher C, Schwöppe C, von Schaewen A (2011) Alternative targeting of Arabidopsis plastidic glucose-6-phosphate dehydrogenase G6PD1 involves cysteine-dependent interaction with G6PD4 in the cytosol. Plant J 66:745–758. doi:10.1111/j.1365-313X.2011.04535.x

Mitsuhara I, Malik KA, Miura M, Ohashi Y (1999) Animal cell-death suppressors Bcl-xL and Ced-9 inhibit cell death in tobacco plants. Curr Biol 9:775–778. doi:10.1016/S0960-9822(99)80341-8

Møller IM (2001) Plant mitochondria and oxidative stress: electron transport, NADPH turnover, and metabolism of reactive oxygen species. Annu Rev Plant Physiol Plant Mol Biol 52:561–591. doi:10.1146/annurev.arplant.52.1.561

Murphy MP (2009) How mitochondria produce reactive oxygen species. Biochem J 417:1–13. doi:10.1042/BJ20081386

Noskov SY, Rostovtseva TK, Bezrukov SM (2013) ATP transport through VDAC and the VDAC–tubulin complex probed by equilibrium and Nonequilibrium MD simulations. Biochemistry 52:9246–9256. doi:10.1021/bi4011495

Ott M, Gogvadze V, Orrenius S, Zhivotovsky B (2007) Mitochondria, oxidative stress and cell death. Apoptosis 12:913–922. doi:10.1007/s10495-007-0756-2

Owsianowski E, Walter D, Fahrenkrog B (2008) Negative regulation of apoptosis in yeast. Biochim Biophys Acta, Mol Cell Res 1783:1303–1310. http://dx.doi.org/10.1016/j.bbamcr.2008.03.006

Peng S, Blachly-Dyson E, Forte M, Colombini M (1992) Large scale rearrangement of protein domains is associated with voltage gating of the VDAC channel. Biophys J 62:123–135

Qiao J, Mitsuhara I, Yazaki Y, Sakano K, Gotoh Y, Miura M, Ohashi Y (2002) Enhanced resistance to salt, cold and wound stresses by overproduction of animal cell death suppressors Bcl-xL and Ced-9 in tobacco cells — their possible contribution through improved function of Organella. Plant Cell Physiol 43:992–1005. doi:10.1093/pcp/pcf122

Radak Z, Zhao Z, Goto S, Koltai E (2011) Age-associated neurodegeneration and oxidative damage to lipids, proteins and DNA. Mol Asp Med 32:305–315. doi:10.1016/j.mam.2011.10.010

Rao RSP, Salvato F, Thal B, Eubel H, Thelen JJ, Møller IM (2016) The proteome of higher plant mitochondria. Mitochondrion In Press. doi:10.1016/j.mito.2016.07.002

Rhoads DM, Umbach AL, Subbaiah CC, Siedow JN (2006) Mitochondrial reactive oxygen species. Contribution to oxidative stress and Interorganellar signaling. Plant Physiol 141:357–366. doi:10.1104/pp.106.079129

Robert N, D'Erfurth I, Marmagne A, Erhardt M, Allot M, Boivin K, Gissot L, Monachello D, Michaud M, Duchêne A-M, Barbier-Brygoo H, Maréchal-Drouard L, Ephritikhine G, Filleur S (2012) Voltage-dependent-anion-channels (VDACs) in Arabidopsis have a dual localization in the cell but show a distinct role in mitochondria. Plant Mol Biol 78:431–446. doi:10.1007/s11103-012-9874-5

Robey RB, Hay N (2006) Mitochondrial hexokinases, novel mediators of the antiapoptotic effects of growth factors and Akt. Oncogene 25:4683–4696. doi:10.1038/sj.onc.1209595

Rostovtseva TK, Bezrukov SM (1998) ATP transport through a single VDAC channel, studied by noise analysis. Biophys J 74:A383–A383

Rostovtseva TK, Bezrukov SM (2008) VDAC regulation: role of cytosolic proteins and mitochondrial lipids. J Bioenerg Biomembr 40:163–170. doi:10.1007/s10863-008-9145-y

Rostovtseva TK, Bezrukov SM (2012) VDAC inhibition by tubulin and its physiological implications. Biochim Biophys Acta Biomembr 1818:1526–1535. doi:10.1016/j.bbamem.2011.11.004

Rostovtseva TK, Colombini M (1996) ATP flux is controlled by a voltage-gated channel from the mitochondrial outer membrane. J Biol Chem 271:28006–28008. doi:10.1074/jbc.271.45.28006

Rostovtseva TK, Colombini M (1997) VDAC channels mediate and gate the flow of ATP: implications for the regulation of mitochondrial function. Biophys J 72:1954–1962. doi:10.1016/S0006-3495(97)78841-6

Rostovtseva TK, Komarov A, Bezrukov SM, Colombini M (2002) Dynamics of nucleotides in VDAC channels: structure-specific noise generation. Biophys J 82:193–205. doi:10.1016/S0006-3495(02)75386-1

Rostovtseva TK, Tan W, Colombini M (2005) On the role of VDAC in apoptosis: fact and fiction. J Bioenerg Biomembr 37:129–142. doi:10.1007/s10863-005-6566-8

Rostovtseva TK, Kazemi N, Weinrich M, Bezrukov SM (2006) Voltage gating of VDAC is regulated by nonlamellar lipids of mitochondrial membranes. J Biol Chem 281:37496–37506

Salinas T, Duchene AM, Delage L, Nilsson S, Glaser E, Zaepfel M, Marechal-Drouard L (2006) The voltage-dependent anion channel, a major component of the tRNA import machinery in plant mitochondria. Proc Natl Acad Sci U S A 103:18362–18367

Salinas T, Duchene AM, Marechal-Drouard L (2008) Recent advances in tRNA mitochondrial import. Trends Biochem 33:320–329

Schein SJ, Colombini M, Finkelstein A (1976) Reconstitution in planar lipid bilayers of a voltage-dependent anion-selective channel obtained from paramecium mitochondria. J Membr Biol 30:99–120. doi:10.1007/BF01869662

Schmid A, Kromer S, Heldt HW, Benz R (1992) Identification of two general diffusion channels in the outer membrane of pea mitochondria. BiochimBiophysActa 1112:174–180. doi:10.1016/0005-2736(92)90389-4

Schneider R, Etzkorn M, Giller K, Daebel V, Eisfeld J, Zweckstetter M, Griesinger C, Becker S, Lange A (2010) The native conformation of the human VDAC1 N terminus. Angew Chemie Int Ed 49:1882–1885. doi:10.1002/anie.200906241

Schredelseker J, Paz A, Lopez CJ, Altenbach C, Leung CS, Drexler MK, Chen J-N, Hubbell WL, Abramson J (2014) High resolution structure and double electron-electron resonance of the zebrafish voltage-dependent Anion Channel 2 reveal an oligomeric population. J Biol Chem 289:12566–12577. doi:10.1074/jbc.M113.497438

Sevilla F, Camejo D, Ortiz-Espín A, Calderón A, Lázaro JJ, Jiménez A (2015) The thioredoxin/peroxiredoxin/sulfiredoxin system: current overview on its redox function in plants and regulation by reactive oxygen and nitrogen species. J Exp Bot 66:2945–2955. doi:10.1093/jxb/erv146
Shanmugavadivu B, Apell HJ, Meins T, Zeth K, Kleinschmidt JH (2007) Correct folding of the beta-barrel of the human membrane protein VDAC requires a lipid bilayer. J Mol Biol 368:66–78. doi:10.1016/j.jmb.2007.01.066
Shao L, Kinnally KW, Mannella CA (1996) Circular dichroism studies of the mitochondrial channel, VDAC, from Neurospora crassa. Biophys J 71:778–786. doi:10.1016/S0006-3495(96)79277-9
Shoshan-Barmatz V, De Pinto V, Zweckstetter M, Raviv Z, Keinan N, Arbel N (2010) VDAC, a multi-functional mitochondrial protein regulating cell life and death. Mol Asp Med 31:227–285. doi:10.1016/j.mam.2010.03.002
Shoshan-Barmatz V, Ben-Hail D, Admoni L, Krelin Y, Tripathi SS (2015) The mitochondrial voltage-dependent anion channel 1 in tumor cells. Biochim Biophys Acta Biomembr 1848:2547–2575. doi:10.1016/j.bbamem.2014.10.040
Šileikytė J, Blachly-Dyson E, Sewell R, Carpi A, Menabò R, Di Lisa F, Ricchelli F, Bernardi P, Forte M (2014) Regulation of the mitochondrial permeability transition pore by the outer membrane does not involve the peripheral benzodiazepine receptor (translocator protein of 18 kDa (TSPO)). J Biol Chem 289:13769–13781. doi:10.1074/jbc.M114.549634
Simamura E, Hirai K-I, Shimada H, Koyama J, Niwa Y, Shimizu S (2006) Furanonaphthoquinones cause apoptosis of cancer cells by inducing the production of reactive oxygen species by the mitochondrial voltage-dependent anion channel. Cancer Biol Ther 5:1523–1529. doi:10.4161/cbt.5.11.3302
Smack DP, Colombini M (1985) Voltage-dependent channels found in the membrane fraction of corn mitochondria. Plant Physiol 79:1094–1097
Sun L, Shukair S, Naik TJ, Moazed F, Ardehali H (2008) Glucose phosphorylation and mitochondrial binding are required for the protective effects of hexokinases I and II. Mol Cell Biol 28:1007–1017. doi:10.1128/MCB.00224-07
Sweetlove LJ, Heazlewood JL, Herald V, Holtzapffel R, Day DA, Leaver CJ, Millar AH (2002) The impact of oxidative stress on Arabidopsis mitochondria. Plant J 32:891–904. doi:10.1046/j.1365-313X.2002.01474.x
Swidzinski JA, Leaver CJ, Sweetlove LJ (2004) A proteomic analysis of plant programmed cell death. Phytochemistry 65:1829–1838. doi:10.1016/j.phytochem.2004.04.020
Takahashi Y, Tateda C (2013) The functions of voltage-dependent anion channels in plants. Apoptosis 18:917–924. doi:10.1007/s10495-013-0845-3
Tateda C, Yamashita K, Takahashi F, Kusano T, Takahashi Y (2009) Plant voltage-dependent anion channels are involved in host defense against Pseudomonas cichorii and in Bax-induced cell death. Plant Cell Rep 28:41–51. doi:10.1007/s00299-008-0630-x
Tateda C, Watanabe K, Kusano T, Takahashi Y (2011) Molecular and genetic characterization of the gene family encoding the voltage-dependent anion channel in Arabidopsis. J Exp Bot 62:4773–4785. doi:10.1093/jxb/err113
Tateda C, Kusano T, Takahashi Y (2012) The Arabidopsis voltage-dependent anion channel 2 is required for plant growth. Plant Signal Behav 7:31–33. doi:10.4161/psb.7.1.18394
Tikunov A, Johnson CB, Pediaditakis P, Markevich N, Macdonald JM, Lemasters JJ, Holmuhamedov E (2010) Closure of VDAC causes oxidative stress and accelerates the Ca2+-induced mitochondrial permeability transition in rat liver mitochondria. Arch Biochem Biophys 495:174–181. doi:10.1016/j.abb.2010.01.008
Tomasello F, Messina A, Lartigue L, Schembri L, Medina C, Reina S, Thoraval D, Crouzet M, Ichas F, De Pinto V, De Giorgi F (2009) Outer membrane VDAC1 controls permeability transition of the inner mitochondrial membrane in cellulo during stress-induced apoptosis. Cell Res 19:1363–1376. doi:10.1038/cr.2009.98
Ujwal R, Cascio D, Colletier JP, Faham S, Zhang J, Toro L, Ping P, Abramson J (2008) The crystal structure of mouse VDAC1 at 2.3 A resolution reveals mechanistic insights into metabolite gating. Proc Natl Acad Sci USA 105:17742–17747. doi:10.1073/pnas.0809634105

Van Aken O, Van Breusegem F (2015) Licensed to kill: mitochondria, chloroplasts, and cell death. Trends Plant Sci 20:754–766. doi:10.1016/j.tplants.2015.08.002

Vanlerberghe G (2013) Alternative oxidase: a mitochondrial respiratory pathway to maintain metabolic and signaling homeostasis during abiotic and biotic stress in plants. Int J Mol Sci 14:6805–6847. doi:10.3390/ijms14046805

Vianello A, Casolo V, Petrussa E, Peresson C, Patui S, Bertolini A, Passamonti S, Braidot E, Zancani M (2012) The mitochondrial permeability transition pore (PTP) — an example of multiple molecular exaptation? Biochim Biophys Acta Bioenerg 1817:2072–2086. http://dx.doi.org/10.1016/j.bbabio.2012.06.620

Villinger S, Giller K, Bayrhuber M, Lange A, Griesinger C, Becker S, Zweckstetter M (2014) Nucleotide interactions of the human voltage-dependent Anion Channel. J Biol Chem 289:13397–13406. doi:10.1074/jbc.M113.524173

Woodson JD, Chory J (2008) Coordination of gene expression between organellar and nuclear genomes. Nat Rev Genet 9:383–395. doi:10.1038/nrg2348

Yan J, He H, Tong S, Zhang W, Li X, Yang Y (2009) Voltage-dependent Anion Channel 2 of *Arabidopsis thaliana* (AtVDAC2) is involved in ABA-mediated early seedling development. Int J Mol Sci 10:2476–2486. doi:10.3390/ijms10062476

Yehezkel G, Hadad N, Zaid H, Sivan S, Shoshan-Barmatz V (2006) Nucleotide-binding sites in the voltage-dependent anion channel – characterization and localization. J Biol Chem 281:5938–5946

Yehezkel G, Abu-Hamad S, Shoshan-Barmatz V (2007) An N-terminal nucleotide-binding site in VDAC1: involvement in regulating mitochondrial function. J Cell Physiol 212:551–561

Young MJ, Bay DC, Hausner G, Court DA (2007) The evolutionary history of mitochondrial porins. BMC Evol Biol 7:31. doi:10.1186/1471-2148-7-31

Zalman LS, Nikaido H, Kagawa Y (1980) Mitochondrial outer membrane contains a protein producing nonspecific diffusion channels. J Biol Chem 255:1771–1774

Zambrowicz EB, Colombini M (1993) Zero-current potentials in a large membrane channel: a simple theory accounts for complex behavior. Biophys J 65:1093–1100. doi:10.1016/S0006-3495(93)81148-2

Zancani M, Casolo V, Petrussa E, Peresson C, Patui S, Bertolini A, De Col V, Braidot E, Boscutti F, Vianello A (2015) The permeability transition in plant mitochondria: the missing link. Front Plant Sci 6:1120. doi:10.3389/fpls.2015.01120

Zhang M, Takano T, Liu S, Zhang X (2015) Arabidopsis mitochondrial voltage-dependent anion channel 3 (AtVDAC3) protein interacts with thioredoxin m2. FEBS Lett 589:1207–1213. doi:10.1016/j.febslet.2015.03.034

Zizi M, Forte M, Blachly-Dyson E, Colombini M (1994) NADH regulates the gating of VDAC, the mitochondrial outer membrane channel. J Biol Chem 269:1614–1616

Chapter 8
Lipids in Regulation of the Mitochondrial Outer Membrane Permeability, Bioenergetics, and Metabolism

Tatiana K. Rostovtseva, David P. Hoogerheide, Amandine Rovini, and Sergey M. Bezrukov

8.1 Introduction

Mitochondrial functions extend far beyond energy production in the form of ATP. Mitochondria play a crucial role in Ca^{2+} homeostasis, in steroid biosynthesis, and in programmed cell death and as a source of the endogenous toxic reactive oxygen species (ROS) in the cell. The involvement of mitochondrial membranes in all these processes is well recognized, especially with regard to lipid synthesis and trafficking or communication with other organelles. The two mitochondrial membranes, the mitochondrial inner membrane (MIM) and the mitochondrial outer membrane (MOM), play rather different roles in mitochondrial function, which are intimately related to their different lipid compositions. The MIM's signature is a high cardiolipin (CL) content (up to 20%), while the MOM has high cholesterol content (up to 10%) (de Kroon et al. 1997; Flis and Daum 2013). The MIM's prominent roles in oxidative phosphorylation, energy production, Ca^{2+} signaling, and mitochondrial metabolism are well established. By contrast, the role of the MOM was previously thought to be the simple confinement of proteins and metabolites in the crowded intermembrane space. This viewpoint changed drastically in the mid-1990s with the discovery of the MOM permeabilization (MOMP) process at the

T.K. Rostovtseva (✉) • A. Rovini • S.M. Bezrukov
Section on Molecular Transport, Eunice Kennedy Shriver National Institute of Child Health and Human Development, National Institutes of Health, Bethesda, MD 20892, USA
e-mail: rostovtt@mail.nih.gov

D.P. Hoogerheide
Center for Neutron Research, National Institute of Standards and Technology, Gaithersburg, MD 20899, USA

T.K. Rostovtseva (ed.), *Molecular Basis for Mitochondrial Signaling*, Biological and Medical Physics, Biomedical Engineering,
DOI 10.1007/978-3-319-55539-3_8

early stage of apoptosis. Since then, the role of the MOM as a platform for apoptosis execution by Bcl-2 family proteins has been well documented (e.g., see Chipuk et al. 2006). The involvement of MOM lipids in formation and promotion of supramolecular openings induced by the pro-apoptotic proteins Bax, Bak, Bid, and others has been demonstrated (Basanez et al. 2002; Kuwana et al. 2002; Terrones et al. 2004; Yethon et al. 2003), leading to the proposal that together with the MOM lipids, the pro-apoptotic Bax/Bak oligomeric complex is directly involved in the formation of the so-called lipidic pores in the MOM (Landeta et al. 2011; Schafer et al. 2009; Terrones et al. 2004). Furthermore, lipid ceramide was itself shown to form large, cytochrome c-permeable pores in the MOM and thus be sufficient to induce MOMP and promote apoptosis (Colombini 2010; Siskind et al. 2002). The unique MIM phospholipid CL is associated with all of the major players in oxidative phosphorylation. Due to its characteristic nonlamellar shape, with four poly-unsaturated acyl chains, CL also plays a pivotal role in maintaining the curvature of the highly bent cristae. In addition to its structural and protein-lipid interaction functions in the MIM, CL is essential for anchoring pro-apoptotic tBid and thus promoting MOMP (Gonzalvez et al. 2005; Shamas-Din et al. 2015). The formation of the apoptotic "activation platform" composed of CL, caspase-8, and full-length Bid at the mitochondrial contact sites has been suggested (Gonzalvez et al. 2008), directly connecting mitochondrial lipids with Barth syndrome, a chromosome X-linked cardioskeletal myopathy and neutropenia that is often fatal in infancy and early childhood due to heart failure and bacterial infections (Schlame and Ren 2006). This is just one of numerous cases in which mitochondrial lipids are involved in human diseases.

The few examples mentioned above represent the conventional understanding of the functional and structural roles of the MOM lipids in the mitochondrial apoptotic pathway. In the current review, we discuss a different, complementary view of the MOM as a platform for cytosolic proteins to regulate MOM permeability by their direct functional interaction with the voltage-dependent anion channel (VDAC). We focus particularly on the role of the mitochondrial lipids in this regulation.

8.2 VDAC Voltage-Gating and Regulation of ATP Flux

As the major channel and most abundant protein in the MOM, VDAC is responsible for most of the metabolite flux in and out of mitochondria. VDAC was shown to be involved in a wide variety of mitochondria-associated pathologies, from various forms of cancer to neurodegeneration (Shoshan-Barmatz et al. 2010). It is therefore not surprising that VDAC has emerged as a promising pharmacological target (Shoshan-Barmatz and Ben-Hail 2012). The uniqueness of this large, passive, weakly selective β-barrel channel arises mainly from its location at the interface between the mitochondria and the cytosol (Colombini 2004), where it serves as a pathway for all mitochondrial water-soluble respiratory substrates, such as ATP,

ADP, and small ions (Colombini et al. 1996; DeHart et al. 2014; Rostovtseva et al. 2005). By contrast, there is a large variety of highly specified carriers and exchangers to translocate these substrates across the MIM. The weak anionic selectivity of the VDAC pore favors transport of negatively charged mitochondrial metabolites (Colombini et al. 1996). It is generally believed that VDAC regulates metabolite fluxes using its conserved ability to "gate" or adopt different conducting and, crucially, ion-selective states (Colombini 2004; Hodge and Colombini 1997; Lemasters and Holmuhamedov 2006; Lemasters et al. 2012; Rostovtseva and Colombini 1996). However, it is not yet known with certainty whether VDAC gating occurs in vivo, since all electrophysiological data on VDAC function have thus far been obtained using an in vitro system of VDAC reconstituted into planar lipid bilayers, where gating is induced by the applied voltage (Colombini 1989).

One major dilemma is that the very existence of a significant potential across MOM in vivo is uncertain. Conventional thinking holds that this potential is essentially zero due to the high abundance of VDAC in the MOM (e.g., see discussions in Colombini 2004; Lemeshko 2006; Rostovtseva and Bezrukov 2012 and Chap. 9 of the current volume). In this view, the main source of the potential across the MOM is the so-called Donnan potential, which arises from the large difference in the concentrations of VDAC-impermeable charged macromolecules (e.g., the 12-kDa cytochrome c protein, which carries nine positive charges) in the mitochondrial intermembrane space and in the cytosol (Colombini 2004). The few available estimates of the MOM transmembrane potential range from 10 mV (Lemeshko 2006) to as high as 46 mV (Porcelli et al. 2005) and even higher (Lemeshko 2014a, b, 2016). Recent work suggests that other biochemical processes can enhance the Donnan potential considerably. Lemeshko (2014a, b, 2016 and Chap. 9 of the current volume) proposes that VDAC complexes with hexokinases at the cytosolic side and/or with mitochondrial creatine kinase in the intermembrane space could play the role of "batteries," generating a potential across the MOM as high as 50 mV. In this picture, the Gibbs free energy of kinase reactions is the driving force. Importantly, the cytoplasmic side of the MOM is negative relative to the potential of the intermembrane space. Regardless of the source and magnitude of the MOM potential in vivo, it has been experimentally demonstrated using reconstituted VDAC that ATP passes readily through the open, weakly anion-selective VDAC state, but essentially does not pass through its low-conducting, weakly cation-selective states induced by the applied potential in the voltage-clamp mode (Rostovtseva and Colombini 1996, 1997). Furthermore, using current noise analysis on a single VDAC channel in the presence of relatively high molecular weight negatively charged metabolites, it was found that VDAC could discriminate between different adenine mono- and dinucleotides and also between similarly charged molecules, such as ATP and UTP (Rostovtseva et al. 2002a, b). These findings suggest that VDAC has been evolutionary selected to regulate fluxes of metabolites by dynamically adapting the electrostatic environment of its pore in favor of adenine nucleotides (Noskov et al. 2016). It is then natural to expect the existence of endogenous cytosolic (and perhaps mitochondrial also) proteins that are able to efficiently regulate VDAC permeability for the major mitochondrial metabolites, especially ATP and ADP.

It was only a matter of time before such endogenous VDAC regulators were found. Two apparently unrelated cytosolic proteins, dimeric tubulin and α-synuclein (α-syn), were discovered to reversibly block reconstituted VDAC with nanomolar (nmol/l) efficiency and thereby control the fluxes of ATP, ADP, and other metabolites through VDAC (Rostovtseva et al. 2008, 2015). Notably, while both α-syn and tubulin are abundant cytosolic proteins with other functions (well known for tubulin, but not as clear for α-syn, as discussed below), they appear to double as membrane-associated proteins, displaying high affinity to mitochondrial membranes. In this second role, they are potent regulators of VDAC.

8.3 VDAC Regulation by Tubulin and α-Synuclein

Tubulin is an abundant cytosolic protein known primarily for its role as the building block of microtubules. It is a heterodimer with α- and β-subunits forming an acidic, water-soluble, compactly folded 110-kDa protein with a well-defined crystal structure (Nogales et al. 1998). Each subunit has an unstructured C-terminal tail (CTT) composed of 11–20 amino acids (Westermann and Weber 2003) exposed at the protein surface (Fig. 8.1a). Both α- and β-tubulin CTTs are highly negatively

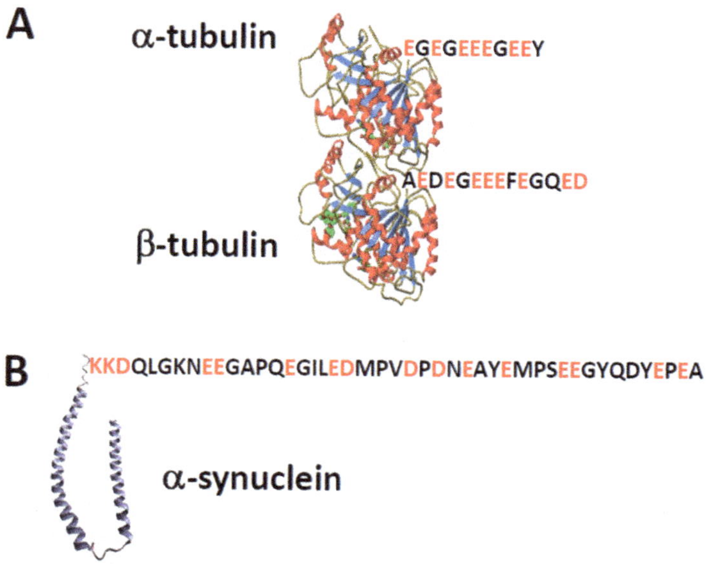

Fig. 8.1 Both VDAC blockers, the α/β-tubulin heterodimer and α-synuclein, possess extended disordered anionic C-terminal tails. Tubulin structure is adapted from (Nogales et al. 1998)

charged and are essentially poly-Glu peptides. In microtubules, dimeric tubulin is assembled in such a way that the exposed, growing end of the microtubule is always the β-subunit. As a result, microtubule-targeting agents (MTAs) used in chemotherapy bind exclusively to the β-subunit (Field et al. 2014). Until recently, any physiological role of unassembled, cytosol-solubilized dimeric tubulin, other than as a supply for polymerization into microtubules, remained unknown.

α-Syn, by contrast, is a small, 14-kDa, 140-amino acid, intrinsically disordered protein (Fig. 8.1b) highly expressed in the central nervous system and constituting up to 1% of total cytosolic protein in normal brain cells (Kruger et al. 2000). It is the major component of the Lewy bodies characteristically observed in the brains of Parkinson's disease (PD) patients (Spillantini et al. 1997). Based on the observation that Lewy bodies consist primarily of fibrillary α-syn, most studies have focused on the pathological role of α-syn aggregates. However, the precise role of α-syn in normal brain cells and in α-synucleinopathies, particularly in its monomeric form, remains largely mysterious.

What do these two genetically, structurally, and physiologically unrelated proteins have in common? In this review, we discuss the following similarities: both proteins (1) possess a disordered highly negative charged C-terminus; (2) induce characteristic rapid reversible blockages of VDAC reconstituted into planar lipid membranes with nanomolar efficiency; (3) are "amphitropic" proteins, i.e., water-soluble proteins that are reversibly associated with cellular membranes under certain physiological conditions; (4) specifically bind to mitochondrial membranes in live cells; and (5) modulate mitochondrial metabolism and respiration by regulating VDAC permeability in a lipid-dependent manner.

8.3.1 *Regulation of VDAC Permeability by Dimeric Tubulin and Monomeric α-Synuclein*

In these experiments, a single VDAC channel spans a planar bilayer lipid membrane that separates and electrically isolates two buffer-filled compartments (Rostovtseva and Bezrukov 2015). The ionic current through the VDAC channel is generated by an applied transmembrane potential and is monitored for changes induced by the addition of dimeric tubulin or α-syn. Single-channel recordings of VDAC reconstituted into planar lipid bilayers yield strikingly similar blockage events induced by addition of dimeric tubulin and α-syn to the membrane-bathing buffer solution at nanomolar concentrations (Fig. 8.2a, b). Two representative experiments in which 50 nM tubulin or α-syn added to one side of the membrane induce fast time-resolved transitions (*insets* at a finer time scale) between VDAC's high-conductance open state and low-conductance blocked state are presented in Fig. 8.2a, b. Without these inhibitors, the VDAC conductance at low applied potentials (<30 mV) is very stable and essentially lacks transitions to the low-conductance state (*left traces* in Fig. 8.2a, b).

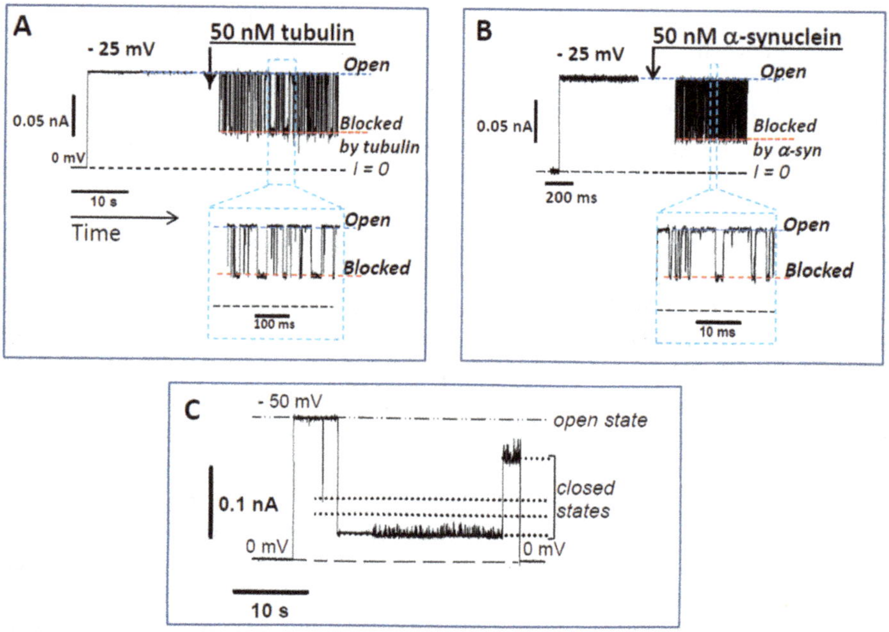

Fig. 8.2 Dimeric tubulin (**a**) and α-synuclein (**b**) induce fast, reversible, partial blockages of VDAC reconstituted into planar lipid bilayers, with kinetics that are dramatically different from those of VDAC closure induced by applied voltage (voltage gating) (**c**). (**a**, **b**) Representative current records of single VDAC channels before and after addition of 50 nM tubulin (**a**) or α-syn (**b**) to one side of the membrane at −25 mV applied voltage (potential is more negative at the side of tubulin or α-syn addition). The channel conductance fluctuates between the high-conducting, "open," and low-conducting, "blocked", state with well time-resolved blockage events shown in *insets* at the finer time scales. *Long-dashed lines* indicate the zero-current level, and *short-dashed lines* show the open and blocked states. (**c**) Typical VDAC voltage-induced gating under −50 mV applied voltage. Under the applied potential, the channel conductance moves from a single open state (*dash-dot-dot lines*) to various low-conducting or "closed" states (*dotted lines*). The medium was 1 M KCl buffered with 5 mM HEPES at pH 7.4. VDAC was isolated from mitochondria of *N. crassa* (**a**, **c**) or rat liver (**b**)

The channel can maintain the open state for up to a few hours at low potentials under the experimental conditions represented by Fig. 8.2a, b. The conductance of the open state in the presence of either inhibitor remains the same as in the control, and the conductance of the blocked state is ~0.4 of the open-state conductance. This is in striking contrast to the typical VDAC voltage-induced gating, which is characterized by long (10–100 s) closures to multiple low-conductance ("closed") states (Fig. 8.2c).

Both tubulin and α-syn induce VDAC blockage in a concentration-dependent manner. The number of blockage events increases with the blocker protein's bulk concentration (Rostovtseva et al. 2012, 2015). The concentration dependence of the blockage on-rate is described by a simple ligand-binding equation with K_d in the 10–50 nM range for both inhibitors, depending on the experimental conditions such as the membrane lipid composition or pH. At low (<50 nM) tubulin or α-syn concentrations, the rates of interaction with VDAC can, to a first approximation, be described by a first-order reaction, where the rate of blockage events (the "on-rate") linearly increases with blocker concentration, while the characteristic duration of the blockages (the inverse "off-rate") is concentration independent.

8.3.2 A General Model of VDAC Blockage, Requiring a Channel Inhibitor to Have an Anionic Disordered Domain and Lipid-Binding Domain

Figure 8.1 illustrates that the existence of a highly negatively charged unstructured CTT is a common feature of tubulin and α-syn. Both tubulin and α-syn block VDAC measurably only when a negative potential is applied from the side of protein addition (Rostovtseva and Bezrukov 2015; Rostovtseva et al. 2015). When the sign of the potential is reversed, no blockage events are observed, and the channel open-state conductance is as steady as in a control experiment without protein addition (see Fig. 1.5 in (Rostovtseva and Bezrukov 2015). This observation, and the fact that tubulin with proteolytically cleaved CTTs does not induce characteristic VDAC blockages even at micromolar concentrations (Rostovtseva et al. 2008), suggest that the negatively charged CTTs are responsible for the channel blockage by tubulin. On the other hand, synthetic peptides of α- and β-tubulin CTTs do not measurably block the channel, up to 10 μM concentration (Rostovtseva et al. 2008). Experiments with a peptide consisting of the 45 C-terminal amino acids of α-syn likewise did not reveal interaction with VDAC (Rostovtseva et al. 2015). Altogether, these observations suggest that while the disordered acidic C-terminal peptide of either tubulin or α-syn is required for VDAC blockage, it should be attached to an "anchor" that holds it at the channel entrance (Fig. 8.3), preventing free translocation through the pore, which is expected to occur at time scales too fast to be experimentally observed. Tubulin's bulky body is about twice as large as the VDAC pore entrance and cannot translocate; in the case of α-syn, its membrane-binding domain is responsible for this anchoring function. In both cases, the increased applied negative voltage keeps the anionic CTT inside the positively charged pore longer (Fig. 8.4c, d). Indeed, experiments show that the dwell time of tubulin- and α-syn-induced blockages is highly voltage dependent and increases exponentially with the

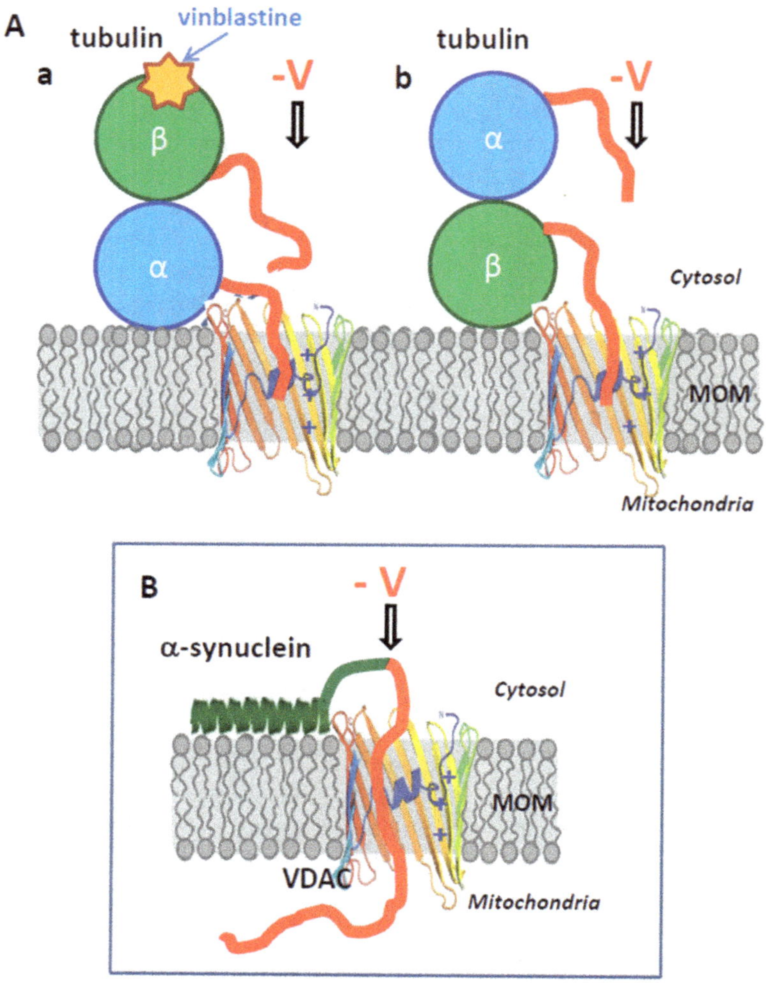

Fig. 8.3 Models of VDAC blockage by tubulin (**a**) and α-syn (**b**), in which the negatively charged C-terminal tails of tubulin or α-syn partially block the channel by entering the net positive VDAC pore in its open state under negative applied potentials. (**a**) Hypothetical structural models of dimeric tubulin binding to the membrane. (*a*) For tubulin bound by the α-subunit, both CTTs of the α- and β-subunits can reach the VDAC pore, while the vinblastine-binding site remains exposed. (*b*) For tubulin bound to the membrane by the β-subunit, only the β-subunit CTT is able to block the channel, while the vinblastine-binding site is inaccessible. Intermediate binding configurations, including the case where both subunits are responsible for membrane binding, are also possible and would in general present both CTTs to the VDAC pore (not shown). (**b**) Structural model of membrane-bound α-syn with its α-helical N-terminal domain bound to the membrane and the acidic disordered CTT blocking the VDAC pore. The VDAC β-barrel (3D model of mouse VDAC1 is adapted from Ujwal et al. (2008)) is shown embedded into lipid bilayer)

applied voltage (Fig. 8.4c) (Rostovtseva et al. 2012). The steep voltage dependence of the open probability yields the effective gating charge of 10–13 elementary charges depending on the salt concentration of the membrane-bathing buffer (Gurnev et al. 2012). Such a high gating charge is characteristic of the most sensitive voltage-gated channels of neurophysiology to date (Swartz 2008).

Additional evidence that the tubulin CTT partitions into the VDAC pore was obtained in experiments designed to probe the tubulin-blocked state using different approaches (Gurnev et al. 2011) in combination with molecular dynamics (MD) simulations (Noskov et al. 2013). It was shown that the small-ion selectivity switches from anionic to cationic when the channel moves from the open state to the tubulin-blocked state (Gurnev et al. 2011). These results indicate that the highly negatively charged tubulin CTT, by entering a net positive pore, reverses the net charge of the pore interior. By measuring nonelectrolyte polymer partitioning into the ion channel, it was found that the dimensions of the blocked pore are significantly smaller than those of the open state (Gurnev et al. 2011). MD simulations of the α-tubulin-VDAC1 complex demonstrate that in the presence of the unstructured CTT of α-tubulin in the VDAC1 pore, the pore conductance decreases by about 60% and switches its selectivity from anion- to cation-preferring channels (Noskov et al. 2013), thus confirming electrophysiological data obtained on reconstituted VDAC. The negatively charged C-terminus of the bound α-tubulin molecule is complemented by the VDAC pore providing matching basic residues that form stable salt bridges involving Arg15, Lys20, Lys12, and Lys32 of VDAC (Noskov et al. 2013). Finally, and most importantly, channel experiments and MD simulations demonstrate that ATP is excluded from the tubulin-blocked state (Gurnev et al. 2011; Noskov et al. 2013). Altogether, these results provide strong evidence that tubulin is a potent regulator of VDAC, which could efficiently modulate fluxes of nucleotides through the channel and thus control mitochondria respiration. Experiments with isolated mitochondria (Monge et al. 2008; Rostovtseva et al. 2008) and human hepatoma HepG2 cells (Maldonado et al. 2011, 2013) confirm that the VDAC-tubulin interaction is functionally important for regulating mitochondrial respiration.

For intrinsically disordered α-syn, where the whole molecule can, in principle, translocate through the channel because a single polypeptide strand fits comfortably into the ~2.5 to 2.7 nm diameter VDAC pore, the relationship between dwell times and applied voltages becomes more complex than for tubulin. In contrast to the tubulin-induced blockages, the dwell time of α-syn-induced blockages exhibits a biphasic dependence on the applied potential (Rostovtseva et al. 2015) (Fig. 8.4d). At voltages lower than a certain value, the so-called turnover potential, (indicated as V^* in Fig. 8.4e) dwell time increases with voltage in a manner characteristic for the blockage regime (Hoogerheide et al. 2016). At voltages higher than the turnover potential, the dwell time decreases with voltage, signifying a predominance of translocation (Fig. 8.4e). Based on these data, a model of α-syn interaction with VDAC was proposed, where the anionic C-terminus partially blocks the positively charged VDAC pore and the membrane-bound N-terminus holds it at the membrane surface preventing complete translocation through the channel at voltages smaller

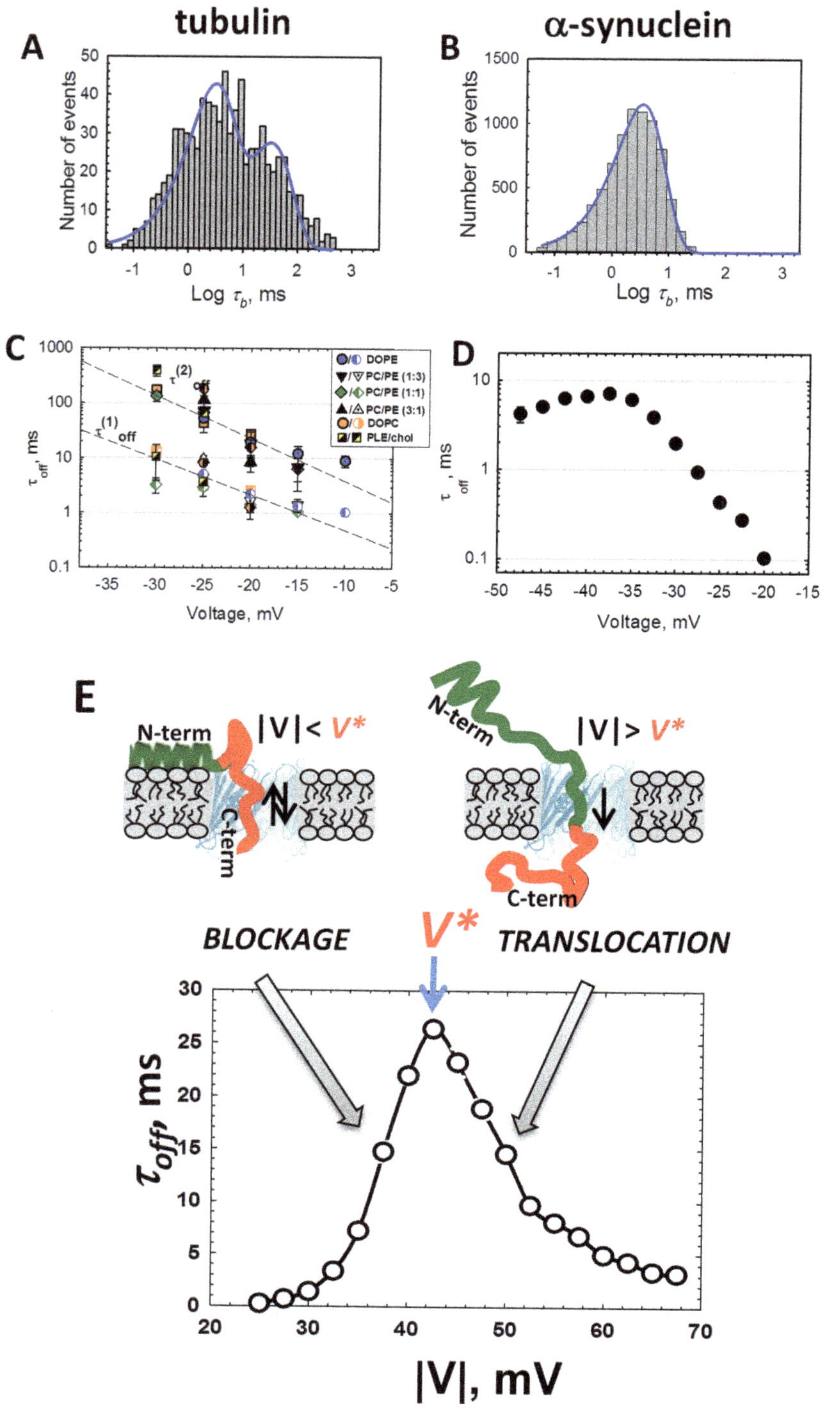

Fig. 8.4 VDAC blockages by tubulin or α-syn are highly voltage dependent. (**a, b**) Log-binned distribution of the time of blockage events induced by tubulin (**a**) or α-syn (**b**) obtained from statistical analysis of current records such as those shown in Fig. 8.2a, b. Distributions of the times

than 30 mV (Fig. 8.4e) (Rostovtseva et al. 2015). Therefore, under certain conditions, such as the applied potential, lipid composition, and ionic strength of buffer solution, α-syn can either block VDAC in qualitatively similar manner to tubulin or translocate through the channel (Fig. 8.4e). The latter process, if confirmed by future experiments in live cells, implies that VDAC serves as a pathway for α-syn into the mitochondrial space, where it could directly target electron transport chain complexes of the inner membrane as in vivo experiments suggest (Devi et al. 2008; Elkon et al. 2002; Ellis et al. 2005; Liu et al. 2009; Luth et al. 2014) as well as other intermembrane space proteins.

Importantly, the proposed model of VDAC inhibition (Fig. 8.3) is quite general and does not require any specific interaction between the VDAC pore and inhibitor. This model explains satisfactorily how two such different proteins as tubulin and α-syn induce qualitatively similar blockages of VDAC and suggests that other cytosolic proteins may also be involved. Generalizing based on these two examples, the architecture of a VDAC inhibitor should fulfil two main requirements: have a negatively charged disordered C- or N-terminus and a membrane-binding domain. Then, it is natural to expect that the membrane lipid composition is an important factor in interaction of both tubulin and α-syn with reconstituted VDAC.

8.4 Tubulin and α-Synuclein Interactions with VDAC Strongly Depend on Membrane Lipid Composition

8.4.1 Effect of Membrane Lipid Composition on VDAC-Tubulin Interaction

The characteristic on-rate constant, k_{on}, of VDAC-tubulin binding depends on both the hydrophobic and polar parts of the phospholipid (Fig. 8.5a–c). The effect of lipid charge is observed primarily at the physiologically low salt concentrations: addition of the negatively charged diphytanoyl-phosphatidylserine (DPhPS) to the neutral diphytanoyl-phosphatidylcholine (DPhPC) (PS:PC = 4:1) resulted in ~200

Fig. 8.4 (continued) spent by the channel in the tubulin-blocked state, τ_b, require at least two exponents for fitting, with characteristic times $\tau^{(1)}_{off}$ and $\tau^{(2)}_{off}$ shown by the *solid line* (**a**). The distribution of blockage events induced by α-syn, on the other hand, is satisfactorily described by single exponential fitting (*solid line*) (**b**). Applied voltages were −25 mV (**a**) and −35 mV (**b**). (**c**) Both characteristic tubulin-induced blockage times exponentially depend on the applied voltage and do not depend on the membrane lipid composition (indicated in the graph). (**d**) Voltage dependence of the blockage time induced by α-syn has a biphasic character. (**e**) The turnover potential V^* separates the blocking regime, where the absolute applied potential $|V| \leq V^*$, from the translocation regime, where $|V| \geq V^*$. At relatively low voltages, when the CTT (*red*) is captured in the pore, increasing the electrical field keeps it inside the pore longer, while the membrane-bound N-terminal domain (*green*) prevents translocation of the molecule. This results in reversible capture of the CTT and an exponential increase in the characteristic blockage time τ_{off}. Voltages higher than V^* detach the N-terminal domain from the membrane surface and allow the whole molecule to translocate through the channel, resulting in a decrease of τ_{off} (Adapted from Rostovtseva et al. (2015))

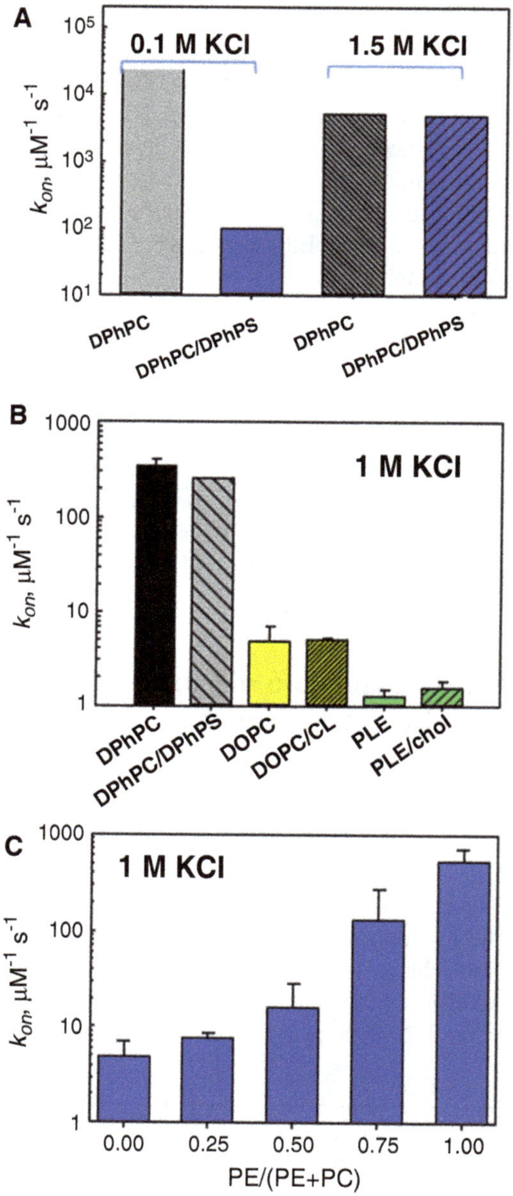

Fig. 8.5 Effect of lipid composition on the on-rate constants of VDAC blockage by tubulin. (**a**) Effect of charged lipids on the on-rate constant of VDAC blockage by tubulin depends on salt concentration. Membranes were formed from pure DPhPC or a DPhPC/DPhPS (1:4) mixture in 0.1 or 1.5 M KCl buffered with 5 mM HEPES at pH 7.4. (**b**) Acyl chain composition affects the k_{on} of VDAC blockage. Membranes were formed from pure DPhPC and DPhPC/DPhPS (1:1), from pure DOPC and DOPC/CL (1:0.08), and from PLE with and without cholesterol (10:1). (**c**) The k_{on} of blockages induced by tubulin increases with PE content in DOPC/DOPE mixture. The data were obtained at –25 mV of applied voltage. Medium consisted of 1 M KCL and 5 mM HEPES, pH 7.4 (**b**, **c**). VDAC was isolated from *N. crassa* mitochondria

times decrease of k_{on} in 0.1 M KCl (Fig. 8.5a) (Rostovtseva et al. 2012), which may suggest that the acidic regions of the tubulin globule may be sufficiently close to the negatively charged membrane surface to be electrostatically repelled. At a high salt concentration of 1.5 or 1 M KCl, the presence of DPhPS does not affect the on-rate (Fig. 8.5a). Salt concentration affects the on-rate of the blockage in both neutral and charged membranes: in 0.1 M KCl, the k_{on} was ~5 times higher than in 1.5 M KCl for the neutral DPhPC membranes but ~50 times lower than in 1.5 M KCl in the negatively charged DPhPC/DPhPS membranes (Fig. 8.5a). The type of lipid hydrocarbon acyl chains also significantly affects the blockage kinetics: when oleoyl chains in DOPC have been replaced with phytanoyl in DPhPC, the on-rate of tubulin blockage increased ~70 times (Fig. 8.5b). Surprisingly, the on-rates appeared to be consistently different when planar bilayers were formed from the same lipid composition but using different organic solvents, such as hexadecane or petroleum jelly (Rostovtseva et al. 2012). A comparison of the k_{on} values obtained in DPhPC membranes, formed after pretreatment of the orifice across which a planar membrane is made from two opposing lipid monolayers (Rostovtseva and Bezrukov 2015), with hexadecane (Fig. 8.5a) (see details in Bezrukov and Vodyanoy 1993) or petroleum jelly (Fig. 8.5b) (see details in Schein et al. 1976) is shown in Fig. 8.5a, b. These data indicate that tubulin-VDAC interaction is sensitive to the hydrophobic core of planar lipid bilayer, thus pointing out to a presence of the non-electrostatic component in this interaction. Additional evidence of the involvement of a hydrophobic component in tubulin-VDAC binding is that k_{on} increases gradually with the DOPE content of DOPC membranes and is ~200 times higher in a pure DOPE membrane than in a pure DOPC membrane (Fig. 8.5c) (Rostovtseva et al. 2012). These observations point to tubulin binding to the membrane as an important requirement for VDAC blockage, in which the on-rate depends on the surface concentration of tubulin.

The natural lipid composition of the rat liver MOM is a mixture of ~44 and 35% of phosphatidylcholine (PC) and phosphatidylethanolamine (PE), respectively, with ~20% of charged lipids consisting primarily of ~14% CL and ~5% of phosphatidylinositol (PI) (de Kroon et al. 1997) with cholesterol (~10% of total MOM lipid). The soybean polar lipid extract (PLE) closely resembles rat liver MOM lipid composition and therefore has been tested in tubulin-VDAC experiments (Rostovtseva et al. 2012). In PLE membranes, tubulin blocked VDAC even less effectively than in DOPC (Fig. 8.5b), perhaps due to the presence of ~25% of the negatively charged PI and PA lipids in the mixture. Addition of 10% of cholesterol to the PLE mixture did not affect the on-rate constant significantly (Fig. 8.5b). CL also did not change VDAC-tubulin binding when added to the DOPC membranes in 1 M KCl (Fig. 8.5b).

For all studied lipid compositions, the off-rate of VDAC blockage by tubulin was not affected (Fig. 8.4c) (Rostovtseva et al. 2012). This is consistent with the picture of VDAC-tubulin binding as a first-order reaction in which the on-rate is concentration dependent and the off-rate is not. In this view, the on-rate depends on the effective concentration of tubulin close to the channel entrance and thus on tubulin binding to the membrane surface, which in turn depends on membrane lipid composition (see Sect. 8.5.2 below). The role of lipid composition, particularly charged headgroups, on the concentration-independent off-rate of capture of the tubulin CTT into the VDAC pore has not yet been determined. When the tubulin CTT enters the VDAC pore, the strength of the binding, in terms of its residence time in the pore, does not depend on the lipid composition. Thus, at a constant applied voltage the equilibrium of VDAC-tubulin binding is predominantly defined by the on-rate. Consistent with the proposed lipid-dependent step of VDAC blockage by tubulin, we can suggest that any modifications of tubulin association with mitochondrial membrane induced by substantial lipid homeostasis in mitochondria in vivo would result in the different on-rates of VDAC-tubulin binding and consequently in different VDAC permeability to ATP or ADP.

8.4.2 Effect of Membrane Lipid Composition on VDAC-α-syn Interaction

When α-syn blocks VDAC (Fig. 8.2b), the lipid membrane composition affects the on-rates of VDAC blockage in a manner qualitatively similar to what was shown for tubulin. Using a lipid mixture mimicking the MOM lipid composition, it was found that the on-rate of α-syn-induced VDAC blockage is higher in non-lamellar DOPE than in lamellar DOPC (Jacobs et al. 2016). In particular, the on-rate increases up to tenfold with the increase of PE content in PC membranes. Remarkably, the off-rate at high transmembrane potentials was also lipid dependent, as seen by a 5 mV increase in the "turnover potential" which separates regimes of blockage and translocation with the increase of PE content (Jacobs et al. 2016). This lipid sensitivity supports a model shown in Fig. 8.3b in which the binding of α-syn's N-terminal domain to the membrane is followed by C-terminal blockage of the VDAC pore governed by transmembrane potential (Rostovtseva et al. 2015). Thus, similar to tubulin blockage of VDAC, the lipid dependence of α-syn-VDAC binding can be primarily attributed to the lipid sensitivity of α-syn molecule binding to the membrane, a phenomenon that has been demonstrated by various in vitro biophysical experiments (see Sect. 8.5.1 below).

At physiologically low salt concentrations of 150 mM KCl, the on-rates increase more than tenfold compared to experiments performed in high salt of 1 M KCl, while translocation of α-syn through VDAC is impeded (Jacobs et al. 2016). The effect of ionic strength on the on-rate of VDAC blockages observed in neutral membranes made of DOPE/DOPC mixture suggests that both electrostatic and hydrophobic components of α-syn-membrane association are involved in α-syn-VDAC binding, thus confirming the previously suggested participation of electrostatic forces in α-syn-membrane binding (Davidson et al. 1998).

It must be noted that despite the striking phenomenological similarity between VDAC blockages by α-syn and tubulin, there are a number of significant quantitative and qualitative differences between the two. For instance, the presence of PE has ~10 times higher effect on the on-rate of tubulin-VDAC binding than on α-syn-induced blockage; the characteristic time of blockage by α-syn is at least ten times smaller than the shortest one for tubulin, $\tau^{(1)}_{\mathrm{off}}$ (compare Fig. 8.4c with Fig. 8.4d); anionic lipids significantly increase the on-rate of VDAC blockage by α-syn in low and high salts but not by tubulin (Fig.8.5a, b).

Overall, these findings provide an example of lipid-controlled protein-protein interactions where the choice of lipid species is able to change the equilibrium binding constant by orders of magnitude. They also suggest a new regulatory role of both charged and neutral mitochondrial lipids in the control of MOM permeability, and hence mitochondria respiration, by modulating VDAC sensitivity to blockage by tubulin and α-syn. Considering the substantial lipid homeostasis in mitochondria during morphological changes such as fusion and fission (Furt and Moreau 2009), under apoptotic stress (Crimi and Esposti 2011), or lipid oxidation by ROS produced by mitochondria (Kagan et al. 2004; Pamplona 2008; Paradies et al. 2011), lipids could be potent regulators of VDAC and therefore of MOM permeability in vivo.

8.5 Membrane-Binding Properties of α-Synuclein and Tubulin

8.5.1 α-Synuclein-Membrane Binding

There are three distinctive regions of α-syn: the slightly net positively charged N-terminal domain (residues 1–60), the central nonpolar "nonamyloid-β component" (NAC) domain (residues 61–95), and the highly acidic C-terminal domain containing 15 negative charges (residues 96–140) (Rochet et al. 2012) (Fig. 8.1b). In solution, α-syn exists in the disordered form but can adopt an α-helical structure in its N-terminal domain upon binding to lipid membranes. The structure of membrane-bound α-syn is the subject of intense investigation and has been comprehensively reviewed (Pfefferkorn et al. 2012b). One of the primary techniques used to study adsorption of α-syn onto membrane surfaces has been circular dichroism (CD). CD reveals that α-syn, while disordered in bulk media, adopts a secondary structure with different helical content upon association with different lipid membranes (Davidson et al. 1998). In particular, anionic lipids stabilize an amphipathic helical structure in bound α-syn (Jiang et al. 2015). CD also shows an increase in helicity for non-lamellar zwitterionic lipids such as PE (Jo et al. 2000).

Complementary techniques find that there is measurable association between α-syn and membranes composed of lipids that do not induce helices. Specifically, fluorescence correlation spectroscopy (FCS) uses fluorescently labeled α-syn to determine the fractions of unbound α-syn, which diffuses freely in bulk solution, and liposome-bound α-syn, which has a diffusion constant similar to the liposomes to which it is bound (Rhoades et al. 2006). By this technique, it has been determined that α-syn does in fact associate with PC lipids, and there is a strong variation in binding to different anionic lipid species (Middleton and Rhoades 2010; Rhoades et al. 2006).

These results suggest a complex correlation between the secondary structure adopted by membrane-bound α-syn and the measured binding affinity. In particular, lipids with small and/or negatively charged headgroups enhance both binding affinity and helix formation; these enhancements are highest for phosphatidic acid (PA) headgroups and weakest for PC. Neutron reflectometry, which reports on membrane structure, protein penetration depth, and protein extension outside the membrane, is a promising technique for examining the structure of α-syn and the membranes to which it binds. Studies on anionic lipids have demonstrated that α-syn resides in the membrane headgroup region without significant penetration into the hydrophobic core of the membrane (Jiang et al. 2015; Pfefferkorn et al. 2012a).

α-Syn has also been demonstrated to show strong curvature sensitivity (Middleton and Rhoades 2010) and in fact appears to have a strong effect on the morphology of membrane surfaces to which it binds (Braun et al. 2014; Jiang et al. 2013; Shi et al. 2015). This is not surprising given α-syn's affinity for binding nonlamellar lipids in planar bilayers; indeed, the two phenomena are difficult to separate conceptually (Cornell 2015). In highly curved micelles, α-syn forms a broken helix (Ulmer et al. 2005); on small unilamellar vesicles, it is thought that the helix can be either broken or continuous (Lokappa and Ulmer 2011). Intriguingly, in membranes with lipid compositions mimicking real systems, a multiplicity of membrane conformations was confirmed in nuclear magnetic resonance (NMR) and computational studies (Bodner et al. 2009; Fusco et al. 2014). Together, these observations suggest that the degree of helix formation depends strongly on the lipid composition and other factors in vivo and may be a mechanism to fine-tune α-syn-lipid-binding affinities.

The propensity of α-syn to target anionic lipids with headgroups that are small relative to the fatty acid chains and as a result to demonstrate strong curvature sensitivity has significant in vivo implications. The mitochondrial outer membrane of *Saccharomyces cerevisiae* is composed of 45% PC lipids; the rest are all small headgroup (PE at 33%, CL at 6%, PA at 4%) and/or charged lipids (CL, PA, phosphoinositol (PI) at 10%, phosphatidylserine (PS) at 1%) (Zinser et al. 1991). In addition, the mitochondrial network is highly dynamic; the fusion/fission processes necessarily create regions of high curvature in MOM that may additionally enhance (or be enhanced by) α-syn binding. In the absence of significant inhibition by proteinaceous membrane components, α-syn can be expected to have substantial interactions with the membranes of mitochondrial composition and topology.

8.5.2 Tubulin-Membrane Association

Tubulin has been known to bind to liposomes and membrane extracts for several decades. These early results have been exhaustively compiled and reviewed (Wolff 2009). Unlike α-syn, however, few definitive statements can be made about the structure of membrane-bound tubulin. CD experiments suggest that both the helical character and the hydrophobic environment of membrane-bound tubulin increase relative to solubilized protein (Kumar et al. 1981). While these results have been interpreted as suggesting tubulin to be integrated into the hydrophobic region of the membrane, they are intriguingly similar to observations for α-syn and may simply indicate the formation of additional secondary structure upon association with the membrane. Other features of tubulin-membrane association, particularly its reversibility (Caron and Berlin 1980) and well-defined binding constant (Bernier-Valentin et al. 1983), support the notion of a peripheral attachment.

The lipid specificity of tubulin binding is not well studied. PE lipids appear to increase microtubule assembly (Hargreaves and McLean 1988), possibly due to a tubulin concentrating effect at surfaces containing PE lipids. This is supported by fluorescence studies showing an increase in tubulin binding to PE-containing large unilamellar vesicles (Rostovtseva et al. 2012). The effect of anionic lipids on microtubule assembly appears to have more to do with microtubule associated proteins than with tubulin itself (Wolff 2009).

Despite the known crystal structure of dimeric tubulin (Nettles et al. 2004; Nogales et al. 1998), the binding mechanism of the tubulin dimer to the membrane surface remains unknown, including the fundamental question of whether the α- or β-subunit (or both) is proximal to the membrane (Fig. 8.3a). The most indicative clue is the unchanged β-tubulin/vinblastine-binding affinity between membrane-bound and solubilized tubulin (Bhattacharyya and Wolff 1975), suggesting that the vinblastine-binding site remains exposed in membrane-bound tubulin (Fig. 8.3a, *a*).

The identification of the membrane-binding surface of tubulin is of critical importance for two reasons. First, MTAs overwhelmingly bind to the β-subunit (Field et al. 2014). If the β-subunit is responsible for membrane binding (Fig. 8.3a, *b*), one might expect that β-tubulin-bound MTAs would interfere with tubulin-membrane binding (see below discussion in 8.6.1). If, on the other hand, the α-subunit binds to membranes (Fig. 8.3a, *a*), new opportunities arise for drug development targeting the membrane-binding surface of tubulin. Second, the detailed mechanism of tubulin's interaction with VDAC depends on the structure and orientation of membrane-bound tubulin on the membrane surface. The relatively short α-tubulin CTT (11 amino acids, ≈4.4 nm long) is attached near one end of the dimer (Fig. 8.1b) and would be unable to reach the membrane surface and be captured into the VDAC pore for a variety of surface configurations, including nearly all in which β-tubulin is bound (Fig. 8.3a, *b*). Conversely, the longer β-tubulin CTT (17 amino acids, ≈6.8 nm long), which is attached near the dimer interface (Fig. 8.1b), is likely to be surface accessible for all reasonable binding configurations, including binding via α-tubulin (Fig. 8.3a, *a*). Binding configurations for which both subunits simultaneously interact with the membrane are also possible. One would expect both CTTs to be accessible to the VDAC pore in this case.

The multiple dwell times of the CTTs in the channel (Rostovtseva et al. 2008) (Fig. 8.4a) argue for a distribution of either CTT isotypes (see discussion in part 8.6.1 below) or of the orientation of surface-bound tubulin dimer (Fig. 8.3a). The simplest explanation is that both the α- and β-CTTs are captured into the channel; however, the wide variety of mammalian CTT isotypes and post-translational modifications which predominantly occur in the CTT (Westermann and Weber 2003), such as polyglutamation (Sirajuddin et al. 2014), as well as the multiplicity of subunits and isoforms of each subunit, may as well be responsible.

The disambiguation of the role of the two tubulin subunits in binding to mitochondrial membranes would be an important step forward in understanding the physiology of the tubulin-mitochondrial interaction and regulation of VDAC in particular. The recent development of systems producing recombinant tubulin (Minoura et al. 2013; Sirajuddin et al. 2014; Vemu et al. 2016) is likely to stimulate research in this direction. Biophysical studies of the lipid composition dependence of tubulin binding to mitochondrial membranes, such as those described for α-syn, would be particularly useful. The large, elongated structure of the tubulin dimer makes it particularly suitable for in vitro structural techniques such as neutron reflectometry (Heinrich and Losche 2014).

To summarize, the membrane-binding properties of tubulin and α-syn have a number of similar features. Both are strongly sensitive to lipid-packing defects caused by the presence of nonlamellar lipids in planar bilayers (PE vs PC). The helical content of both proteins increases with binding, with α-syn's amphipathic helical binding domains stabilized by anionic lipids. These properties are well suited for targeting the MOM and enhancing the availability of these molecules for voltage-induced interaction with VDAC. In addition, if tubulin is shown to bind peripherally, they appear to form a new and interesting class of peripheral membrane proteins that target lipid-packing defects and/or curvature without (as best we know) being directly involved in lipid-trafficking pathways (Cornell 2015). It remains to be seen if this is a general property of peripheral membrane proteins associated with the MOM.

8.6 Physiological Relevance of Tubulin and α-Synuclein Interaction with Mitochondrial Membranes

8.6.1 Tubulin Association with Mitochondrial Membranes

The year 2017 marks the 50th anniversary of the discovery of tubulin, yet some topics related to this protein remain underappreciated. One such topic is the nature of tubulin association with cell membranes. As pointed out in Sect. 8.5.2, the presence of tubulin in membranes isolated from various types of cells and tissues has been known since the 1970s (Bhattacharyya and Wolff 1975; Feit and Barondes 1970) followed a few years later by the demonstration of tubulin's association with mitochondrial membranes (Bernier-Valentin et al. 1983). A clear co-localization of β-tubulin with MOM has been shown by (Saetersdal et al. 1990). Since then, only a few studies have been dedicated to establish the nature of this "mitochondrial" tubulin.

Tubulin is found to be associated with mitochondrial membranes from a wide range of cancerous and noncancerous human cell lines, namely, neuroblastoma (SK-N-SH and IMR32), lung carcinoma (A549), breast adenocarcinoma (MCF-7), nasal septum adenocarcinoma (RPMI 2650), cervix carcinoma (HeLa), ovarian carcinoma (A2780 and OVCAR-3, Cicchillitti et al. 2008), and noncancerous breast cells (HBL-100) (Carre et al. 2002). β-Tubulin was also found associated with mitochondria from rat cardiomyocytes (Guzun et al. 2011). While the amount of mitochondria-associated tubulin appears to differ between cell lines from which mitochondria were isolated, Carre et al. estimated that it represents ~2% of total cellular tubulin. Interestingly, in this study α- and β-tubulins were present at comparable levels in mitochondria, suggesting that mitochondrial tubulin is a heterodimer. Noteworthy differences were found between cytosolic and mitochondrial tubulins: mitochondria contain more tyrosinated and acetylated α-tubulin and also more βIII isotype compared to cytosolic tubulin. In rat cardiomyocytes, βII-tubulin was associated with mitochondria, while βIII-tubulin was colocalized with sarcomeric Z-lines and βIV with polymerized microtubules (Guzun et al. 2011). Interestingly, Cicchillitti et al. (2008) analyzed the migration pattern of βIII-tubulin from ovarian cancer cells on two-dimensional gels and observed that βIII isotype exists under two distinct isoforms that differ in their isoelectric point (pI). While one βIII isoform was detectable only in the mitochondrial compartment, the other βIII isoform was present in both cytoskeletal and mitochondrial preparations. Further characterization of βIII isoforms by separation of polymerized and free tubulin fractions allowed finding the "mitochondrial" isoform in the pool of free tubulin, while the "cytoskeletal" isoform was found mainly associated with the pellet containing assembled tubulin. Altogether these data suggest that tubulin associated with mitochondria is a dimer consisting of α-tubulin and one of the β-tubulin isotypes, depending on cellular tissue. Whether there is βIII isotype specificity for mitochondria has yet to be proven.

The fact that tubulin and VDAC could be co-immunoprecipitated (Carre et al. 2002) is of particular relevance for the in vitro findings described in previous sections. Tubulin interaction with mitochondrial membranes and VDAC is a promising direction for chemotherapy and, more specifically, for understanding MTA's mechanism of action. MTAs like vinorelbine and paclitaxel have been shown to have a direct effect on mitochondria isolated from human neuroblastoma. They notably induce cytochrome c release, PTP opening, and mitochondria swelling (Andre et al. 2000). Mitochondria-associated tubulin has been proposed to be responsible for the mitochondrial dysfunctions induced by both stabilizing and destabilizing MTAs. Yet, the exact underlying mechanisms remain to be elucidated, requiring a clear depiction of how MTAs bind to dimeric mitochondrial tubulin and whether this binding modifies tubulin dimer conformation, as suggested by in vitro experiments (Arnal and Wade 1995), and consequently affects tubulin-membrane association (Fig. 8.3a). Another related question to explore is the impact of MTA-tubulin binding on tubulin interaction with VDAC. This would bring a better understanding of MTA direct targets in cancer cells and also in neuronal cells, since MTA treatment is associated with the development of peripheral neuropathies (Ballatore et al. 2012).

Mitochondrial membrane composition and its maintenance are essential for normal mitochondrial function, structure, and biogenesis. Thus, changes in the phospholipid composition affect mitochondrial functions and dynamics and have been linked to a variety of human diseases such as Barth syndrome (Schlame and Ren 2006), heart failure (Sparagna et al. 2007), neurodegenerative diseases (Aufschnaiter et al. 2016), and, more recently, cancer (Ribas et al. 2016). While lipid metabolism in apoptosis and cancer together with lipid replacement therapies are emerging fields (Huang and Freter 2015; Monteiro et al. 2013; Nicolson and Ash 2014), the way mitochondrial tubulin could be affected by MOM lipid content changes is still an open avenue to be explored. Nowadays, profiling of mitochondrial lipids in pathological conditions is made possible through "lipidomics" studies. This term, coined in the early 2000s (Wilson 2003), relates to novel mass spectrometry analytical approaches to quantitatively define lipid compositions from many different sources. Search for reliable procedures to standardize mitochondria isolation and minimize contaminations by membranes from other organelles is now a point of focus (Kappler et al. 2016). A future perspective is to discern whether tubulin binding to mitochondria is affected by MTA treatment in cancer and neuronal cells.

Potentially, the key factor in modulating tubulin binding to mitochondrial membranes is the increased generation of ROS, a common feature in aging and cancer (Brieger et al. 2012; Minelli et al. 2009). Mitochondria are the source and also the first target of oxidative products. Oxidation of mitochondrial phospholipids and associated proteins triggers structural changes in lipid bilayer organization and consequently alters membrane fluidity and permeability, leading inevitably to mitochondrial dysfunction (Paradies et al. 2014).

Future research will show if lipid peroxidation affects tubulin binding to mitochondria. So far, it is known that products of lipid peroxidation and especially 4-hydroxy-2-nonenal (HNE) disrupt the soluble/polymer tubulin equilibrium in favor of free tubulin (Kokubo et al. 2008; Neely et al. 1999; Olivero et al. 1990). Chemotherapy treatments, including taxanes and vinca alkaloids to a lesser extent, generate oxidative stress and lipid peroxidation products (Conklin 2004; Panis et al. 2012). Cellular data confirmed a mitochondrial origin of ROS after MTAs treatment (Le Grand et al. 2014) as well as an increased level of lipid peroxidation in an animal model of the chemotherapy-induced peripheral neuropathy (Greeshma et al. 2015). βIII-tubulin lacks the highly oxidizable Cys-239 found in βI, βII, and βIV isotypes (Joe et al. 2008), thus making βIII more resistant to free radicals. This could be one of the plausible explanations why mitochondria are enriched for this specific isotype. βIII-tubulin is of high interest in the cancer research field since this isotype has been associated with tumor development and aggressiveness and also in resistance to chemotherapy in tumors with poor prognosis (Mariani et al. 2015; Seve and Dumontet 2010). In this context, it would be beneficial to investigate if MTA-induced lipid peroxidation could regulate VDAC-tubulin interaction and thus participate in affecting physiopathological conditions.

8.6.2 α-Synuclein Association with Mitochondrial Membranes

Since the identification of dominant mutations in the SNCA gene, α-syn became the subject of numerous investigations. It has been reported that α-syn localizes at both the MOM and MIM from diverse models of dopaminergic neurons (Cole et al. 2008; Devi et al. 2008; Li et al. 2007; Parihar et al. 2008; Shavali et al. 2008) and that accumulation of α-syn in mitochondria impairs complex I resulting in increased oxidative stress (Devi et al. 2008; Parihar et al. 2008; Pennington et al. 2010). However, other group showed the absence of the inhibition effect on complex I by

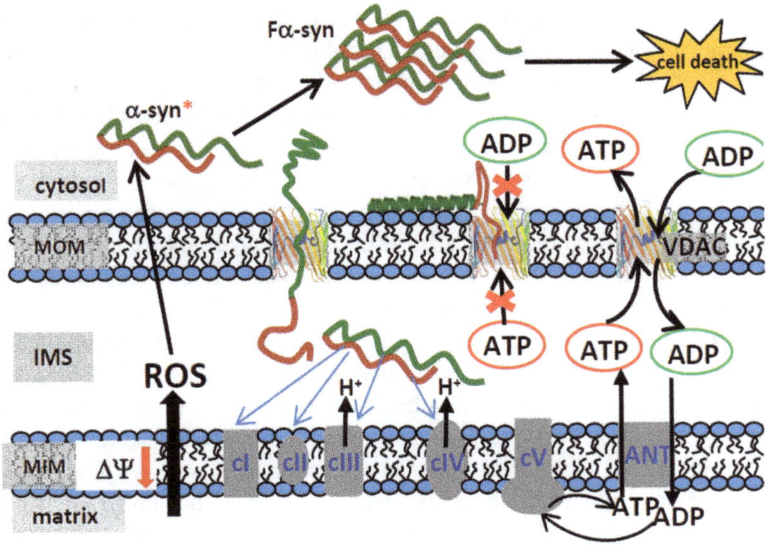

Fig. 8.6 Proposed physiological implications of α-syn blockage and translocation through VDAC. By reversibly blocking the VDAC pore, α-syn temporarily disrupts ATP/ADP fluxes between mitochondria and the cytosol and thus regulates them. By translocating through VDAC, α-syn reaches complexes of the electron transport chain (cI, cII, cIII, and cIV) in the MIM and impairs their function. This leads to the loss of mitochondrial potential ΔΨ and enhanced ROS production. ROS induces mitochondrial lipid peroxidation, which may modulate α-syn binding to the MOM as well as monomeric α-syn oxidation, leading to amplification of toxic fibrillary α-syn (Fα-syn) in the cytosol (Reprinted with permission from Rostovtseva et al. (2015))

α-syn accumulated in mitochondria isolated from mouse brain (Banerjee et al. 2010). While most of the studies agree on the inhibitory effect of α-syn on the oxidative phosphorylation capacity and on the promotion of oxidative stress, the exact mechanism(s) of α-syn effect on the mitochondrial bioenergetics remains largely unknown. There is still no consensus about the molecular identity of the pathway for α-syn to cross the MOM. A few studies suggested the translocase of the outer membrane complex 40 (TOM40) as the α-syn pathway (Bender et al. 2013; Devi et al. 2008). However, it seems very unlikely that acidic protein without a mitochondria-specific precursor would translocate through the highly substrate-specific and cation-selective Tom40 pore (Kuszak et al. 2015). On the contrary, VDAC would nicely account for the translocation pathway of α-syn into the mitochondria and its access to complex I and probably to other electron transport complexes (Fig. 8.6). Toxicity associated with such interaction has been shown in a yeast PD model (Rostovtseva et al. 2015) and is currently under investigation in a neuronal cell model.

α-Syn's connection to mitochondrial lipid homeostasis in cells was proposed in a study showing that a decreased level of mitochondrial PE found in yeast and worm models of PD was correlated with accumulation of α-syn into cytoplasmic foci

(Wang and Witt 2014). An earlier study showed a decreased level of the total PE and PC content in brains from PD patients (Riekkinen et al. 1975). These results are consistent with the in vitro data and model of α-syn-VDAC interaction discussed above in Sect. 8.4.2. The role of CL in α-syn-induced mitochondrial dysfunctions is now quite well documented (Ghio et al. 2016). In mouse models, the lack of α-syn was associated with decrease in CL (Ellis et al. 2005). Remarkably, both interactions between CL and α-syn and changes in CL content have been reported to have deleterious effects on mitochondria. In addition, the direct association of α-syn and CL has been correlated with disruption of mitochondrial dynamics in favor of fragmentation (Bueler 2009). On the other hand, altered CL content and expression levels of α-syn have generated seemingly contradictory results. Some data report that overexpression of the N-terminal domain of α-syn in neuronal models of PD is correlated with a decline in mitochondrial CL content, alterations in mitochondrial morphology, bioenergetics deficits, and decrease in mitochondrial membrane potential (Shen et al. 2014), while another study reported that α-syn knockout mice showed a decrease in mitochondrial CL and CL precursors and consequently reduction of complex I/III activity (Ellis et al. 2005). The number of similarities between the effects produced by the decrease of CL content and changes of α-syn expression levels points to an intricate relationship between α-syn and CL that requires further investigation.

There is strong evidence that oxidative stress is associated with PD pathology (Borza 2014) (Dias et al. 2013) as well as with numerous other pathologies. Several cellular models showed that oligomeric but not monomeric α-syn significantly increases the rate of ROS production, subsequently inducing lipid peroxidation in both primary co-cultures of neurons and astrocytes. Since inhibition of lipid peroxidation protects cells from cell death induced by oligomeric α-syn, the authors concluded that lipid peroxidation induced by misfolding of α-syn may play an important role in the cellular mechanism of neuronal cell loss observed in PD (Angelova et al. 2015). But so far, details about the specific phospholipid species undergoing oxidation and leading to mitochondrial dysfunction in PD are lacking. Given the high vulnerability of CL to peroxidation and the overall role of CL oxidation products in apoptosis and metabolic signaling (Kagan et al. 2014), CL is a promising candidate for further investigation in a PD context. Recently, Tyurina and coauthors (Tyurina et al. 2015) used a rat model of PD and lipidomics approach to identify oxygenated molecular species of CL formed in dysfunctional mitochondria. This particular study showed an increase in oxidized CL; therefore, a crucial point that remains unanswered is to what extent α-syn still binds to a modified CL. More in vitro data are needed to unravel the connections between α-syn and lipid peroxidation products in regard to α-syn conformational states and their relative binding to mitochondrial membranes.

The cartoon presented in Fig. 8.6 (Rostovtseva et al. 2015) illustrates how the previously discussed array of data on ROS production, mitochondrial dysfunction, lipid peroxidation, α-syn oxidation and fibrillation, and α-syn expression level can be reconciled within a model of MOM permeability regulation by α-syn interaction with VDAC and translocation through VDAC. By crossing the MOM through

VDAC, α-syn is able to directly target complexes of the electron transport chain in the MIM, which could lead to the loss of the mitochondrial potential, enhanced ROS production, and mitochondrial lipid peroxidation followed by mitochondrial dysfunction. ROS in turn could oxidize monomeric α-syn in the cytosol (and most likely mitochondrial bound α-syn), causing α-syn oligomerization (Hashimoto et al. 1999) and consequently amplification of α-syn toxicity. ROS-induced mitochondrial lipid peroxidation could change monomeric α-syn binding to mitochondria (Ruiperez et al. 2010) and thus affect the α-syn-induced mitochondrial toxicity cycle. Depending on physiological conditions in the cell, such as α-syn expression level (Devi et al. 2008), cytosolic pH (Cole et al. 2008), MOM lipid composition (especially its CL and PE content Ellis et al. 2005), and potential across MOM, α-syn could regulate a normal ATP/ADP exchange through VDAC or cause mitochondrial dysfunction. In general, the model in Fig. 8.6 illustrates how mitochondrial lipids could be intimately involved in regulation of MOM permeability and mitochondrial function by cytosolic proteins as long as these proteins have the characteristics identified in Sect. 8.3.

8.7 Future Perspectives

We have attempted to show that the voltage-activated interaction between VDAC and charged cytosolic proteins is not specific in the traditional sense of ion channel regulation. Rather, the complexity of the dependences of the interaction rate on salt concentration, lipid composition, and protein concentration appears to arise entirely from the crucial involvement of peripheral binding of the cytosolic proteins to the lipid membrane. Thus, a complete characterization of this phenomenon requires a dissection of each step of interaction by applying a number of biophysical and biochemical techniques that report on various aspects of peripheral protein binding. As we have shown here, the VDAC channel itself is a sensitive single-molecule probe of the membrane-bound tubulin and α-syn. The planar bilayer approach is also well suited for study of protein binding using the nonlinear electrical properties (e.g. "second harmonics" techniques) of lipid bilayers (Peterson et al. 2002; Sokolov and Kuzmin 1980). Due to the absence of intrinsic curvature, planar bilayers are complementary to the liposome-based techniques (CD, FCS, gravimetric, electrokinetic mobility). Other useful platforms include solid supported (Castellana and Cremer 2006) or tethered (He et al. 2005; Lang et al. 1994) bilayer lipid membrane systems. Though these membranes have the disadvantages of steric hindrances and membrane stresses due to proximity to a planar substrate, they nonetheless feature unparalleled stability and have led to significant advances using neutron reflectometry, surface plasmon resonance, and optical spectroscopy. We expect these experimental techniques to be especially informative when complemented by in silico platforms, including both atomistic and coarse-grained (e.g., MARTINI MD simulation Marrink et al. 2007; Monticelli et al. 2008) models for protein binding to MOM-mimicking membranes. Visualization of proteins interacting with VDAC in biological samples through the use of a dual-color super-resolution microscopy of protein distribution in the MOM would provide a translation of in vitro data and models into a live-cell context. Another important element to assay in cells is the mitochondrial lipidomics and its modifications (remodeling, variation of content, oxidative forms) since lipid changes have been related to numerous pathological conditions and are even considered in some cases as biomarkers. Altogether, the combination of electrophysiology with an array of biophysical, biochemical, and computational methods is needed to reveal the role(s) of lipids in regulation of MOM permeability and, consequently, in mitochondrial function.

Acknowledgments These studies were supported by the Intramural Research Program of the NIH, *Eunice Kennedy Shriver* National Institute of Child Health and Human Development.

References

Andre N, Braguer D, Brasseur G, Goncalves A, Lemesle-Meunier D, Guise S, Jordan MA, Briand C (2000) Paclitaxel induces release of cytochrome c from mitochondria isolated from human neuroblastoma cells. Cancer Res 60(19):5349–5353

Angelova PR, Horrocks MH, Klenerman D, Gandhi S, Abramov AY, Shchepinov MS (2015) Lipid peroxidation is essential for alpha-synuclein-induced cell death. J Neurochem 133(4):582–589

Arnal I, Wade RH (1995) How does taxol stabilize microtubules? Curr Biol 5(8):900–908

Aufschnaiter A, Kohler V, Diessl J, Peselj C, Carmona-Gutierrez D, Keller W, Buttner S (2016) Mitochondrial lipids in neurodegeneration. Cell Tissue Res 367(1):125–140

Ballatore C, Brunden KR, Huryn DM, Trojanowski JQ, Lee VM, Smith AB 3rd (2012) Microtubule stabilizing agents as potential treatment for Alzheimer's disease and related neurodegenerative tauopathies. J Med Chem 55(21):8979–8996

Banerjee K, Sinha M, Pham Cle L, Jana S, Chanda D, Cappai R, Chakrabarti S (2010) Alpha-synuclein induced membrane depolarization and loss of phosphorylation capacity of isolated rat brain mitochondria: implications in Parkinson's disease. FEBS Lett 584(8):1571–1576

Basanez G, Sharpe JC, Galanis J, Brandt TB, Hardwick JM, Zimmerberg J (2002) Bax-type apoptotic proteins porate pure lipid bilayers through a mechanism sensitive to intrinsic monolayer curvature. J Biol Chem 277(51):49360–49365

Bender A, Desplats P, Spencer B, Rockenstein E, Adame A, Elstner M, Laub C, Mueller S, Koob AO, Mante M et al (2013) TOM40 mediates mitochondrial dysfunction induced by alpha-synuclein accumulation in Parkinson's disease. PLoS ONE 8(4):e62277

Bernier-Valentin F, Aunis D, Rousset B (1983) Evidence for tubulin-binding sites on cellular membranes – plasma-membranes, mitochondrial-membranes, and secretory granule membranes. J Cell Biol 97(1):209–216

Bezrukov SM, Vodyanoy I (1993) Probing alamethicin channels with water-soluble polymers. Effect on conductance of channel states. Biophys J 64(1):16–25

Bhattacharyya B, Wolff J (1975) Membrane-bound tubulin in brain and thyroid tissue. J Biol Chem 250(19):7639–7646

Bodner CR, Dobson CM, Bax A (2009) Multiple tight phospholipid-binding modes of α-synuclein revealed by solution NMR spectroscopy. J Mol Biol 390(4):775–790

Borza LR (2014) A review on the cause-effect relationship between oxidative stress and toxic proteins in the pathogenesis of neurodegenerative diseases. Rev Med Chir Soc Med Nat Iasi 118(1):19–27

Braun AR, Lacy MM, Ducas VC, Rhoades E, Sachs JN (2014) Alpha-synuclein-induced membrane remodeling is driven by binding affinity, partition depth, and interleaflet order asymmetry. J Am Chem Soc 136(28):9962–9972
Brieger K, Schiavone S, Miller FJ Jr, Krause KH (2012) Reactive oxygen species: from health to disease. Swiss Med Wkly 142:w13659
Bueler H (2009) Impaired mitochondrial dynamics and function in the pathogenesis of Parkinson's disease. Exp Neurol 218(2):235–246
Caron JM, Berlin RD (1980) Reversible adsorption of microtubule protein to phospholipid-vesicles. J Cell Biol 87(2):A255–A255
Carre M, Andre N, Carles G, Borghi H, Brichese L, Briand C, Braguer D (2002) Tubulin is an inherent component of mitochondrial membranes that interacts with the voltage-dependent anion channel. J Biol Chem 277(37):33664–33669
Castellana ET, Cremer PS (2006) Solid supported lipid bilayers: from biophysical studies to sensor design. Surf Sci Rep 61(10):429–444
Chipuk JE, Bouchier-Hayes L, Green DR (2006) Mitochondrial outer membrane permeabilization during apoptosis: the innocent bystander scenario. Cell Death Differ 13(8):1396–1402
Cicchillitti L, Penci R, Di Michele M, Filippetti F, Rotilio D, Donati MB, Scambia G, Ferlini C (2008) Proteomic characterization of cytoskeletal and mitochondrial class III beta-tubulin. Mol Cancer Ther 7(7):2070–2079
Cole NB, Dieuliis D, Leo P, Mitchell DC, Nussbaum RL (2008) Mitochondrial translocation of alpha-synuclein is promoted by intracellular acidification. Exp Cell Res 314(10):2076–2089
Colombini M (1989) Voltage gating in the mitochondrial channel, VDAC. J Membr Biol 111(2):103–111
Colombini M (2004) VDAC: the channel at the interface between mitochondria and the cytosol. Mol Cell Biochem 256(1–2):107–115
Colombini M (2010) Ceramide channels and their role in mitochondria-mediated apoptosis. Biochim Biophys Acta 1797(6–7):1239–1244
Colombini M, Blachly-Dyson E, Forte M (1996) VDAC, a channel in the outer mitochondrial membrane. In: Narahashi T (ed) In ion channels. Plenum Press, New York, pp 169–202
Conklin KA (2004) Chemotherapy-associated oxidative stress: impact on chemotherapeutic effectiveness. Integr Cancer Ther 3(4):294–300
Cornell RB (2015) Membrane lipid compositional sensing by the inducible amphipathic helix of CCT. Biochim Biophys Acta. doi:10.1016/j.bbalip.2015.1012.1022
Crimi M, Esposti MD (2011) Apoptosis-induced changes in mitochondrial lipids. Biochim Biophys Acta 1813(4):551–557
Davidson WS, Jonas A, Clayton DF, George JM (1998) Stabilization of alpha-synuclein secondary structure upon binding to synthetic membranes. J Biol Chem 273(16):9443–9449
de Kroon AI, Dolis D, Mayer A, Lill R, de Kruijff B (1997) Phospholipid composition of highly purified mitochondrial outer membranes of rat liver and Neurospora crassa. Is cardiolipin present in the mitochondrial outer membrane? Biochim Biophys Acta 1325(1):108–116
DeHart DN, Gooz M, Rostovtseva TK, Sheldon KL, Lemasters JJ, Maldonado EN (2014) Antagonists of the inhibitory effect of free tubulin on VDAC induce oxidative stress and mitochondrial dysfunction. Biophys J 106(2):591a–591a
Devi L, Raghavendran V, Prabhu BM, Avadhani NG, Anandatheerthavarada HK (2008) Mitochondrial import and accumulation of alpha-synuclein impair complex I in human dopaminergic neuronal cultures and Parkinson disease brain. J Biol Chem 283(14):9089–9100
Dias V, Junn E, Mouradian MM (2013) The role of oxidative stress in Parkinson's disease. J Park Dis 3(4):461–491
Elkon H, Don J, Melamed E, Ziv I, Shirvan A, Offen D (2002) Mutant and wild-type alpha-synuclein interact with mitochondrial cytochrome C oxidase. J Mol Neurosci 18(3):229–238
Ellis CE, Murphy EJ, Mitchell DC, Golovko MY, Scaglia F, Barcelo-Coblijn GC, Nussbaum RL (2005) Mitochondrial lipid abnormality and electron transport chain impairment in mice lacking alpha-synuclein. Mol Cell Biol 25(22):10190–10201

Feit H, Barondes SH (1970) Colchicine-binding activity in particulate fractions of mouse brain. J Neurochem 17(9):1355–1364

Field JJ, Waight AB, Senter PD (2014) A previously undescribed tubulin binder. Proc Natl Acad Sci U S A 111(38):13684–13685

Flis VV, Daum G (2013) Lipid transport between the endoplasmic reticulum and mitochondria. Cold Spring Harb Perspect Biol 5(6):1–22

Furt F, Moreau P (2009) Importance of lipid metabolism for intracellular and mitochondrial membrane fusion/fission processes. Int J Biochem Cell B 41(10):1828–1836

Fusco G, De Simone A, Gopinath T, Vostrikov V, Vendruscolo M, Dobson CM, Veglia G (2014) Direct observation of the three regions in alpha-synuclein that determine its membrane-bound behaviour. Nat Commun 5:3827

Ghio S, Kamp F, Cauchi R, Giese A, Vassallo N (2016) Interaction of alpha-synuclein with biomembranes in Parkinson's disease – role of cardiolipin. Prog Lipid Res 61:73–82

Gonzalvez F, Pariselli F, Dupaigne P, Budihardjo I, Lutter M, Antonsson B, Diolez P, Manon S, Martinou JC, Goubern M et al (2005) tBid interaction with cardiolipin primarily orchestrates mitochondrial dysfunctions and subsequently activates Bax and Bak. Cell Death Differ 12(6):614–626

Gonzalvez F, Schug ZT, Houtkooper RH, MacKenzie ED, Brooks DG, Wanders RJ, Petit PX, Vaz FM, Gottlieb E (2008) Cardiolipin provides an essential activating platform for caspase-8 on mitochondria. J Cell Biol 183(4):681–696

Greeshma N, Prasanth KG, Balaji B (2015) Tetrahydrocurcumin exerts protective effect on vincristine induced neuropathy: behavioral, biochemical, neurophysiological and histological evidence. Chem Biol Interact 238:118–128

Gurnev PA, Rostovtseva TK, Bezrukov SM (2011) Tubulin-blocked state of VDAC studied by polymer and ATP partitioning. FEBS Lett 585(14):2363–2366

Gurnev PA, Queralt-Martin M, Aguilella VM, Rostovtseva TK, Bezrukov SM (2012) Probing tubulin-blocked state of VDAC by varying membrane surface charge. Biophys J 102(9):2070–2076

Guzun R, Karu-Varikmaa M, Gonzalez-Granilo M, Kuznetsov A, Michel L, Cottet-Rousselle C, Saaremae M, Kaam T, Metsis M, Grimm M et al (2011) Mitochondria-cytoskeleton interaction: distribution of β-tubulins in cardiomyocytes and HL-1 cells. Biochim Biophys Acta 1807:458–469

Hargreaves AJ, McLean WG (1988) The characterization of phospholipids associated with microtubules, purified tubulin and microtubule associated proteins in vitro. Int J Biochem 20(10):1133–1138

Hashimoto M, Hsu LJ, Xia Y, Takeda A, Sisk A, Sundsmo M, Masliah E (1999) Oxidative stress induces amyloid-like aggregate formation of NACP/alpha-synuclein in vitro. Neuroreport 10(4):717–721

He L, Robertson JW, Li J, Karcher I, Schiller SM, Knoll W, Naumann R (2005) Tethered bilayer lipid membranes based on monolayers of thiolipids mixed with a complementary dilution molecule. 1. Incorporation of channel peptides. Langmuir 21(25):11666–11672

Heinrich F, Losche M (2014) Zooming in on disordered systems: neutron reflection studies of proteins associated with fluid membranes. Biochim Biophys Acta 1838(9):2341–2349

Hodge T, Colombini M (1997) Regulation of metabolite flux through voltage-gating of VDAC channels. J Membr Biol 157(3):271–279

Hoogerheide DP, Gurnev PA, Rostovtseva TK, Bezrukov SM (2016) Mechanism of alpha-synuclein translocation through a VDAC nanopore revealed by energy landscape modeling of escape time distributions. Nanoscale 9(1):183–192

Huang C, Freter C (2015) Lipid metabolism, apoptosis and cancer therapy. Int J Mol Sci 16(1):924–949

Jacobs D, Hoogerheide DP, Rovini A, Gurnev PA, Bezrukov SM, Rostovtseva TK (2016) Membrane lipid composition regulates alpha-synuclein blockage of and translocation through the mitochondrial voltage-dependent anion channel. Biophys J 110(3):19a–20a

Jiang Z, de Messieres M, Lee JC (2013) Membrane remodeling by alpha-synuclein and effects on amyloid formation. J Am Chem Soc 135(43):15970–15973

Jiang Z, Hess SK, Heinrich F, Lee JC (2015) Molecular details of alpha-synuclein membrane association revealed by neutrons and photons. J Phys Chem B 119(14):4812–4823

Jo E, McLaurin J, Yip CM, St George-Hyslop P, Fraser PE (2000) Alpha-synuclein membrane interactions and lipid specificity. J Biol Chem 275(44):34328–34334

Joe PA, Banerjee A, Luduena RF (2008) The roles of cys 124 and ser239 in the functional properties of human betaIII tubulin. Cell Motil Cytoskeleton 65(6):476–486

Kagan VE, Borisenko GG, Tyurina YY, Tyurin VA, Jiang JF, Potapovich AI, Kini V, Amoscato AA, Fujii Y (2004) Oxidative lipidomics of apoptosis: redox catalytic interactions of cytochrome C with cardiolipin and phosphatidylserine. Free Radic Biol Med 37(12):1963–1985

Kagan VE, Chu CT, Tyurina YY, Cheikhi A, Bayir H (2014) Cardiolipin asymmetry, oxidation and signaling. Chem Phys Lipids 179:64–69

Kappler L, Li J, Haring HU, Weigert C, Lehmann R, Xu G, Hoene M (2016) Purity matters: a workflow for the valid high-resolution lipid profiling of mitochondria from cell culture samples. Sci Rep 6:21107

Kokubo J, Nagatani N, Hiroki K, Kuroiwa K, Watanabe N, Arai T (2008) Mechanism of destruction of microtubule structures by 4-hydroxy-2-nonenal. Cell Struct Funct 33(1):51–59

Kruger R, Muller T, Riess O (2000) Involvement of alpha-synuclein in Parkinson's disease and other neurodegenerative disorders. J Neural Transm (Vienna) 107(1):31–40

Kumar N, Klausner RD, Weinstein JN, Blumenthal R, Flavin M (1981) Interaction of tubulin with phospholipid-vesicles. 2. Physical changes of the protein. J Biol Chem 256(11):5886–5889

Kuszak AJ, Jacobs D, Gurnev PA, Shiota T, Louis JM, Lithgow T, Bezrukov SM, Rostovtseva TK, Buchanan SK (2015) Evidence of distinct channel conformations and substrate binding affinities for the mitochondrial outer membrane protein translocase pore Tom40. J Biol Chem 290(43):26204–26217

Kuwana T, Mackey MR, Perkins G, Ellisman MH, Latterich M, Schneiter R, Green DR, Newmeyer DD (2002) Bid, Bax, and lipids cooperate to form supramolecular openings in the outer mitochondrial membrane. Cell 111(3):331–342

Landeta O, Landajuela A, Gil D, Taneva S, Di Primo C, Sot B, Valle M, Frolov VA, Basanez G (2011) Reconstitution of proapoptotic BAK function in liposomes reveals a dual role for mitochondrial lipids in the BAK-driven membrane permeabilization process. J Biol Chem 286(10):8213–8230

Lang H, Duschl C, Vogel H (1994) A new class of thiolipids for the attachment of lipid bilayers on gold surfaces. Langmuir 10(1):197–210

Le Grand M, Rovini A, Bourgarel-Rey V, Honore S, Bastonero S, Braguer D, Carre M (2014) ROS-mediated EB1 phosphorylation through Akt/GSK3beta pathway: implication in cancer cell response to microtubule-targeting agents. Oncotarget 5(10):3408–3423

Lemasters JJ, Holmuhamedov E (2006) Voltage-dependent anion channel (VDAC) as mitochondrial governator – thinking outside the box. Biochim Biophys Acta 1762(2):181–190

Lemasters JJ, Holmuhamedov EL, Czerny C, Zhong Z, Maldonado EN (2012) Regulation of mitochondrial function by voltage dependent anion channels in ethanol metabolism and the Warburg effect. Biochim Biophys Acta 1818(6):1536–1544

Lemeshko VV (2006) Theoretical evaluation of a possible nature of the outer membrane potential of mitochondria. Eur Biophys J 36(1):57–66

Lemeshko VV (2014a) VDAC electronics: 1. VDAC-hexo(gluco)kinase generator of the mitochondrial outer membrane potential. Biochim Biophys Acta 1838(5):1362–1371

Lemeshko VV (2014b) VDAC electronics: 2. A new, anaerobic mechanism of generation of the membrane potentials in mitochondria. Biochim Biophys Acta 1838(7):1801–1808

Lemeshko VV (2016) VDAC electronics: 3. VDAC-creatine kinase-dependent generation of the outer membrane potential in respiring mitochondria. BBA Biomembr 1858(7):1411–1418

Li WW, Yang R, Guo JC, Ren HM, Zha XL, Cheng JS, Cai DF (2007) Localization of alpha-synuclein to mitochondria within midbrain of mice. Neuroreport 18(15):1543–1546

Liu G, Zhang C, Yin J, Li X, Cheng F, Li Y, Yang H, Ueda K, Chan P, Yu S (2009) Alpha-synuclein is differentially expressed in mitochondria from different rat brain regions and dose-dependently down-regulates complex I activity. Neurosci Lett 454(3):187–192

Lokappa SB, Ulmer TS (2011) α-synuclein populates both elongated and broken helix states on small unilamellar vesicles. J Biol Chem 286(24):21450–21457

Luth ES, Stavrovskaya IG, Bartels T, Kristal BS, Selkoe DJ (2014) Soluble, prefibrillar alpha-synuclein oligomers promote complex I-dependent, Ca2+-induced mitochondrial dysfunction. J Biol Chem 289(31):21490–21507

Maldonado EN, Patnaik J, Mullins MR, Lemasters JJ (2011) Free tubulin modulates mitochondrial membrane potential in cancer cells. Cancer Res 70(24):10192–10201

Maldonado EN, Sheldon KL, DeHart DN, Patnaik J, Manevich Y, Townsend DM, Bezrukov SM, Rostovtseva TK, Lemasters JJ (2013) Voltage-dependent anion channels modulate mitochondrial metabolism in cancer cells: regulation by free tubulin and erastin. J Biol Chem 288(17):11920–11929

Mariani M, Karki R, Spennato M, Pandya D, He S, Andreoli M, Fiedler P, Ferlini C (2015) Class III beta-tubulin in normal and cancer tissues. Gene 563(2):109–114

Marrink SJ, Risselada HJ, Yefimov S, Tieleman DP, de Vries AH (2007) The MARTINI force field: coarse grained model for biomolecular simulations. J Phys Chem B 111(27):7812–7824

Middleton ER, Rhoades E (2010) Effects of curvature and composition on alpha-synuclein binding to lipid vesicles. Biophys J 99(7):2279–2288

Minelli A, Bellezza I, Conte C, Culig Z (2009) Oxidative stress-related aging: a role for prostate cancer? Biochim Biophys Acta 1795(2):83–91

Minoura I, Hachikubo Y, Yamakita Y, Takazaki H, Ayukawa R, Uchimura S, Muto E (2013) Overexpression, purification, and functional analysis of recombinant human tubulin dimer. FEBS Lett 587(21):3450–3455

Monge C, Beraud N, Kuznetsov AV, Rostovtseva T, Sackett D, Schlattner U, Vendelin M, Saks VA (2008) Regulation of respiration in brain mitochondria and synaptosomes: restrictions of ADP diffusion in situ, roles of tubulin, and mitochondrial creatine kinase. Mol Cell Biochem 318(1–2):94–1562

Monteiro JP, Oliveira PJ, Jurado AS (2013) Mitochondrial membrane lipid remodeling in pathophysiology: a new target for diet and therapeutic interventions. Prog Lipid Res 52(4):513–528

Monticelli L, Kandasamy SK, Periole X, Larson RG, Tieleman DP, Marrink SJ (2008) The MARTINI coarse-grained force field: extension to proteins. J Chem Theory Comput 4(5):819–834

Neely MD, Sidell KR, Graham DG, Montine TJ (1999) The lipid peroxidation product 4-hydroxynonenal inhibits neurite outgrowth, disrupts neuronal microtubules, and modifies cellular tubulin. J Neurochem 72(6):2323–2333

Nettles JH, Li HL, Cornett B, Krahn JM, Snyder JP, Downing KH (2004) The binding mode of epothilone A on alpha, beta-tubulin by electron crystallography. Science 305(5685):866–869

Nicolson GL, Ash ME (2014) Lipid replacement therapy: a natural medicine approach to replacing damaged lipids in cellular membranes and organelles and restoring function. Biochim Biophys Acta 1838(6):1657–1679

Nogales E, Wolf SG, Downing KH (1998) Structure of the alpha beta tubulin dimer by electron crystallography. Nature 391(6663):199–203

Noskov SY, Rostovtseva TK, Bezrukov SM (2013) ATP transport through VDAC and the VDAC-tubulin complex probed by equilibrium and nonequilibrium MD simulations. Biochemistry 52(51):9246–9256

Noskov SY, Rostovtseva TK, Chamberlin AC, Teijido O, Jiang W, Bezrukov SM (2016) Current state of theoretical and experimental studies of the voltage-dependent anion channel (VDAC). Biochim Biophys Acta 1858(7 Pt B):1778–1790

Olivero A, Miglietta A, Gadoni E, Gabriel L (1990) 4-hydroxynonenal interacts with tubulin by reacting with its functional -SH groups. Cell Biochem Funct 8(2):99–105

Pamplona R (2008) Membrane phospholipids, lipoxidative damage and molecular integrity: a causal role in aging and longevity. Biochim Biophys Acta 1777(10):1249–1262

Panis C, Herrera AC, Victorino VJ, Campos FC, Freitas LF, De Rossi T, Colado Simao AN, Cecchini AL, Cecchini R (2012) Oxidative stress and hematological profiles of advanced breast cancer patients subjected to paclitaxel or doxorubicin chemotherapy. Breast Cancer Res Treat 133(1):89–97

Paradies G, Petrosillo G, Paradies V, Ruggiero FM (2011) Mitochondrial dysfunction in brain aging: role of oxidative stress and cardiolipin. Neurochem Int 58(4):447–457

Paradies G, Paradies V, Ruggiero FM, Petrosillo G (2014) Oxidative stress, cardiolipin and mitochondrial dysfunction in nonalcoholic fatty liver disease. World J Gastroenterol 20(39):14205–14218

Parihar MS, Parihar A, Fujita M, Hashimoto M, Ghafourifar P (2008) Mitochondrial association of alpha-synuclein causes oxidative stress. Cell Mol Life Sci 65(7–8):1272–1284

Pennington K, Peng J, Hung CC, Banks RE, Robinson PA (2010) Differential effects of wild-type and A53T mutant isoform of alpha-synuclein on the mitochondrial proteome of differentiated SH-SY5Y cells. J Proteome Res 9(5):2390–2401

Peterson U, Mannock DA, Lewis RN, Pohl P, McElhaney RN, Pohl EE (2002) Origin of membrane dipole potential: contribution of the phospholipid fatty acid chains. Chem Phys Lipids 117(1–2):19–27

Pfefferkorn CM, Heinrich F, Sodt AJ, Maltsev AS, Pastor RW, Lee JC (2012a) Depth of α-synuclein in a bilayer determined by fluorescence, neutron reflectometry, and computation. Biophys J 102(3):613–621

Pfefferkorn CM, Jiang Z, Lee JC (2012b) Biophysics of alpha-synuclein membrane interactions. Biochim Biophys Acta 1818(2):162–171

Porcelli AM, Ghelli A, Zanna C, Pinton P, Rizzuto R, Rugolo M (2005) pH difference across the outer mitochondrial membrane measured with a green fluorescent protein mutant. Biochem Biophys Res Commun 326(4):799–804

Rhoades E, Ramlall TF, Webb WW, Eliezer D (2006) Quantification of alpha-synuclein binding to lipid vesicles using fluorescence correlation spectroscopy. Biophys J 90(12):4692–4700

Ribas V, Garcia-Ruiz C, Fernandez-Checa JC (2016) Mitochondria, cholesterol and cancer cell metabolism. Clin Transl Med 5(1):22

Riekkinen P, Rinne UK, Pelliniemi TT, Sonninen V (1975) Interaction between dopamine and phospholipids. Studies of the substantia nigra in Parkinson disease patients. Arch Neurol 32(1):25–27

Rochet JC, Hay BA, Guo M (2012) Molecular insights into Parkinson's disease. Prog Mol Biol Transl Sci 107:125–188

Rostovtseva TK, Bezrukov SM (2012) VDAC inhibition by tubulin and its physiological implications. Biochim Biophys Acta 1818(6):1526–1535

Rostovtseva TK, Bezrukov SM (2015) Function and regulation of mitochondrial voltage-dependent anion channel. In: Delcour AH (ed) Electrophysiology of unconventional channels and pores. Springer, Switzerland, pp 3–31

Rostovtseva T, Colombini M (1996) ATP flux is controlled by a voltage-gated channel from the mitochondrial outer membrane. J Biol Chem 271(45):28006–28008

Rostovtseva T, Colombini M (1997) VDAC channels mediate and gate the flow of ATP: implications for the regulation of mitochondrial function. Biophys J 72(5):1954–1962

Rostovtseva TK, Komarov A, Bezrukov SM, Colombini M (2002a) Dynamics of nucleotides in VDAC channels: structure-specific noise generation. Biophys J 82(1):193–205

Rostovtseva TK, Komarov A, Bezrukov SM, Colombini M (2002b) VDAC channels differentiate between natural metabolites and synthetic molecules. J Membr Biol 187(2):147–156

Rostovtseva TK, Tan WZ, Colombini M (2005) On the role of VDAC in apoptosis: fact and fiction. J Bioenerg Biomembr 37(3):129–142

Rostovtseva TK, Sheldon KL, Hassanzadeh E, Monge C, Saks V, Bezrukov SM, Sackett DL (2008) Tubulin binding blocks mitochondrial voltage-dependent anion channel and regulates respiration. Proc Natl Acad Sci U S A 105(48):18746–18751

Rostovtseva TK, Gurnev PA, Chen MY, Bezrukov SM (2012) Membrane lipid composition regulates tubulin interaction with mitochondrial voltage-dependent anion channel. J Biol Chem 287(35):29589–29598
Rostovtseva TK, Gurnev PA, Protchenko O, Hoogerheide DP, Yap TL, Philpott CC, Lee JC, Bezrukov SM (2015) Alpha-synuclein shows high affinity interaction with voltage-dependent anion channel, suggesting mechanisms of mitochondrial regulation and toxicity in Parkinson disease. J Biol Chem 290(30):18467–18477
Ruiperez V, Darios F, Davletov B (2010) Alpha-synuclein, lipids and Parkinson's disease. Prog Lipid Res 49(4):420–428
Saetersdal T, Greve G, Dalen H (1990) Associations between beta-tubulin and mitochondria in adult isolated heart myocytes as shown by immunofluorescence and immunoelectron microscopy. Histochemistry 95(1):1–10
Schafer B, Quispe J, Choudhary V, Chipuk JE, Ajero TG, Du H, Schneiter R, Kuwana T (2009) Mitochondrial outer membrane proteins assist bid in Bax-mediated lipidic pore formation. Mol Biol Cell 20(8):2276–2285
Schein SJ, Colombini M, Finkelstein A (1976) Reconstitution in planar lipid bilayers of a voltage-dependent anion-selective channel obtained from paramecium mitochondria. J Membr Biol 30(2):99–120
Schlame M, Ren M (2006) Barth syndrome, a human disorder of cardiolipin metabolism. FEBS Lett 580(23):5450–5455
Seve P, Dumontet C (2010) Class III beta tubulin expression in nonsmall cell lung cancer. Rev Mal Respir 27(4):383–386
Shamas-Din A, Bindner S, Chi X, Leber B, Andrews DW, Fradin C (2015) Distinct lipid effects on tBid and Bim activation of membrane permeabilization by pro-apoptotic Bax. Biochem J 467(3):495–505
Shavali S, Brown-Borg HM, Ebadi M, Porter J (2008) Mitochondrial localization of alpha-synuclein protein in alpha-synuclein overexpressing cells. Neurosci Lett 439(2):125–128
Shen J, Du T, Wang X, Duan C, Gao G, Zhang J, Lu L, Yang H (2014) Alpha-synuclein amino terminus regulates mitochondrial membrane permeability. Brain Res 1591:14–26
Shi Z, Sachs JN, Rhoades E, Baumgart T (2015) Biophysics of alpha-synuclein induced membrane remodelling. Phys Chem Chem Phys 17(24):15561–15568
Shoshan-Barmatz V, Ben-Hail D (2012) VDAC, a multi-functional mitochondrial protein as a pharmacological target. Mitochondrion 12(1):24–34
Shoshan-Barmatz V, De Pinto V, Zweckstetter M, Raviv Z, Keinan N, Arbel N (2010) VDAC, a multi-functional mitochondrial protein regulating cell life and death. Mol Asp Med 31(3):227–285
Sirajuddin M, Rice LM, Vale RD (2014) Regulation of microtubule motors by tubulin isotypes and post-translational modifications. Nat Cell Biol 16(4):335–344
Siskind LJ, Kolesnick RN, Colombini M (2002) Ceramide channels increase the permeability of the mitochondrial outer membrane to small proteins. J Biol Chem 277(30):26796–26803
Sokolov VS, Kuzmin VG (1980) Study of surface-potential difference in bilayer-membranes according to the 2nd harmonic response of capacitance current. Biofizika 25(1):170–172
Sparagna GC, Chicco AJ, Murphy RC, Bristow MR, Johnson CA, Rees ML, Maxey ML, McCune SA, Moore RL (2007) Loss of cardiac tetralinoleoyl cardiolipin in human and experimental heart failure. J Lipid Res 48(7):1559–1570
Spillantini MG, Schmidt ML, Lee VMY, Trojanowski JQ, Jakes R, Goedert M (1997) Alpha-synuclein in Lewy bodies. Nature 388(6645):839–840
Swartz KJ (2008) Sensing voltage across lipid membranes. Nature 456(7224):891–897
Terrones O, Antonsson B, Yamaguchi H, Wang HG, Liu JH, Lee RM, Herrmann A, Basanez G (2004) Lipidic pore formation by the concerted action of proapoptotic BAX and tBID. J Biol Chem 279(29):30081–30091
Tyurina YY, Polimova AM, Maciel E, Tyurin VA, Kapralova VI, Winnica DE, Vikulina AS, Domingues MR, McCoy J, Sanders LH et al (2015) LC/MS analysis of cardiolipins in substantia nigra and plasma of rotenone-treated rats: implication for mitochondrial dysfunction in Parkinson's disease. Free Radic Res 49(5):681–691

Ujwal R, Cascio D, Colletier J-P, Faham S, Zhang J, Toro L, Ping P, Abramson J (2008) The crystal structure of mouse VDAC1 at 2.3 A resolution reveals mechanistic insights into metabolite gating. Proc Natl Acad Sci U S A 105(46):17742–17747
Ulmer TS, Bax A, Cole NB, Nussbaum RL (2005) Structure and dynamics of micelle-bound human alpha-synuclein. J Biol Chem 280(10):9595–9603
Vemu A, Atherton J, Spector JO, Szyk A, Moores CA, Roll-Mecak A (2016) Structure and dynamics of single-isoform recombinant neuronal human tubulin. J Biol Chem:12907–12915
Wang S, Witt SN (2014) The Parkinson's disease-associated protein α-synuclein disrupts stress signaling–a possible implication for methamphetamine use? Microb Cell 1(4):131
Westermann S, Weber K (2003) Post-translational modifications regulate microtubule function. Nat Rev Mol Cell Biol 4(12):938–947
Wilson JF (2003) Long-suffering lipids gain respect: technical advances and enhanced understanding of lipid biology fuel a trend toward lipidomics. (Lab consumer). Scientist 17(5):34–37
Wolff J (2009) Plasma membrane tubulin. Biochim Biophys Acta 1788(7):1415–1433
Yethon JA, Epand RF, Leber B, Epand RM, Andrews DW (2003) Interaction with a membrane surface triggers a reversible conformational change in Bax normally associated with induction of apoptosis. J Biol Chem 278(49):48935–48941
Zinser E, Sperka-Gottlieb CD, Fasch EV, Kohlwein SD, Paltauf F, Daum G (1991) Phospholipid synthesis and lipid composition of subcellular membranes in the unicellular eukaryote *Saccharomyces cerevisiae*. J Bacteriol 173(6):2026–2034

Chapter 9
The Mitochondrial Outer Membrane Potential as an Electrical Feedback Control of Cell Energy Metabolism

Victor V. Lemeshko

9.1 Introduction

Mitochondria, intracellular structures composed of two membranes, represent the main source of energy in aerobic eukaryotic cells. The mitochondrial inner membrane (MIM) is responsible for the oxidative phosphorylation process coupled to respiration, while the mitochondrial outer membrane (MOM) seems to control energy flux between mitochondria and the cytosol (Rostovtseva and Colombini 1997; Colombini 2004; Lemeshko 2002, 2014a, 2016; Lemasters and Holmuhamedov 2006; Saks et al. 2010; Colombini and Mannella 2012; Rostovtseva and Bezrukov 2008, 2012; Messina et al. 2012; Maldonado et al. 2013), cell energy homeostasis in general, and cell survival (Shoshan-Barmatz and Gincel 2003; Rostovtseva et al. 2008; Shoshan-Barmatz and Ben-Hail 2012; Maldonado et al. 2013; Lemeshko 2014a, 2016; Guzun et al. 2015). The MOM permeabilization induced by various proapoptotic signals is known to liberate apoptogenic factors from the mitochondrial intermembrane space (MIMS) into the cytosol, hence starting the apoptotic cascade in mammalian cells, finally leading to cell death (Rostovtseva et al. 2005; Báthori et al. 2006; Shoshan-Barmatz et al. 2010; Smilansky et al. 2015). In general, a broad range of pathologies is directly related to various mitochondrial dysfunctions at the level of both membranes.

The voltage-dependent anion channel (VDAC), constituting more than 50% of the MOM total protein (Mannella 1982; Colombini and Mannella 2012), is highly permeable in its open state for negatively charged metabolites, including ATP, ADP, AMP, CrP, and inorganic phosphate (P_i), and it is almost impermeable in its closed state for most of them (Hodge and Colombini 1997; Vander Heiden et al. 2000; Colombini 2016). This allows the assumption that the outer membrane potential

V.V. Lemeshko (✉)
Escuela de Física, Facultad de Ciencias, Universidad Nacional de Colombia, Sede Medellín, Calle 59A, No 63-20, Medellín, Colombia
e-mail: vvasilie@unal.edu.co

T.K. Rostovtseva (ed.), *Molecular Basis for Mitochondrial Signaling*, Biological and Medical Physics, Biomedical Engineering,
DOI 10.1007/978-3-319-55539-3_9

(OMP) of mitochondria, presumably generated under physiological conditions by various metabolically dependent mechanisms, suggested earlier (Lemeshko and Lemeshko 2004; Lemeshko 2002, 2006, 2014a, 2016), might be a powerful regulator of MOM permeability and thus of the cell energy metabolism in general.

VDAC is characterized by well-conserved voltage-gating properties (Colombini and Mannella 2012) that may be additionally controlled by its phosphorylation-dephosphorylation (Pastorino et al. 2005; Sheldon et al. 2011; Kerner et al. 2012) and SH/SS redox state (Okazaki et al. 2015; De Pinto et al. 2016), various proteins (Liu and Colombini 1992; Colombini et al. 1996; Rostovtseva and Bezrukov 2012; Kuznetsov et al. 2013; Rostovtseva et al. 2015; Chernoivanenko et al. 2015), metabolites, and other factors (Lee et al. 1996; Colombini et al. 1996; Lemasters and Holmuhamedov 2006; Báthori et al. 2006; Stein and Colombini 2008; Shoshan-Barmatz et al. 2010; Maldonado et al. 2013; Sheldon et al. 2015). That is why VDAC has been considered a very important target for the development of new drugs, including anticancer medications (Shoshan-Barmatz and Ben-Hail 2012; Mathupala and Pedersen 2010; Rimmerman et al. 2013; Head et al. 2015; Zhang et al. 2016).

Many cases of apparently anomalous behavior of mitochondria have been related to a possible MOM permeability restriction, such as global suppression of mitochondria under ischemia, anoxia, sepsis, and aerobic glycolysis in cancer cells, among others (Lemasters and Holmuhamedov 2006). According to these authors, a closure of VDAC, then a block of MOM permeability, could lead to a global suppression of mitochondrial metabolism. Nevertheless, the possibility of electrical closure of VDACs by OMP under physiological conditions has been questioned, because even in its closed state VDAC is highly permeable to small ions such as potassium, sodium, and chloride (Hodge and Colombini 1997), assuming to prevent the possibility of OMP generation (Lemasters and Holmuhamedov 2006). With this respect, Benz et al. (1990) concluded earlier that "the existence of an electrochemical potential across the outer membrane can't be expected." However, zero electrochemical potential does not mean that the electrical membrane potential cannot exist. In the case of the Donnan potential, for example, generally understood as the electrical potential generated across porous membranes by membrane impermeable charged colloids, the electrochemical potentials of small permeable ions are equal to zero. Even in the case of a relatively high steady-state resting potential in the giant axon, generated by the well-known Na^+/K^+ mechanism, the electrochemical potential of highly permeable chloride ions across the excitable membrane is known to be almost zero.

OMP of the Donnan nature has been suggested to influence VDAC's conductance under physiological conditions (Liu and Colombini 1992; Porcelli et al. 2005), but such a possibility has been questioned because the average charge of cytosolic proteins has to be different from that of MIMS proteins (Maldonado and Lemasters 2014). On the other hand, a Donnan potential of even a small value might be superimposed on variable metabolically derived OMP, as described earlier (Lemeshko and Lemeshko 2000).

A Donnan potential of high value across MOM has been assumed by Porcelli et al. (2005) on the basis of experiments with living cells, first mentioned by Colombini (2004). In the experiments of Porcelli et al. (2005), the estimated OMP achieved up to −43 mV. The authors also demonstrated that rotenone and oligomycin, mitochondrial inhibitors, caused the complete disappearance of the monitored potential that seems to contradict the Donnan potential nature of OMP. Nevertheless, if we consider the mitochondrial matrix as one "giant macromolecule" with a very high and variable negative charge inside the energized mitochondria, it might be understood as a Donnan potential. Along these lines, it has been demonstrated that electrophoretic mobility of the energized mitochondria is significantly higher than that of the de-energized mitochondria (Kamo et al. 1976). This new aspect of a possible nature of OMP seems to be worthy of further theoretical consideration.

The origin of the potential monitored by Porcelli et al. (2005) is not yet clear, because the pH-sensitive fluorescent probe has been covalently attached to the glycerol-3-phosphate dehydrogenase. This is an integral membrane protein on the external side of MIM, thus allowing the sensing of relatively high negative surface potential of MIM, depending on the mitochondrial metabolic state (see discussion by Lemeshko 2006). OMP, monitored by Porcelli et al. (2005), might also reflect, at least partially, metabolically derived OMP generated by some of the steady-state mechanisms suggested earlier (Lemeshko and Lemeshko 2000, 2004; Lemeshko 2002, 2014a, 2016). The experimental approach undertaken by Porcelli et al. (2005) is promising for the monitoring of OMP in living cells in real time using a pH-sensitive fluorescent probe covalently attached to a MIMS protein, which does not have an affinity to MIM.

Metabolically dependent generation of OMP, although of relatively low value, might arise from the difference in the MOM permeability to charged metabolites. This has been demonstrated earlier using a theoretical model of a steady-state energy flux from mitochondria into the cytosol, maintained by cycling of creatine phosphate (CrP)/creatine (Cr) and P_i (Lemeshko and Lemeshko 2000). High values of OMP, positive and negative, were predicted by another model, where OMP generation is coupled to phosphoryl group transfer from the matrix ATP into the cytosol through the bi-transmembrane contact sites formed by the adenine nucleotide translocator (ANT) and VDAC (Lemeshko 2002). In this model, inorganic phosphate returns into MIMS through VDACs not bound to other proteins in MOM ("free" VDACs). Such energy-supported cycling of one net negative charge allows the presentation of OMP as a part of IMP applied to MOM through the ANT-VDAC or ANT-VDAC-HK electrogenic contact sites (Lemeshko 2002). The possibility of induction of OMP by the inner membrane electrochemical gradient, at least at the intermembrane contact sites, to regulate the open-closed states of the VDACs, has been also assumed by Mathupala and Pedersen (2010).

A very powerful and, at the same time, the simplest active mechanism of OMP generation has been suggested recently on the basis of the thermodynamic analysis of the VDAC-kinase electrogenic complexes in MOM, which might be considered as generators of metabolically dependent OMP (Lemeshko 2002, 2014a, 2016).

According to the VDAC-HK model of OMP generation, the Warburg effect might be described as an electrical suppression of mitochondrial ATP transfer to the cytosol (Lemeshko 2015). The exception, in this case, is the initial step of glycolysis associated with the VDAC-HK complexes of MOM (Lemeshko 2002, 2014a; Lemeshko and Lemeshko 2004). As it has been shown by Warburg, cancer cells display very high glycolysis, in which fully active mitochondria (not damaged) are relatively inactive in terms of respiration and ATP generation (see Lemeshko 2002; Lemasters and Holmuhamedov 2006).

Although various possibilities exist for generation of metabolically derived OMP, allowing a fast and wide range of VDAC conductance modulation by an electrical feedback control mechanism, the most accepted concept might be described as a "molecular corking up" of open VDAC pores, based on many experimental observations (Colombini et al. 1996; Lemasters and Holmuhamedov 2006; Rostovtseva et al. 2008, 2015; Rostovtseva and Bezrukov 2012; Maldonado et al. 2013; Simson et al. 2016). Recently, it has been assumed, on the basis of a model that does not consider any possibility of the electrical closure of VDACs by OMP, that only 2% of mitochondrial VDACs in cardiomyocytes are open, resulting in up to 98% of VDACs being blocked by tubulin and/or wrapped by sarcoplasmic reticulum (Simson et al. 2016). Alternatively, these observations support the possibility of OMP generation causing up to 98% electrical suppression of MOM permeability to ADP and to other charged metabolites.

It has been demonstrated that the cytosolic protein tubulin significantly decreases the conductance of VDAC reconstituted into planar lipid membrane (Rostovtseva and Bezrukov 2012, 2015; Maldonado et al. 2013). The most important finding is that VDAC blockage by tubulin is impressively voltage-dependent (Rostovtseva and Bezrukov 2015). Thus, if the metabolically dependent positive OMP (which corresponds to the negative potential on the cytosolic side of the MOM) is generated by some mechanism, it will increase VDAC blockage by tubulin. Vice versa, VDAC blockage by tubulin and other proteins might increase the probability of OMP generation, as it has been demonstrated in previous theoretical works (Lemeshko 2014a, 2016). In general, the fast electrical modulation of VDACs together with their "molecular corking up" might represent a very powerful synergistic signaling mechanism controlling cell energy metabolism.

In the presented work, we describe three types of possible steady-state mechanisms of OMP generation coupled to phosphoryl group transfer through the mitochondrial membranes (Lemeshko 2002, 2014a, 2016). The performed simplest thermodynamic estimations of a possible range of generated OMP support the suggestion of a new metabolically dependent electrical signaling system associated with MOM, allowing physiological regulation of cell energy metabolism. In cancer cells, the metabolically dependent OMP might cause electrical suppression of mitochondria, underlying the Warburg and Crabtree effects. In addition, OMP-dependent expulsion/capture of Ca^{2+} from/into MIMS seems to also be fundamental in controlling mitochondrial permeability transition and cell death resistance.

9.2 The Mitochondrial Outer Membrane Potential Arising from Steady-State Passive Diffusion of Charged Metabolites

There are various possibilities for generation of metabolically dependent OMP in cells that may be tissue specific, from the normal contractile and noncontractile cells to cancer cells. OMP at relatively low levels, not high enough to restrict MOM permeability by electrical closing of VDACs, may still modulate steady-state concentration levels of organic and inorganic ions in MIMS, such as ADP^{3-} and Ca^{2+}, for example, influencing their transportation into the mitochondrial matrix and thus modulating the mitochondrial metabolic state.

One type of the mechanisms of OMP generation may be based on the well-known difference in VDAC's permeability to the most important anionic metabolites, such as ATP, ADP, AMP, CrP, and P_i, in both open and closed states of the channel, at different steady-state electrochemical gradients of these metabolites across MOM (Fig. 9.1). The generation of negative OMP, resulting from the steady-state return of P_i into MIMS, after ATP hydrolysis in the cytosol, is expected due to a relatively high MOM permeability to P_i in comparison with that to ATP (Hodge and Colombini 1997; Vander Heiden et al. 2000; Colombini 2016). To recover ATP, P_i from MIMS is transported into the mitochondrial matrix through the electrically neutral pH-dependent P_i transporter (PT) located in MIM (Fig. 9.1a).

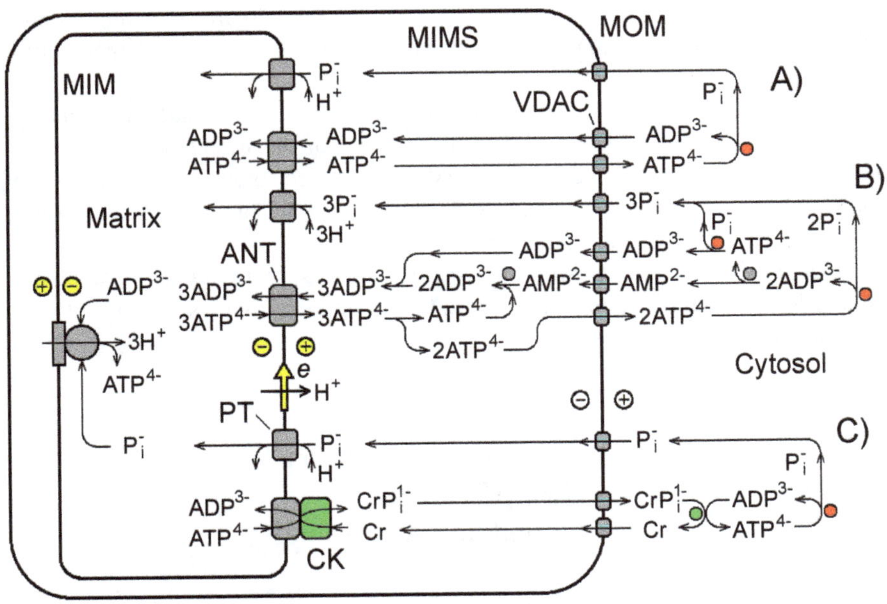

Fig. 9.1 Possible passive diffusion mechanisms of OMP generation due to a difference in the VDAC's permeability to various charged metabolites under their steady-state circulation through the mitochondrial outer membrane. *Circles*: *red*, ATPase; *green*, creatine kinase; *gray*, adenylate kinase

For the most general case (Fig. 9.1b), ADP production in the cytosol by ATP hydrolysis results in a local increase of ADP concentration that in turn leads to a recovery of 1 ATP per 2 ADP in the adenylate kinase reaction, also producing 1 AMP. To close the cycle in a steady-state process, 3 P_i, 1 ADP, and 1 AMP return into MIMS in exchange for 2 ATP released from MIMS into the cytosol (Fig. 9.1b).

In the cardiac and muscle cells, the largest portion of mitochondrial energy has been suggested to be channeled into the cytosol through the creatine kinase system (Kottke et al. 1991; Schlattner et al. 2006; Saks et al. 2010; Guzun et al. 2015; Wallimann 2015), the simplified part of which is presented in Fig. 9.1c. Earlier we have estimated the possibility of generation of OMP for this case on the basis of computational modeling (Lemeshko and Lemeshko 2000). Metabolically derived negative OMP up to −5 mV has been calculated using Goldman's equations for steady-state fluxes of CrP^{2-}, P_i^{1-} and P_i^{2-} through MOM, varying the maximal rate of CrP hydrolysis in the cytosol at a fixed rate of CrP^{2-} production in MIMS and essentially arbitrary concentrations of these metabolites.

For these calculations, the voltage sensitivity of VDAC has been taken as very high, assuming the presence of some VDAC's modulators in the cytosol, for example, a protein factor "X" (Saks et al. 1995) or some MIMS proteins (Liu and Colombini 1992; Holden and Colombini 1993). The protein factor "X" has been subsequently identified as the cytosolic protein tubulin (Rostovtseva and Bezrukov 2008) that has been earlier shown to reversibly interact with mitochondria through protein component(s) in an ionic strength-insensitive manner (Bernier-Valentin and Rousset 1982). This first model (Fig. 9.1c) also showed that the upper values of generated OMP depend on the cell workload at the given maximal rate of mitochondrial energy production and on the MOM permeability to charged metabolites (Lemeshko and Lemeshko 2000).

Theoretically, all circuits presented in Fig. 9.1 might be included in a general computational model of the passive diffusion mechanism of metabolically dependent generation of OMP. The value of OMP is assumed to depend on the rates of corresponding reactions and steady-state concentrations of charged metabolites (Fig. 9.1). In addition, the voltage-gating properties of various VDAC isoforms, modulated by corking up proteins and chemical factors, seem to play a crucial role in the MOM signaling system controlling the mitochondria-cytosol energy flux by generated OMP.

9.3 Outer Membrane Potential Generation Associated with Phosphoryl Group Transfer Through the Mitochondrial Membranes

According to endosymbiotic models of the evolution of eukaryotic cells, their mitochondria are considered to originate from former primitive aerobic bacteria that were captured by large anaerobic eukaryotic host cells as a new, aerobic energy source (Martin et al. 2001; Gray 2012). The acquired prokaryotic bacteria were compartmentalized due to their wrapping by the plasma membrane of host cells containing VDAC, thus converting them into mitochondria (Grimm and Brdiczka 2007). Relatively permeable MOM allowed fast metabolic communication between the cytosol and the mitochondrial matrix. Such communication includes hydrogen shuttling from cytosolic NADH into the matrix or into MIMS, as well as ATP^{4-}/ADP^{3-} turnover with continuous return of inorganic phosphate into the mitochondrial matrix after ATP hydrolysis in the cytosol.

It has been suggested that the smart mechanism of mitochondrial energy export to the cytosol is not to transfer ATP but rather to transfer phosphoryl group from ATP produced in the mitochondrial matrix to either glucose or creatine through ANT coupled to hexokinase or creatine kinase, respectively (Grimm and Brdiczka 2007). It has also been reported that P_i may play a role in the regulation of cell respiration, as a substrate feedback control (Scheibye-Knudsen and Quistorff 2009), and that the mechanism of permeation of monovalent P_i through VDAC remarkably differs from those of divalent P_i, AMP, and ATP (Krammer et al. 2015). Thus, P_i seems to be a very important anionic metabolite of the mitochondria-cytosol energy transfer system and of respiratory feedback control (Jeneson et al. 2011).

We have considered the structural and functional organization of VDAC-mediated phosphoryl group transfer to be crucial for the suggested active mechanisms of metabolically dependent generation of OMP (Lemeshko 2002, 2014a, 2016). As shown in Fig. 9.2, the mitochondria-cytosol transfer of phosphoryl groups might be realized through the VDAC-HK electrogenic complexes of MOM (Fig. 9.2a), through the bi-transmembrane ANT-VDAC (Fig. 9.2b) and ANT-CK-VDAC (Fig. 9.2c) electrogenic contact sites. The most general characteristic of all of these pathways is the continuous return of P_i from the cytosol into MIMS through free, unbound VDACs in MOM.

The concept of the channeling of phosphoryl groups of mitochondrial ATP to the specific cytosolic energy consumers (Schlattner et al. 2006; Saks et al. 2010; Guzun et al. 2015) represents a great interest. In order to coordinate mitochondrial and cytosolic ATP producing systems, aimed to maintain cell energy homeostasis, such energy channeling requires metabolic feedback control. It could be realized through the metabolically dependent generation of OMP as an electrical signaling system modulating MOM permeability to charged macroergic compounds (Fig. 9.2). This type of regulation is consistent with the highly conserved voltage-gating properties of VDACs (Shoshan-Barmatz et al. 2010; Colombini and Mannella 2012) and with the existence of many physiological modulators of the VDAC's conductance and voltage sensitivity (Colombini et al. 1996; Lemasters and Holmuhamedov 2006; Shoshan-Barmatz et al. 2010; Kerner et al. 2012; Rostovtseva and Bezrukov 2008, 2012, 2015; Sheldon et al. 2011, 2015; Maldonado et al. 2013; Rostovtseva et al. 2015; Okazaki et al. 2015; De Pinto et al. 2016), in addition to the cell-specific structural organization of mitochondrial energy channeling systems through the VDAC-kinase complexes (Kottke et al. 1991; Lemeshko 2002, 2014a, 2016; Schlattner et al. 2006; Saks et al. 2010; Mathupala and Pedersen 2010; Guzun et al. 2015).

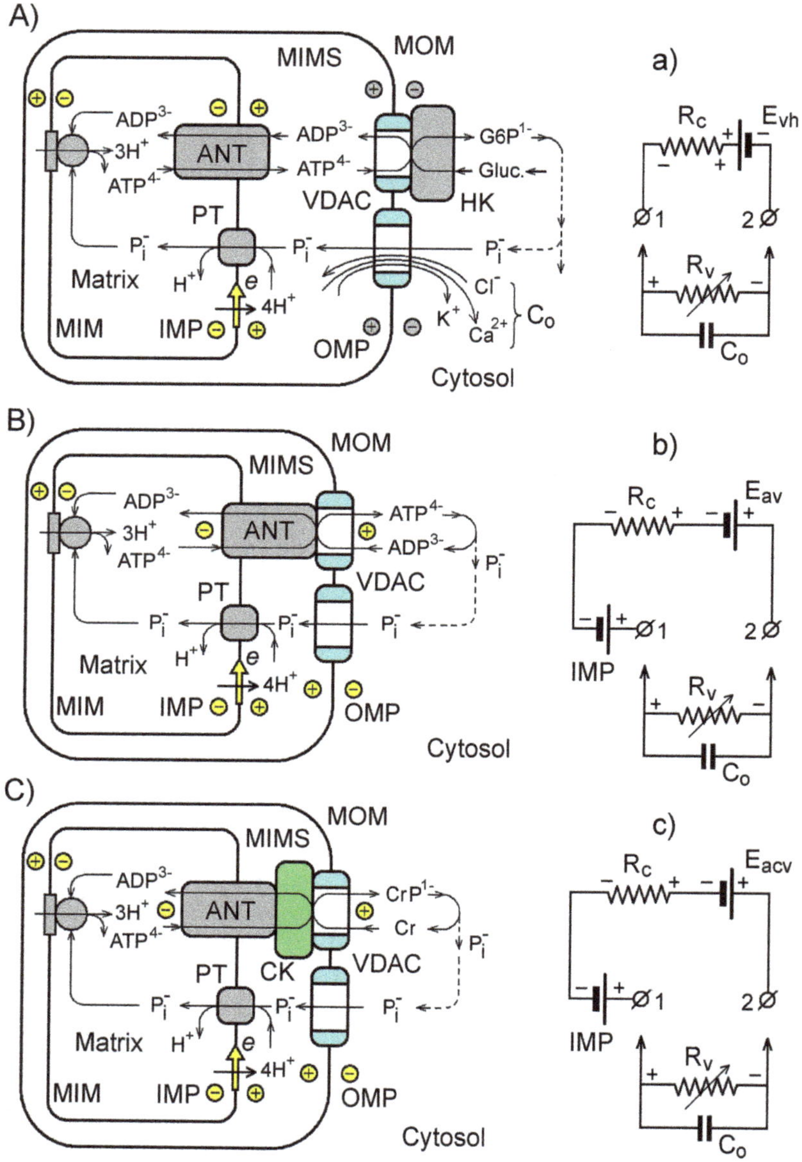

Fig. 9.2 Possible active mechanisms of OMP generation mediated by steady-state phosphoryl group transfer through the mitochondrial membranes. **a** Direct generation of OMP by the VDAC-HK electrogenic complexes using the Gibbs free energy of the HK reaction; **b** OMP generation due to an application of a part of IMP to MOM through the ANT-VDAC electrogenic contact sites; **c** OMP generation due to an application of a part of IMP to MOM through the ANT-CK-VDAC electrogenic contact sites and using the Gibbs free energy of the CK reaction associated with them; **a–c** equivalent electrical circuits of the models **a–c**, in which points 1 and 2 are the MIMS and cytosolic sides of MOM, R_v is the resistance of free, unbound VDACs in MOM, and C_o represents MOM as an electrolytic capacitor, resulting in charge separation (K^+, Cl^-, Ca^{2+}, etc.) in an electric field of OMP

9.3.1 Generation of the Outer Membrane Potential by the VDAC-HK Electrogenic Complexes

It has been reported in the literature that the binding of HK to VDACs (Pastorino and Hoek 2008; Abu-Hamad et al. 2008; Mathupala and Pedersen 2010; John et al. 2011) leads to a decrease in the conductance of VDAC reconstituted into planar lipid membrane (Azoula-Zohar et al. 2004) and that the kinetic properties of HK bound to mitochondria are very different to those of free HK (Wilson 2003). A very interesting phenomenon of the energy metabolism in cancer cells is the extremely high quantity of HK bound to mitochondria, more than two orders of magnitude higher in liver cancer cells than in normal hepatocytes (Marín-Hernández et al. 2006). The preferential high-affinity binding of HK to the ANT-VDAC contact sites has also been reported (Brdiczka et al. 2006). Although the intermembrane contact sites have not been detected for some types of cancer cells, these cells contain a very high quantity of mitochondrial HK (Denis-Pouxviel et al. 1987), which is known to bind to VDACs beyond the contact sites with less affinity, forming VDAC-HK complexes (Brdiczka et al. 2006), as shown in Fig. 9.2a.

For the computational analysis of three models of OMP generation, shown in Fig. 9.2, all quantities of VDACs in MOM for each model were normalized to one arbitrary unit, representing the sum of the fraction N_c of VDACs forming VDAC-HK (Fig. 9.2a), ANT-VDAC (Fig. 9.2b), or ANT-CK-VDAC (Fig. 9.2c) complexes and of the fraction of free, unbound VDACs. For the majority of calculations, the fraction of unbound VDACs has been considered as only the fraction of voltage-sensitiveVDAC1 and VDAC2 isoforms, N_{vs}:

$$N_c + N_{vs} = 1 \tag{9.1}$$

The fraction N_c of VDACs forming VDAC-kinase complexes of MOM or bi-transmembrane contact sites (Fig. 9.2a–c) represents numerically the relative conductance g_c to transfer phosphoryl groups through them (Fig. 9.2a–c, where $R_c = 1/g_c$):

$$g_c = N_c \tag{9.2}$$

For the steady-state return of P_i^- into MIMS through MOM, the P_i^- conductance g_{vs} of the fraction N_{vs} of voltage-sensitive unbound VDACs in MOM was presented as a bell-shaped function of OMP (Fig. 9.3):

$$g_{vs} = N_{vs} \cdot \left[G_c + (1 - G_c) \cdot \exp - \left(S \cdot |OMP| \right)^3 \right]. \tag{9.3}$$

Here, the parameter G_c is the relative P_i^- conductance of unbound VDACs in the closed state as a fraction of that in the open state. It was ranged from $G_c = 0.5$ to $G_c = 0.1$ (Hodge and Colombini 1997). S is the voltage sensitive parameter. For

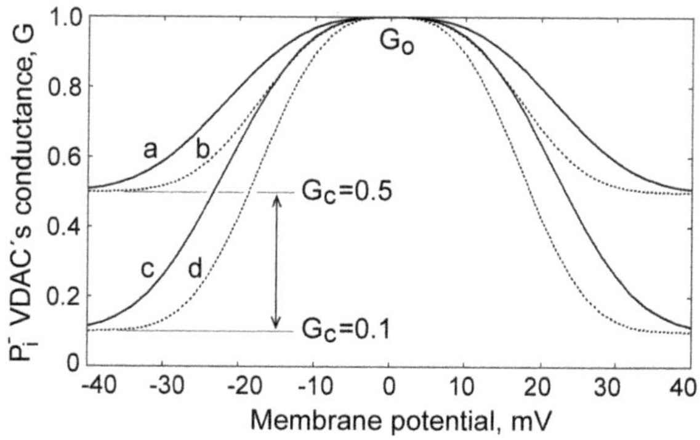

Fig. 9.3 VDAC voltage-gating characteristics used for computational analysis of the models. G_c relative P_i- conductance of VDAC in the closed state, *S* voltage sensitivity parameter; **a, b** $G_c = 0.5$; **c, d** $G_c = 0.1$; *Solid lines* at $S = 40$ V-1; *Dashed lines* at $S = 50$ V-1; $G_c = 0.1–0.5$ were used for calculations

calculations, the latter was taken at $S = 40\ V^{-1}$ or $S = 50\ V^{-1}$ (Fig. 9.3). The shape of the VDAC conductance-voltage curve with the power index of 3 is somewhat different to that with the power index of 2 (Lemeshko 2014a):

$$g_{vs} = N_{vs} \cdot \left[G_c + (1 - G_c) \cdot \exp -(S \cdot OMP)^2 \right], \tag{9.4}$$

With the aim to estimate the influence of the shape of the VDAC conductance-voltage function on the probability of OMP generation and MOM permeability modulation, some calculations were also performed using Eq. 9.4 in comparison to those performed using Eq. 9.3.

Actually, there is significant variation among the reported VDAC conductance-voltage plots, obtained under different experimental conditions (Mangan and Colombini 1987; Rostovtseva et al. 2008; Shoshan-Barmatz et al. 2010, 2015; Teijido et al. 2014; Okazaki et al. 2015; Maurya and Mahalakshmi 2015). The shape of these plots may be fitted with Eq. 9.4 using the power index ranging from 2 (or less) to 3 (or greater) and changing the voltage-sensitive parameter *S*. A physical mechanism that would allow the description of VDAC voltage-gating properties with an equation like Eq. 9.4, depending on the experimental conditions, such as pH changes (Teijido et al. 2014), the presence of polyvalent anions, or some proteins (Mangan and Colombini 1987; Rostovtseva et al. 2008), SH/SS redox state of various VDAC isoforms (Okazaki et al. 2015), has to be determined yet.

On the basis of experiments with knockdown of various VDAC isoforms, the minor VDAC3 isoform, having very low voltage sensitivity, has been reported as the most important for maintenance of mitochondrial metabolism, at least in HepG2 cancer cells (Maldonado et al. 2013). That is why we also estimated the influence of the presence of VDAC3 fraction and of its theoretical knockout on generation of OMP by the VDAC-HK-complexes. The free, unbound VDAC3 isoform was included as the fraction N_{vi} of voltage-insensitive VDACs, for simplicity, (always open at least in the range of ±50 mV for OMP) by replacing a part of the fraction N_{vs} of voltage-sensitive VDACs, thus transforming Eq. 9.3 into

$$g_{vs} = \left(N_{vs} - N_{vi}\right)\cdot\left[G_c + \left(1-G_c\right)\cdot \exp -\left(S\cdot\left|OMP\right|\right)^3\right] + N_{vi}. \tag{9.5}$$

If in addition, we assume a knockout of the fraction N_{vi}, Eq. 9.5 will be transformed into

$$g_{vs} = \left(N_{vs} - N_{vi}\right)\cdot\left[G_c + \left(1-G_c\right)\cdot \exp -\left(S\cdot\left|OMP\right|\right)^3\right]. \tag{9.6}$$

The fraction of VDAC3 was taken at $N_{vi} = 0.1$, i.e., 10% of all VDACs, very close to 11% reported by Maldonado et al. (2013).

The VDAC-HK electrogenic complexes of MOM may be considered direct voltage generators (Fig. 9.2a, *a*), which use the Gibbs free energy of the essentially irreversible HK reaction of energized mitochondria at high ratio $[\mathrm{ATP}]_s/[\mathrm{ADP}]_s$ in MIMS. The voltage V_{vh} of such VDAC-HK battery E_{vh} (Fig. 9.2a, *a*) may be estimated (Lemeshko 2014a) as

$$V_{vh} = -\left(\frac{G_h^{o'}}{F} + \frac{RT}{F}\ln\frac{[G6P]_c\cdot[ADP]_s}{[\mathrm{Gluc}]_c\cdot[ATP]_s}\right). \tag{9.7}$$

Here, $[G6P]_c$ and $[\mathrm{Gluc}]_c$ are the cytosolic concentrations of glucose-6-phosphate and glucose, respectively, and $G_h^{o'} = -16.7\,kJ/mol$ is the standard Gibbs free energy of the HK reaction at pH = 7. For calculations, the ratio $[\mathrm{Gluc}]_c/[G6P]_c$ was varied from 0.1 to 100. Taking into account that under normal physiological conditions $[G6P]_c$ = 0.1 mM, approximately, it means that the calculations were performed at glucose concentrations of up to 10 mM (at $[\mathrm{Gluc}]_c/[G6P]_c = 100$).

The ratio $[\mathrm{ATP}]_s/[\mathrm{ADP}]_s$ in MIMS (TD_s) may be determined from a given value of the mitochondrial phosphorylation potential, ΔG_a, defined as

$$G_a = \ G_a^{o'} + \mathrm{RT}\ln\frac{\left[P_{i,s}\right]}{TD_s}, \tag{9.8}$$

where $G_a^{o'} = -31\,kJ/mol$ is the standard Gibbs free energy of ATP hydrolysis at pH = 7.0. For calculations, we used $\Delta G_a = -61$ kJ/mol that is near the values reported in the literature for the resting state of cells (Wallis et al. 2005; Pinz et al. 2008). The concentration of P_i in MIMS was considered constant, at $[P_{i,s}]$ = 5 mM, in the range for moderate workloads of isolated perfused mouse hearts (Spindler et al. 2002).

The voltage V_{vh} of VDAC-HK complexes, calculated according to Eqs. 9.7 and 9.8 at $\Delta G_a = -61$ kJ/mol, partially drops on the internal resistance R_c ($1/g_c$) of the battery E_{vh}, as well as on the resistance R_v ($1/g_{vs}$) of free, unbound VDACs in MOM, thus providing direct generation of OMP (Fig. 9.2a, *a*) that can be estimated on the basis of Ohm's law:

$$OMP = V_{vh} \cdot \frac{g_c}{g_c + g_{vs}}. \tag{9.9}$$

OMP was calculated for various fractions N_c of VDACs forming such VDAC-HK complexes, and also as a function of the $[\text{Gluc}]_c/[G6P]_c$ ratio, using the system of Eqs. 9.1, 9.2, and 9.7–9.9 and one of Eqs. 9.3, 9.5, and 9.6, depending on the presence of VDAC3 isoform and on its theoretical knockout (see figure legends). The software Mathcad Professional was used for all calculations.

First, the computational analysis was performed at the voltage sensitivity parameter $S = 40\ V^{-1}$ and the P_i^- conductance parameter $G_c = 0.4$ for voltage-sensitive VDACs. At a high concentration of glucose, the calculations demonstrated generation of a remarkable OMP if only 3.5% of all VDACs in MOM form VDAC-HK complexes (i.e., at $N_c = 0.035$) or even at very small glucose concentrations, if N_c was increased to $N_c = 0.04$ (Fig. 9.4a). The calculated OMP is high enough to electrically restrict MOM permeability to P_i^- through voltage-sensitive VDACs, depending on the $[\text{Gluc}]_c/[G6P]_c$ ratio and on the percentage V_{HK} of VDACs forming VDAC-HK complexes in MOM ($V_{\text{HK}} = 2$–4% in Fig. 9.4b, where $V_{\text{HK}} = N_c \cdot 100\%$).

Many factors have been reported to modulate VDAC's voltage sensitivity and conductance, as referred above. One of such most powerful factors seems to be free tubulin, suggested to maintain VDAC1 and VDAC2 isoforms in a mostly closed state and to increase VDAC's voltage sensitivity (Rostovtseva et al. 2008; Rostovtseva and Bezrukov 2012, 2015; Maldonado et al. 2013). The great importance of such properties of free tubulin to modulate the VDAC-HK-dependent generation of OMP was evaluated by increasing the parameter S from $S = 40$ to $S = 50\ V^{-1}$, and by decreasing the parameter G_c from $G_c = 0.4$ to only $G_c = 0.3$ (Eq. 9.3, Fig. 9.3). The calculations demonstrated a very strong increase in OMP due to both an increase in S (Fig. 9.4c) and a decrease in G_c (Fig. 9.4e), in comparison to OMP values calculated for the case of $S = 40$ and $G_c = 0.4$ (Fig. 9.4a). This increase in OMP resulted in corresponding significant electrical restriction of MOM permeability to P_i^- (Fig. 9.4d, f) compared with the data determined for the case of $S = 40$ and $G_c = 0.4$ (Fig. 9.4b).

Under the same conditions, significantly lower values of OMP were calculated when a relatively small fraction of unbound voltage-sensitive VDACs was replaced by the fraction N_{vi} of voltage-insensitive VDACs, taking $N_{\text{vi}} = 0.1$ (Fig. 9.5a). In this case, the fraction of N_{vs} was reduced (Eq. 9.5), for example, from $N_{\text{vs}} = 0.96$ to $N_{\text{vs}} = 0.96\text{-}N_{\text{vi}}$. Because of the lower values of OMP (Fig. 9.5a in comparison to Fig. 9.4a), the electrical restriction of MOM permeability to P_i^- was also significantly reduced (Fig. 9.5b in comparison to Fig. 9.4b).

To estimate the possible effects of knockout of the fraction N_{vi} of VDACs, Eq. 9.5 was replaced by Eq. 9.6, conserving the reduced fraction N_{vs}, i.e., the same N_{vs} value that was before N_{vi} knockout. The calculations demonstrated that the theoretical knockout of $N_{\text{vi}} = 0.1$ strongly increased the probability of generation of high OMP (Fig. 9.5c) and of electrical restriction of MOM permeability to P_i^- (Fig. 9.5d).

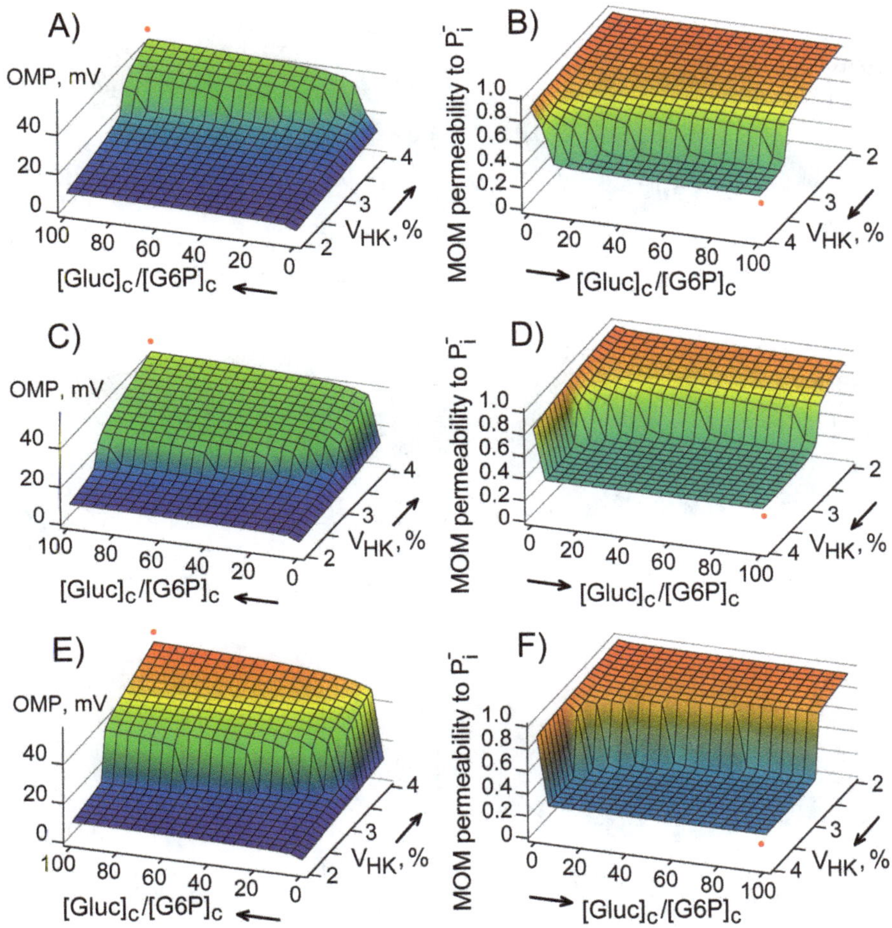

Fig. 9.4 OMP generation (**a**, **c**, **e**) and electrical restriction of the MOM permeability to P_i^- (g_{vs}) (**b**, **d**, **f**) according to the VDAC-HK model (Fig. 9.2a) in mitochondria with the phosphorylation potential $\Delta G_a = -61$ kJ/mol (Eq. 9.8 at $[P_{i,s}] = 5$ mM) as functions of the ratio $[\text{Gluc}]_c/[G6P]_c$ in the cytosol (Eq. 9.7) and of the fraction N_c of VDACs forming VDAC-HK complexes ($V_{HK,\%} = N_c \cdot 100\%$). **a**, **b** At $S = 40$ V^{-1} and $G_c = 0.4$; **c**, **d** at $S = 50$ V^{-1} and $G_c = 0.4$; **e**, **f** at $S = 40$ V^{-1} and $G_c = 0.3$. The calculations were performed using the system of Eqs. 9.1–9.3, 9.7–9.9

Interestingly, even in the presence of voltage-insensitive fraction N_{vi} of VDAC's, at $N_{vi} = 0.1$, high OMP (Fig. 9.5e) and pronounced electrical restriction of MOM permeability to P_i^- (Fig. 9.5f) may be expected on increase in the voltage sensitivity parameter S and on decrease in the parameter G_c (Eq. 9.5). Such effects might be caused by free tubulin (Rostovtseva et al. 2008) interacting mainly with the voltage-sensitive VDAC1 and VDAC2, but not with the voltage-insensitive VDAC3 isoform (Maldonado et al. 2013).

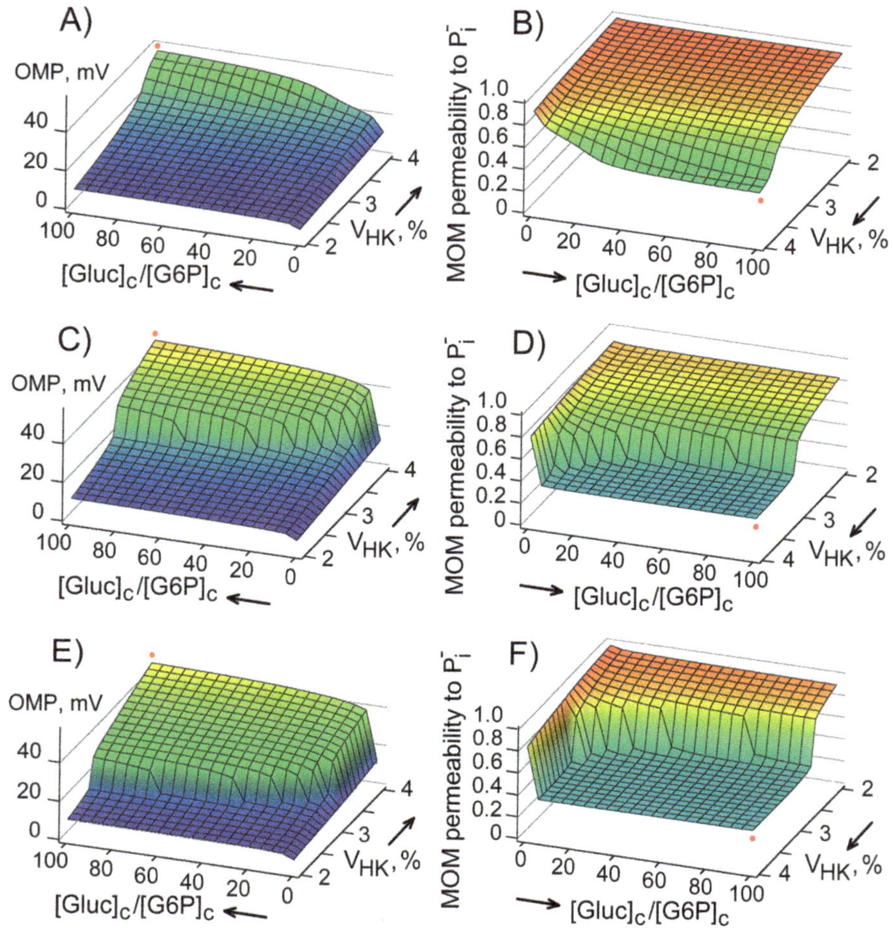

Fig. 9.5 The influence of a small fraction of voltage-insensitive VDACs, N_{vi} and of its theoretical knockout on OMP generation (**a, c, e**) and electrical restriction of the MOM permeability to P_i^- (g_{vs}) (**b, d, f**) according to the VDAC-HK model (Fig. 9.2a) in mitochondria with the phosphorylation potential $\Delta G_a = -61$ kJ/mol (Eq. 9.8 at $[P_{i,s}] = 5$ mM) as functions of the ratio $[\text{Gluc}]_c/[\text{G6P}]_c$ in the cytosol (Eq. 9.7) and of the fraction N_c of VDACs forming VDAC-HK complexes ($V_{HK,\%} = N_c \cdot 100\%$). **a, b** At $G_c = 0.4$, $S = 40\ \text{V}^{-1}$, $N_{vi} = 0.1$ (Eq. 9.5); **c, d** at $G_c = 0.4$, $S = 40\ \text{V}^{-1}$ and knockout of $N_{vi} = 0.1$ (Eq. 9.6); **e, f** at $G_c = 0.3$, $S = 50\ \text{V}^{-1}$, $N_{vi} = 0.1$ (Eq. 9.5). The system of Eqs. 9.1, 9.2, and 9.7–9.9, with Eq. 9.5 or Eq. 9.6, was used for respective calculations

In the presence of small inorganic ions, the generated OMP has been assumed to dissipate due to a high permeability to these ions of unbound VDACs, even in their closed state (Lemasters and Holmuhamedov 2006). Nevertheless, as the suggested mechanisms of OMP generation are steady state (Fig. 9.2), due to the continued circulation of phosphoryl groups and P_i, OMP should be recovered after reaching the electrochemical equilibrium for small ions, such as K^+, Cl^-, and Ca^{2+}, that can be presented as a process of the charging of an electrolytic capacitor C_o (Fig. 9.2a–c).

9.3.2 ANT-VDAC-Mediated Generation of the Outer Membrane Potential

One of the mechanisms of OMP generation might be coupled to the transfer of negative phosphoryl groups from the mitochondrial matrix ATP to the cytosol through the ANT-VDAC complexes (Fig. 9.2b), as it has been suggested earlier (Lemeshko 2002). This mechanism may be explained by the simplest equivalent electrical circuit shown in Fig. 9.2b, *b*.

For the electrochemical equilibrium of ATP^{4-}/ADP^{3-} turnover through the ANT-VDAC contact sites (Fig. 9.2b), the voltage V_{av} of the battery E_{av} (Fig. 9.2b, *b*) may be described by Nernst's equation as

$$V_{av} = -\frac{RT}{F}\ln\frac{TD_c}{TD_m}, \tag{9.10}$$

where R is the universal gas constant, $T = 310\ K$ is normal body temperature, and F is the Faraday constant, giving $RT/F = 26.7$ mV. Here $TD_c = [ATP]_c/[ADP]_c$ and $TD_m = [ATP]_m/[ADP]_m$ are corresponding concentration ratios in the cytosol and in the mitochondrial matrix, respectively. TD_m for respiring mitochondria was considered fixed at $TD_m = 3$ (Korzeniewski and Mazat 1996). Thus, the ANT-VDAC electrogenic complexes may be presented as a battery E_{av} with voltage V_{av}, which depends on TD_c that in turn may be defined by the ATP phosphorylation potential in the cytosol. The internal resistance R_c (Fig. 9.2b, *b*) of the battery E_{av} is inversely proportional to the conductance g_c of the fraction N_c of mitochondrial VDACs forming the intermembrane contact sites ($R_c = 1/g_c$, taking into account that $g_c = N_c$, Eq. 9.2).

The steady-state ATP^{4-}/ADP^{3-} electrogenic turnover and the P_i^- return into MIMS will depend on IMP and on the TD_c ratio immediately at the external side of MOM, providing OMP generation as a voltage drop on the resistance R_v of unbound VDACs (Fig. 9.2b, *b*), as described earlier (Lemeshko 2002). The sign and value of OMP depend on both V_{av} (Eq. 9.10) and IMP. According to the electrical circuit (Fig. 9.2b, *b*), OMP will be zero at V_{av} = IMP (where V_{av} is the voltage of the battery E_{av}), or OMP will be positive if the value of IMP is greater than that of V_{av}. A negative OMP is expected when the IMP value is less than that of V_{av}, in the case of a relatively high TD_c ratio in the cytosol, maintained by glycolysis, and a low phosphorylation potential of mitochondria, ΔG_a defined by Eq. 9.8.

On the other hand, at high cell workloads, accompanied by a relatively low steady-state TD_c ratio, the greatest portion of IMP in respiring mitochondria will be applied to the ANT-VDAC contact sites through the unbound VDACs of MOM (Fig. 9.2b), thus functioning as a driving force directly moving the phosphoryl groups from the mitochondrial matrix ATP toward the cytosol.

OMP, generated as the voltage drop on MOM (Fig. 9.2b), on the resistance $R_v = 1/g_{vs}$, may be presented according to Ohm's law (Fig. 9.2b, *b*) as

$$OMP = \left(V_{av} - IMP\right) \cdot \frac{g_c}{g_c + g_{vs}}. \tag{9.11}$$

Here, $g_c = 1/R_c$. For the simplicity of thermodynamic estimations, we took IMP at a fixed value of −140 mV (Gerencser et al. 2012; Bagkos et al. 2014) and the concentration of P_i in the cytosol at $[P_{i,c}]$ = 5 mM. In accordance with the well-known ratio of four protons returning into the matrix from MIMS per each synthesized ATP molecule, the phosphorylation potential maintained by respiring mitochondria, ΔG_a, may be presented as

$$G_a = 4F\left(IMP - \frac{R \cdot T}{F}\Delta pH\right). \tag{9.12}$$

Under the described conditions, taking ΔpH = 0.3 across MIM (relatively alkaline matrix), Eq. 9.12 gives $\Delta G_a = -61$ kJ/mol, the same as the values used above for the computational analysis of the VDAC-HK model.

In the cytosol, the free energy of ATP hydrolysis, i.e., the cytosolic ATP phosphorylation potential, $\Delta G_{a,c}$, is the function of the ratio $TD_c = [ATP]_c/[ADP]_c$, at a fixed $[P_{i,c}]$ = 5 mM:

$$G_{a,c} = \ G_a^{o'} + RT\ln\frac{\left[P_{i,c}\right]}{TD_c}. \tag{9.13}$$

Here, $G_a^{o'} = -31 kJ/mol$ is the standard Gibbs free energy of ATP hydrolysis at pH = 7.0. The thermodynamic analysis of the model of ANT-VDAC contact sites (Fig. 9.2b, *b*) was performed by solving the system of Eqs. 9.1–9.3 and 9.10–9.13.

First, OMP and relative permeability of MOM to P_i^- (g_{vs}) (Fig. 9.6a, b, respectively) were estimated at the phosphorylation potential of mitochondria $\Delta G_a = -61$ kJ/mol as functions of the ATP phosphorylation potential in the cytosol, $\Delta G_{a,c}$, at various fractions N_{vs} of unbound VDACs that was varied in the range of N_{vs} = 0.4–0.8, corresponding to the fraction of VDACs forming ANT-VDAC contact sites N_c = 0.6–0.2, respectively.

Zero OMP was obtained at $\Delta G_{a,c} = \Delta G_a$, independently of other conditions (Fig. 9.6a). Calculations showed positive OMP at $|\Delta G_{a,c}| < |\Delta G_a|$, and vice versa; negative OMP was demonstrated at $|\Delta G_{a,c}| > |\Delta G_a|$, at $G_c = 0.4$ (the P_i^- conductance of unbound VDACs in the closed state), and $S = 40\ V^{-1}$ (the voltage sensitivity parameter). The absolute values of the calculated OMP were higher when the difference between $\Delta G_{a,c}$ and ΔG_a was larger and also when the fraction of unbound VDACs, N_{vs}, was smaller (Fig. 9.6a).

Positive and negative OMPs, generated at relatively small fractions N_{vs} of unbound VDACs, were high enough to cause the voltage-dependent restriction of the MOM permeability to P_i^- (Fig. 9.6b). These results indicate the possibility of generation of high OMP (positive in MIMS) at increased cell workloads, accompanied with a

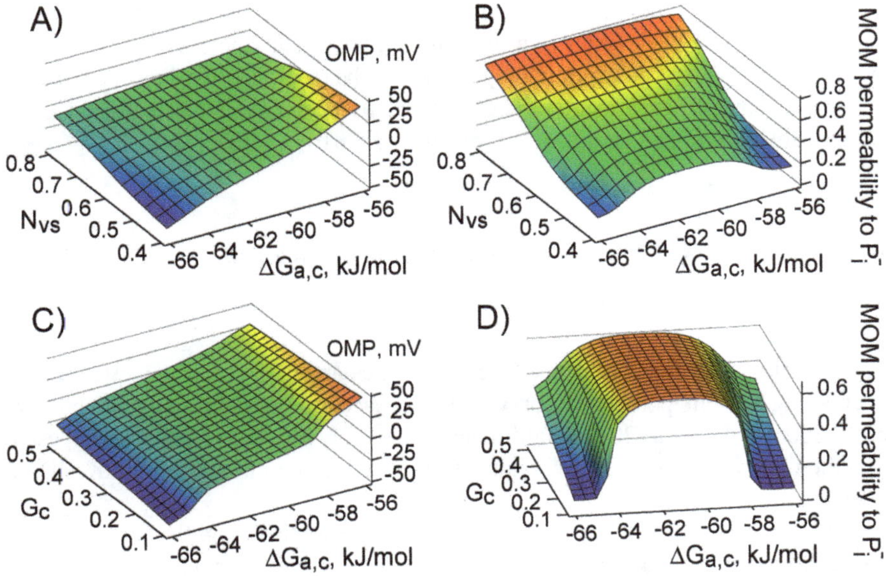

Fig. 9.6 OMP generation (**a**, **c**) and electrical restriction of the MOM permeability to P_i^- (g_{vs}) (**b**, **d**) according to the ANT-VDAC model (Fig. 9.2b) in mitochondria with the phosphorylation potential $\Delta G_a = -61$ kJ/mol (Eq. 9.12) as a function of cytosolic ATP phosphorylation potential $\Delta G_{a,c}$ (Eq. 9.13 at $[P_{i,c}] = 5$ mM), depending on the fraction N_{vs} of unbound VDACs (**a**, **b**) and on the conductance of unbound VDACs in the closed state (G_c) (**c**, **d**). **a**, **b** At $G_c = 0.4$ and $S = 40\ V^{-1}$; **c**, **d** at $N_{vs} = 0.6$ and $S = 50\ V^{-1}$. The calculations were performed using the system of Eqs. 9.1–9.3 and 9.10–9.13

decreasing steady-state $[ATP]_c/[ADP]_c$ ratio in the cytosol (TD_c), while negative OMP might be expected at relatively high ATP phosphorylation potential in the cytosol ($\Delta G_{a,c}$) with respect to mitochondria, maintained, for example, by glycolysis under hypoxic/anoxic conditions.

We also estimated the possible dependence of generated OMP on the conductance G_c of unbound VDACs in their closed state, at the voltage sensitivity parameter $S = 50\ V^{-1}$, that might result from the action of some VDAC's modulators. The calculations performed for $N_{vs} = 0.6$ as a function of $\Delta G_{a,c}$ showed the possibility of generation of positive and negative OMPs, which were of higher values at lower G_c (Fig. 9.6c). In addition, it revealed the very interesting phenomenon of energy transfer optimization, showing higher MOM permeability to P_i^- in a relatively narrow range of $\Delta G_{a,c}$ changes in the cytosol, around the given mitochondrial phosphorylation potential $\Delta G_a = -61$ kJ/mol (Fig. 9.6d).

Thus, the performed thermodynamic estimations showed the possibility of electrical suppression of mitochondria by positive OMP generated at very high cell workloads, i.e., at relatively low values of the ATP phosphorylation potential in the cytosol near mitochondria, $\Delta G_{a,c}$, as well as of the electrical suppression by negative OMP generated at $\Delta G_{a,c}$ significantly higher than the phosphorylation potential of mitochondria (Fig. 9.6d).

Note, these calculations were performed assuming the changes of the steady-state ratio $TD_c = [ATP]_c/[ADP]_c$ in the cytosol immediately near the mitochondrial ANT-VDAC contact sites, required by Eq. 9.10. The ratio TD_c may be very different at some distance from mitochondria due to a restriction of ADP electro-diffusion in the cytosol (Wallimann et al. 2011; Simson et al. 2016).

9.3.3 ANT-CK-VDAC-Mediated Generation of the Outer Membrane Potential

The mitochondrial and cytosolic CK reactions are parts of the most powerful energy channeling system postulated for various types of mammalian cells (Schlattner et al. 2006; Saks et al. 2010; Wallimann et al. 2011, 2015). It has been suggested that the octameric CK of mitochondria forms the intermembrane ANT-CK-VDAC contact sites.

Here we estimated the possibility of OMP generation by tightly coupled mitochondrial bi-transmembrane ANT-CK-VDAC contact sites allowing direct usage of mitochondrial matrix ATP for the CK reaction associated with the ANT-CK-VDAC contact sites to produce cytosolic CrP from cytosolic Cr (Fig. 9.2c), as it has been suggested by various authors (Brdiczka et al. 2006; Saks et al. 2010; Wallimann et al. 2011; Guzun et al. 2015).

The functioning of ANT-CK-VDAC contact sites might be presented as a battery E_{acv} (Fig. 9.2c, *c*) with a voltage V_{acv} that depends on the Gibbs free energy of the CK reaction associated with these contact sites:

$$V_{acv} = \frac{G_{CK}^{o'}}{F} + \frac{RT}{F}\ln\frac{[Cr]_c \cdot [ATP]_m}{[CrP]_c \cdot [ADP]_m}. \tag{9.14}$$

Here, $G_{CK}^{o'} = -12.7\,kJ/mol$ is the standard Gibbs free energy of CK reaction at pH = 7.0, and $[ATP]_m$ and $[ADP]_m$ are ATP and ADP concentrations, respectively, in the mitochondrial matrix, with the ratio $[ATP]_m/[ADP]_m = 3$ (Korzeniewski and Mazat 1996), considered invariable. $[Cr]_c$ and $[CrP]_c$ are cytosolic concentrations of Cr and CrP, respectively, with the ratio $[CrP]_c/[Cr]_c$ defined by a given CrP phosphorylation potential in the cytosol, $\Delta G_{c,c}$:

$$G_{c,c} = G_c^{o'} + RT\ln\frac{[Cr]_c \cdot [P_{i,c}]}{[CrP]_c}, \tag{9.15}$$

where $G_c^{o'} = -43.7\,kJ/mol$ is the standard Gibbs free energy of CrP hydrolysis at pH = 7.0 and $[P_{i,c}]$ is the concentration of inorganic phosphate in the cytosol fixed at 5 mM, as defined above.

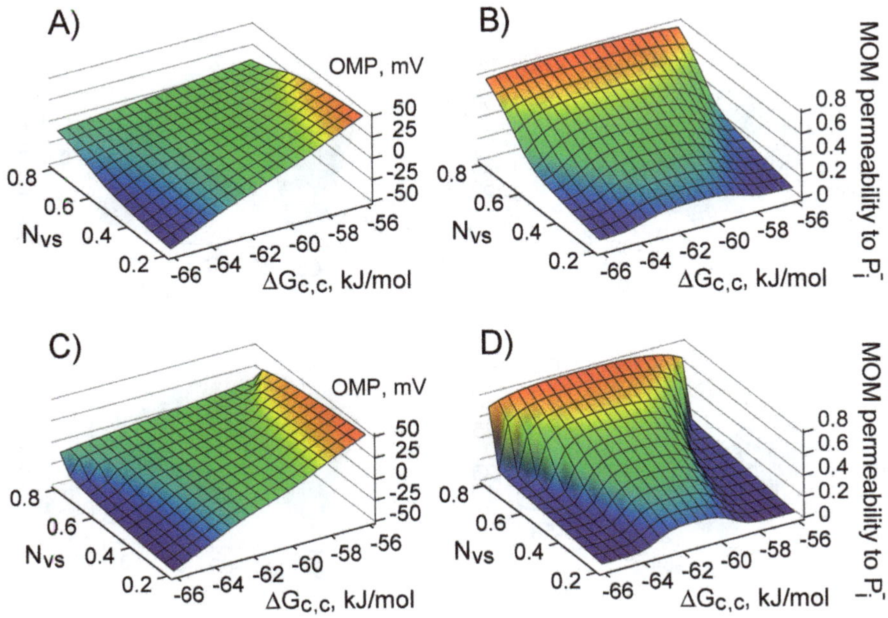

Fig. 9.7 OMP generation (**a, c**) and electrical restriction of MOM permeability to P_i^- (g_{vs}) (**b, d**) according to the ANT-CK-VDAC model (Fig. 9.2c) in mitochondria with the phosphorylation potential $\Delta G_a = -61$ kJ/mol (Eq. 9.12) as a function of the CrP phosphorylation potential in the cytosol $\Delta G_{c,c}$ (Eq. 9.15 at $[P_{i,c}] = 5$ mM), depending on the fraction N_{vs} of unbound VDACs. **a, b** At $G_c = 0.4$ and $S = 40$ V^{-1}; **c, d** at $G_c = 0.2$ and $S = 50$ V^{-1}. The calculations were performed using the system of Eqs. 9.1–9.3, 9.12, and 9.14–9.16

A part of the voltage difference between V_{acv} and IMP will drop on the internal resistance R_c ($1/g_c$) of the battery E_{acv} (Fig. 9.2c, *c*) and the other part will drop on the resistance R_v ($1/g_{vs}$) of unbound VDACs as OMP that may be estimated on the bases of Ohm's law:

$$OMP = \left(V_{acv} - IMP\right) \cdot \frac{g_c}{g_c + g_{vs}}. \tag{9.16}$$

For calculations, IMP was set at a fixed −140 mV, as mentioned above.

The calculations (solving the system of Eqs. 9.1–9.3, 9.12, 9.14–9.16) of OMP and of the electrical restriction of MOM permeability to P_i^- (g_{vs}) for the model of ANT-CK-VDAC contact sites (Fig. 9.2c) were performed as a function of the CrP phosphorylation potential in the cytosol, $\Delta G_{c,c}$. The fraction N_c of VDACs forming such bi-transmembrane contact sites was varied in the range of N_c = 0.2–0.8 (Fig. 9.7), with the respective changes in the fraction N_{vs} of unbound VDACs, N_{vs} = 0.8–0.2. The voltage sensitivity parameter S for these calculations was taken at $S = 40$ V^{-1} (Fig. 9.7a, b) and $S = 50$ V^{-1} (Fig. 9.7c, d).

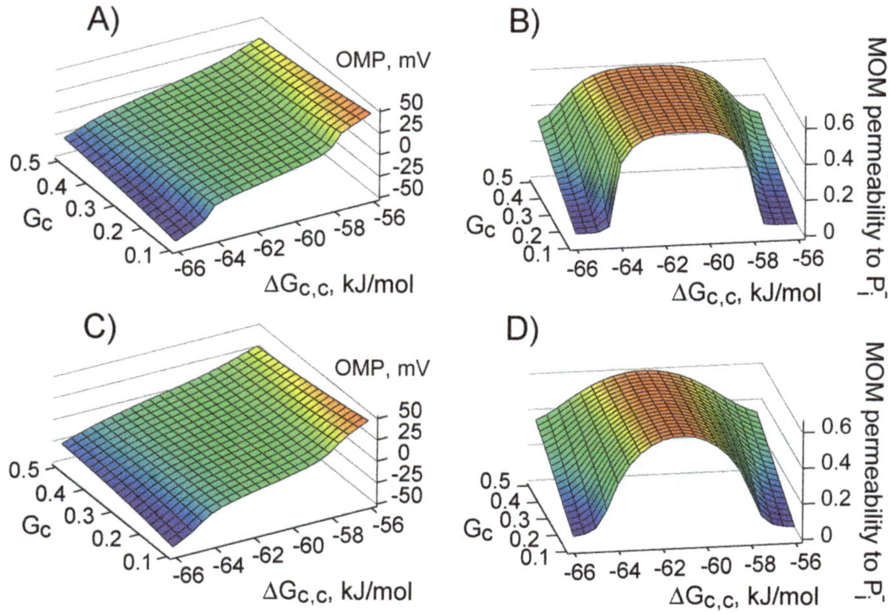

Fig. 9.8 OMP generation (**a**, **c**) and electrical restriction of MOM permeability to P_i^- (g_{vs}) (**b**, **d**) according to the ANT-CK-VDAC model (Fig. 9.2c) in mitochondria with the phosphorylation potential $\Delta G_a = -61$ kJ/mol (Eq. 9.12) as a function of the CrP phosphorylation potential in the cytosol $\Delta G_{c,c}$ (Eq. 9.15 at $[P_{i,c}] = 5$ mM), depending on the conductance of unbound VDACs in the closed state (G_c). The calculations were performed at fixed $N_{vs} = 0.6$ and $S = 50$ V^{-1} using the system of Eqs. 9.1–9.3, 9.12, and 9.14–9.16 (**a**, **b**) or the system of Eqs. 9.1, 9.2, 9.4, 9.12, and 9.14–9.16 (**c**, **d**)

The thermodynamic estimations, performed for the range of CrP phosphorylation potential in the cytosol between $\Delta G_{c,c} = -56$ kJ/mol and $\Delta G_{c,c} = -66$ kJ/mol, at $S = 40$ V^{-1} and $G_c = 0.4$, demonstrated the dependence of calculated OMP (Fig. 9.7a) and of the MOM permeability to P_i^- (Fig. 9.7b) on $\Delta G_{c,c}$, very similar to that obtained for the model of ANT-VDAC contact sites, in the range of $N_{vs} = 0.4$–0.8 (Figs. 9.6a, b, respectively). Relatively high values of OMP were calculated outside of the relatively narrow range of $\Delta G_{c,c}$ changes around the fixed value of the mitochondrial phosphorylation potential $\Delta G_a = -61$ kJ/mol, for the conditions of $G_c = 0.2$ and $S = 50$ V^{-1} (Fig. 9.7c), also showing a significant electrical restriction of MOM permeability to P_i^- (Fig. 9.7d).

We also estimated the influence of the P_i^- conductance of unbound VDACs in the closed state, G_c, changing it in the range of $G_c = 0.1$–0.5, at fixed parameters $N_{vs} = 0.6$ and $S = 50$ V^{-1} (Fig. 9.8). The calculations were performed by solving the system of Eqs. 9.1, 9.2, 9.12, and 9.14–9.16 together with Eq. 9.3 (with the power index of 3) (Fig. 9.8a, b) or with Eq. 9.4 (with the power index of 2) (Fig. 9.8c, d). One of the goals was to estimate the influence of possible changes in the shape of the conductance-voltage function of VDAC on the probability of OMP generation and MOM permeability modulation (Fig. 9.8).

The calculations performed using Eq. 9.3, in the abovementioned system of equations, for the fraction $N_{vs} = 0.6$ of unbound VDACs, at different values of the parameter G_c and the voltage-sensitivity parameter $S = 50\ V^{-1}$, allowed demonstration of a strong dependence of the sign and value of calculated OMP on $\Delta G_{c,c}$ (Fig. 9.8a). The range of $\Delta G_{c,c}$ changes, around $\Delta G_a = -61$ kJ/mol, for almost unrestricted, optimum energy transfer between mitochondria and the cytosol, was more narrow at the lowest values of the parameter G_c (Fig. 9.8b). Qualitatively, similar results were obtained using Eq. 9.4, instead of Eq. 9.3, in the system of equations for calculation of OMP (Fig. 9.8c) and MOM permeability to P_i^- (Fig. 9.8d). Nevertheless, the steeper slopes of the functions of OMP and MOM permeability to P_i^- on $\Delta G_{c,c}$ were revealed using Eq. 9.3 (Fig. 9.8a, b, respectively) than those for Eq. 9.4 (Fig. 9.8c, d, respectively). With this respect, various factors have been reported to change the steepness of the VDAC voltage dependence (Mangan and Colombini 1987; Lee et al. 1996; Rostovtseva et al. 2008; Teijido et al. 2014; Okazaki et al. 2015).

The most interesting results of the performed computational analysis of the model of electrogenic ANT-CK-VDAC contact sites (Fig. 9.2c) are the possibility of electrical suppression of mitochondria by generated OMP when the cytosolic CrP phosphorylation potential is significantly different from the phosphorylation potential of mitochondria (Figs. 9.7a, c and 9.8a, c). This effect is strongly increased when the parameter G_c of the P_i^- conductance of unbound VDACs in the closed state is decreased and when the voltage sensitivity parameter S is increased. Both parameters were reported to be controlled by tubulin and other physiological modulators of VDAC's voltage-gating properties (Liu and Colombini 1992; Colombini et al. 1996; Lee et al. 1996; Lemasters and Holmuhamedov 2006; Rostovtseva et al. 2008; Shoshan-Barmatz et al. 2010; Rostovtseva and Bezrukov 2012, 2015; Maldonado et al. 2013).

9.4 Possible Physiological Consequences of Generation of the Mitochondrial Outer Membrane Potential

The flux of macroergic compounds and respiratory metabolites between mitochondria and the cytosol is provided by VDACs in MOM (Rostovtseva and Colombini 1997; Colombini 2004; Lemasters and Holmuhamedov 2006; Shoshan-Barmatz et al. 2010; Colombini and Mannella 2012). The main unresolved question in this respect is the physiological role of the highly conserved VDAC's voltage-gating properties (Colombini and Mannella 2012). In any case, in order for these VDAC's voltage-gating properties to play a crucial role in cell energy homeostasis, a high enough OMP has to be generated as an electrical feedback control of mitochondrial and cell energy metabolism. According to Mannella et al. (1992), it seems to be unlikely that VDACs simply convert MOM into a coarse sieve. We may add that it seems to also be unlikely that the VDAC permeability is regulated by only molecular corks, without voltage gating of VDACs by metabolically derived OMP.

Actually, various possibilities exist for OMP generation. One group of the mechanisms of OMP generation might be classified as passive diffusion mechanisms (Fig. 9.1), based on the difference in the VDAC's permeability to charged metabolites, mainly to ATP^{4-}, ADP^{3-}, AMP^{2-}, $CrP^{1-}(CrP^{2-})$, and $P_i^{1-}(P_i^{2-})$ (Hodge and Colombini 1997; Vander Heiden et al. 2000; Colombini 2016), in addition to the Donnan potential suggested by various authors (Liu and Colombini 1992; Porcelli et al. 2005 and references therein) as a possible regulator of MOM permeability that may be superimposed on the metabolically derived OMP (Lemeshko and Lemeshko 2000; Lemeshko 2006).

Another group represents active mechanisms of OMP generation, based on the phosphoryl group transfer through the mitochondrial membranes coupled to various energy sources (Lemeshko 2002, 2014a, 2016). The three possible mechanisms of this type were described here (Fig. 9.2) and analyzed using simplified computational models by performing thermodynamic estimations of a possible range of changes of generated OMP at various VDAC's voltage-gating properties.

The simplest and the most powerful mechanism of OMP generation, based on the phosphoryl group transfer through MOM, seems to be the energy-dependent transmembrane separation of charges by the electrogenic VDAC-HK complexes (Fig. 9.2a) (Lemeshko 2002, 2014a; Lemeshko and Lemeshko 2004). The Gibbs free energy of the HK reaction is used by these complexes, functioning as a direct MOM voltage generator. Computational analysis of the VDAC-HK model (Fig. 9.2a) demonstrated the generation of OMP of only positive sign (in MIMS), due to the essentially irreversible HK reaction. High values of OMP were calculated even at relatively low concentrations of glucose (at low ratios $[Gluc]_c/[G6P]_c$ in the cytosol) and low percentage of VDACs forming VDAC-HK complexes (Figs. 9.4 and 9.5). Similar results were obtained for the model of OMP generation involving electrogenic bi-transmembrane ANT-VDAC-HK contact sites (data not shown). The calculated OMPs were high enough to cause a significant electrical restriction of the MOM permeability of respiring mitochondria to P_i^- (Figs. 9.4 and 9.5) and to other charged metabolites (Rostovtseva and Colombini 1997; Hodge and Colombini 1997; Vander Heiden et al. 2000; Colombini 2016).

A very important characteristic of most cancer cells is a high quantity of HK bound to the mitochondrial VDACs, more than two orders of magnitude higher than in normal cells (Marín-Hernández et al. 2006; John et al. 2011 and references therein). The computational analysis of the VDAC-HK model demonstrated that OMP generation is potentiated by even a relatively small decrease in the VDAC's closed-state conductance G_c or by an increase in the VDAC's voltage sensitivity parameter S (Figs. 9.4 and 9.5). Based on the high probability of OMP generation by the VDAC-HK complexes, we have suggested an explanation of the Warburg effect as an electrical suppression of mitochondria (Lemeshko 2002, 2014a, 2015), under which the mitochondrial ATP is mainly accessible to initiate glycolysis through the VDAC-HK complexes of MOM. The generated OMP should inhibit glucose phosphorylation by the mitochondria-associated HK using external ATP, and it should also suppress the release of ADP from MIMS to recover ATP in the cytosol, thus preventing glycolysis acceleration by glycolitically produced ATP. This

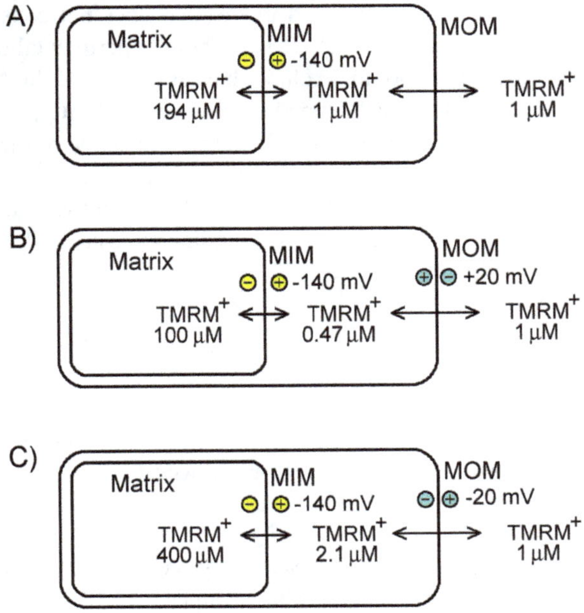

Fig. 9.9 The possible influence of OMP on apparent values of IMP monitored with cationic fluorescent probes like TMRM. **a** Zero influence at OMP = 0 mV; **b** decreased TMRM capture at OMP = +20 mV, as it would show a supposed mitochondrial "depolarization"; **c** increased TMRM capture at OMP = −20 mV, as it would show a supposed mitochondrial "hyperpolarization." Estimations were performed using Eq. 9.18 at fixed IMP = −140 mV, assuming 1 μM TMRM concentration in the external space

hypothetical OMP-dependent anti-turbo mechanism (Lemeshko 2014a) is consistent with the observations that the kinetic properties of the unbound HK are very different from those of mitochondria-associated HK, which preferentially uses intramitochondrially generated ATP (see Wilson 2003).

The direct monitoring of OMP, and even quantitative estimation of only IMP of mitochondria in living cells (Lemasters and Ramshesh 2007; Gerencser et al. 2012), presents certain difficulties. An experimental approach for the quantitative monitoring of IMP in cultured cells using cationic fluorescent probes has been recently suggested taking into account the contribution of the plasma membrane potential (Gerencser et al. 2012) but ignoring the possible existence of OMP and thus its influence on the measured values of IMP. The method to measure IMP, based on the quantitative imaging analysis of cells with a fluorescent cationic probe TMRM, has also been proposed (Lemasters and Ramshesh 2007). The authors assumed that IMP is directly related to the logarithmic function of the ratio of the TMRM concentration within the mitochondrial matrix (F_{in}) to that in the cytosol/nucleus (F_{out}), according to Nernst's equation:

$$IMP = -\frac{RT}{F}\ln\frac{F_{in}}{F_{out}}, \tag{9.17}$$

It is true, if we follow the most general concept that OMP = 0, illustrated in Fig. 9.9a and based on the assumption that MOM is highly permeable to charged metabolites and small ions. Nevertheless, it has been recently suggested that only 2% of all VDACs in cardiomyocytes are open that followed from a description of the experimental data by a model that does not consider the possibility of OMP generation (Simson et al. 2016). These experimental data (Simson et al. 2016), on the other hand, do not exclude that up to 98% of the MOM permeability might be blocked by the OMP-dependent closure of VDACs, if we assume that OMP is generated (Figs. 9.4, 9.5, 9.6, 9.7, and 9.8) by some of the mechanisms described here (Fig. 9.2). Thus, if OMP is generated, it should influence TMRM distribution between the cytosol and MIMS and finally between the cytosol and the mitochondrial matrix, thus influencing the mitochondrial TMRM fluorescence in accordance to the following equation:

$$IMP + OMP = -\frac{RT}{F}\ln\frac{F_{\text{in}}}{F_{\text{out}}}. \tag{9.18}$$

The effects of OMP on TMRM capture by respiring mitochondria may be significant even at unchanged IMP, as shown in Fig. 9.9 for OMP = +20 mV (Fig. 9.9b) and OMP = −20 mV (Fig. 9.9c). These calculations were performed using Eq. 9.18 at a fixed IMP = −140 mV (Fig. 9.9).

The experimental approach suggested by Lemasters and Ramshesh (2007) has been used to estimate possible changes of IMP in cancer cells with the fluorescent cationic probe TMRM in experiments with knockdown of various VDAC isoforms (Maldonado et al. 2013), as well as after cell treatments leading to a change in the cytosolic concentration of free tubulin (Maldonado et al. 2010). On the other hand, according to the VDAC-HK model (Fig. 9.2a), knockdown of VDAC isoforms and changes in the concentration of free tubulin should influence OMP generation and thus TMRM capture by mitochondria (Fig. 9.9), according to Eq. 9.18.

Only for theoretical estimations, let us calculate a possible range of OMP changes on the basis of the reported, very informative experimental data of Maldonado et al. (2010, 2013), assuming, for simplicity, that IMP of respiring mitochondria within the cells is maintained constant at all performed cell treatments, excepting VDAC3 knockout. Assuming unchanged IMP, the changes of OMP, ΔOMP, may be estimated (based on Eq. 9.18) as

$$\Delta OMP = -\frac{RT}{F}\ln\frac{F_{\text{in},t}}{F_{\text{in},o}}, \tag{9.19}$$

where $F_{\text{in},t}$ is the TMRM fluorescence of mitochondria after a cell treatment, hypothetically influenced only by OMP, and $F_{\text{in},o}$ is the TMRM fluorescence before the treatment.

Interestingly, in HepG2 cancer cells, knockdown of VDAC1 or VDAC2 caused a decrease in mitochondrial TMRM fluorescence by nearly 42% and 58%, respectively, without any influence on the mitochondrial NAD(P)H level (Maldonado et al. 2013, Fig. 9.1), as an indirect indicator of IMP changes (see Lemeshko 2014b). Thus, the mentioned decrease in TMRM fluorescence might be attributed to only an increase in positive OMP by 15 mV and 23 mV, respectively, according to Eq. 9.19.

On the other hand, knockdown of the least abundant VDAC3 isoform (11%) in HepG2 cells decreased mitochondrial TMRM fluorescence by nearly 80% and also caused remarkable decrease in the mitochondrial NAD(P)H level (Maldonado et al. 2013, Fig. 9.1). In this case, the mentioned significant decrease in the TMRM capture by mitochondria might be explained by both MIM depolarization and an increase in positive OMP (Eq. 9.18).

According to the VDAC-HK model (Fig. 9.2a), the replacement of a small part of voltage-sensitive VDAC isoforms by voltage-insensitive VDACs (10% of all VDACs) significantly decreased the calculated OMP (Fig. 9.5a in comparison to Fig. 9.4a). On the other hand, a hypothetical knockout of this fraction (as it would be knock-out of VDAC3) should increase OMP by 20 mV, calculated at $V_{HK} = 4\%$ and $[\text{Gluc}]_c/[G6P]_c = 10$ (Fig. 9.5c) with respect to that before the knockout (Fig. 9.5a). These estimations are very consistent with the conclusion of Maldonado et al. (2013) that the minor VDAC3 isoform is the most important for maintenance of mitochondrial metabolism, at least in HepG2 cells. According to the authors, VDAC3 knockdown might lead to a restriction of the influx of respiratory substrates that reduce $NAD(P)^+$ in the mitochondrial matrix, finally resulting in a remarkable decrease in the $NAD(P)H/NAD(P)^+$ ratio (Maldonado et al. 2013). In addition, a knockdown of VDAC3, but not of VDAC1 and/or VDAC2, caused a significant impairment of mitochondria in cells.

Our model predicts that OMP increase by 20 mV after VDAC3 knockout (Fig. 9.5c) leads to the electrical closure of voltage-sensitive VDAC isoforms in MOM (Fig. 9.5d). In other words, the fraction of VDAC3 isoform, having very low voltage sensitivity, might be considered as an "emergency exit-entrance" factor, preventing complete electrical suppression of mitochondrial MOM permeability caused by high OMPs due to the closure of VDAC1 and VDAC2. The small fraction of constitutively open VDAC3 isoform in MOM would still allow the influx of respiratory substrates into the mitochondria and the return of P_i^- to recover ATP in mitochondria even at high OMP. The effect similar to a VDAC3 knockdown might also be caused by a disulfide group formation, leading to inhibition of the VDAC3 conductance (Okazaki et al. 2015; De Pinto et al. 2016). This demonstrates the possibility of additional, redox-signaling potentiation of OMP generation and of electrical regulation of MOM permeability to charged metabolites and even of electrical suppression of mitochondria by oxidative stress.

It has been demonstrated that the treatment of HepG2 cells with colchicine and nocodazole, microtubule destabilizers increasing the concentration of free cytosolic tubulin, caused a decrease in TMRM fluorescence by 60% and 70%, respectively (Maldonado et al. 2010, Fig. 9.3), that according to Eq. 9.19 might result from an increase in OMP by 24–32 mV, due to, for example, a decrease in the parameter G_c (Fig. 9.4e in comparison to Fig. 9.4a) and/or an increase in the parameter S (Fig. 9.4c in comparison to Fig. 9.4a) by free tubulin, as the modulator of the VDAC's voltage-gating properties (Rostovtseva et al. 2008; Rostovtseva and Bezrukov 2008, 2012, 2015; Maldonado et al. 2013). On the other hand, cell treatment with the microtubule stabilizer paclitaxel, known to decrease cellular free tubulin, increased TMRM fluorescence by nearly 65% (Maldonado et al. 2010, Fig. 9.3) that could be interpreted as a consequence of a decrease of the positive OMP by 13 mV (Eq. 9.19). The total range of 37–45 mV for possible tubulin-mediated modulation of OMP in HepG2 cancer cells, estimated on the basis of the reported experimental data (Maldonado et al. 2010), is in a good agreement with the results predicted by the VDAC-HK model (Figs. 9.4a, c, f and 9.5a, e). This analysis demonstrates that OMP should be considered at least as a possible additional factor influencing the mitochondrial TMRM fluorescence changes reported by Maldonado et al. (2010, 2013).

The treatment of HepG2 cells with erastin, a compound hypothesized to target VDACs (Yagoda et al. 2007), thus disrupting the tubulin-dependent inhibition of VDAC conductance, antagonizing Warburg-type metabolism (Maldonado et al. 2013) and preventing formation of VDAC-HK complexes (see Lemeshko 2014a), has been demonstrated to increase the mitochondrial TMRM fluorescence by 46% (Maldonado et al. 2013; DeHart et al. 2014). According to Eq. 9.19, this corresponds to a possible decrease in OMP by 10 mV. The effect of such magnitude is predicted by the VDAC-HK model, if we assume a decrease in the percentage of VDAC-HK complexes from 4% to 2% at $[\mathrm{Gluc}]_c/[G6P]_c = 10$, or from 3.5% to 2% at $[\mathrm{Gluc}]_c/[\mathrm{G6P}]_c = 100$) (Fig. 9.5a). An even more significant effect on OMP might be expected (Fig. 9.5a, e) assuming that erastin decreases the binding of free tubulin to voltage-sensitive VDACs, thus leading to an increase in the parameter G_c and to a decrease in the parameter S.

The VDAC-HK model predicts that tubulin and/or HK binding to voltage-sensitive unbound VDACs favors OMP generation, thus increasing the electrical suppression of mitochondria in the cells with the Warburg-type metabolism, as it has been suggested earlier (Lemeshko 2002, 2014a, 2015). The performed estimations indicate that the changes in TMRM fluorescence of cancer cells, caused by various cell treatments, known to influence the voltage-gating properties of VDACs and/or formation of VDAC-HK complexes, might reflect a contribution from generated OMP (Eq. 9.18) in addition to IMP changes, suggested by various authors (Vander Heiden et al. 1999; Maldonado et al. 2010, 2013; Sheldon et al. 2011; Gerencser et al. 2012; Zhang et al. 2016). In this respect, our model predicts that factors preventing formation of VDAC-HK complexes might cause an anti-Warburg effect by decreasing the probability of the glucose-dependent generation of positive OMP and thus of electrical suppression of mitochondria (Lemeshko 2014a, 2015). In addition, the lowered capacity to decrease the Ca^{2+} concentration in MIMS at low positive OMP might significantly decrease the cell death resistance to factors causing an increase in the cytosolic concentration of Ca^{2+} and leading to mitochondrial permeability transition, as discussed earlier (Lemeshko 2014a, 2016).

The models based on the bi-transmembrane transfer of phosphoryl groups through the ANT-VDAC (Fig. 9.2b) and ANT-CK-VDAC (Fig. 9.2c) electrogenic complexes demonstrated the possibility of generation of both positive and negative OMPs. The generated OMP depends on the ATP and CrP phosphorylation potentials in the cytosol with respect to the ATP phosphorylation potential of mitochondria (Figs. 9.6 and 9.7, respectively). Interestingly, thermodynamic estimations demonstrated very similar results for both ANT-VDAC and ANT-CK-VDAC models (Figs. 9.6, 9.7, and 9.8, respectively), apparently questioning the physiological importance of the CK system

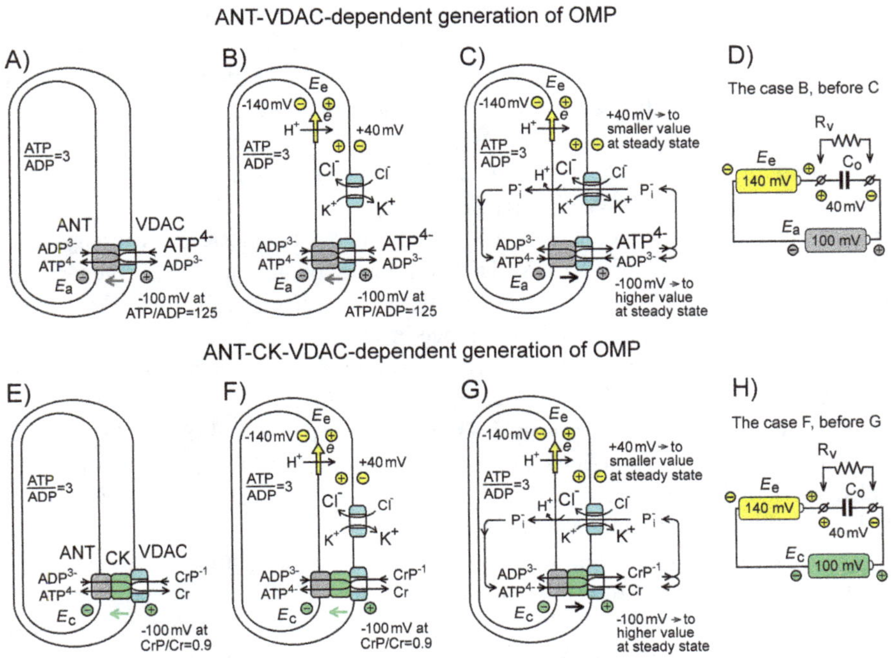

Fig. 9.10 Main principles of ANT-VDAC- (**a–d**) and ANT-CK-VDAC-dependent (**e–h**) generation of positive OMP. **a**, **e** bi-transmembrane potential is generated by the electrogenic ANT-VDAC contact sites in a hypothetical bi-liposomal structure at the fixed internal ratio [ATP]/[ADP] = 3 and the external ratio [ATP]/[ADP] = 125 (**a**), or by the electrogenic ANT-CK-VDAC contact sites at the external ratio [CrP[/[Cr] = 0.9 (**e**). **b**, **f** OMP of +40 mV is generated at the thermodynamic equilibrium if a fixed internal membrane potential of −140 mV is generated by an electron transport chain. A decrease in OMP is expected at starting the P_i^- circulation through unbound VDACs and the phosphoryl group transfer through the ANT-VDAC (**c**) and ANT-CK-VDAC (**g**) electrogenic contact sites coupled to energy use in the external space. **d**, **h** The equivalent electrical circuits of **b**, **f**, respectively, where E_e, E_a, and E_c are the batteries corresponding to an electron transport system (E_e), ANT-VDAC (E_a) and ANT-CK-VDAC (E_c) contact sites. C_o-MOM as an electrolytic capacitor (**d**, **h**) resulting of charge separation (K^+, Cl^-) in an electric field of equilibrium state OMP (**b**, **f**) or steady-state OMP (**c**, **g**)

for mitochondria-cytosol energy channeling in cells with high and fluctuating energy demands. Nevertheless, note that both calculations were performed for the cytosolic ratios [ATP]/[ADP] and [CrP]/[Cr] in the immediate proximity to MOM.

The general principles for ANT-VDAC- and ANT-CK-VDAC-mediated generation of OMP are presented in Fig. 9.10a–d, e–h, respectively. Let us assume first that the internal space of a bi-membrane structure contains adenine nucleotides at the ratio $[ATP]_m/[ADP]_m = 3$, as in the mitochondrial matrix (Fig. 9.10a, e), and the inner and outer membranes are impermeable to ions. At the electrochemical equilibrium, the bi-transmembrane potential of −100 mV, for example, will be generated by the bi-transmembrane ANT-VDAC contact sites at the external ratio [ATP]/[ADP] = 125 (Fig. 9.10a) or by the ANT-CK-VDAC contact sites at the ratio [CrP]/[Cr] = 0.9 (Fig. 9.10e).

Now, let us assume that the inner membrane contains an electron transport chain pumping protons through the inner membrane, thus maintaining the membrane potential at −140 mV (Fig. 9.10b, f). As a result, the outer membrane potential of +40 mV will be maintained, although the outer membrane is highly permeable to small ions due to the presence of VDAC. The OMP will be maintained, because the flow of small ions will cease after achieving the electrochemical equilibrium for these ions, at fixed [ATP]/[ADP] = 125 (Fig. 9.10b) or [CrP]/[Cr] = 0.9 (Fig. 9.10f). In this case, the outer membrane may be considered as an electrolytic capacitor charged to the voltage of +40 mV. This simplest explanation of OMP generation (Fig. 9.10b, f) is demonstrated with two batteries connected as shown in Fig. 9.10d, h, respectively, where the voltage difference of the batteries E_e and E_a (Fig. 9.10d), or of the batteries E_e, and E_c (Fig. 9.10h), is applied to the electrolytic capacitor, without a current through it after its charging is finished.

Next, let us assume ATP hydrolysis (Fig. 9.10c) and CrP hydrolysis (Fig. 9.10g) in the external space with a rate that is equal to the rate of ATP recovery in the internal space of a bi-membrane structure coupled to the protonic electrochemical potential of the inner membrane, accompanied with the return of P_i^{1-} into the internal space, as shown in more detail in Fig. 9.2c, g. The outer membrane potential at such steady state will be less than +40 mV, and will depend on the percentage of free, unbound VDACs, on their voltage-gating properties and on steady-state ratios [ATP]/[ADP] (Fig. 9.10c) and [CrP]/[Cr] (Fig. 9.10g). We have to note that the thermodynamic analysis was performed here without taking into account the possible restriction of electro-diffusion/diffusion of ATP, ADP, CrP, and Cr in the cytosol.

Our data demonstrate similar results for a very low cytosolic concentration of ADP, two orders of magnitude less than the concentrations of ATP (Fig. 9.10a–c) and for a relatively high concentration of electrically neutral Cr, comparable with that of CrP (Fig. 9.10e–h). We have to highlight that the [ATP]/[ADP] ratios near the mitochondria under physiological conditions might be very different from those in the cytosolic sites of ATP hydrolysis, especially in spermatozoids, due to a restriction of ADP^{3-} electro-diffusion (Wallimann et al. 2011; Simson et al. 2016), which is at least three orders of magnitudes less than the rate of Cr diffusion (Wallimann et al. 2011 and references therein). Thus, the mitochondrial-cytosolic CK kinase system is very important for cells with high and fluctuating energy demands, functioning not only as the cytosol energy buffer (Schlattner et al. 2006; Wallimann et al. 2011; Wallimann 2015; Guzun et al. 2015) but also as a system allowing the fast recovery of the cytosolic CrP phosphorylation potential by mitochondria (Wallimann 2015). As we showed (Lemeshko 2016), this mitochondria-cytosol CrP flux may be controlled by an electrical signaling system of MOM coupled to the generation of the metabolically dependent OMP (Figs. 9.7 and 9.8).

Recently we have explained the protective effects of creatine against mitochondrial permeability transition (Lemeshko 2016), induced by an elevated concentration of Ca^{2+} in the cytosol, as a result of pushing out of Ca^{2+} ions from MIMS by positive OMP, directly generated by the CK-VDAC electrogenic complexes of MOM, or by the ANT-CK-VDAC complexes coupled to IMP (Fig. 9.2c).

In addition, even the generation of positive OMP that is not yet high enough to essentially inhibit MOM permeability to ADP might contribute in a decrease of $K_{m,\mathrm{ADP}}$ for mitochondrial state 3 respiration in the presence of creatine (Lemeshko 2016) by more than one order of magnitude (Saks et al. 2010; Guzun et al. 2015). Thus, even OMP of relatively small values, not yet sufficient to close VDACs, might influence steady-state concentrations of charged metabolites and inorganic ions in MIMS, such as ADP^{3-} and Ca^{2+}, thus modulating the mitochondrial metabolic state, cell energetics, and cell death resistance.

9.5 Conclusions

Up to now, the main concept of the MOM permeability regulation has been the "molecular corking up" of MOM's porins, although there is a high probability that the mechanism controlling cell energy metabolism is a combination of electrical and "corking up" components, when relatively high OMP is generated as an electrical feedback control of the energy flux between mitochondria and the cytosol. In addition to the natural modulators of VDAC's voltage-gating properties, like tubulin (Rostovtseva et al. 2008; Maldonado et al. 2013; Rostovtseva and Bezrukov 2015), synuclein (Rostovtseva et al. 2015) and other allosteric factors, this combined regulation might be also supplemented with the redox-dependent modulation of VDAC conductance and with other chemical modifications of various VDAC isoforms. In general, the combined VDAC electrical and "molecular corking up" regulation of the MOM permeability to charged metabolites seem to represent a novel physiological mechanism of coordination of mitochondrial and cytosolic energy metabolism, thus playing a crucial role in cell energy homeostasis. The real values of the metabolically derived OMP may not be as high as calculated for the models described here because the voltage-gating properties of VDACs under natural conditions might be different from those known from the studies of VDACs in lipid bilayers and also because the rates of phosphoryl group transfer through the VDAC-HK complexes and through the bi-transmembrane ANT-VDAC and ANT-CK-VDAC contact sites depend on local concentrations of corresponding metabolites. All these require further theoretical and experimental studies. On the other hand, even the simplest thermodynamic estimations demonstrate that the generated metabolically dependent OMP may be high enough to control cell energy metabolism at the level of MOM, representing a fast electrical signaling system to control VDAC's conductance, especially if combined with the "molecular corking up," redox signaling, and other mechanisms of modulation of the VDAC's voltage-gating properties.

Acknowledgments The author thanks Dr. Andriy Anishkin and Dr. Sergy V. Lemeshko for their critical reading of the manuscript, discussion, and valuable observations.

References

Abu-Hamad S, Zaid H, Israelson A, Nahon E, Shoshan-Barmatz V (2008) Hexokinase-I protection against apoptotic cell death is mediated via interaction with the voltage-dependent anion channel-1: mapping the site of binding. J Biol Chem 283(19):13482–13490

Azoula-Zohar H, Israelson A, Abu-Hamad S, Shoshan-Barmatz V (2004) In self-defence: hexokinase promotes VDAC closure and prevents mitochondria-mediated apoptotic cell death. Biochem J 377:347–355

Bagkos G, Koufopoulos K, Piperi C (2014) A new model for mitochondrial membrane potential production and storage. Med Hypotheses 83(2):175–181

Báthori G, Csordás G, Garcia-Perez C, Davies E, Hajnóczky G (2006) Ca2+-dependent control of the permeability properties of the mitochondrial outer membrane and voltage-dependent anion-selective channel (VDAC). J Biol Chem 281(25):17347–17358

Benz R, Kottke M, Brdiczka D (1990) The cationically selective state of the mitochondrial outer membrane pore: a study with intact mitochondria and reconstituted mitochondrial porin. Biochim Biophys Acta 1022:311–318

Bernier-Valentin F, Rousset B (1982) Interaction of tubulin with rat liver mitochondria. J Biol Chem 257(12):7092–7099

Brdiczka DG, Zorov DB, Sheu SS (2006) Mitochondrial contact sites: their role in energy metabolism and apoptosis. Biochim Biophys Acta 1762:148–163

Chernoivanenko IS, Matveeva EA, Gelfand VI, Goldman RD, Minin AA (2015) Mitochondrial membrane potential is regulated by vimentin intermediate filaments. FASEB J 29(3):820–827

Colombini M (2004) VDAC: the channel at the interface between mitochondria and the cytosol. Mol Cell Biochem 256(1–2):107–115

Colombini M (2016) The VDAC channel: molecular basis for selectivity. Biochim Biophys Acta 1863(10):2498–2502

Colombini M, Mannella CA (2012) VDAC, the early days. Biochim Biophys Acta 1818(6):1438–1443

Colombini M, Blachly-Dyson E, Forte M (1996) VDAC, a channel in the outer mitochondrial membrane. Ion Channels 4:169–202

De Pinto V, Reina S, Gupta A, Messina A, Mahalakshmi R (2016) Role of cysteines in mammalian VDAC isoforms' function. Biochim Biophys Acta 1857(8):1219–1227

DeHart DN, Gooz M, Rostovtseva TK, Sheldon KL, Lemasters JJ, Maldonado EN (2014) Antagonists of the inhibitory effect of free tubulin on VDAC induce oxidative stress and mitochondrial dysfunction. Biophys J 106(Issue 2):p591a

Denis-Pouxviel C, Riesinger I, Buhler C, Brdiczka D, Murat J-C (1987) Regulation of mitochondrial hexokinase in cultured HT 29 human cancer cells: an ultrastructural and biochemical study. Biochim Biophys Acta 902:335–348

Gerencser AA, Chinopoulos C, Birket MJ, Jastroch M, Vitelli C, Nicholls DG, Brand MD (2012) Quantitative measurement of mitochondrial membrane potential in cultured cells: calcium-induced de- and hyperpolarization of neuronal mitochondria. J Physiol 590(12):2845–2871

Gray MW (2012) Mitochondrial evolution. Cold Spring Harb Perspect Biol 4(9):a011403

Grimm S, Brdiczka D (2007) The permeability transition pore in cell death. Apoptosis 12(5):841–855

Guzun R, Kaambre T, Bagur R, Grichine A, Usson Y, Varikmaa M, Anmann T, Tepp K, Timohhina N, Shevchuk I, Chekulayev V, Boucher F, Dos Santos P, Schlattner U, Wallimann T, Kuznetsov AV, Dzeja P, Aliev M, Saks V (2015) Modular organization of cardiac energy metabolism: energy conversion, transfer and feedback regulation. Acta Physiol (Oxford) 213(1):84–106

Head SA, Shi W, Zhao L, Gorshkov K, Pasunooti K, Chen Y, Deng Z, Li RJ, Shim JS, Tan W, Hartung T, Zhang J, Zhao Y, Colombini M, Liu JO (2015) Antifungal drug itraconazole targets VDAC1 to modulate the AMPK/mTOR signaling axis in endothelial cells. Proc Natl Acad Sci U S A 112(52):E7276–E7285

Hodge T, Colombini M (1997) Regulation of metabolite flux through voltage-gating of VDAC channels. J Membr Biol 157(3):271–279

Holden MJ, Colombini M (1993) The outer mitochondrial channel, VDAC, is modulated by a protein localized in the intermembrane space. Biochim Biophys Acta 1144:396–402

Jeneson JA, ter Veld F, Schmitz JP, Meyer RA, Hilbers PA, Nicolay K (2011) Similar mitochondrial activation kinetics in wild-type and creatine kinase-deficient fast-twitch muscle indicate significant Pi control of respiration. Am J Physiol Regul Integr Comp Physiol 300(6):R1316–R1325

John S, Weiss JN, Ribalet B (2011) Subcellular localization of hexokinases I and II directs the metabolic fate of glucose. PLoS One 6(3):e17674

Kamo N, Muratsugu M, Kurihara K, Kobatake Y (1976) Change in surface charge density and membrane potential of intact mitochondria during energization. FEBS Lett 72(2):247–250

Kerner J, Lee K, Tandler B, Hoppel CL (2012) VDAC proteomics: post-translation modifications. Biochim Biophys Acta 1818(6):1520–1525

Korzeniewski B, Mazat J-R (1996) Theoretical studies of the control of oxidative phosphorylation in muscle mitochondria: application to mitochondria deficiencies. Biochem J 319:143–148

Kottke M, Adams V, Wallimann T, Nalam VK, Brdiczka D (1991) Location and regulation of octameric mitochondrial creatine kinase in the contact sites. Biochim Biophys Acta 1061:215–225

Krammer EM, Vu GT, Homblé F, Prévost M (2015) Dual mechanism of ion permeation through VDAC revealed with inorganic phosphate ions and phosphate metabolites. PLoS One 10(4):e0121746

Kuznetsov AV, Javadov S, Guzun R, Grimm M, Saks V (2013) Cytoskeleton and regulation of mitochondrial function: the role of beta-tubulin II. Front Physiol 4:82

Lee AC, Xu X, Colombini M (1996) The role of pyridine dinucleotides in regulating the permeability of the mitochondrial outer membrane. J Biol Chem 271(43):26724–26731

Lemasters JJ, Holmuhamedov E (2006) Voltage-dependent anion channel (VDAC) as mitochondrial governator – thinking outside the box. Biochim Biophys Acta 1762(2):181–190

Lemasters JJ, Ramshesh VK (2007) Imaging of mitochondrial polarization and depolarization with cationic fluorophores. Methods Cell Biol 80:283–295

Lemeshko VV (2002) Model of the outer membrane potential generation by the inner membrane of mitochondria. Biophys J 82:684–692

Lemeshko VV (2006) Theoretical evaluation of a possible nature of the outer membrane potential of mitochondria. Eur Biophys J 36:57–66

Lemeshko VV (2014a) VDAC electronics: 1. VDAC-hexo(gluco)kinase generator of the mitochondrial outer membrane potential. Biochim Biophys Acta 1838:1362–1371

Lemeshko VV (2014b) Competitive interactions of amphipathic polycationic peptides and cationic fluorescent probes with lipid membrane: experimental approaches and computational model. Arch Biochem Biophys 545:167–178

Lemeshko V (2015) The Warburg effect as a VDAC-hexokinase-mediated electrical suppression of mitochondrial energy metabolism. FASEB J 29(Suppl 1):725.27

Lemeshko VV (2016) VDAC electronics: 3. VDAC-creatine kinase-dependent generation of the outer membrane potential in respiring mitochondria. Biochim Biophys Acta 1858(7 PtA):1411–1418

Lemeshko SV, Lemeshko VV (2000) Metabolically derived potential on the outer membrane of mitochondria: a computational model. Biophys J 79:2785–2800

Lemeshko SV, Lemeshko VV (2004) Energy flux modulation on the outer membrane of mitochondria by metabolically-derived potential. Mol Cell Biochem 256–257:127–139

Liu MY, Colombini M (1992) A soluble mitochondrial protein increases the voltage dependence of the mitochondrial channel, VDAC. J Bioenerg Biomembr 24:41–46

Maldonado EN, Lemasters JJ (2014) ATP/ADP ratio, the missed connection between mitochondria and the Warburg effect. Mitochondrion 19(Pt A):78–84

Maldonado EN, Patnaik J, Mullins MR, Lemasters JJ (2010) Free tubulin modulates mitochondrial membrane potential in cancer cells. Cancer Res 70(24):10192–10201

Maldonado EN, Sheldon KL, DeHart DN, Patnaik J, Manevich Y, Townsend DM, Bezrukov SM, Rostovtseva TK, Lemasters JJ (2013) Voltage-dependent anion channels modulate mitochondria metabolism in cancer cells: regulation by free tubulin and erastin. J Biol Chem 288(17):11920–11929

Mangan PS, Colombini M (1987) Ultrasteep voltage dependence in a membrane channel. Proc Natl Acad Sci U S A 84(14):4896–4900

Mannella CA (1982) Structure of the outer mitochondrial membrane: ordered arrays of porelike subunits in outer-membrane fractions from Neurospora crassa mitochondria. J Cell Biol 94:680–687

Mannella CA, Forte M, Colombini M (1992) Toward the molecular structure of the mitochondrial channel, VDAC. J Bioenerg Biomembr 24:7–19

Marín-Hernández A, Rodríguez-Enríquez S, Vital-González PA, Flores-Rodríguez FL, Macías-Silva M, Sosa-Garrocho M, Moreno-Sánchez R (2006) Determining and understanding the control of glycolysis in fast-growth tumor cells. Flux control by an over-expressed but strongly product-inhibited hexokinase. FEBS J 273(9):1975–1988

Martin W, Hoffmeister M, Rotte C, Henze K (2001) An overview of endosymbiotic models for the origins of eukaryotes, their ATP-producing organelles (mitochondria and hydrogenosomes), and their heterotrophic lifestyle. Biol Chem 382:1521–1539

Mathupala SP, Pedersen PL (2010) Voltage dependent anion channel-1 (VDAC-1) as an anti-cancer target. Cancer Biol Ther 9(12):1053–1056

Maurya SR, Mahalakshmi R (2015) N-helix and cysteines inter-regulate human mitochondrial VDAC-2 function and biochemistry. J Biol Chem 290(51):30240–30252

Messina A, Reina S, Guarino F, De Pinto V (2012) VDAC isoforms in mammals. Biochim Biophys Acta 1818:1466–1476

Okazaki M, Kurabayashi K, Asanuma M, Saito Y, Dodo K, Sodeoka M (2015) VDAC3 gating is activated by suppression of disulfide-bond formation between the N-terminal region and the bottom of the pore. Biochim Biophys Acta 1848(12):3188–3196

Pastorino JG, Hoek JB (2008) Regulation of hexokinase binding to VDAC. J Bioenerg Biomembr 40(3):171–182

Pastorino JG, Hoek JB, Shulga N (2005) Activation of glycogen synthase kinase 3beta disrupts the binding of hexokinase II to mitochondria by phosphorylating voltage-dependent anion channel and potentiates chemotherapy-induced cytotoxicity. Cancer Res 65(22):10545–10554

Pinz I, Ostroy SE, Hoyer K, Osinska H, Robbins J, Molkentin JD, Ingwall JS (2008) Calcineurin-induced energy wasting in a transgenic mouse model of heart failure. Am J Physiol Heart Circ Physiol 294(3):H1459–H1466

Porcelli AM, Ghelli A, Zanna C, Pinton P, Rizzuto R, Rugolo M (2005) pH difference across the outer mitochondrial membrane measured with a green fluorescent protein mutant. Biochem Biophys Res Commun 326:799–804

Rimmerman N, Ben-Hail D, Porat Z, Juknat A, Kozela E, Daniels MP, Connelly PS, Leishman E, Bradshaw HB, Shoshan-Barmatz V, Vogel Z (2013) Direct modulation of the outer mitochondrial membrane channel, voltage-dependent anion channel 1 (VDAC1) by cannabidiol: a novel mechanism for cannabinoid-induced cell death. Cell Death Dis 4:e949

Rostovtseva TK, Bezrukov SM (2008) VDAC regulation: role of cytosolic proteins and mitochondrial lipids. J Bioenerg Biomembr 40(3):163–170

Rostovtseva TK, Bezrukov SM (2012) VDAC inhibition by tubulin and its physiological implications. Biochim Biophys Acta 1818(6):1526–1535

Rostovtseva TK, Bezrukov SM (2015) Function and regulation of mitochondrial voltage-dependent anion channel. Springer Series in Biophysics. In: Delcour AH (ed) Electrophysiology of unconventional channels and pores, vol 18. Springer International Publishing Cham Springer Series in Biophysics pp 3–31

Rostovtseva T, Colombini M (1997) VDAC channels mediate and gate the flow of ATP: implications for the regulation of mitochondrial function. Biophys J 72(5):1954–1962

Rostovtseva TK, Tan W, Colombini M (2005) On the role of VDAC in apoptosis: fact and fiction. J Bioenerg Biomembr 37(3):129–142

Rostovtseva TK, Sheldon KL, Hassanzadeh E, Monge C, Saks V, Bezrukov SM, Sackett DL (2008) Tubulin binding blocks mitochondrial voltage-dependent anion channel and regulates respiration. Proc Natl Acad Sci U S A 105(48):18746–18751

Rostovtseva TK, Gurnev PA, Protchenko O, Hoogerheide DP, Yap TL, Philpott CC, Lee JC, Bezrukov SM (2015) α-synuclein shows high affinity interaction with voltage-dependent anion channel, suggesting mechanisms of mitochondrial regulation and toxicity in Parkinson disease. J Biol Chem 290(30):18467–18477

Saks VA, Kuznetsov AV, Khuchua ZA, Vasilyeva EV, Belikova JO, Kesvatera T, Tiivel T (1995) Control of cellular respiration in vivo by mitochondrial outer membrane and by creatine kinase. A new speculative hypothesis: possible involvement of mitochondrial-cytoskeleton interactions. J Mol Cell Cardiol 27:625–645

Saks V, Guzun R, Timohhina N, Tepp K, Varikmaa M, Monge C, Beraud N, Kaambre T, Kuznetsov A, Kadaja L, Eimre M, Seppet E (2010) Structure-function relationships in feedback regulation of energy fluxes in vivo in health and disease: mitochondrial interactosome. Biochim Biophys Acta 1797(6–7):678–697

Scheibye-Knudsen M, Quistorff B (2009) Regulation of mitochondrial respiration by inorganic phosphate; comparing permeabilized muscle fibers and isolated mitochondria prepared from type-1 and type-2 rat skeletal muscle. Eur J Appl Physiol 105(2):279–287

Schlattner U, Tokarska-Schlattner M, Wallimann T (2006) Mitochondrial creatine kinase in human health and disease. Biochim Biophys Acta 1762:164–180

Sheldon KL, Maldonado EN, Lemasters JJ, Rostovtseva TK, Bezrukov SM (2011) Phosphorylation of voltage-dependent anion channel by serine/threonine kinases governs its interaction with tubulin. PLoS One 6(10):e25539

Sheldon KL, Gurnev PA, Bezrukov SM, Sackett DL (2015) Tubulin tail sequences and post-translational modifications regulate closure of mitochondrial voltage-dependent anion channel (VDAC). J Biol Chem 290(44):26784–26789

Shoshan-Barmatz V, Ben-Hail D (2012) VDAC, a multi-functional mitochondrial protein as a pharmacological target. Mitochondrion 12(1):24–34

Shoshan-Barmatz V, Gincel D (2003) The voltage-dependent anion channel – characterization, modulation, and role in mitochondrial function in cell life and death. Cell Biochem Biophys 39:279–292

Shoshan-Barmatz V, De Pinto V, Zweckstetter M, Raviv Z, Keinan N, Arbel N (2010) VDAC, a multi-functional mitochondrial protein regulating cell life and death. Mol Asp Med 31(3):227–285

Shoshan-Barmatz V, Ben-Hail D, Admoni L, Krelin Y, Tripathi SS (2015) The mitochondrial voltage-dependent anion channel 1 in tumor cells. Biochim Biophys Acta 1848(10 Pt B):2547–2575

Simson P, Jepihhina N, Laasmaa M, Peterson P, Birkedal R, Vendelin M (2016) Restricted ADP movement in cardiomyocytes: cytosolic diffusion obstacles are complemented with a small number of open mitochondrial voltage-dependent anion channels. J Mol Cell Cardiol 97:197–203

Smilansky A, Dangoor L, Nakdimon I, Ben-Hail D, Mizrachi D, Shoshan-Barmatz V (2015) The voltage-dependent anion channel 1 mediates amyloid β toxicity and represents a potential target for Alzheimer disease therapy. J Biol Chem 290(52):30670–30683

Spindler M, Niebler R, Remkes H, Horn M, Lanz T, Neubauer S (2002) Mitochondrial creatine kinase is critically necessary for normal myocardial high-energy phosphate metabolism. Am J Physiol Heart Circ Physiol 283:H680–H687

Stein CA, Colombini M (2008) Specific VDAC inhibitors: phosphorothioate oligonucleotides. J Bioenerg Biomembr 40(3):157–162

Teijido O, Rappaport SM, Chamberlin A, Noskov SY, Aguilella VM, Rostovtseva TK, Bezrukov SM (2014) Acidification asymmetrically affects voltage-dependent anion channel implicating the involvement of salt bridges. J Biol Chem 289(34):23670–23682

Vander Heiden MG, Chandel NS, Schumacker PT, Thompson CB (1999) Bcl-xL prevents cell death following growth factor withdrawal by facilitating mitochondrial ATP/ADP exchange. Mol Cell 3(2):159–167

Vander Heiden MG, Chandel NS, Li XX, Schumacker PT, Colombini M, Thompson CB (2000) Outer mitochondrial membrane permeability can regulate coupled respiration and cell survival. Proc Natl Acad Sci U S A 97(9):4666–4671

Wallimann T (2015) The extended, dynamic mitochondrial reticulum in skeletal muscle and the creatine kinase (CK)/phosphocreatine (PCr) shuttle are working hand in hand for optimal energy provision. J Muscle Res Cell Motil 36(4–5):297–300

Wallimann T, Tokarska-Schlattner M, Schlattner U (2011) The creatine kinase system and pleiotropic effects of creatine. Amino Acids 40:1271–1296

Wallis J, Lygate CA, Fischer A, ten Hove M, Schneider JE, Sebag-Montefiore L, Dawson D, Hulbert K, Zhang W, Zhang MH, Watkins H, Clarke K, Neubauer S (2005) Supranormal myocardial creatine and phosphocreatine concentrations lead to cardiac hypertrophy and heart failure: insights from creatine transporter-overexpressing transgenic mice. Circulation 112:3131–3139

Wilson JE (2003) Isozymes of mammalian hexokinase: structure, subcellular localization and metabolic function. J Exp Biol 206(Pt 12):2049–2057

Yagoda N, von Rechenberg M, Zaganjor E, Bauer AJ, Yang WS, Fridman DJ, Wolpaw AJ, Smukste I, Peltier JM, Boniface JJ, Smith R, Lessnick SL, Sahasrabudhe S, Stockwell BR (2007) RAS-RAF-MEK-dependent oxidative cell death involving voltage-dependent anion channels. Nature 447:864–868

Zhang YX, Zhao W, Tang YJ (2016) Multilevel induction of apoptosis by microtubule-interfering inhibitors 4β-S-aromatic heterocyclic podophyllum derivatives causing multi-fold mitochondrial depolarization and PKA signaling pathways in HeLa cells. Oncotarget 7(17):24303–24313

Part IV
Mitochondria and Cellular Signaling Network

Chapter 10
New Insights on the Regulation of Programmed Cell Death by Bcl-2 Family Proteins at the Mitochondria: Physiological and Pathophysiological Implications

Laurent Dejean and Stéphen Manon

10.1 Introduction

Since the first description, in 1972, that cell death spontaneously occurred in otherwise healthy tissues (Kerr et al. 1972), apoptosis has grown as one of the most exciting fields of modern cell biology. Biology is defined as the science of life, and it is rather fascinating that so much efforts are made by investigators around the world to understand how a beautifully organized cell is returning to the inanimate. Beyond our obvious interest to understand the mechanistic and regulatory aspects of a process that is crucial for major health-related issues (such as embryonic development, cancer, and neurodegenerative diseases), the study of apoptosis is a constant reminder to us that nothing is eternal. Scientist or not, every human being wants to understand death, and our differences in perceiving this concept have inspired the most remarkable artistic works, as well as the ugliest wars. For us, biologists, this is not so much the concept of cell death that is fascinating, since we know too well that life is dependent on such a fragile equilibrium and that death is the "normal" status while life is the "exceptional" one. Our fascination might come from the concept of *programmed* cell death that life evolution, in all its complexity, has also included death evolution. A bacterium does not die like a mammalian neuron, for instance, and this mammalian neuron does not die like a mammalian lymphocyte. All of this is obviously without considering the (somewhat exaggerated) concept of immortal cancer cell; cells which

L. Dejean (✉)
Department of Chemistry, California State University, Fresno, CA 93740, USA
e-mail: ldejean@csufresno.edu

S. Manon (✉)
CNRS, Université de Bordeaux, UMR5095,
1 Rue Camille Saint-Saëns, 33077 Bordeaux, France
e-mail: stephen.manon@ibgc.cnrs.fr

T.K. Rostovtseva (ed.), *Molecular Basis for Mitochondrial Signaling*,
Biological and Medical Physics, Biomedical Engineering,
DOI 10.1007/978-3-319-55539-3_10

are halfway between the dream of eternal life and the nightmare of uncontrolled proliferation incompatible with the maintenance of organismal structural integrity. Underlying these somewhat epistemological considerations exist molecular mechanisms, which investigators have begun to understand in the 1980s and 1990s from two perspectives: (i) that of development, with studies using the nematode *Caenorhabditis elegans* as model (Ellis and Horvitz 1986), and (ii) that of cancer research, with a special focus on mammalian/human Bcl-2 family proteins (Tsujimoto et al. 1984). It turns out that these two fields met each other, when it was found that the worm protein *ced-9* was a Bcl-2 family member (Hengartner and Horvitz 1994). An additional degree of excitement was reached in these studies when it was found that the function(s) of Bcl-2 family members had something to do with mitochondria (Hockenbery et al. 1993; Zamzami et al. 1995), organelles which, as cellular energy converters, symbolize the power of life.

In this chapter, we will review several recent advances on the study of the molecular links between Bcl-2 family members and mitochondria. Specifically, we will emphasize on the molecular aspects underlying these links and show that some concepts which are commonly accepted nowadays are bound to be reevaluated. It is worth noting that if mitochondria have for long been considered as "passive" players in apoptosis, it is now considered that their function in bioenergetic metabolism is playing an important part in the apoptotic response. This recent knowledge may provide new means to manipulate apoptosis in pathological situations and proves itself useful in the development of new therapeutics against cancer and/or neurodegenerative diseases.

10.2 Presentation of Structure-Function Features of Bcl-2 Family Proteins

10.2.1 Characteristics of Bcl-2 Family Members

Bcl-2 was first identified in 1984 as an oncogenic protein that was overexpressed in B-cell lymphomas (Tsujimoto et al. 1984). Sequence analyzes later indicated that Bcl-2 was the first discovered member of a large family of proteins that play a major role in the regulation of cell death, most specifically – but not exclusively – by regulating apoptosis. As such, the molecular mechanisms underlying their function have been the subject of intense scrutiny. Indeed, alterations of their function are involved in major cancer-related issues such as tumorigenesis or chemoresistance, and targeting these proteins is a promising approach for cancer therapy (Czabotar et al. 2014; Kvansakul and Hinds 2015). Bcl-2 family members are present in the whole animal reign: the *Caenorhabditis elegans* protein *Ced9* is a Bcl-2 family member (Hengartner and Horvitz 1994). Bcl-2 family members have actually been found in all vertebrates, including zebrafish (Arnaud et al. 2006; Chen et al. 2000; Kratz et al. 2006), and in invertebrates, such as Drosophila (Colussi et al. 2000; Quinn et al. 2003). A summary of the main structural features of Bcl-2 family proteins is presented below and summarized in Fig. 10.1.

Bcl-2 family members have been once characterized by the presence of Bcl-2-homology (BH) domains (Kelekar and Thompson 1998). Structural data later showed that homologies in the primary structure were associated to homologies in the tertiary structure. Indeed, anti-apoptotic proteins Bcl-2 (Petros et al. 2001) and Bcl-xL (Muchmore et al. 1996) and pro-apoptotic proteins Bax (Suzuki et al. 2000), Bak (Moldoveanu et al. 2006), and Bid (Chou et al. 1999) all shared common structural features (Petros et al. 2004).

Besides these proteins that can be considered as "strict" Bcl-2 homologs, other proteins only having the BH3 domain have been identified, namely, Bad (Yang et al. 1995), Bim (O'Connor et al. 1998), and Puma (Yu et al. 2001). It should be noted that Bid, although sharing the global structure of Bcl-2 family members, has been classified for long as a BH3-only protein, since it does not have obvious BH1 and BH2 domains. All these proteins, termed "BH3-only proteins," interact with Bcl-2 family members, because of the ability of their BH3 domain to interact with the BH1/BH2 domains of anti-apoptotic proteins, thus preventing the interaction between anti-apoptotic and pro-apoptotic proteins (Kelekar and Thompson 1998). Other "non-strict Bcl-2 homologs" containing a BH3 domain have been found in mammals [e.g., the pro-autophagic proteins Beclin-1 (Oberstein et al. 2007) and Nix (Farooq et al. 2001)], but also in organisms from other life domains (e.g., the yeast protein Bxi1/Ybh3 (Buttner et al. 2011; Cebulski et al. 2011)). Phylogenetic studies suggest that the presence of BH3-like domains in these proteins might be the result of convergent evolution instead of the evolution from a common ancestor (Aouacheria et al. 2013). This suggests that genuine Bcl-2 family members are limited to the animal reign. However, under certain conditions, these noncanonical BH3-containing proteins can regulate genuine Bcl-2 family members, e.g., between Beclin-1 and Bcl-2 (Liang et al. 1999). Thus, these studies led to the proposal of an alternative phylogenetic classification of these proteins (Fig. 10.1).

10.2.2 Different Roles of Bcl-2 Family Members in Apoptosis

Multi-domain pro-apoptotic proteins, such as Bax and Bak, are the central effectors of cell death. They are the direct effectors of the permeabilization of the outer mitochondrial membrane. Although defects on Bax/Bak have rarely been identified in a tumor context, stimulating their activity (Gavathiotis et al. 2012) and mimicking their action (Valero et al. 2011) are feasible alternatives to increase the death of cancer cells. The mechanistic aspects of the permeabilization process will be detailed below.

Anti-apoptotic proteins Bcl-2 and Bcl-xL inhibit apoptosis through a physical interaction with pro-apoptotic proteins. Early studies supported the view that the formation of complexes between anti-apoptotic proteins Bcl-2/Bcl-xL and pro-apoptotic protein Bax favored the cytosolic retention of Bax, thus preventing its mitochondrial localization and/or activation. When Bcl-2/Bcl-xL are overexpressed, they therefore cause the resistance of tumor cells to signals or to chemotherapy that

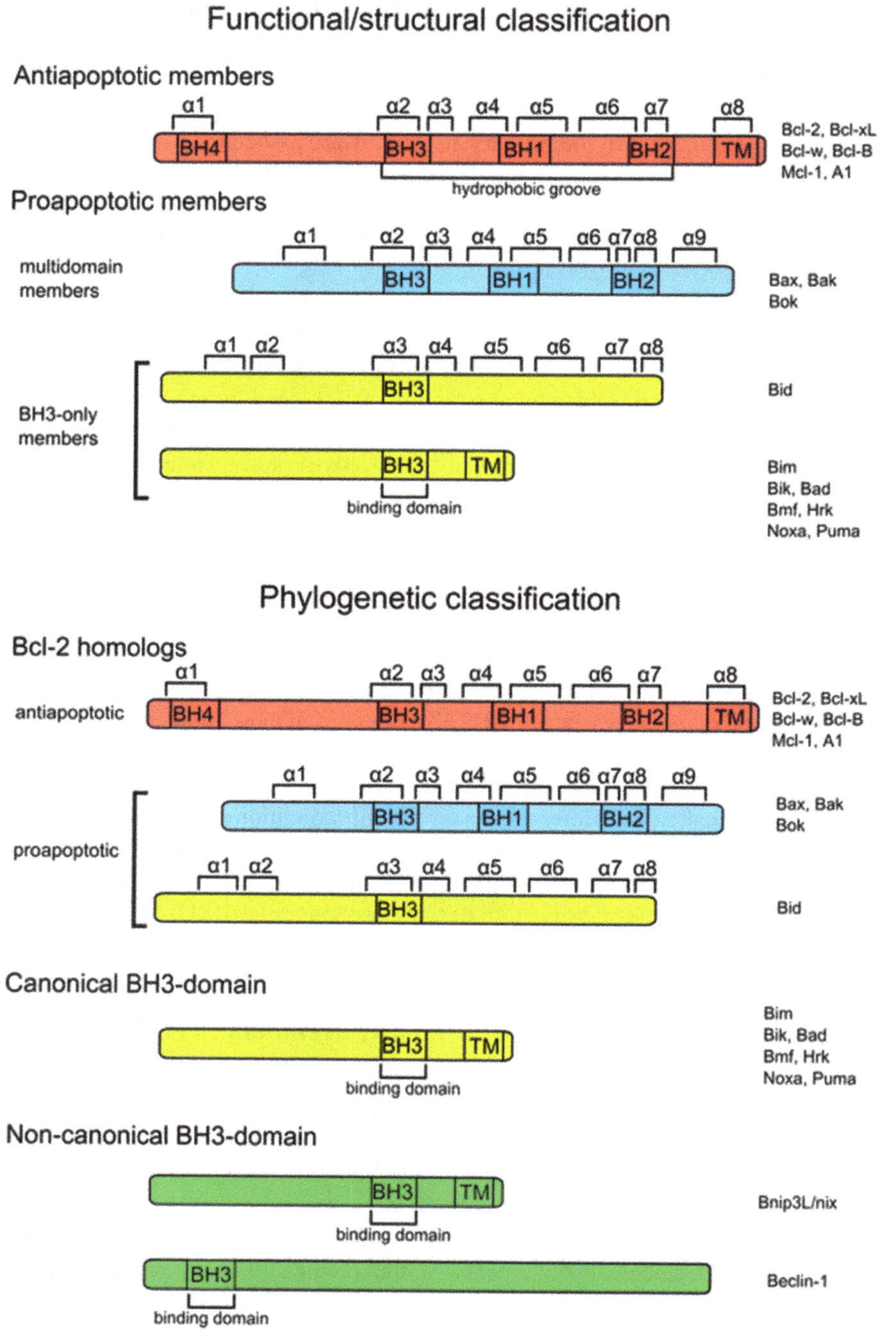

Fig. 10.1 Classifications of Bcl-2 family members. (*Top*) Classical classification based on structure/function studies. Bcl-2 family members were characterized by the presence of BH domains and sub-characterized by their function. Anti-apoptotic proteins contain a hydrophobic groove that can interact with BH3 domains of pro-apoptotic proteins. Multi-domain pro-apoptotic proteins share structural features with anti-apoptotic proteins. They can organize as a large-sized pore

lead to cell death (Adams and Cory 2007). They might also help the survival of cancer cells through other mechanisms such as regulation of autophagy (Pattingre et al. 2005; Maiuri et al. 2007; Priault et al. 2010).

As noted above, BH3-only proteins form a heterogeneous group of proteins that only share the BH3 domain of homology with Bcl-2 family members. Although they are mostly pro-apoptotic, they display different functions in the process. Some of them, like Bad, are acting by breaking the interaction between pro- and anti-apoptotic proteins. They are then called "derepressors." Others, like tBid, are mostly, if not exclusively, "activators" of multi-domain pro-apoptotic proteins. A third group, including Bim or Puma, may have both functions, depending on the cellular context. These proteins have emerged as suitable templates to design molecules able to trigger apoptosis in cancer cells (Delbridge and Strasser 2015).

10.2.3 Structure-Function Studies of Bcl-2 Family Members

Since most cells are able to express the vast majority of Bcl-2 family members, the mitochondrial outer membrane permeabilization (MOMP) leading to cytochrome c release during apoptosis is regulated by a complex network of interactions (Renault et al. 2012; Willis et al. 2007 for review).

It has been early established that homology domains BH1, BH2, and BH3 are mainly responsible for these interactions. Site-directed mutagenesis of the BH3 domain of Bax (Zha et al. 1996a) or the BH1 domain of Bcl-2/Bcl-xL (Sedlak et al. 1995; Yin et al. 1994) is sufficient to prevent the interactions between pro- and anti-apoptotic proteins. Similarly, mutations of the BH3 domain of BH3-only proteins prevent their interaction with anti-apoptotic proteins. It is noteworthy that the binding affinities of Bcl-2 and Bcl-xL for the BH3 domains of different proteins may differ significantly (Chen et al. 2005). Even though the conclusion of the experiments done with isolated BH3 domains should be extrapolated to the whole proteins with caution, the BH3 domain of Bak has successfully served as a model to design ABT-737, a very specific and high-affinity inhibitor of anti-apoptotic proteins Bcl-2 and Bcl-xL (Oltersdorf et al. 2005). Further studies have since allowed to refine the

Fig. 10.1 (continued) responsible for mitochondria permeabilization. BH3-only proteins can interact with anti-apoptotic proteins, inhibiting them (Bad, Noxa), or with pro-antiapoptotic proteins, activating them (tBid, Bim, Puma). A part from the BH3 domain, they do not share structural features with anti- or pro-apoptotic proteins, with the exception of Bid. (*Bottom*) Anti-apoptotic and pro-apoptotic proteins, including Bid, have been generated from gene duplication and divergent evolution, and all together form a group of Bcl-2 homologs. The canonical BH3 domain is a protein-protein interaction domain than can be found in these Bcl-2 homologs and in other proteins that regulate their function. Closely resembling noncanonical BH3 domains can be found in a large number of proteins making them potentially able to modulate the activity of Bcl-2 homologs. It is unlikely that these noncanonical domains there generated through duplication of the canonical BH3 domain: as a matter of fact, they are also present in organisms where the canonical domain is absent (Figure from Renault et al. 2016)

process by designing ABT-199 that is highly specific to Bcl-2 but not to Bcl-xL (Souers et al. 2013). Conversely, other studies have permitted to design Bcl-xL-selective inhibitors, such as WEHI-539 (Lessene et al. 2013).

As already noted, both anti- and pro-apoptotic Bcl-2 family members share several common structural features, beyond the homology domains. One of them is the presence of a C-terminal hydrophobic α-helix, long enough to form a membrane anchor (Kelekar and Thompson 1998; Garcia-saez et al. 2004). It has been unambiguously demonstrated that the removal of this C-terminal domain prevented both in vitro and in vivo membrane localization of both Bcl-2 and Bcl-xL (Kaufmann et al. 2003). As a matter of fact, this domain alone can spontaneously insert into membranes (Del Mar Martinez-Senac et al. 2000). Interestingly, the C-terminal domain of Bcl-xL was sufficient to promote the mitochondrial localization of Bax deprived of its own C-terminus, in vitro (Tremblais et al. 1999), in mammalian cells (Oliver et al. 2000), or following its heterologous expression in yeast.

Conversely, it was and is still unclear if the C-terminal hydrophobic α-helix of Bax is involved in its anchoring into the outer mitochondrial membrane. The removal of this domain leads to contradictory results on its localization both in vitro and in vivo. The domain alone is not able to anchor a reporter protein, unless the Ser184 residue is deleted or substituted by a Val residue (Suzuki et al. 2000). It is not able to restore the mitochondrial localization of Bcl-xL deprived of its own C-terminal domain (Tremblais et al. 1999).

A series of experiments have suggested that the mitochondrial receptor translocase of outer mitochondrial membrane 22 (TOMM22) helped the mitochondrial localization of Bax (Bellot et al. 2007). However, the absence of TOMM22 did not affect the mitochondrial localization of Bax carrying a deletion (Cartron et al. 2008) or a substitution (Ross et al. 2009) of Ser184. By using the heterologous expression of Bax in yeast, it has been observed that the overexpression of the cytosolic domain of human TOMM22 could interfere with Bax interaction with mitochondria: it increased Bax mitochondrial localization, but decreased its ability to permeabilize the outer mitochondrial membrane (Renault et al. 2012), suggesting that Bax was relocalized under a poorly active conformation.

Even though the C-terminal α-helix of Bax is not a genuine membrane anchor, like its counterparts in Bcl-2 and Bcl-xL, mutations that are expected to move it away from the core of the protein are activating Bax. The introduction of a negative charge in position 174 (Thr174>Asp), facing a negative charge in position 69 (glu69), is sufficient to promote a high mitochondrial localization (Arokium et al. 2004). Also, favoring the movement of the domain by substituting the proline residue located in the loop between α8 and α9 helices (Pro168>Ala, Val, or Gly) was found to strongly increase the mitochondrial localization of Bax in several biological models (Arokium et al. 2004; Cartron et al. 2005) but not in a universal fashion (Schinzel et al. 2004). The consequences of these substitutions may likely depend on other factors modulating Bax conformation. Indeed, the stimulating effect of a substitution Pro168>Ala was abolished by a substitution Ser60>Ala, but further restored by a substitution Ser163>Asp (Arokium et al. 2007) suggesting that conformational changes in several different domains of Bax may participate to its full activation.

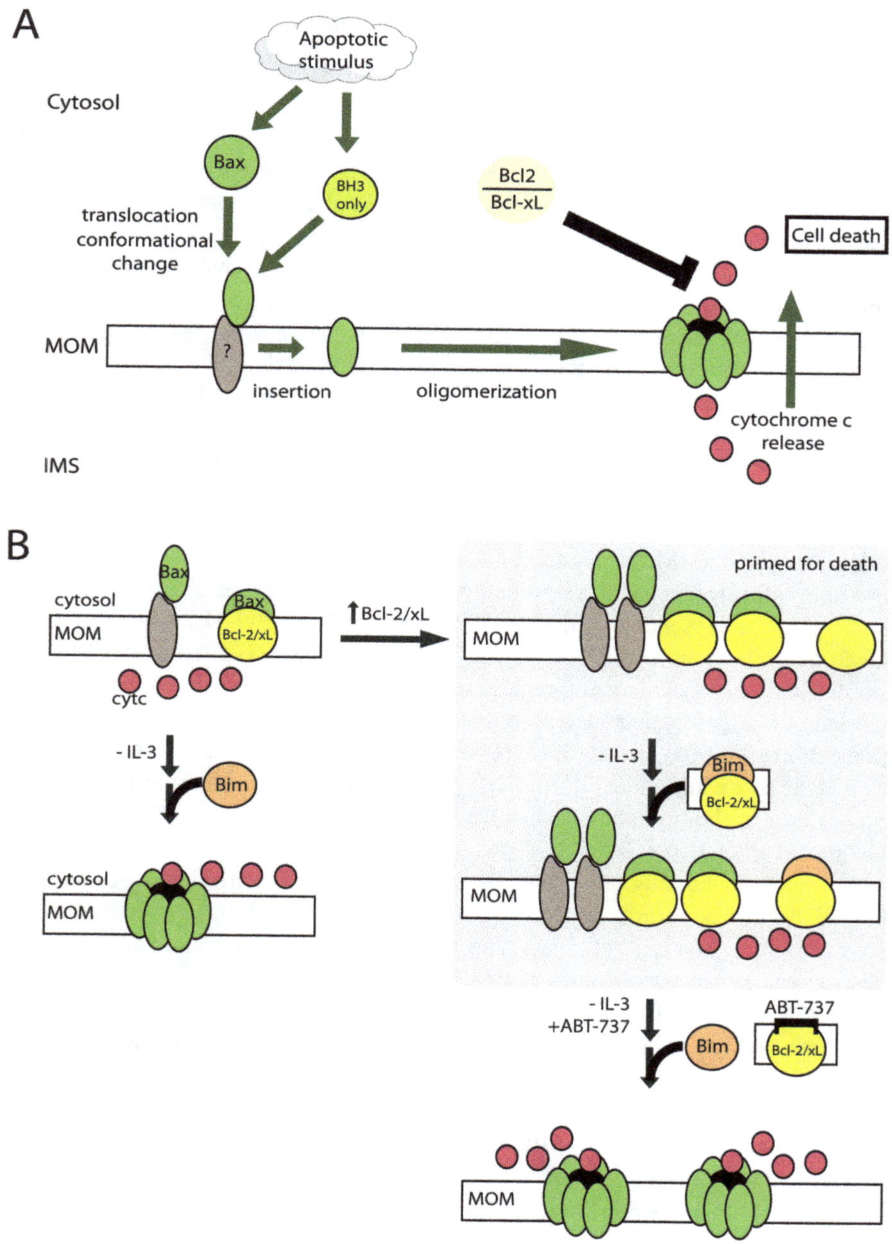

Fig. 10.2 Bax signaling pathway leading to MOMP in cells which are normal or primed for death. (**a**) Induction of apoptosis results in Bax translocation and change of conformation (pre-activation) at the mitochondrial outer membrane (*MOM*) (*white rectangle*). During this step, Bax docking may be favored by its interaction with native MOM protein such as TOMM22 (labeled "?").

10.2.4 Relationship Between Bax Structure and Permeabilization of the Outer Mitochondrial Membrane

As already mentioned, the canonical pathway of Bax-mediated mitochondrial permeabilization depends on several conformational changes of the protein leading sequentially to its translocation, insertion, and oligomerization in the outer membrane; the last of these steps being the cause of MOMP and cytochrome c release (Fig. 10.2a). Following the first structural data on Bcl-2 family members, a parallel has been drawn between the presence of two amphipathic α-helices (α5–α6) forming a hairpin structure, with those found in several bacterial toxins known to form pores in membranes (Minn et al. 1997). When added to artificial membranes, Bcl-2, Bcl-xL, and Bax could form low-conductance channels (several tenths of pS) (Minn et al. 1997; Schendel et al. 1997; Schlesinger et al. 1997). However, the extrapolated size of the pores of these putative channels was too small to explain neither the release of large molecules such as cytochrome c observed in apoptotic cells or to constitute a mechanistic basis of the antagonistic function of anti- and pro-apoptotic proteins. It was next demonstrated that, under certain conditions and, namely, in the presence of tBid, Bax (but not Bcl-2 and Bcl-xL) was able to permeabilize mitochondria to cytochrome c (Desagher et al. 1999). Electrophysiological studies showed that this membrane permeabilization was associated with the formation of a large-conductance pore (Pavlov et al. 2001) containing Bax oligomers (Dejean et al. 2005). Recent progresses in imaging techniques have allowed to show that Bax could indeed form large pore-like structures both in artificial membranes and in the mitochondrial outer membrane (Grosse et al. 2016; Salvador-Gallego et al. 2016).

Although some alternative models, such as the Bax-induced activation of the so-called mitochondrial permeability transition pore (mPTP), have been popular for some time (Marzo et al. 1998) and might still be involved in other death processes such as necrosis (Baines et al. 2005; Nakagawa et al. 2005), it is now readily established that Bax can genuinely form a pore in the outer mitochondrial membrane which is large enough to allow the release of, at least, cytochrome c. The remaining question however was how a globally hydrophilic and globular protein such as Bax is able to reorganize itself as a large membrane-inserted pore. Recent structural and structure/function studies have allowed to partially answer this crucial question.

The structure of soluble, supposedly inactive, Bax has been determined by NMR in 2000 (Suzuki et al. 2000). At this date, most data described above had yet to be obtained. This structure has been a template to study which domains/residues of Bax were of interest for structure/function studies. However, it was not sufficient to predict the structure of the fully active/membrane-associated protein. Worthy information has been obtained from the structure of antibodies directed against the N-terminal domain of Bax (Peyerl et al. 2007). This domain, corresponding to the residues 3–16 of human Bax, is highly immunogenic and has been used to generate both polyclonal and monoclonal antibodies. One monoclonal antibody, 6A7, was found to selectively recognize the active form of Bax, opposite to another antibody, 2D2, that recognizes all Bax conformations (Hsu and Youle 1998). The differences in the structures of the recognition sites of 2D2 and 6A7 showed that part of the domain supported a large amplitude movement between the two conformations (Peyerl et al. 2007).

A breakthrough advance has been done when the structure of Bax activated by a BH3 domain has been resolved by X-ray diffraction of crystals (Czabotar et al. 2013). It should be noted, however, that, for technical reasons, the authors used a version of Bax deprived of the C-terminal hydrophobic domain, making the questions about the actual role of this domain (see above) still relevant. This structure revealed a symmetrical Bax dimer. Opposite to what had been postulated before, the amphipathic helices α5–α6 were not organized as hairpins, like in the inactive monomer, but were unfolded to form a head-to-tail interface between the two monomers. This led to the hypothesis that, instead of being inserted in the membrane, α5–α6 laid flat at the surface of the membrane. This hypothesis was further supported by experiments of residue accessibility (Westphal et al. 2014). Models were then proposed to explain how these Bax dimers could further self-organize as oligomers (Bleicken et al. 2014; Subburaj et al. 2015). It is noteworthy that such a self-organization is fully compatible with experiments showing that once several Bax molecules are inserted in the outer mitochondrial membrane, they can serve as "receptors" for additional Bax molecules (Cartron et al. 2008). This self-organizing model is also compatible with experiments showing that the lipid composition, more specifically the presence of ceramides and cardiolipin (Zhang and Saghatelian 2013 for a review), membrane curvature (Renault et al. 2015a), and the overall mitochondrial cristae structure (Yamaguchi et al. 2008) are crucial factors for Bax-induced permeabilization. Indeed, this model requires Bax oligomers to be able to reorganize lipids around them to form large-sized pores (Bleicken et al. 2014).

Fig. 10.2 (continued) Induction of apoptosis also increases the expression and translocation of BH3-only proteins at the MOM which trigger Bax insertion and oligomerization. These Bax oligomers are essential components of the MOMP machinery. MOMP allows the release of cytochrome c (*red*) from the intermembrane space to the cytosol, ultimately leading to cell death. Bcl-2/xL inhibits cytochrome c release and physical interactions between anti-apoptotic Bcl-2 family members are required to inhibit MOMP (**b**) *Left*, Upon IL-3 withdrawal, Bim catalyzes the activation of loosely bound Bax into a functional MOMP machinery leading to the release of cytochrome c. *Right*, Bcl-2/xL overexpression leads to accumulation of activated Bax in the mitochondrial outer membrane (*MOM*) even in the absence of an apoptotic stimulus (i.e., prior to IL-3 withdrawal). Withdrawal of IL-3 induces accumulation of Bim-Bcl-2/xL complexes in the MOM, but the MOMP machinery is still in its precursor form and cytochrome c is not released (compare *Left* and *Right* side on the figure). *Right*, Bim is accumulated at the MOM in response to IL-3 withdrawal. Bcl-2/xL overexpression leads to a mitochondrial accumulation of activated Bax present in noncompetent MOMP structure. ABT-737 addition leads to a release of the interaction between Bax and Bim and Bcl-2/Bcl-xL and therefore to a MOMP with higher intensity than in a normal context

10.3 Linking Signaling Pathways to Bax-Dependent MOMP

Structural and structure/function studies described above have allowed to get a better understanding about how a monomeric, globular, and soluble protein like Bax is able to be reorganized in a large-sized oligomer inserted in the membranes. We are now going to detail how this dramatic structural change might be connected to upstream death signaling pathways.

While protein phosphorylation is widely established as the major conveyor of short-term intracellular signaling, how phosphorylation could possibly regulate Bcl-2 family members function remains elusive. The first regulation by phosphorylation of a Bcl-2 family member has been shown on the BH3-only protein Bad (Zha et al. 1996b). Bad strictly interacts with the anti-apoptotic proteins Bcl-2 and Bcl-xL, inhibiting their ability to interact with Bax. On the other hand, Bad does not directly interact with Bax, opposite to other BH3-only proteins such as Bim or Puma (Willis et al. 2007). In living cells, Bad is normally phosphorylated on two serine residues by the survival kinase protein kinase B (AKT) (Del Peso et al. 1997). When phosphorylated, Bad cannot interact with Bcl-2/Bcl-xL. If the activity of AKT is decreased, Bad is less phosphorylated and can interact with Bcl-2/Bcl-xL, thus releasing the inhibition of Bax (Datta et al. 1997).

The BH3-only Bcl-2 homolog Bid is a major regulator of apoptosis. The protein is normally inactive but can be activated through a cleavage by caspase-8 (Luo et al. 1998). The active form of the protein, called tBid, has been shown to favor apoptosis in multiple ways; i.e., not only tBid directly favors Bax oligomerization (Desagher et al. 1999), but it also may cause mitochondrial structure rearrangements which help the release of cytochrome c and other apoptogenic factors during apoptosis (Kim et al. 2004). It was shown that Bid can be phosphorylated by casein kinases I and II, preventing its cleavage and thus activation, by caspase-8 (Desagher et al. 2001).

Anti-apoptotic proteins can also be phosphorylated under certain conditions by different kinases, namely, c-Jun N-terminal kinase (JNK) (Maundrell et al. 1997), mitogen-activated protein kinases 1 and 2 (MAPK1/2) (Torcia et al. 2001), and cyclin-dependent kinase 1 (CDK1) (Vantieghem et al. 2002). Multiple phosphorylation by JNK and MAPK1/MAPK2 has been shown to stabilize Bcl-2 and prevent its ubiquitin-dependent degradation (Breitschopf et al. 2000). On the other hand, the transient phosphorylation of both proteins by CDK1 has been shown to disable their anti-apoptotic activity (Terrano et al. 2010). Note however that these kinases have multiple targets in the death and survival signaling processes, a characteristic which does not allow to completely eliminate the possibility that the effects of the mutations mentioned above may be indirect.

The phosphorylation of specific residues might be the basis of conformational changes underlying the activation of Bax, including its mitochondrial translocation and its oligomerization. Because of its potential role in regulating the conformation of the C-terminal helix α9, the putative phosphorylation of the Ser184 residue within this helix has been the focus of several studies. In this context, it has been shown that Bax Ser184 can be phosphorylated by several kinases such as AKT (Gardai et al. 2004) and PKCζ (Xin et al. 2007). These protein kinases are involved in survival signaling pathways; it was therefore not surprising to find out that the phosphorylation of Ser184 was associated to a decrease of the mitochondrial translocation of Bax, both in mammalian cells (Gardai et al. 2004) and after heterologous expression in yeast (Simonyan et al. 2016). Conversely, the substitution of Ser184 by a non-phosphorylatable residue (Ala or Val) has reproducibly been associated with a constitutive mitochondrial localization of Bax (Arokium et al. 2007; Gardai et al. 2004). However, the consequences on Bax activity might not be so clear. Indeed, when expressed alone in the yeast model system, the mutant carrying a phosphomimetic substitution Ser184>Asp displayed a high capacity to release cytochrome c in spite of its weak mitochondrial localization, while non-phosphorylatable substitutions Ser184>Ala or >Val exhibited relatively low activities in spite of high mitochondrial localizations (Simonyan et al. 2016). Besides, the phosphomimetic mutant was less resistant to proteases and more sensitive to the inhibition by Bcl-xL than the non-phosphorylatable mutants (Simonyan et al. 2016). This may indicate that phosphorylation of Ser184 very finely regulates Bax stability, membrane localization, membrane permeabilization activity, and capacity to bind to anti-apoptotic partners.

An additional level of regulation of Bax activity might be provided by the phosphorylation of Ser163. This residue was shown to be phosphorylated by glycogen synthase kinase-3β (GSK3β); this phosphorylation event leads to an increase of Bax mitochondrial localization (Linseman et al. 2004), which was relevant to the role of this kinase in death signaling pathways (Forde and Dale 2007 for review). Remarkably, GSK3β is also known to inactivate AKT, through the phosphorylation of two of its core residues (Gold et al. 2000 for review). GSK3β would consequently be able to increase Bax mitochondrial relocalization through two mechanisms: by directly phosphorylating Ser16, and by indirectly preventing the phosphorylation of Ser184. However, like for Ser184, the consequence(s) of the phosphorylation of Ser163 might not be so straightforward. Indeed, the phosphomimetic substitution Ser163>Asp is not sufficient to increase the mitochondrial localization; however, it dramatically increased the mitochondrial localization of Bax substituted on other residues (Arokium et al. 2007), suggesting that concerted conformational changes of different Bax domains might participate to its mitochondrial relocation. It is worth noting that GSK3β phosphorylates the first Ser/Thr residue in a sequence Ser/Thr-X-X-X-Ser/Thr if the second Ser or Thr residue is already phosphorylated (Jho et al. 1999). Bax has a Thr at position 167, and it has been shown that this residue can be phosphorylated by extracellular signal-regulated kinases 1 and 2 (Erk1/2) and is close to the residue Pro168 (Schinzel et al. 2004), which was shown to be crucial to maintain the position of helix α9 (Arokium et al. 2004; Cartron et al. 2005). Incremental phosphorylation and conformational changes of this part of the protein are therefore likely to initiate its mitochondrial relocation.

Other residues of Bax are susceptible to be phosphorylated, and other kinases might contribute to the regulation of Bax location and function. For example, during anoikis, the 38 kDa mitogen-activated protein kinase (p38MAPK) has been shown to increase the mitochondrial localization of Bax without, however, fully activating it (Owens et al. 2009). Like for Bcl-2 and Bcl-xL, the phosphorylation of Bax could simply stabilize the protein, like it has been shown in yeast when Bax is phosphorylated by the protein kinase C alpha (PKCα) (Silva et al. 2011).

These observations, which have been made in different cellular models, might obviously not be extrapolated to every biological system. The more intricate is a regulation, the more likely it is to be specific of the model in which it is observed. However, the molecular mechanisms underlying these regulations, such as the fact that the phosphorylation of Ser184 might help to move helix α9 away from the core of the protein, are intrinsically present. In the next paragraph, we will summarize reports demonstrating that the regulation of Bax relocation and activation is more dynamic than what was previously proposed.

10.4 Regulation of Bax Mitochondrial Location by BH3-Only and Multi-domain Anti-apoptotic Bcl-2 Family Proteins

10.4.1 BH3-Only Proteins Favor Bax Translocation to Mitochondria

The BH3-only proteins are the sentinels that translate the survival and apoptotic signals emanating from throughout the cell. There are two functional classes of sentinels; they are direct activators or sensitizers (Fig. 10.1). Bid, Bim, and most probably Puma directly interact with the multi-domain pro-apoptotic Bax and Bak to cause a conformational change, which presumably triggers their oligomerization leading to MOMP (Luna-Vargas and Chipuk 2016; Sarosiek and Letai 2016). On the other end, Bad and Noxa are sensitizers which bind the hydrophobic pocket of the anti-apoptotic proteins like Bcl-2 to displace the normally sequestered direct activators, like Bim, or even possibly the pro-apoptotic effectors Bax and Bak (Luna-Vargas and Chipuk 2016; Sarosiek and Letai 2016). The elevation in liberated pro-apoptotic proteins shifts the balance within the family toward apoptosis. Nevertheless, Bax and Bak presumably remain inert until a direct activator sentinel induces the conformational change that allows their oligomerization. Yet another layer of regulation exists within the BH3-only group. For example, Bad must be dephosphorylated while Bid needs to be cleaved in order to assume their sentinel status as sensitizers and direct activators, respectively (Luna-Vargas and Chipuk 2016).

During apoptosis, inactive Bid is cleaved to generate a C-terminal truncated form referred to as tBid, which functions as a direct activator. The fragment tBid triggers oligomerization of both Bax and Bak in the mitochondrial outer membrane which causes cytochrome c release (Desagher et al. 1999; Wei et al. 2001). Furthermore, tBid can trigger oligomerization of recombinant monomeric Bax in artificial membranes (Dejean et al. 2005; Roucou et al. 2002). The oligomerization results in formation of voltage independent and slightly cationic channels with conductances of 1.5–10 nS, which can be detected by patch clamp techniques (Dejean et al. 2005). Moreover, cytochrome c is transported through these tBid-induced Bax channels, which makes them very similar to the electrophysiological manifestation of MOMP, i.e., the mitochondrial apoptosis-induced channel or MAC (Dejean et al. 2005). We used tBid to induce MAC activity in mitochondria in order to visualize formation of the pore of MAC in real time (Martinez-Caballero et al. 2009). Nanomolar concentrations of tBid catalyzed MAC formation and cytochrome *c* release in a matter of minutes. Interestingly, the amount of tBid needed to form MAC and release cytochrome *c* in mitochondria lacking Bax and/or Bak was different. MAC formed from Bak with an EC_{50} of about 20 nM tBid; but MAC needed >200 nM tBid to be formed from Bax. It is however possible that the requirement of less tBid to induce MAC was due to an increase of Bak expression levels in the Bax −/− KO cells used during this study (Martinez-Caballero et al. 2009). As expected, MAC did not form in mitochondria lacking both Bax and Bak. Quantitative analysis of the stepwise changes in conductance associated with tBid-induced MAC formation was consistent with pore assembly by a barrel-stave model. Assuming the stave is composed of two transmembrane α-helices in Bax and Bak, mature MAC pores would typically contain ~9 monomers and have diameters of 5.5–6 nm (Martinez-Caballero et al. 2009).

10.4.2 Bax Retrotranslocation from Mitochondria to the Cytosol

For long, it had been considered that Bax translocation from the cytosol to the mitochondria was a "one-way" process. This came from the observation that, in non-apoptotic cells, Bax localization is essentially diffuse in the cytosol while, following an apoptotic signal, it gets relocated to mitochondria (Wolter et al. 1997) and forms membrane-inserted oligomers that are responsible for MOMP. Since this process involves major conformational changes (Peyerl et al. 2007; Suzuki et al. 2000), it was hardly conceivable that these oligomers could be disassembled and that Bax could go back to its previously inactive state in the cytosol. However, significant observations suggested that the passage from soluble/monomeric Bax to membrane-inserted/oligomeric Bax is not a one-step process. For example, it has been shown that, during anoikis, Bax could be relocalized to mitochondria, but that the process was reversible (Owens et al. 2009). This, and other experiments, demonstrated that "mitochondrial Bax" was not obligatorily "membrane-inserted" Bax (Renault et al. 2013).

Structural studies have allowed designing a complex Bax mutant that has a constitutive mitochondrial localization but that cannot support the conformational change associated with the oligomerization, thus remaining incompetent for cytochrome c release. Under oxidative conditions, this mutant displays two internal disulfide bonds that constraint helices α1 and α2 and α1–α2 loop and α6, therefore blocking the "opening" of Bax structure associated to its activation (Edlich et al. 2011). When expressed in HCT-116 or HeLa cells, a GFP-tagged version of this mutant displayed a mitochondrial localization and, as expected, did not trigger apoptosis. Following the photobleaching of GFP fluorescence in cell nuclei, both nuclear and cytosolic fluorescence disappeared while mitochondrial fluorescence remained. This reflected the rapid dynamics of exchange between the nucleus and the cytosol and its absence for mitochondrial Bax. However, by following the reappearance of the fluorescence in the cytosol, the authors showed that a fraction of mitochondrial Bax could be retrotranslocated from the mitochondria to the cytosol. Most interestingly, this retrotranslocation process was greatly accelerated when the anti-apoptotic protein Bcl-xL was overexpressed (Edlich et al. 2011). This led the authors to conclude that, in non-apoptotic cells, (1) Bax subcellular localization followed a dynamic equilibrium between mitochondria and cytosol and that (2) anti-apoptotic proteins could displace this equilibrium toward a more cytosolic localization. An additional interesting observation was that a mutant of Bcl-xL deleted of the C-terminal α-helix (Bcl-xLΔC) was unable to promote Bax retrotranslocation, and further experiments demonstrated that the deletion of the four last residues of Bcl-xL was sufficient to prevent it (Todt et al. 2013). The question remained whether this process was only dependent on the intrinsic properties of interaction between the two proteins or also involved external factors (such as the presence of other Bcl-2 family members). Since yeast does not contain endogenous Bcl-2 family members, it is a powerful tool to answer this question. Bax was expressed alone, or co-expressed with full-length Bcl-xL, or truncated Bcl-xLΔC in a yeast mutant inactivated for the major vacuolar protease Pep4p (the yeast ortholog of cathepsin D) and in the presence of the general inhibitor of serine proteases phenylmethylsulfonyl fluoride (PMSF). Under these conditions, following the addition of an inhibitor of protein synthesis (cycloheximide), the cellular Bax content remained stable for at least 4 h. Mitochondria were then isolated to measure the evolution of Bax mitochondrial content following cycloheximide addition. Since the total content was stable, any decrease of the mitochondrial content would indicate retrotranslocation. When expressed alone, mitochondrial Bax levels remained constant. When co-expressed with Bcl-xL, mitochondrial Bax levels were decreased by 50% within 2 h. When co-expressed with Bcl-xLΔC, mitochondrial Bax levels remained constant. Similar experiments were done in mouse prolymphocytic cells FL5.12, with the same results, showing that the experiment in yeast actually led to observations relevant to mammalian physiology. These results were therefore consistent with the hypothesis that Bcl-xL (but not Bcl-xLΔC) was able to promote the retrotranslocation of Bax, when the two proteins are co-expressed in yeast. It also demonstrated that the retrotranslocation of Bax by Bcl-xL was only dependent on molecular properties of Bax and Bcl-xL, without interference from factors present in mammalian cells and absent from yeast, such as other Bcl-2 family members (Renault et al. 2015b). Importantly, no retrotranslocation was observed when a constitutively active mutant of Bax, carrying a substitution P168A (Arokium et al. 2004), was used. This supports the view that, once it is inserted and oligomerized (since it is able to promote the release of cytochrome c), Bax cannot be retrotranslocated.

10.4.3 Steady-State Bax Subcellular Localization

Intuitively, the existence of Bax retrotranslocation suggests that, in the presence of Bcl-xL, the steady-state Bax localization should be more cytosolic than when Bax is expressed alone. This can be correlated with the classical view of overexpressed Bcl-xL (or Bcl-2) sequestering Bax away from mitochondria, thus preventing apoptosis by restraining Bax ability to reach and permeabilize MOM. However, opposite to this prediction, we observed that the overexpression of Bcl-xL or of Bcl-2 increased the mitochondrial localization of endogenous Bax in FL5.12 cells (Renault et al. 2015b; Teijido and Dejean 2010). Strikingly, Bax mitochondrial content reached the same level as the one measured in parental cells committed to apoptosis through IL-3 removal. Although less marked, a similar effect was observed in HCT-116 cells overexpressing Bcl-xL, showing that it was not a particular property of FL5.12 cells. We then investigated if this apparently paradoxical effect of Bcl-xL on Bax mitochondrial localization was dependent only on the interaction between Bax and Bcl-xL or if other factors could be involved, by doing the experiments in yeast. We observed that Bax mitochondrial content was increased when Bcl-xL was co-expressed with Bax in the yeast model. This last result showed that the presence of BH3-only proteins was not essential to Bcl-xL-mediated Bax increase at the mitochondria and that this relocation was a general phenomenon depending only in the presence of Bax and Bcl-xL.

Most interestingly, we observed that the co-expression of Bcl-xLΔC induced an even higher increase of Bax mitochondrial localization in FL5.12 cells, in HCT-116 cells, and in yeast (Renault et al. 2015b). As discussed above, truncated Bcl-xLΔC is not able to promote Bax retrotranslocation. This suggested that Bcl-xL regulates Bax mitochondrial localization through two opposed processes: retrotranslocation, for which Bcl-xLΔC is impaired, and translocation, promoted both by full-length Bcl-xL and truncated Bcl-xLΔC.

10.4.4 Bcl-xL and Bcl-2 Differentially Affect MOMP Mechanisms in Ways Which Eventually Favor Bax Association to Mitochondrial Membranes

Previous to the observation that Bcl-xL was an inhibitor of Bax retrotranslocation, the overexpression of Bcl-2 had been shown to increase the mitochondrial localization of Bax in non-apoptotic FL5.12 cells (Teijido and Dejean 2010). However, in spite of this apparent similarity, the consequences of Bcl-xL or Bcl-2 overexpression on Bax conformation were clearly distinct. Both proteins induced the conformational change of the N-terminus of Bax, leading to recognition by the 6A7 antibody, and reveal the association of Bax to the MOM. But, while the overexpression of Bcl-2 also favored the formation of Bax oligomers (Renault et al. 2013), the overexpression of Bcl-xL did not (Renault et al. 2015b). It is yet difficult to draw any definitive conclusion about the consequences of this difference. Indeed, under both conditions, Bax remains inactive, being inhibited by its anti-apoptotic partner, and no MOMP was observed. But these observations emphasize the view that the effects of Bcl-2 and Bcl-xL overexpression on Bax apoptotic signaling are very distinct. This might be a crucial consideration to take for attempting to reactivate apoptosis in Bcl-2- or Bcl-xL-overexpressing tumoral cells.

Recent structural data brought major advances in the knowledge of Bax organization in the outer mitochondrial membrane. The interaction of BaxΔC (deprived of its C-terminal hydrophobic α-helix) with the BH3 domain of Bid favors the formation of a symmetrical Bax dimer, in which helices $\alpha 5$ and $\alpha 6$ would first lay flat on the membrane instead of being inserted as a hairpin and then would arrange a large pore by tilting in the membrane (Czabotar et al. 2014; Westphal et al. 2014). The formation of the pore as a juxtaposition of dimers has been proposed as an alternative to the membrane-inserted α-helices (Bleicken et al. 2014; Subburaj et al. 2015). This model is compatible with the recent microscopy images showing the ringlike organization of Bax in mitochondria of apoptotic cells (Grosse et al. 2016; Salvador-Gallego et al. 2016). It is also compatible with electrophysiological data suggesting that the pore formed by Bax may have different sizes (Martinez-Caballero et al. 2009).

In non-apoptotic FL5.12 cells overexpressing Bcl-2, the presence of oligomers similar in size to those observed in apoptotic cells has been observed (Renault et al. 2013), while no oligomer was detected in the same cells overexpressing Bcl-xL (Renault et al. 2015b). By correlating these observations to the structural model of Bax oligomers, one could speculate that Bcl-2 would favor the formation of Bax oligomers that remain laid down on, but only loosely bound to the membrane, while Bcl-xL prevents the formation of oligomers, may be by stabilizing Bax dimers. In any case, the disruption of the interaction between Bax and its anti-apoptotic partner would be sufficient to promote the organization of the pore-forming oligomer(s).

10.5 Bax Mitochondrial Relocation: Apoptotic and Metabolic Implications

10.5.1 The Upregulation of Anti-apoptotic Multi-domain Bcl-2 Family Proteins Primes Cancer Cells to Programmed Cell Death

Importantly, the anti-apoptotic proteins like Bcl-2 cause an accumulation of sentinel proteins, like Bim, in mitochondria (Certo et al. 2006). While overexpression of Bcl-2 suppresses apoptosis, the excess of sentinel proteins may be an Achilles heel that might be exploited to selectively target cancer cells to die. Imbalances in the interaction network of Bcl-2 family proteins, e.g., through the variation of the expression level of various members, likely underlie some degenerative diseases and are known to cause cancer. But these imbalances may also reveal therapeutic targets. For example, genetic events leading to overexpression of the Bcl-2 proto-oncogene suppress apoptosis and are associated with tumor formation and, more particularly, B-cell non-Hodgkin's lymphoma (Egle et al. 2004; Meijerink et al.

2005; Swanson et al. 2004). Prolymphocytic cell lines overexpressing Bcl-2 exhibit a resistance to mitochondrial apoptosis and induce lymphoma upon injection in mice (Goping et al. 1998; Gross et al. 1998; Meijerink et al. 2005; Pavlov et al. 2001). Hence, overexpression of Bcl-2 is associated with cancer. While these cells are more resistant to death, apoptotic stimuli cause the sequestration of sentinel proteins such as Bim in mitochondria (Certo et al. 2006; Del Gaizo Moore et al. 2008). This accumulation has been coined sensitization or "primed for death" and is illustrated in Fig. 10.2b. Small molecules that mimic the important BH3 domain can function as competitive inhibitors and bind to the hydrophobic cleft in Bcl-2. In doing so, these BH3 mimetics release sequestered pro-apoptotic proteins and the primed cells now die. Like most apoptotic stimuli, BH3 mimetics alone do not kill these cells. However, the BH3 mimetics act synergistically with chemotherapeutic agents (which provides the apoptosis signal) to trigger an increase in MAC formation to selectively kill cancer cells overexpressing anti-apoptotic proteins like Bcl-2. That is, BH3 mimetics, represented by ABT-737, are able to selectively kill these cancer cells by inhibiting the interactions, e.g., between Bcl-2 and Bim (Certo et al. 2006; Letai 2008; Oltersdorf et al. 2005) and triggering an increase of MOMP (Fig. 10.2b), a property potentially optimized by the activity of direct MAC activators which are yet to be found. It is important to note that ABT-737 inhibits all the known anti-apoptotic proteins except Mcl-1. Naturally, this primed to die approach will only work on cancer cells that survive because they overexpress anti-apoptotic proteins and accumulate pro-apoptotic proteins in mitochondria. Hence, development of personalized chemotherapy using BH3 mimetics will likely require prediagnosis of such an anti-apoptotic addiction through BH3 profiling (Certo et al. 2006; Letai 2008; Oltersdorf et al. 2005).

10.5.2 Bax Priming Is Enhanced in Cells Overexpressing Bcl-xLΔC or Bcl-xL

The immediate consequence of the increase in the steady-state level of mitochondrial Bax is a potentially higher ability to initiate apoptosis. However, the presence of Bcl-xL, and the interaction between the two proteins, obviously prevents Bax activation. This can be verified by the fact that IL-3 removal induced a lower level of apoptosis in Bcl-xL-overexpressing FL5.12 cells than in parental cells (Renault et al. 2015b). Here again, the inhibition only depends on the interaction between Bax and Bcl-xL since, in yeast, the effect of a constitutive mitochondrial and active mutant of Bax on mitochondrial permeabilization is also prevented by the co-expression of Bcl-xL. On the other hand, the overexpression of Bcl-xLΔC does not prevent the apoptosis of FL5.12 cells induced by the removal of IL-3 and does not prevent the permeabilization of yeast mitochondrial membrane by a constitutively active Bax mutant. In line with these observations, co-immunoprecipitation assays showed that the interaction between Bax and Bcl-xLΔC could not be depicted under conditions where the one between Bax and full-length Bcl-xL was detected. This demonstrates that the presence of the C-terminal α-helix of Bcl-xL is required for a stable interaction and subsequent efficient inhibition of Bax (Renault et al. 2015b).

Furthermore, we observed that the expression of Bcl-xLΔC promoted Bax-driven MOMP in yeast. This rather unexpected effect was dependent on the interaction between the two proteins, since it was abrogated by the canonical G138A mutation in the BH1 domain (Renault et al. 2015b), which is known to fully prevent the interaction between Bax and Bcl-xL (Ottilie et al. 1997). These results showed that, although it was not detectable by co-immunoprecipitation, a transient interaction between Bax and Bcl-xLΔC occurred and was sufficient to increase mitochondrial Bax content, but was not sufficient to prevent Bax-induced permeabilization. This was not observed with full-length Bcl-xL since its interaction with Bax, while inducing translocation of the pro-apoptotic protein, was stable enough to also keep it inactive. Since Bcl-xLΔC is not a natural variant of the protein, we searched for experimental conditions which could reveal that full-length Bcl-xL is also able to prime Bax. Experiments on pure recombinant proteins had previously shown that the binding, then release, of Bax and Bcl-xL could initiate the active conformation of Bax that can be depicted with the 6A7 monoclonal antibody (Hsu and Youle 1997). The release could be induced by peptides corresponding to the BH3 domain of canonical BH3-only proteins, such as Puma, or by BH3-mimetic molecules such as ABT-737 (Gautier et al. 2011). Therefore we used ABT-737 to disrupt Bax/Bcl-xL interaction both in the FL5.12 and the yeast models. The addition of ABT-737 to parental FL5.12 cells did not induce apoptosis, in agreement with the observations from (Certo et al. 2006). However, the addition of ABT-737 to FL5.12 cells overexpressing Bcl-xL did induce a significant level of apoptosis (Renault et al. 2015b). We also observed that ABT-737 had no effect on the biomass formation of yeast cells expressing Bax alone, or Bcl-xL alone, but that it significantly decreased the biomass formation of cells co-expressing Bax and Bcl-xL. Although this result cannot be simply interpreted in terms of Bax-induced MOMP, for the obvious reason that yeast does not naturally express canonical Bcl-2 family proteins (Kissova et al. 2006), it is nevertheless in agreement with the effect of ABT-737 we observed in Bcl-xL-overexpressing FL5.12 cells (Renault et al. 2015b).

These experiments show that, both in FL5.12 cells and in yeast, the transient interaction between Bax and Bcl-xL followed by the release of the interaction promotes the activation of Bax, like previously reported using pure recombinant proteins. This has interesting consequences on the potential efficiency of BH3-mimetic molecules in antitumoral therapy. These molecules were designed to prevent the inhibition of apoptosis by anti-apoptotic Bcl-2 family members (essentially Bcl-2, Bcl-xL, and Mcl-1), which are overexpressed in most cancer cells (see, e.g., Castilla et al. 2006; Pena et al. 1998; Tang et al. 1998). The fact that, by breaking the interaction between Bax and Bcl-2/xL, not only they liberate Bax, but they further activate it, renders these molecules even more desirable as antitumor therapy, provided that Bax is still expressed and not mutated.

10.5.3 A Possible Correlation Between Bax Phosphorylation and Its Interaction with Bcl-xL

Combined together, the results presented above indicate that Bcl-xL is able to stimulate Bax translocation to the MOM, but also Bax retrotranslocation to the cytosol. However, this putative futile cycle would appear to be shifted toward a net increase of Bax activation and mitochondrial relocation when the expression levels of either Bcl-2 or Bcl-xL are high (Renault et al. 2015b; Teijido and Dejean 2010). The consequence of these effects is the establishment of a steady state for which the mitochondrial membranes of Bcl-2/xL-overexpressing cells are "overloaded" with a high mitochondrial content of "primed" Bax, ready to trigger apoptosis. This might make the system more reactive than if Bax was only sequestered away from mitochondria, since Bax is already on target, ready to be activated after the release of its interaction with Bcl-2/Bcl-xL. However, the efficiency of such a "dynamic cycling" model implies that the interaction between Bax and Bcl-xL has different properties depending if it happens in the cytosol or in the MOM. Interestingly, these different properties might be directly related with differences in the chemical compositions of the cytosolic vs mitochondrial form of Bax (Fig. 10.3).

The study of the phosphorylation of S184 has attracted much interest because this residue is located in Bax C-terminal α-helix, a structure for which the conformational change is crucial to initiate Bax relocalization (see Sect. 10.2). It had been shown that the phosphorylation of S184 retained Bax away from mitochondria, while its dephosphorylation has the opposite effect (Gardai et al. 2004). This could be reproduced in

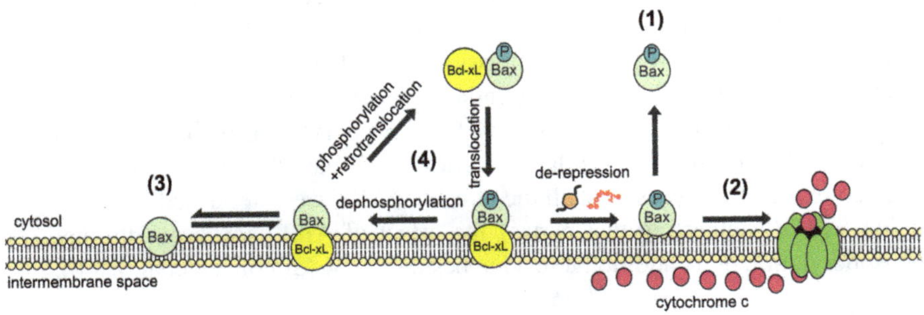

Fig. 10.3 Regulation of Bax mitochondrial localization through phosphorylation and interaction with Bcl-xL (*1*) When Bax is phosphorylated on S184, it is spontaneously mostly located in the cytosol. However, the small fraction that remains in the mitochondrial membrane is able to oligomerize to form a pore that promotes the release of cytochrome c (*2*). (*3*) When Bax is not phosphorylated, it is spontaneously mostly located in the mitochondrial membrane, but is unable to oligomerize to form the pore. (*4*) In the presence of Bcl-xL, phosphorylated Bax and Bcl-xL are conveyed together to the membrane where the high stability of the interaction prevents the activation of Bax. Although the process is reversible through retrotranslocation, the system is favored toward Bax mitochondrial localization through the possible dephosphorylation of Bax. Conversely, in the presence of de-repressors of the interaction between Bax and Bcl-xL (BH3-only proteins, such as Bim, or BH3-mimetic drugs, such as ABT-737), Bax is able to form a pore with great efficiency because it is already present in great amount in the membrane

yeast where the mutants S184A/V have a higher mitochondrial localization than the mutant S184D (Arokium et al. 2007; Simonyan et al. 2016). However, this did not fully correlate with the ability of these mutants to induce MOMP. Indeed, the mutant S184D was surprisingly more efficient than mutants S184A/V (and wild-type Bax) to promote the release of cytochrome c. These results suggest that the intrinsic capacity of the phosphomimetic mutant to induce MOMP was higher than that of the non-phosphorylatable mutant and that was apparently in contradiction with the fact that Bax phosphorylated on S184 is less able to induce apoptosis that its non-phosphorylated counterpart (Gardai et al. 2004). Part of the answer to this paradox might be related to the fact that, in mammalian cells, Bax is not expressed alone, but in the presence of, namely, anti-apoptotic proteins. Bax phosphomimetic and non-phosphorylatable mutants were then co-expressed with full-length Bcl-xL or truncated Bcl-xLΔC (Simonyan et al. 2016). It was observed that the mitochondrial localization of the phosphomimetic mutant S184D did not differ from that of wild-type Bax, with an increase induced by Bcl-xL and a further increase induced by Bcl-xLΔC, linked to the absence of retrotranslocation, as described above. Interestingly, the behavior of the non-phosphorylatable mutant S184A differed: its mitochondrial content was increased by Bcl-xL, like that of wild-type Bax and of the phosphomimetic mutant S184D, but it was not further increased, and was even about twice less increased, by Bcl-xLΔC. This suggests that retrotranslocation did not have any effect on the mitochondrial localization of non-phosphorylatable Bax (Simonyan et al. 2016).

In mammalian cells, a correlation between the phosphorylation of S184 and the retrotranslocation of Bax by Bcl-xL had also been observed (Schellenberg et al. 2013): in this context, a mutant GFP-Bax-S184V was retrotranslocated slower than GFP-BaxWT, paralleling the observations in yeast discussed above. Also, the overexpression of AKT stimulated the retrotranslocation of GFP-BaxWT. However, this study did not mention if the overexpression of AKT had also an effect on the retrotranslocation of GFP-Bax-S184V or not. It cannot be then yet ascertained that this effect of AKT and the consequence of the phosphorylation of S184 on Bax retrotranslocation are strictly correlated. It should be noted that, when Bax and AKT were co-expressed in yeast, not all the effects of AKT were abolished by the substitution S184V, suggesting that part of the effect of AKT may occur through the phosphorylation of an additional residue besides S184 (Simonyan et al. 2016).

10.5.4 Is There a Direct Link Between Bax Mitochondrial Relocation and Metabolic Regulation?

There are many hallmarks associated with cancer cells. They include "sustaining proliferative signaling, evading growth suppressors, resisting cell death, enabling replicative immortality, inducing angiogenesis, metastasis, evading immune destruction and reprogramming of energy metabolism" (Hanahan and Weinberg 2011). A cancer hallmark that has always been of great interest are metabolic shifts which can be observed in most tumor cells (Hsu and Sabatini 2008). Otto Warburg's group discovered in the early 1920s that fuel metabolism in cancer cells tends to rely more on lactic fermentation, a phenomenon which happened to be later named "the Warburg effect" (Vander Heiden et al. 2009). This in return helps the cancer cells survive in various environments and more particularly hypoxia (Moreno-Sanchez et al. 2007). However, this effect does not seem to be due only to a strict increase of lactic fermentation vs oxidative phosphorylation in all cancer cells: for instance, if a clear fermentative shift may be observed in insulinoma RINm5F cells, a parallel increase in both glycolytic and oxidative phosphorylation fluxes is observed in C6 glioma cells (Martin et al. 1998; Moreno-Sanchez et al. 2007). Finally, no clear picture has yet emerged in terms of the molecular definition of metabolic shifts in lymphoma cells. Overexpression of the anti-apoptotic Bcl-2 has been reported in many tumor types and is highly correlated with poor survival and resistance to chemotherapy (Lessene et al. 2008). It is worth noting that the BH3-only Bcl-2 family protein Bad was recently shown to be a master regulator of the balance between oxidative glycolysis and gluconeogenesis in normal hepatocytes (Gimenez-Cassina et al. 2014). It is actually worth noting that Bad exerts its regulatory function in the form of a glucokinase-containing complex loosely bound to the mitochondria and that phosphorylation of Bad by survival kinases (such as AKT) is essential in this process (Gimenez-Cassina and Danial 2015). It was also observed that Bcl-xL overexpression in normal neurons leads to a decrease of mitochondrial proton leak, a phenomenon which was attributed to physical interactions between the c subunit of the ATP synthase and an inner membrane embedded form of Bcl-xL (Jonas 2014). The activity of the mitochondrial enzyme cytochrome c oxidase was also positively correlated with Bcl-2 expression levels in isolated mitochondria from several cancer models (Chen and Pervaiz 2010; Moreno-Sanchez et al. 2007). However, there is surprisingly only a very few molecular and mechanistic data on the effects of Bcl-2 expression levels on whole-cell bioenergetics and metabolism. Our laboratory previously showed that, in pre-lymphocytes B, Bax translocation to mitochondria increased in response to an overexpression of Bcl-2 (Teijido and Dejean 2010). We also recently observed that Bcl-2 or Bcl-xL overexpression caused a significant change to the ^{1}H-NMR-defined metabolomic pattern of FL5.12 cells (Fig. 10.4). Interestingly, both this Bcl-2-dependent Bax relocation and metabolic changes were strongly attenuated if a Bcl-2 mutant of interaction with Bax (Bcl-2 G145E) was overexpressed instead of native Bcl-2 (Fig. 10.4 and Teijido and Dejean 2010). All these studies lead to our hypothesis that carbohydrate metabolism can be directly regulated by anti-apoptotic multi-domain Bcl-2 family proteins. They also suggest that the ability of Bcl-2 to interact with Bax and the increase of the abundance of mitochondrial membrane loosely bound Bax-containing complexes in Bcl-2-overexpressing cells are essential during this regulatory process.

10.6 Concluding Remarks

Bcl-2/xL are anti-apoptotic proteins that eventually insert into the MOM during apoptosis. It is well established that an excess of Bcl-2/xL inhibits apoptosis; this inhibition is requiring the ability of Bcl-2/xL to physically interact with Bax and to

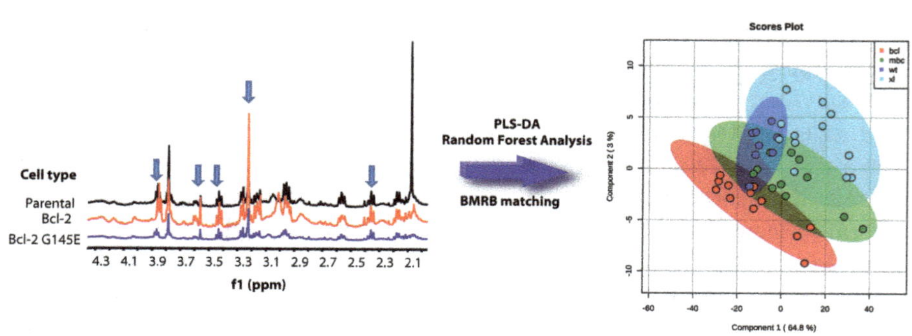

Fig. 10.4 Effects of Bcl-2/xL on ^{1}H-NMR-defined metabolomes 400 MHz ^{1}H NMR spectra of three different intracellular metabolites samples from whole FL5.12 cell lines obtained after acetonitrile extraction (Odunsi et al. 2005). f1 (in ppm) is relative to the reference peak of TSP (set arbitrarily at 0). Regions of the spectrum showing difference in the measured metabolites are shown by *vertical blue arrows*. Data are then binned, cluster analyzed, and matched for metabolite identification using BMRB database (Izquierdo-Garcia et al. 2011). Comparison of metabolomic cloud profiles show significant differences between the metabolomics profiles of Parental (wt), Bcl-2-overexpressing (bcl), and Bcl-xL-overexpressing (xl) FL5.12 cell lines. Of note, these differences tend to be cancelled out if Bcl-2-G145E (mbc), a Bcl-2 mutant of interaction with Bax, is overexpressed instead of native Bcl-2 (compare the *red* and *dark blue* cloud patterns on the *right* side of the figure)

sequester Bim at the mitochondria (Letai 2008). Bax oligomerization and cytochrome c release through the mitochondrial channel MAC are also inhibited by Bcl-2/xL overexpression (Antonsson et al. 2001; Pavlov et al. 2001). We recently found that Bax translocation in mitochondria was increased in response to Bcl-2 or Bcl-xL overexpression (Fig. 10.3). This observation suggests that an excess of Bcl-2/xL triggers not only a mitochondrial accumulation of pro-apoptotic sentinel proteins such as Bim but also of pro-apoptotic effector proteins such as Bax. Considering that Bcl-2 overexpression does not inhibit Bax interaction with its putative receptor Tom22 (Renault et al. 2012), these results strongly suggest that Bcl-2/xL upregulation leads to the accumulation of MAC precursors as a consequence of inhibiting MAC formation. Finally, Bcl-2 and Bcl-xL are positive effectors of mitochondrial respiration and mass, respectively (Berman et al. 2009; Chen and Pervaiz 2010). Our recent results show that Bcl-2 effects on Bax translocation and carbohydrate metabolomics depend on the ability of Bcl-2 to bind Bax (Teijido and Dejean 2010 and Fig. 10.4). Taken together, these findings reinforce the hypothesis that several members of the Bcl-2 family proteins can behave as master co-regulators of both cell death and energy metabolism. Part of the molecular mechanisms underlying this co-regulation involves shifts in steady states of Bax locations; i.e., shifts which tend toward a net increase of Bax mitochondrial content (and change of conformation) are observed in prolymphocytes and yeast cells when Bcl-2 or Bcl-xL is constitutively overexpressed (Renault et al. 2013, 2015b). A model illustrating the details of these possible chain of events is shown in Fig. 10.5. In summary, Bcl-2 and Bcl-xL have long been associated with inhibition of cell

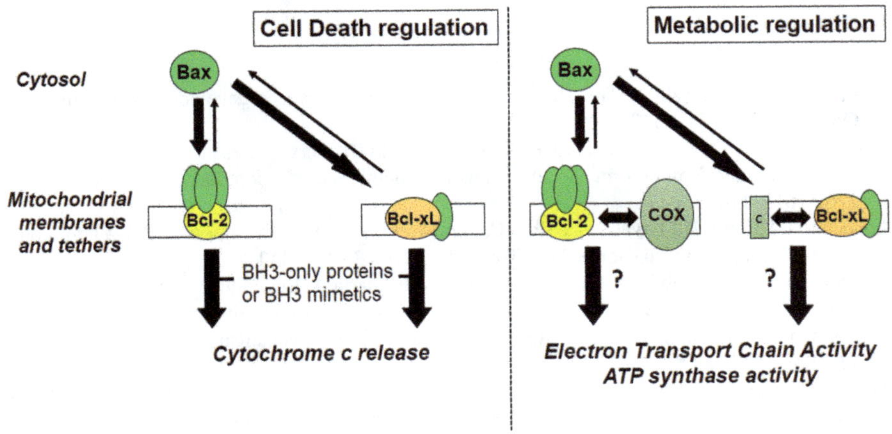

Fig. 10.5 Bcl-2 and Bcl-xL inhibit apoptosis and regulate metabolism via different molecular mechanisms *Left*. Bcl-2/xL overexpression results in Bax translocation and change of conformation (pre-activation) at the mitochondrial membrane level. During this step, Bax, in combination with Bcl-2, forms MOMP-machinery precursors if Bcl-2 is overexpressed; when Bcl-xL overexpression leads to the accumulation of membrane-embedded heterodimers Bax/Bcl-xL which are cytochrome c release incompetent. When the interactions between Bax and Bcl-2 or Bcl-xL are inhibited by BH3-only proteins or the drug ABT-737, cytochrome c release occurs and the cell dies by apoptosis. *Right*. Potential mechanistic link between apoptotic and metabolic regulations by anti-apoptotic Bcl-2 family proteins. Bcl-2/COX and Bcl-xL/ATP synthase c subunit interactions at the mitochondrial membranes and/or tethers are observed in response to Bcl-2 and Bcl-xL overexpression. Bcl-2/COX interaction and Bcl-xL/ATP synthase c subunit interaction have been associated with an increase of cytochrome c oxidase and efficiency of the ATP synthase activities, respectively. We hypothesize that Bax relocation triggered by Bcl-2/xL overexpression is important to Bcl-2/xL-mediated metabolic regulation. The rationale of our hypothesis reposes on the facts that both Bax accumulation and metabolomics pattern are cancelled if Bcl-2 overexpression is replaced by the overexpression of a Bcl-2 mutant which has lost its ability to interact with Bax (see Fig. 10.4)

death. However, the study of the molecular mechanisms associated with their functions in the definition of both cell metabolism and cell fate is still in its infancy. Therefore, a deeper understanding of these cross regulatory processes should provide essential clues to improve and/or define therapeutic strategies against selective forms of cancer associated with abnormal increased amounts of Bcl-2 and/or Bcl-xL (e.g., certain forms of blood cancer such as non-Hodgkin's lymphoma (Meijerink et al. 2005)).

Acknowledgments The work in the labs of the authors has been supported by the CNRS, the ANR, and the Université de Bordeaux (to S.M.) and by the CSUPERB and the California State University of Fresno (to L.D.). L.D. would also like to thank Dr. Krish Krishnan and Rhaul Llanos from the California State University of Fresno for their help with the NMR experiments presented in Fig. 10.4.

References

Adams JM, Cory S (2007) The Bcl-2 apoptotic switch in cancer development and therapy. Oncogene 26:1324–1337

Antonsson B, Montessuit S, Sanchez B, Martinou JC (2001) Bax is present as a high molecular weight oligomer/complex in the mitochondrial membrane of apoptotic cells. J Biol Chem 276:11615–11623

Aouacheria A, Rech De Laval V, Combet C, Hardwick JM (2013) Evolution of Bcl-2 homology motifs: homology versus homoplasy. Trends Cell Biol 23:103–111

Arnaud E, Ferri KF, Thibaut J, Haftek-Terreau Z, Aouacheria A, Le Guellec D, Lorca T, Gillet G (2006) The zebrafish Bcl-2 homologue Nrz controls development during somitogenesis and gastrulation via apoptosis-dependent and -independent mechanisms. Cell Death Differ 13:1128–1137

Arokium H, Camougrand N, Vallette FM, Manon S (2004) Studies of the interaction of substituted mutants of BAX with yeast mitochondria reveal that the C-terminal hydrophobic alpha-helix is a second ART sequence and plays a role in the interaction with anti-apoptotic BCL-xL. J Biol Chem 279:52566–52573

Arokium H, Ouerfelli H, Velours G, Camougrand N, Vallette FM, Manon S (2007) Substitutions of potentially phosphorylatable serine residues of Bax reveal how they may regulate its interaction with mitochondria. J Biol Chem 282:35104–35112

Baines CP, Kaiser RA, Purcell NH, Blair NS, Osinska H, Hambleton MA, Brunskill EW, Sayen MR, Gottlieb RA, Dorn GW, Robbins J, Molkentin JD (2005) Loss of cyclophilin D reveals a critical role for mitochondrial permeability transition in cell death. Nature 434:658–662

Bellot G, Cartron PF, Er E, Oliver L, Juin P, Armstrong LC, Bornstein P, Mihara K, Manon S, Vallette FM (2007) TOM22, a core component of the mitochondria outer membrane protein translocation pore, is a mitochondrial receptor for the proapoptotic protein Bax. Cell Death Differ 14:785–794

Berman SB, Chen YB, Qi B, Mccaffery JM, Rucker EB 3rd, Goebbels S, Nave KA, Arnold BA, Jonas EA, Pineda FJ, Hardwick JM (2009) Bcl-x L increases mitochondrial fission, fusion, and biomass in neurons. J Cell Biol 184:707–719

Bleicken S, Jeschke G, Stegmueller C, Salvador-Gallego R, Garcia-Saez AJ, Bordignon E (2014) Structural model of active Bax at the membrane. Mol Cell 56:496–505

Breitschopf K, Zeiher AM, Dimmeler S (2000) Ubiquitin-mediated degradation of the proapoptotic active form of bid. A functional consequence on apoptosis induction. J Biol Chem 275:21648–21652

Buttner S, Ruli D, Vogtle FN, Galluzzi L, Moitzi B, Eisenberg T, Kepp O, Habernig L, Carmona-Gutierrez D, Rockenfeller P, Laun P, Breitenbach M, Khoury C, Frohlich KU, Rechberger G, Meisinger C, Kroemer G, Madeo F (2011) A yeast BH3-only protein mediates the mitochondrial pathway of apoptosis. EMBO J 30:2779–2792

Cartron PF, Arokium H, Oliver L, Meflah K, Manon S, Vallette FM (2005) Distinct domains control the addressing and the insertion of Bax into mitochondria. J Biol Chem 280:10587–10598

Cartron PF, Bellot G, Oliver L, Grandier-Vazeille X, Manon S, Vallette FM (2008) Bax inserts into the mitochondrial outer membrane by different mechanisms. FEBS Lett 582:3045–3051

Castilla C, Congregado B, Chinchon D, Torrubia FJ, Japon MA, Saez C (2006) Bcl-xL is overexpressed in hormone-resistant prostate cancer and promotes survival of LNCaP cells via interaction with proapoptotic Bak. Endocrinology 147:4960–4967

Cebulski J, Malouin J, Pinches N, Cascio V, Austriaco N (2011) Yeast Bax inhibitor, Bxi1p, is an ER-localized protein that links the unfolded protein response and programmed cell death in *Saccharomyces cerevisiae*. PLoS One 6:e20882

Certo M, Del Gaizo Moore V, Nishino M, Wei G, Korsmeyer S, Armstrong SA, Letai A (2006) Mitochondria primed by death signals determine cellular addiction to antiapoptotic BCL-2 family members. Cancer Cell 9:351–365

Chen ZX, Pervaiz S (2010) Involvement of cytochrome c oxidase subunits Va and Vb in the regulation of cancer cell metabolism by Bcl-2. Cell Death Differ 17:408–420

Chen MC, Gong HY, Cheng CY, Wang JP, Hong JR, Wu JL (2000) Cloning and characterization of a novel nuclear Bcl-2 family protein, zfMcl-1a, in zebrafish embryo. Biochem Biophys Res Commun 279:725–731

Chen L, Willis SN, Wei A, Smith BJ, Fletcher JI, Hinds MG, Colman PM, Day CL, Adams JM, Huang DC (2005) Differential targeting of prosurvival Bcl-2 proteins by their BH3-only ligands allows complementary apoptotic function. Mol Cell 17:393–403

Chou JJ, Li H, Salvesen GS, Yuan J, Wagner G (1999) Solution structure of BID, an intracellular amplifier of apoptotic signaling. Cell 96:615–624

Colussi PA, Quinn LM, Huang DC, Coombe M, Read SH, Richardson H, Kumar S (2000) Debcl, a proapoptotic Bcl-2 homologue, is a component of the *Drosophila melanogaster* cell death machinery. J Cell Biol 148:703–714

Czabotar PE, Westphal D, Dewson G, Ma S, Hockings C, Fairlie WD, Lee EF, Yao S, Robin AY, Smith BJ, Huang DC, Kluck RM, Adams JM, Colman PM (2013) Bax crystal structures reveal how BH3 domains activate Bax and nucleate its oligomerization to induce apoptosis. Cell 152:519–531

Czabotar PE, Lessene G, Strasser A, Adams JM (2014) Control of apoptosis by the BCL-2 protein family: implications for physiology and therapy. Nat Rev Mol Cell Biol 15:49–63

Datta SR, Dudek H, Tao X, Masters S, Fu H, Gotoh Y, Greenberg ME (1997) Akt phosphorylation of BAD couples survival signals to the cell-intrinsic death machinery. Cell 91:231–241

Dejean LM, Martinez-Caballero S, Guo L, Hughes C, Teijido O, Ducret T, Ichas F, Korsmeyer SJ, Antonsson B, Jonas EA, Kinnally KW (2005) Oligomeric Bax is a component of the putative cytochrome c release channel MAC, mitochondrial apoptosis-induced channel. Mol Biol Cell 16:2424–2432

Del Gaizo Moore V, Schlis KD, Sallan SE, Armstrong SA, Letai A (2008) BCL-2 dependence and ABT-737 sensitivity in acute lymphoblastic leukemia. Blood 111:2300–2309

Del Mar Martinez-Senac M, Corbalan-Garcia S, Gomez-Fernandez JC (2000) Study of the secondary structure of the C-terminal domain of the antiapoptotic protein bcl-2 and its interaction with model membranes. Biochemistry 39:7744–7752

Del Peso L, Gonzalez-Garcia M, Page C, Herrera R, Nunez G (1997) Interleukin-3-induced phosphorylation of BAD through the protein kinase Akt. Science 278:687–689

Delbridge AR, Strasser A (2015) The BCL-2 protein family, BH3-mimetics and cancer therapy. Cell Death Differ 22:1071–1080

Desagher S, Osen-Sand A, Nichols A, Eskes R, Montessuit S, Lauper S, Maundrell K, Antonsson B, Martinou JC (1999) Bid-induced conformational change of Bax is responsible for mitochondrial cytochrome c release during apoptosis. J Cell Biol 144:891–901

Desagher S, Osen-Sand A, Montessuit S, Magnenat E, Vilbois F, Hochmann A, Journot L, Antonsson B, Martinou JC (2001) Phosphorylation of bid by casein kinases I and II regulates its cleavage by caspase 8. Mol Cell 8:601–611

Edlich F, Banerjee S, Suzuki M, Cleland MM, Arnoult D, Wang C, Neutzner A, Tjandra N, Youle RJ (2011) Bcl-x(L) retrotranslocates Bax from the mitochondria into the cytosol. Cell 145:104–116

Egle A, Harris AW, Bath ML, O'Reilly L, Cory S (2004) VavP-Bcl2 transgenic mice develop follicular lymphoma preceded by germinal center hyperplasia. Blood 103:2276–2283

Ellis HM, Horvitz HR (1986) Genetic control of programmed cell death in the nematode *C. elegans*. Cell 44:817–829

Farooq M, Kim Y, Im S, Chung E, Hwang S, Sohn M, Kim M, Kim J (2001) Cloning of BNIP3h, a member of proapoptotic BNIP3 family genes. Exp Mol Med 33:169–173

Forde JE, Dale TC (2007) Glycogen synthase kinase 3: a key regulator of cellular fate. Cell Mol Life Sci 64:1930–1944

Garcia-Saez AJ, Mingarro I, Perez-Paya E, Salgado J (2004) Membrane-insertion fragments of Bcl-xL, Bax, and Bid. Biochemistry 43:10930–10943

Gardai SJ, Hildeman DA, Frankel SK, Whitlock BB, Frasch SC, Borregaard N, Marrack P, Bratton DL, Henson PM (2004) Phosphorylation of Bax Ser184 by Akt regulates its activity and apoptosis in neutrophils. J Biol Chem 279:21085–21095

Gautier F, Guillemin Y, Cartron PF, Gallenne T, Cauquil N, Le Diguarher T, Casara P, Vallette FM, Manon S, Hickman JA, Geneste O, Juin P (2011) Bax activation by engagement with, then release from, the BH3 binding site of Bcl-xL. Mol Cell Biol 31:832–844

Gavathiotis E, Reyna DE, Bellairs JA, Leshchiner ES, Walensky LD (2012) Direct and selective small-molecule activation of proapoptotic BAX. Nat Chem Biol 8:639–645

Gimenez-Cassina A, Danial NN (2015) Regulation of mitochondrial nutrient and energy metabolism by BCL-2 family proteins. Trends Endocrinol Metab 26:165–175

Gimenez-Cassina A, Garcia-Haro L, Choi CS, Osundiji MA, Lane EA, Huang H, Yildirim MA, Szlyk B, Fisher JK, Polak K, Patton E, Wiwczar J, Godes M, Lee DH, Robertson K, Kim S, Kulkarni A, Distefano A, Samuel V, Cline G, Kim YB, Shulman GI, Danial NN (2014) Regulation of hepatic energy metabolism and gluconeogenesis by BAD. Cell Metab 19:272–284

Gold MR, Ingham RJ, Mcleod SJ, Christian SL, Scheid MP, Duronio V, Santos L, Matsuuchi L (2000) Targets of B-cell antigen receptor signaling: the phosphatidylinositol 3-kinase/Akt/glycogen synthase kinase-3 signaling pathway and the Rap1 GTPase. Immunol Rev 176:47–68

Goping IS, Gross A, Lavoie JN, Nguyen M, Jemmerson R, Roth K, Korsmeyer SJ, Shore GC (1998) Regulated targeting of BAX to mitochondria. J Cell Biol 143:207–215

Gross A, Jockel J, Wei MC, Korsmeyer SJ (1998) Enforced dimerization of BAX results in its translocation, mitochondrial dysfunction and apoptosis. EMBO J 17:3878–3885

Grosse L, Wurm CA, Bruser C, Neumann D, Jans DC, Jakobs S (2016) Bax assembles into large ring-like structures remodeling the mitochondrial outer membrane in apoptosis. EMBO J 35:402–413

Hanahan D, Weinberg RA (2011) Hallmarks of cancer: the next generation. Cell 144:646–674

Hengartner MO, Horvitz HR (1994) *C. elegans* cell survival gene ced-9 encodes a functional homolog of the mammalian proto-oncogene bcl-2. Cell 76:665–676

Hockenbery DM, Oltvai ZN, Yin XM, Milliman CL, Korsmeyer SJ (1993) Bcl-2 functions in an antioxidant pathway to prevent apoptosis. Cell 75:241–251

Hsu PP, Sabatini DM (2008) Cancer cell metabolism: Warburg and beyond. Cell 134:703–707

Hsu YT, Youle RJ (1997) Nonionic detergents induce dimerization among members of the Bcl-2 family. J Biol Chem 272:13829–13834

Hsu YT, Youle RJ (1998) Bax in murine thymus is a soluble monomeric protein that displays differential detergent-induced conformations. J Biol Chem 273:10777–10783

Izquierdo-Garcia JL, Villa P, Kyriazis A, Del Puerto-Nevado L, Perez-Rial S, Rodriguez I, Hernandez N, Ruiz-Cabello J (2011) Descriptive review of current NMR-based metabolomic data analysis packages. Prog Nucl Magn Reson Spectrosc 59:263–270

Jho E, Lomvardas S, Costantini F (1999) A GSK3beta phosphorylation site in axin modulates interaction with beta-catenin and Tcf-mediated gene expression. Biochem Biophys Res Commun 266:28–35

Jonas EA (2014) Contributions of Bcl-xL to acute and long term changes in bioenergetics during neuronal plasticity. Biochim Biophys Acta 1842:1168–1178

Kaufmann T, Schlipf S, Sanz J, Neubert K, Stein R, Borner C (2003) Characterization of the signal that directs Bcl-x(L), but not Bcl-2, to the mitochondrial outer membrane. J Cell Biol 160:53–64

Kelekar A, Thompson CB (1998) Bcl-2-family proteins: the role of the BH3 domain in apoptosis. Trends Cell Biol 8:324–330

Kerr JF, Wyllie AH, Currie AR (1972) Apoptosis: a basic biological phenomenon with wide-ranging implications in tissue kinetics. Br J Cancer 26:239–257

Kim TH, Zhao Y, Ding WX, Shin JN, He X, Seo YW, Chen J, Rabinowich H, Amoscato AA, Yin XM (2004) Bid-cardiolipin interaction at mitochondrial contact site contributes to mitochondrial cristae reorganization and cytochrome C release. Mol Biol Cell 15:3061–3072

Kissova I, Plamondon LT, Brisson L, Priault M, Renouf V, Schaeffer J, Camougrand N, Manon S (2006) Evaluation of the roles of apoptosis, autophagy, and mitophagy in the loss of plating efficiency induced by Bax expression in yeast. J Biol Chem 281:36187–36197

Kratz E, Eimon PM, Mukhyala K, Stern H, Zha J, Strasser A, Hart R, Ashkenazi A (2006) Functional characterization of the Bcl-2 gene family in the zebrafish. Cell Death Differ 13:1631–1640

Kvansakul M, Hinds MG (2015) The Bcl-2 family: structures, interactions and targets for drug discovery. Apoptosis 20:136–150

Lessene G, Czabotar PE, Colman PM (2008) BCL-2 family antagonists for cancer therapy. Nat Rev Drug Discov 7:989–1000

Lessene G, Czabotar PE, Sleebs BE, Zobel K, Lowes KN, Adams JM, Baell JB, Colman PM, Deshayes K, Fairbrother WJ, Flygare JA, Gibbons P, Kersten WJ, Kulasegaram S, Moss RM, Parisot JP, Smith BJ, Street IP, Yang H, Huang DC, Watson KG (2013) Structure-guided design of a selective BCL-X(L) inhibitor. Nat Chem Biol 9:390–397

Letai AG (2008) Diagnosing and exploiting cancer's addiction to blocks in apoptosis. Nat Rev Cancer 8:121–132

Liang XH, Jackson S, Seaman M, Brown K, Kempkes B, Hibshoosh H, Levine B (1999) Induction of autophagy and inhibition of tumorigenesis by beclin 1. Nature 402:672–676

Linseman DA, Butts BD, Precht TA, Phelps RA, Le SS, Laessig TA, Bouchard RJ, Florez-Mcclure ML, Heidenreich KA (2004) Glycogen synthase kinase-3beta phosphorylates Bax and promotes its mitochondrial localization during neuronal apoptosis. J Neurosci 24:9993–10002

Luna-Vargas MP, Chipuk JE (2016) Physiological and pharmacological control of BAK, BAX, and beyond. Trends Cell Biol 26:906

Luo X, Budihardjo I, Zou H, Slaughter C, Wang X (1998) Bid, a Bcl2 interacting protein, mediates cytochrome c release from mitochondria in response to activation of cell surface death receptors. Cell 94:481–490

Maiuri MC, Le Toumelin G, Criollo A, Rain JC, Gautier F, Juin P, Tasdemir E, Pierron G, Troulinaki K, Tavernarakis N, Hickman JA, Geneste O, Kroemer G (2007) Functional and physical interaction between Bcl-X(L) and a BH3-like domain in Beclin-1. EMBO J 26:2527–2539

Martin M, Beauvoit B, Voisin PJ, Canioni P, Guerin B, Rigoulet M (1998) Energetic and morphological plasticity of C6 glioma cells grown on 3-D support; effect of transient glutamine deprivation. J Bioenerg Biomembr 30:565–578

Martinez-Caballero S, Dejean LM, Kinnally MS, Oh KJ, Mannella CA, Kinnally KW (2009) Assembly of the mitochondrial apoptosis-induced channel, MAC. J Biol Chem 284:12235–12245

Marzo I, Brenner C, Zamzami N, Jurgensmeier JM, Susin SA, Vieira HL, Prevost MC, Xie Z, Matsuyama S, Reed JC, Kroemer G (1998) Bax and adenine nucleotide translocator cooperate in the mitochondrial control of apoptosis. Science 281:2027–2031

Maundrell K, Antonsson B, Magnenat E, Camps M, Muda M, Chabert C, Gillieron C, Boschert U, Vial-Knecht E, Martinou JC, Arkinstall S (1997) Bcl-2 undergoes phosphorylation by c-Jun N-terminal kinase/stress-activated protein kinases in the presence of the constitutively active GTP-binding protein Rac1. J Biol Chem 272:25238–25242

Meijerink JP, Van Lieshout EM, Beverloo HB, Van Drunen E, Mensink EJ, Macville M, Pieters R (2005) Novel murine B-cell lymphoma/leukemia model to study BCL2-driven oncogenesis. Int J Cancer 114:917–925

Minn AJ, Velez P, Schendel SL, Liang H, Muchmore SW, Fesik SW, Fill M, Thompson CB (1997) Bcl-x(L) forms an ion channel in synthetic lipid membranes. Nature 385:353–357

Moldoveanu T, Liu Q, Tocilj A, Watson M, Shore G, Gehring K (2006) The X-ray structure of a BAK homodimer reveals an inhibitory zinc binding site. Mol Cell 24:677–688

Moreno-Sanchez R, Rodriguez-Enriquez S, Marin-Hernandez A, Saavedra E (2007) Energy metabolism in tumor cells. FEBS J 274:1393–1418

Muchmore SW, Sattler M, Liang H, Meadows RP, Harlan JE, Yoon HS, Nettesheim D, Chang BS, Thompson CB, Wong SL, Ng SL, Fesik SW (1996) X-ray and NMR structure of human Bcl-xL, an inhibitor of programmed cell death. Nature 381:335–341

Nakagawa T, Shimizu S, Watanabe T, Yamaguchi O, Otsu K, Yamagata H, Inohara H, Kubo T, Tsujimoto Y (2005) Cyclophilin D-dependent mitochondrial permeability transition regulates some necrotic but not apoptotic cell death. Nature 434:652–658

O'Connor L, Strasser A, O'Reilly LA, Hausmann G, Adams JM, Cory S, Huang DC (1998) Bim: a novel member of the Bcl-2 family that promotes apoptosis. EMBO J 17:384–395

Oberstein A, Jeffrey PD, Shi Y (2007) Crystal structure of the Bcl-XL-Beclin 1 peptide complex: Beclin 1 is a novel BH3-only protein. J Biol Chem 282:13123–13132

Odunsi K, Wollman RM, Ambrosone CB, Hutson A, Mccann SE, Tammela J, Geisler JP, Miller G, Sellers T, Cliby W, Qian F, Keitz B, Intengan M, Lele S, Alderfer JL (2005) Detection of epithelial ovarian cancer using 1H-NMR-based metabonomics. Int J Cancer 113:782–788

Oliver L, Priault M, Tremblais K, Lecabellec M, Meflah K, Manon S, Vallette FM (2000) The substitution of the C-terminus of Bax by that of Bcl-xL does not affect its subcellular localization but abrogates its pro-apoptotic properties. FEBS Lett 487:161–165

Oltersdorf T, Elmore SW, Shoemaker AR, Armstrong RC, Augeri DJ, Belli BA, Bruncko M, Deckwerth TL, Dinges J, Hajduk PJ, Joseph MK, Kitada S, Korsmeyer SJ, Kunzer AR, Letai A, Li C, Mitten MJ, Nettesheim DG, NG S, Nimmer PM, O'Connor JM, Oleksijew A, Petros AM, Reed JC, Shen W, Tahir SK, Thompson CB, Tomaselli KJ, Wang B, Wendt MD, Zhang H, Fesik SW, Rosenberg SH (2005) An inhibitor of Bcl-2 family proteins induces regression of solid tumours. Nature 435:677–681

Ottilie S, Wang Y, Banks S, Chang J, Vigna NJ, Weeks S, Armstrong RC, Fritz LC, Oltersdorf T (1997) Mutational analysis of the interacting cell death regulators CED-9 and CED-4. Cell Death Differ 4:526–533

Owens TW, Valentijn AJ, Upton JP, Keeble J, Zhang L, Lindsay j, Zouq NK, Gilmore AP (2009) Apoptosis commitment and activation of mitochondrial Bax during anoikis is regulated by p38MAPK. Cell Death Differ 16:1551–1562

Pattingre S, Tassa A, Qu X, Garuti R, Liang XH, Mizushima N, Packer M, Schneider MD, Levine B (2005) Bcl-2 antiapoptotic proteins inhibit Beclin 1-dependent autophagy. Cell 122:927–939

Pavlov EV, Priault M, Pietkiewicz D, Cheng EH, Antonsson B, Manon S, Korsmeyer SJ, Mannella CA, Kinnally KW (2001) A novel, high conductance channel of mitochondria linked to apoptosis in mammalian cells and Bax expression in yeast. J Cell Biol 155:725–731

Pena JC, Rudin CM, Thompson CB (1998) A Bcl-xL transgene promotes malignant conversion of chemically initiated skin papillomas. Cancer Res 58:2111–2116

Petros AM, Medek A, Nettesheim DG, Kim DH, Yoon HS, Swift K, Matayoshi ED, Oltersdorf T, Fesik SW (2001) Solution structure of the antiapoptotic protein bcl-2. Proc Natl Acad Sci U S A 98:3012–3017

Petros AM, Olejniczak ET, Fesik SW (2004) Structural biology of the Bcl-2 family of proteins. Biochim Biophys Acta 1644:83–94

Peyerl FW, Dai S, Murphy GA, Crawford F, White J, Marrack P, Kappler JW (2007) Elucidation of some Bax conformational changes through crystallization of an antibody-peptide complex. Cell Death Differ 14:447–452

Priault M, Hue E, Marhuenda F, Pilet P, Oliver L, Vallette FM (2010) Differential dependence on Beclin 1 for the regulation of pro-survival autophagy by Bcl-2 and Bcl-xL in HCT116 colorectal cancer cells. PLoS One 5:e8755

Quinn L, Coombe M, Mills K, Daish T, Colussi P, Kumar S, Richardson H (2003) Buffy, a drosophila Bcl-2 protein, has anti-apoptotic and cell cycle inhibitory functions. EMBO J 22:3568–3579

Renault TT, Grandier-Vazeille X, Arokium H, Velours G, Camougrand N, Priault M, Teijido O, Dejean LM, Manon S (2012) The cytosolic domain of human Tom22 modulates human Bax mitochondrial translocation and conformation in yeast. FEBS Lett 586:116–121

Renault TT, Teijido O, Antonsson B, Dejean LM, Manon S (2013) Regulation of Bax mitochondrial localization by Bcl-2 and Bcl-x(L): keep your friends close but your enemies closer. Int J Biochem Cell Biol 45:64–67

Renault TT, Floros KV, Elkholi R, Corrigan KA, Kushnareva Y, Wieder SY, Lindtner C, Serasinghe MN, Asciolla JJ, Buettner C, Newmeyer DD, Chipuk JE (2015a) Mitochondrial shape governs BAX-induced membrane permeabilization and apoptosis. Mol Cell 57:69–82

Renault TT, Teijido O, Missire F, Ganesan YT, Velours G, Arokium H, Beaumatin F, Llanos R, Athane A, Camougrand N, Priault M, Antonsson B, Dejean LM, Manon S (2015b) Bcl-xL stimulates Bax relocation to mitochondria and primes cells to ABT-737. Int J Biochem Cell Biol 64:136–146

Renault TT, Dejean LM, Manon S (2016) A brewing understanding of the regulation of Bax function by Bcl-xL and Bcl-2. Mech Ageing Dev 477:33

Ross K, Rudel T, Kozjak-Pavlovic V (2009) TOM-independent complex formation of Bax and Bak in mammalian mitochondria during TNFalpha-induced apoptosis. Cell Death Differ 16:697–707

Roucou X, Montessuit S, Antonsson B, Martinou JC (2002) Bax oligomerization in mitochondrial membranes requires tBid (caspase-8-cleaved Bid) and a mitochondrial protein. Biochem J 368:915–921

Salvador-Gallego R, Mund M, Cosentino K, Schneider J, Unsay J, Schraermeyer U, Engelhardt J, Ries J, Garcia-Saez AJ (2016) Bax assembly into rings and arcs in apoptotic mitochondria is linked to membrane pores. EMBO J 35:389–401

Sarosiek KA, Letai A (2016) Directly targeting the mitochondrial pathway of apoptosis for cancer therapy using BH3 mimetics – recent successes, current challenges and future promise. FEBS J 283:3523

Schellenberg B, Wang P, Keeble JA, Rodriguez-Enriquez R, Walker S, Owens TW, Foster F, Tanianis-Hughes J, Brennan K, Streuli CH, Gilmore AP (2013) Bax exists in a dynamic equilibrium between the cytosol and mitochondria to control apoptotic priming. Mol Cell 49:959–971

Schendel SL, Xie Z, Montal MO, Matsuyama S, Montal M, Reed JC (1997) Channel formation by antiapoptotic protein Bcl-2. Proc Natl Acad Sci U S A 94:5113–5118

Schinzel A, Kaufmann T, Schuler M, Martinalbo J, Grubb D, Borner C (2004) Conformational control of Bax localization and apoptotic activity by Pro168. J Cell Biol 164:1021–1032

Schlesinger PH, Gross A, Yin XM, Yamamoto K, Saito M, Waksman G, Korsmeyer SJ (1997) Comparison of the ion channel characteristics of proapoptotic BAX and antiapoptotic BCL-2. Proc Natl Acad Sci U S A 94:11357–11362

Sedlak TW, Oltvai ZN, Yang E, Wang K, Boise LH, Thompson CB, Korsmeyer SJ (1995) Multiple Bcl-2 family members demonstrate selective dimerizations with Bax. Proc Natl Acad Sci U S A 92:7834–7838

Silva RD, Manon S, Goncalves J, Saraiva L, Corte-Real M (2011) Modulation of Bax mitochondrial insertion and induced cell death in yeast by mammalian protein kinase Calpha. Exp Cell Res 317:781–790

Simonyan L, Renault TT, Novais MJ, Sousa MJ, Corte-Real M, Camougrand N, Gonzalez C, Manon S (2016) Regulation of Bax/mitochondria interaction by AKT. FEBS Lett 590:13–21

Souers AJ, Leverson JD, Boghaert ER, Ackler SL, Catron ND, Chen J, Dayton BD, Ding H, Enschede SH, Fairbrother WJ, Huang DC, Hymowitz SG, Jin S, Khaw SL, Kovar PJ, Lam LT, Lee J, Maecker HL, Marsh KC, Mason KD, Mitten MJ, Nimmer PM, Oleksijew A, Park CH, Park CM, Phillips DC, Roberts AW, Sampath D, Seymour JF, Smith ML, Sullivan GM, Tahir SK, Tse C, Wendt MD, Xiao Y, Xue JC, Zhang H, Humerickhouse RA, Rosenberg SH, Elmore SW (2013) ABT-199, a potent and selective BCL-2 inhibitor, achieves antitumor activity while sparing platelets. Nat Med 19:202–208

Subburaj Y, Cosentino K, Axmann M, Pedrueza-Villalmanzo E, Hermann E, Bleicken S, Spatz J, Garcia-Saez AJ (2015) Bax monomers form dimer units in the membrane that further self-assemble into multiple oligomeric species. Nat Commun 6:8042

Suzuki M, Youle RJ, Tjandra N (2000) Structure of Bax: coregulation of dimer formation and intracellular localization. Cell 103:645–654

Swanson PJ, Kuslak SL, Fang W, Tze L, Gaffney P, Selby S, Hippen KL, Nunez G, Sidman CL, Behrens TW (2004) Fatal acute lymphoblastic leukemia in mice transgenic for B cell-restricted bcl-xL and c-myc. J Immunol 172:6684–6691

Tang L, Tron VA, Reed JC, Mah KJ, Krajewska M, Li G, Zhou X, Ho VC, Trotter MJ (1998) Expression of apoptosis regulators in cutaneous malignant melanoma. Clin Cancer Res 4:1865–1871
Teijido O, Dejean L (2010) Upregulation of Bcl2 inhibits apoptosis-driven BAX insertion but favors BAX relocalization in mitochondria. FEBS Lett 584:3305–3310
Terrano DT, Upreti M, Chambers TC (2010) Cyclin-dependent kinase 1-mediated Bcl-xL/Bcl-2 phosphorylation acts as a functional link coupling mitotic arrest and apoptosis. Mol Cell Biol 30:640–656
Todt F, Cakir Z, Reichenbach F, Youle RJ, Edlich F (2013) The C-terminal helix of Bcl-x(L) mediates Bax retrotranslocation from the mitochondria. Cell Death Differ 20:333–342
Torcia M, De Chiara G, Nencioni L, Ammendola S, Labardi D, Lucibello M, Rosini P, Marlier LN, Bonini P, Dello Sbarba P, Palamara AT, Zambrano N, Russo T, Garaci E, Cozzolino F (2001) Nerve growth factor inhibits apoptosis in memory B lymphocytes via inactivation of p38 MAPK, prevention of Bcl-2 phosphorylation, and cytochrome c release. J Biol Chem 276:39027–39036
Tremblais K, Oliver L, Juin P, Le Cabellec TM, Meflah K, Vallette FM (1999) The C-terminus of bax is not a membrane addressing/anchoring signal. Biochem Biophys Res Commun 260:582–591
Tsujimoto Y, Finger LR, Yunis J, Nowell PC, Croce CM (1984) Cloning of the chromosome breakpoint of neoplastic B cells with the t(n18) chromosome translocation. Science 226:1097–1099
Valero JG, Sancey L, Kucharczak J, Guillemin Y, Gimenez D, Prudent J, Gillet G, Salgado J, Coll JL, Aouacheria A (2011) Bax-derived membrane-active peptides act as potent and direct inducers of apoptosis in cancer cells. J Cell Sci 124:556–564
Vander Heiden MG, Cantley LC, Thompson CB (2009) Understanding the Warburg effect: the metabolic requirements of cell proliferation. Science 324:1029–1033
Vantieghem A, Xu Y, Assefa Z, Piette J, Vandenheede JR, Merlevede W, De Witte PA, Agostinis P (2002) Phosphorylation of Bcl-2 in G2/M phase-arrested cells following photodynamic therapy with hypericin involves a CDK1-mediated signal and delays the onset of apoptosis. J Biol Chem 277:37718–37731
Wei MC, Zong WX, Cheng EH, Lindsten T, Panoutsakopoulou V, Ross AJ, Roth KA, Macgregor GR, Thompson CB, Korsmeyer SJ (2001) Proapoptotic BAX and BAK: a requisite gateway to mitochondrial dysfunction and death. Science 292:727–730
Westphal D, Kluck RM, Dewson G (2014) Building blocks of the apoptotic pore: how Bax and Bak are activated and oligomerize during apoptosis. Cell Death Differ 21:196–205
Willis SN, Fletcher JI, Kaufmann T, Van Delft MF, Chen L, Czabotar PE, Ierino H, Lee EF, Fairlie WD, Bouillet P, Strasser A, Kluck RM, Adams JM, Huang DC (2007) Apoptosis initiated when BH3 ligands engage multiple Bcl-2 homologs, not Bax or Bak. Science 315:856–859
Wolter KG, Hsu YT, Smith CL, Nechushtan A, Xi XG, Youle RJ (1997) Movement of Bax from the cytosol to mitochondria during apoptosis. J Cell Biol 139:1281–1292
Xin M, Gao F, May WS, Flagg T, Deng X (2007) Protein kinase Czeta abrogates the proapoptotic function of Bax through phosphorylation. J Biol Chem 282:21268–21277
Yamaguchi R, Lartigue L, Perkins G, Scott RT, Dixit A, Kushnareva Y, Kuwana T, Ellisman MH, Newmeyer DD (2008) Opa1-mediated cristae opening is Bax/Bak and BH3 dependent, required for apoptosis, and independent of Bak oligomerization. Mol Cell 31:557–569
Yang E, Zha J, Jockel J, Boise LH, Thompson CB, Korsmeyer SJ (1995) Bad, a heterodimeric partner for Bcl-XL and Bcl-2, displaces Bax and promotes cell death. Cell 80:285–291
Yin XM, Oltvai ZN, Korsmeyer SJ (1994) BH1 and BH2 domains of Bcl-2 are required for inhibition of apoptosis and heterodimerization with Bax. Nature 369:321–323
Yu J, Zhang L, Hwang PM, Kinzler KW, Vogelstein B (2001) PUMA induces the rapid apoptosis of colorectal cancer cells. Mol Cell 7:673–682
Zamzami N, Marchetti P, Castedo M, Decaudin D, Macho A, Hirsch T, Susin SA, Petit PX, Mignotte B, Kroemer G (1995) Sequential reduction of mitochondrial transmembrane potential and generation of reactive oxygen species in early programmed cell death. J Exp Med 182:367–377

Zha H, Aime-Sempe C, Sato T, Reed JC (1996a) Proapoptotic protein Bax heterodimerizes with Bcl-2 and homodimerizes with Bax via a novel domain (BH3) distinct from BH1 and BH2. J Biol Chem 271:7440–7444
Zha J, Harada H, Yang E, Jockel J, Korsmeyer SJ (1996b) Serine phosphorylation of death agonist BAD in response to survival factor results in binding to 14-3-3 not BCL-X(L). Cell 87:619–628
Zhang T, Saghatelian A (2013) Emerging roles of lipids in BCL-2 family-regulated apoptosis. Biochim Biophys Acta 1831:1542–1554

Chapter 11
The 18 kDa Translocator Protein (TSPO): Cholesterol Trafficking and the Biology of a Prognostic and Therapeutic Mitochondrial Target

Michele Frison, Anna Katherina Mallach, Emma Kennedy, and Michelangelo Campanella

Abbreviations

Å	Ångström
ACBD	Acyl-coenzyme A binding domain containing
ACTH	Adrenocorticotropin
AD	Alzheimer's disease
ALAS-1	5′-aminolevulinate synthase 1
ANT	Adenine nucleotide transporter
AP	Activator protein
Atg8	Autophagy related protein 8
Bcl-2	B-cell lymphoma 2
*Bc*TSPO	*Bacillus cereus* TspO
cAMP	Cyclic adenosine monophosphate
CARC	Cholesterol consensus domain
CNS	Central nervous system
CRAC	Cholesterol recognition amino acid consensus
CRF	Corticotropin-releasing factor
DBI	Diazepam-binding inhibitor

M. Frison • A.K. Mallach • E. Kennedy
Department of Comparative Biomedical Sciences, The Royal Veterinary College, University of London, Royal College Street, NW1 0TU London, UK

M. Campanella (✉)
Department of Comparative Biomedical Sciences, The Royal Veterinary College, University of London, Royal College Street, NW1 0TU London, UK

University College London Consortium for Mitochondrial Research, Gower Street, WC1E 6BT London, UK
e-mail: mcampanella@rvc.ac.uk

T.K. Rostovtseva (ed.), *Molecular Basis for Mitochondrial Signaling*, Biological and Medical Physics, Biomedical Engineering,
DOI 10.1007/978-3-319-55539-3_11

EM	Electron microscopy
ERα	Oestrogen receptor α
Ets	E26 oncogene homolog
FCCP	Carbonyl cyanide p-(trifluoromethoxy) phenylhydrazone
GABA	Gamma-aminobutyric acid
HIV	Human immunodeficiency virus
HPA	Hypothalamic-pituitary-adrenal
KO	Knockout
I/R	Ischemia/reperfusion
LC3	Microtubule-associated protein light chain 3
mPTP	Mitochondrial permeability transition pore
MS	Multiple sclerosis
mTSPO	Mammalian TSPO
NAT	Natural antisense transcript
NLRP3	Nod-like receptor family, pyrin domain containing 3
NMR	Nuclear magnetic resonance
NOX	NADPH oxidase
OMM	Outer mitochondrial membrane
PBR	Peripheral-type benzodiazepine receptor
PD	Parkinson's disease
PEPCK	Phosphoenolpyruvate carboxykinase
PET	Positron emission tomography
PGC-1α	PPAR-gamma coactivator-1 α
PINK1	PTEN-induced putative kinase 1
PKA	Protein kinase A
PKCε	Protein kinase Cε
PMA	Phorbol-12-myristate 13-acetate
PPAR	Peroxisome proliferator-activated receptor
PpIX	Protoporphyrin IX
ROS	Reactive oxygen species
RsTspO	*R. sphaeroides* TspO
Sp	Specificity protein
StAR	Steroidogenic acute response protein
STAT	Signal transducer and activator of transcription
TM	Transmembrane (number assignment for helices in TSPO structure)
TSPO	Translocator protein
TspO	Tryptophan-rich sensory protein
VDAC	Voltage gated anion channel
WT	Wild type

11.1 A Historical Perspective

TSPO was discovered in studies determining the physiological targets of benzodiazepines. This group of small molecules bears psychoactive properties which have made them the standard treatment of anxiety disorders and seizures in which they inhibit neuronal excitation through their binding to gamma amino butyric acid (GABA) receptors (Saari et al. 2011). Initial studies in 1977 had found that radiolabelled benzodiazepines bind to human cortical areas of the brain, with a specific member of the group, named Ro5-4864, having preferential affinity for peripheral areas (Braestrup et al. 1977). Experiments on fractionated rat brain membranes then proved that Ro5-4864 displayed superior binding to the mitochondrion, observed beyond the brain in the kidneys, liver and lungs (Braestrup and Squires 1977), thus providing evidence of a separate benzodiazepine-binding site.

The second site, named peripheral benzodiazepine receptor (PBR), was characterised as being specifically localised in olfactory nerves (Anholt et al. 1984), with much greater affinities found in endocrine organs, like the pituitary gland, adrenal gland and testis (De Souza et al. 1985). Finally, the discovery of an isoquinoline carboxamide ligand, named PK11195, with a high affinity to PBR (Benavides et al. 1983a, b, 1985), allowed researchers in 1986 to identify this protein as being subcellularly localised on the outer mitochondrial membrane (OMM) (Anholt et al. 1986).

The work carried out over the subsequent years largely utilised pharmacological tools to infer the function of PBR. The abundance of PBR in steroidogenic tissues led to focussed studies on its involvement in endocrinology. In the 1990s, the role of PBR in steroid hormone pathways was investigated (Gavish et al. 1999). The hypothalamic-pituitary-adrenal (HPA) axis, an important stress-induced pathway which controls adrenal corticosteroid production from the central nervous system (CNS) (Smith and Vale 2006), was linked with PBR. PK1195 and Ro5-4864 were found to, respectively, stimulate the production of adrenocorticotropin (ACTH) and corticotropin-releasing factor (CRF) when administered onto the hypothalamus and pituitary, whilst PBR levels were modulated in the adrenal gland by the action of ACTH (Bar-Ami et al. 1989; Calogero et al. 1990). Studies on the sexual hormone pathways gave similar outcomes, as PBR expression was observed to increase in genital organs during the oestrous cycle (Fares et al. 1987, 1988), testes maturation (Mercer et al. 1992) and also in the adrenal gland controlled by the action of testosterone (Amiri et al. 1991; Weizman et al. 1992). At this time, speculations of the pleiotropic nature of PBR emerged. The protein was found to interact with multiple drugs and targets, including porphyrins (Verma et al. 1987), and diazepam-binding inhibitor (DBI) (Alho et al. 1985; Gray et al. 1986), as well as observed change in expression in various pathologies, such as stress disorders and neurodegenerative diseases (Gavish et al. 1999).

The greatest clarification for its role however came from concomitant cell biology studies. By using cell lines derived from testicular Leydig cells and adrenocortical tissues, PBR was discovered to hold the specific function of mediating cholesterol transport across the mitochondrial membranes (Krueger and Papadopoulos 1990). This finding fundamentally explained the involvement of PBR in steroid hormone pathways as steroidogenesis requires cholesterol to be transported from the cytoplasm, where it is produced or deposited, into the mitochondrion, to be catalysed into steroids.

This developed understanding of PBR laid the groundwork for the use of more systematic studies, with genetic disruption techniques solidifying its molecular function (Papadopoulos et al. 1997b). PBR was soon after realised to adopt multiple oligomeric states (Papadopoulos et al. 1994), and benzodiazepine binding was found to be dependent on the heteromeric (Garnier et al. 1994) association with other membrane-bound proteins, such as VDAC1 and adenine nucleotide translocator (ANT) (McEnery et al. 1992). However, at the same time the idea that PBR was associated with other aspects of mitochondrial physiology began to manifest, as more processes, such as apoptosis (Hirsch et al. 1998), mitochondrial permeability transition pore (mPTP) (Chelli et al. 2001) and oxidative stress (Jayakumar et al. 2002), were linked to the protein. Furthermore, associations with various pathological conditions were strengthened, including; (i) cancer (Austin et al. 2013), (ii) neurodegeneration (Rogers et al. 2007) and (iii) cardiovascular diseases (Veenman and Gavish 2006).

Henceforth, in 2006, PBR was renamed TSPO to better reflect its molecular function (Papadopoulos et al. 2006).

Notably, consequent studies linked TSPO to functions in mammals, plants (Vanhee et al. 2011) and prokaryotes (Yeliseev et al. 1997). Notable publications provided the nuclear magnetic resonance (NMR) structure of the mammalian TSPO (Jaremko et al. 2014) and showed that a newer, more specific ligand produced anxiolytic effects by increasing the microglial production of neurosteroids (Rupprecht et al. 2009).

From its discovery, TSPO has historically been of great interest for neuroscientists and neurologists. Since early studies were mostly carried out with radiopharmacological techniques, it did not take long for researchers to observe an increase in TSPO expression upon damage to the nervous system. Towards the end of the 1980s, substantial evidence was already gathered on the connection between levels of TSPO and neuronal damage (Benavides et al. 1987; Dubois et al. 1988), as well as the presence in gliomas (Junck et al. 1989). The protein is currently targeted by a range of ligands utilised to extrapolate data on its levels and localisation in the CNS. However, newer generation ligands, more advanced techniques and knowledge of the behaviour of the protein in neuroglia have permitted preclinical use in patients. Targeting TSPO is in fact exploited by positron emission tomography (PET) to non-invasively visualise gliomas and patterns of CNS neurological damage (Janczar et al. 2015; Turkheimer et al. 2015). In the former, this is possible due to the aberrant expression of TSPO observed in cancerous glial cells. As for neurological damage, the technique exploits the properties of neuroinflammation, a CNS-specific inflammatory response. When neurological damage is inflicted upon the CNS, microglia and astrocytes undergo a switch in metabolism and proliferate to take on a new set of functions (Ransohoff and Perry 2009). These glial cells have a complex and yet not fully known physiology. However, microglia have been shown to adopt different states, each harbouring functional responsibilities like phagocytosis, neuromodulation, synaptic remodelling and inflammation (Gomez-Nicola and Perry 2014). The latter can be triggered by damaged neurons in a very localised and sensitive way, which can be detected within minutes (Graeber 2010). Neuroinflammation is accompanied by a vast upregulation in TSPO in microglia and astrocytes, which is detected by the radioligands (Banati 2002; Cagnin et al. 2007). This technique allows early detection of neurological damage of a variety of types, from multiple sclerosis (MS) to Parkinson's disease (PD), and does not suffer from the uncertainties surrounding the function of TSPO, as it does not require a functional involvement in the neuroinflammatory processes.

11.2 The Structure of Mammalian TSPO and Bacterial Homologs

TSPO was identified early on to be a small 18 kDa protein residing in the OMM (Anholt et al. 1986; McEnery et al. 1992). Yet despite the attention given to the protein, the structure of TSPO has only been recently resolved at atomic resolution (>1.6 Å) (Guo et al. 2015; Li et al. 2015). This is mainly due to the naturally flexible, membrane-bound nature of the protein, requiring a lipidic environment to provide structural stability and native conformations, needed for X-ray crystallographic and NMR techniques. Nevertheless, various structural characteristics and low resolution models were defined before 2014.

A molecular dynamics simulation predicted TSPO to have five α-helices, each long enough to span a single lipid monolayer, with the N-terminus in the intermembrane space and the C-terminus in the cytoplasm (Bernassau et al. 1993). This hypothesis was perfected by topological analysis using immunolabelling and amino acid tagging to reveal membrane-enclosed α-helical regions (Joseph-Liauzun et al. 1998), demonstrated to span the entire lipid bilayer, known as transmembrane (TM) helices, and extend outside of the membrane via loops. Successive works showed that the previously identified cholesterol recognition amino acid consensus (CRAC) motif, a region responsible for cholesterol binding (Li and Papadopoulos 1998), was located on the C-terminal cytoplasmic side, suggested to attract cholesterol and successively allow the molecule to translocate across TSPO (Lacapère et al. 2001; Murail et al. 2008). It was additionally demonstrated by NMR that cholesterol stabilises the tertiary helical bundle structure of TSPO, binding with a 1:1 stoichiometry.

Cryo-electron microscopy (EM) has provided the first 10 Å resolution structure of the bacterial ortholog of TSPO (Korkhov et al. 2010), known as tryptophan-rich sensory protein (TspO) in *Rhodobacter sphaeroides*, a close ancestor of the mitochondrion (Yeliseev and Kaplan 2000). This was possible due to the fact that TSPO is an ancient protein, with remarkably conserved sequence found in bacteria to mammals (Gatliff and Campanella 2016; Selvaraj and Stocco 2015), with notable exceptions of *Escherichia coli* (Li and Papadopoulos 1998) and yeast (Joseph-Liauzun et al. 1998). *R. sphaeroides* TspO (RsTspO) has proved to be a valuable model for structural studies due to its structural and functional similarities, such as its ability to porphyrin molecules and PK11195 at high affinity (Li et al. 2015) and bacterially non-native cholesterol. Korkhov and colleagues modelled a dimeric structure with the 5 TMs, slightly inclined with respect to the vertical axis, arranged in a sequential N-to-C terminus order forming a bundle or cylinder. The interface between the two TspO dimers was shown to consist of TM1 and TM2, whilst the arrangement of TSPO in its native multimeric state remains under discussion. Studies in prokaryotes suggested that TspO may exist in a dimeric state, while in mammalian cells functional data has suggested that TSPO forms complexes with VDAC1 in a stoichiometry of 5:1 (Papadopoulos et al. 1994), to which reactive oxygen species (ROS) are able to regulate this association (Delavoie et al. 2003). Interestingly, differences in TSPO states of oligomerisation conserved between phyla and conferred by the lack of TM1 sequence indicated this region of the protein to be involved in oligomeric interactions.

The NMR study of mammalian TSPO (mTSPO) instead showed that monomers of the protein are indeed stable when reconstituted in dodecylphosphocholine (DPC) micelles (Jaremko et al. 2014). The resulting structure, obtained in complex with PK1195, exhibited striking similarity in comparison to that complex structure obtained with *Rs*TspO, indicating conserved structural importance of TSPO. The high-resolution model obtained also provided additional information. Firstly, the helical bundle does not exactly follow the anticlockwise sequential arrangement seen from cytosolic view, as previously suggested; the circle starts from TM2, continues with TM1 and then continues with the arrangement. Furthermore, a loop (LP1) between TM1 and TM2 forms a small α-helix on the cytosolic side of the membrane, near the C-terminus (Fig. 11.1a). This small helix has a flexible structure which assumes a more steady state upon PK11195 binding, acting as a lid to stabilising ligand binding into a solvent-filled cavity. This study determined 61 molecular contacts between PK11195 in TSPO's binding pocket, made by ten conserved residues contributed by all five TMs. Notably, the CRAC motif was found to be located on TM5's C-terminus, in contact with the cytosolic side of the lipid layer (Fig. 11.1b). This result contradicted the previous models, in which the CRAC motif was suggested to constitute a cholesterol-docking site, which then allows passage across a "TSPO channel". Jaremko et al. also showed in a following publication that mTSPO can dynamically adopt a range of different conformations in the absence of ligand binding (Jaremko et al. 2015a).

In 2015, two X-ray crystallography studies were published using prokaryotic TspO: an *R. sphaeroides* model of the human polymorphism (Li et al. 2015a) and a *Bacillus cereus* structure to support the hypothesis of an ancient function of the TSPO proteins in porphyrin metabolism (Guo et al. 2015). The human alanine (147) to threonine (A147T) substitution is a commonly found polymorphism which has raised interest due to its inhibitory effect on PET ligand binding (Turkheimer et al. 2015). However, the importance of this polymorphism extends beyond its preclinical exploitation. Patients carrying the A147T mutation have in fact been shown to

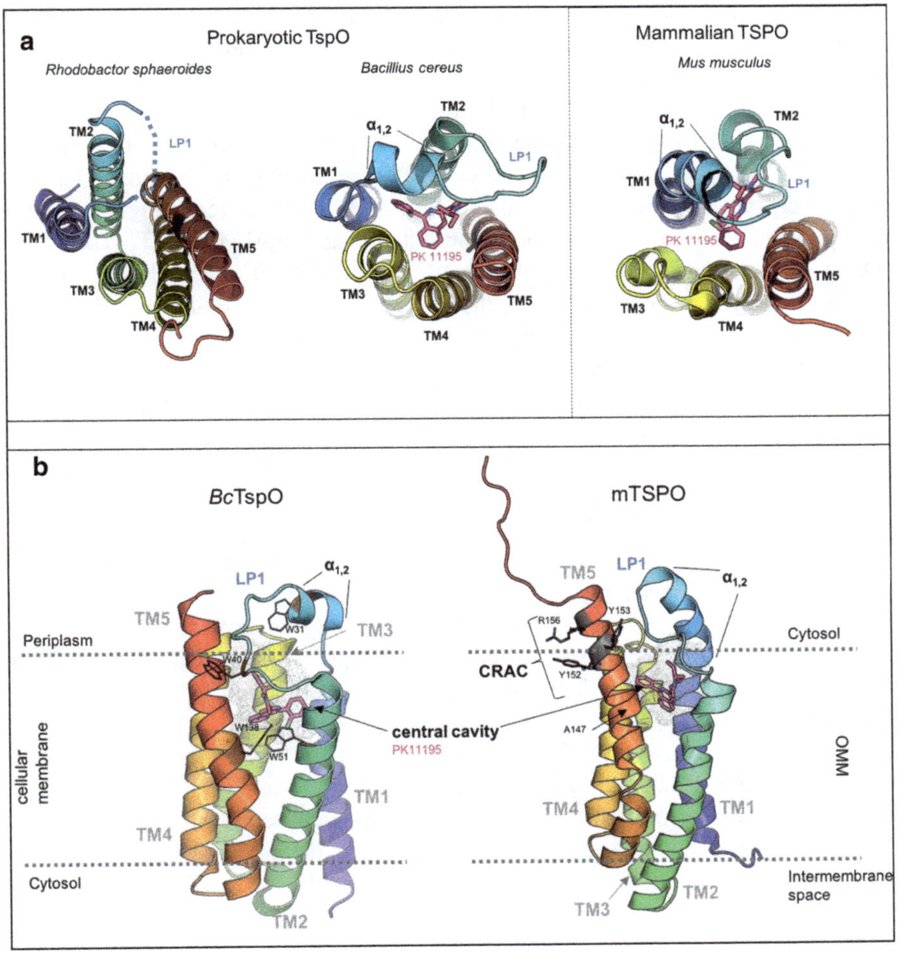

Fig. 11.1 Structure of prokaryotic and mammalian TSPO homologs. Ribbon structure of current single unit mammalian TSPO, achieved through NMR solution structure, and prokaryotic homolog structures obtained through X-ray crystallography. Arrangement of TM bundle of single unit TSPO is constricted by structural data obtained from homodimeric interactions. Sequence is progressively coloured from the N-terminus (*dark blue*) to C-terminus (*red*), whilst PK11195 moiety, located in TSPO central cavity, is *pink*. **a** Cytosolic view of *Rs*TspO, *Bc*TspO-PK11195 and mTSPO-PK11195, highlighting TM arrangement in membrane unanimously conferring to TM1, TM2, TM5, TM3, TM4 clockwise topological order. Particular differences can be seen for *Bc*TspO and mTSPO small $\alpha_{1,2}$ helix length, with missing LP1 electron density data in apo-*Rs*TspO due to its flexible nature in solution. **b** *Bc*TspO and mTSPO viewed from the membrane plane, spanning lipid bilayer. *Bc*TspO-PK11195 with conserved tryptophan residues drawn as sticks (*brown*), known to interact with PpIX. mTSPO, located on OMM, with essential CRAC residues drawn in sticks (*black*), and indicated A147 SNP preceding CRAC motif. Models of *Bc*TspO, *Rs*TspO and mTSPO monomers adopted from the following publications, respectively (Guo et al. 2015; Jaremko et al. 2014; Li et al. 2015a), PDB accession codes – *BcTspO-PK11195*, 4RYI; *RsTspO*, 4UC3; *mTSPO-PK11195*, 2MGY

display predisposition to a spectrum of anxiety-related disorders as a consequence of decreased neurosteroids production, thereby creating the only solid causative link between the protein and disease.

Structural investigation into the polymorphism was presented by Li and colleagues which compared to the crystal structures of WT *Rs*TspO and the polymorphism model A139T, both without the presence of pharmacological ligands. The WT *Rs*TspO crystal structures support the findings of Korkhov et al. regarding the dimeric arrangement, albeit with an interface contributed mainly by TM3 and TM1, consisting on ~15% of the monomer's surface area. This interface is efficiently interlocked, lacking hydrogen bond interactions, thereby refuting the possibility of this region providing a channel for cholesterol passage. The major differences between WT and A139T *Rs*TspO structures are attributed to the LP1 α-helix and the nearby CRAC motif conformation. Due to its flexible nature, the structure of LP1 was not fully resolved in WT *Rs*TspO; however the A139T mutation allowed stabilisation and therefore modelling of this helix. Additionally, A139T *Rs*TspO resulted in a 7.7° tilt of TM2 towards TM5 and therefore tighter bundle packing, in turn observing a distorted CRAC motif region. Nearby TM1 and TM2 observes a groove beginning from the porphyrin binding site and extending onto the external face of the protein, suggesting an external transport pathway requiring protein partners. Lastly, experiments with protoporphyrin IX (PpIX), cholesterol and PK11195 show decreased binding affinity towards A139T compared to wildtype *Rs*TspO. The publication in the same year of the structure of A147T mTSPO in complex with PK11195 however showed contradictory results to those explained above (Jaremko et al. 2015b). This study shows that the structure of TSPO is only perturbed on TM1 and in the loop helix. The CRAC motif is instead seen to be unaffected by the mutation, both in ligand bound and apo-TSPO structures. Jaremko and colleagues criticise the crystal structure of WT RsTspO as showing artificial characteristics due to crystal packing and justify this by observing that BcTspO, A139T RsTspO, WT and A147T mTSPO all share the same topology with regard to certain conserved residues involved in ligand binding. The differences in the other conserved residues are attributed to evolutionary modifications, validated by the fact that PK11195 has a 1,000-fold lower affinity to bacterial TspO compared to mTSPO. Moreover, its binding affinity to mTSPO is not perturbed by the polymorphism unlike with second-generation ligands (Owen et al. 2011), which validates Jaremko's findings. It is also fundamental to consider that bacterial TspO proteins are less relevant models compared to mTSPO with regard to cholesterol binding. This is due to the CRAC motif being not strongly conserved, consistent with the functional irrelevance of cholesterol binding in bacterial membranes, where this particular lipid is not present (Gut et al. 2015). In conclusion, the authors make the hypothesis for which the structure perturbations caused by the A147T mutations impair cholesterol binding not by disrupting the CRAC structure but by impeding interactions with other proteins. However, alternative studies of cholesterol interacting motifs have not yet been considered in regard to TSPO structure. A cholesterol consensus domain (CARC) has been identified as being structurally next to, and functionally cooperating with, CRAC domains present in TSPO, at residues 139–146 (Fantini et al. 2016). Additionally a cholesterol binding enhancement motif containing the A139 residue (A147 in mTSPO) has been identified in *Rs*TspO (Li et al. 2015b). It is therefore tempting to hypothesise that the loss of affinity to cholesterol is due to the structural distortion of the alternative CARC motif. Nevertheless, this hypothesis is yet to be corroborated.

The last structural biology work of relevance to be published is *Bacillus cereus* TspO (*Bc*TspO) (Guo et al. 2015). This publication provides data regarding two different features of TSPO: its structure and function. Structurally, BcTspO exists in at least three oligomeric states, with Guo et al. (2015) providing crystal structure of monomeric and dimeric states, with the dimerisation interactions mostly located on TM2, and not TM1 or TM3. The general topology of the protein is maintained with regard to whole tertiary structure, with the highest structural homology found with A139T RsTspO, however showing significant changes in the location of conserved residues in remaining determined models. A noteworthy difference comes from the absence of significant structural changes upon binding of PK11195 (Fig. 11.1b). However, the focus of this work lies on the structural features interacting with PpIX, which was previously found to be degraded in a light- and oxygen-dependent manner by both prokaryotes and eukaryotes (Ginter et al. 2013). This phenomenon was measured by a shift in absorbance spectrum produced by the porphyrin chemical group. Mutations of key tryptophan residues and a model of the A147T and addition of PK11195 were shown to abolish this process. The consequential suggestion made by this paper is that the tryptophan residues mediate the transfer of high-energy electrons, originating from free radicals to PpIX. This molecular mechanism could fundamentally provide the basis for the role of TSPO, not only linking the processes in porphyrin metabolism and oxidative stress regulation but also in other functions associated with the protein in higher metazoans.

11.3 Expression and Distribution of a Pleiotropic Protein

TSPO is expressed across the whole mammalian body, albeit with a tissue and cell-specific mode. As mentioned, the highest expressing tissues are those that produce steroids: the adrenal glands and the gonads. Within the adrenal gland, TSPO expression is limited to the cortex, the layer specifically producing steroids. In decreasing order of transcription, the organs are listed as follows: lung, bone marrow, kidney, spleen, liver, bladder, heart, pancreas, eyes, muscle, bone and brain, with the latter containing 2% of the adrenal's TSPO mRNA at resting condition in mice (Anholt et al. 1985; Banati et al. 2014). Within the CNS, TSPO is expressed at higher levels in microglia and the endothelium (Cosenza-Nashat et al. 2009; Turkheimer et al. 2015), with the localisation in neurons still being controversial. Initial findings showed significant binding of Ro5-4846 to olfactory neurons (Anholt et al. 1984). In vitro cell cultures of primary neurons (Jayakumar et al. 2002; Jordà et al. 2005; Karchewski et al. 2004) and neuronal cell lines (Wu et al. 2015) express TSPO at resting conditions. However, detailed immunohistochemical analysis demonstrated that during development, expression is significant in neuronal precursors but

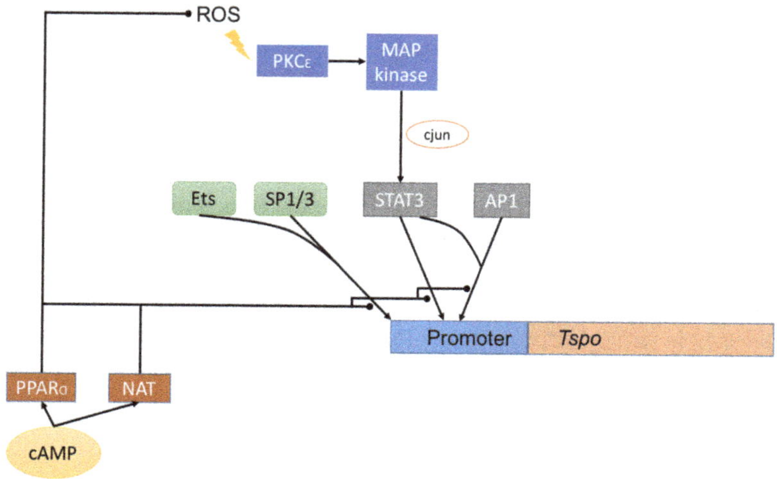

Fig. 11.2 Transcriptional control of TSPO. Ets, SP1/SP3, STAT3 and AP1 can all activate (*arrow*) the promoter of TSPO. Ets and SP1/SP2 are required for basal transcription. STAT3 is activated downstream of the MAP kinase pathway, most notably characterised as being induced by ROS via PKCε. Below PPARα and NAT, activated by cAMP, can work together to inhibit binding of the SP1/SP3-Ets complex (*round arrow*), STAT3 or API to the promoter. Additionally, PPARα can further inhibit ROS

strongly decreases in adult neurons, still allowing for upregulation following neuronal injury (Cosenza-Nashat et al. 2009; Varga et al. 2009). Finally, the presence of TSPO has been confirmed in the various lineages of the immune system and is believed to have a strong functional role within their physiology (Cosenza-Nashat et al. 2009; Costa et al. 2009a; Lee et al. 2015; Zhou et al. 2014).

Subcellular localisation of TSPO has primarily found it to be located in the OMM, but in particular cases, it has also been observed in the nuclear membrane (Hardwick et al. 1999) and in erythrocytic plasma membranes (Olson et al. 1988). In addition, phylogenetic studies have also identified a paralogous protein, TSPO2. This protein most likely arose from gene duplication before the divergence between mammals and avians. TSPO2 is expressed in the endoplasmic reticulum and nuclear membranes and appears to have a much more restricted function as it has only been found to be responsible for cholesterol redistribution from stores to nucleus, an essential step in erythropoiesis (Fan et al. 2009).

The regulatory mechanisms behind the expression and activity of TSPO still remain to be elucidated. Nevertheless, two main pathways of activation have been described, as well as different negative feedback pathways. Despite the absence of a TATA box or CCAAT element, the promoter of *Tspo*, the gene coding for TSPO, contains tandem binding sites for the specificity proteins 1/3 (Sp1/Sp3), regulating the basal transcription of the gene in cooperation with the E26 oncogene homolog (Ets) (Giatzakis and Papadopoulos 2004; Giatzakis et al. 2007) (Fig. 11.2). However, various other, confirmed and putative, transcription factor binding sites have been identified in the promoter, in particular for signal transducer and activator of transcription (STAT) 3 or activator protein (AP) 1 (Batarseh et al. 2012). TSPO's best characterised transcriptional activation pathway is mediated by protein kinase Cε (PKCε) and controlled by ROS levels. This pathway is constitutively activated in MA-10 Leydig cells and inducible in nonsteroidogenic cell lines, like NIH-3T3 fibroblasts and COS-7 kidney cells (Batarseh et al. 2008). Phorbol-12-myristate 13-acetate (PMA), a PKCε inducer (Fay et al. 2006; Huang et al. 2014; Keshari et al. 2013), was shown to activate the MAP kinase pathway and induce nuclear translocation of the transcription factors c-Jun and STAT3, which effected an AP1-mediated induction of transcription (Batarseh et al. 2010). Furthermore, other ROS inducers have been shown to upregulate TSPO in various cell types such as cytokine induction in human pancreatic islets (Trincavelli et al. 2002) and hypoosmotic swelling in astrocytes (Kruczek et al. 2009).

Interestingly, astrocytes provide an example of differential TSPO regulation, as they are also susceptible to estradiol-induced regulation of the protein's activity (Chen et al. 2014). As mentioned earlier, hormones can modulate TSPO expression. This can take place in a wide variety of contexts, for example, in the activation of the HPA axis, during the oestrous cycle or under the control of sex hormones like estradiol (Bitran et al. 1998; Mazurika et al. 2009). However, the precise nature of this process still remains unclear, with a characterised activation pathway of TSPO expression that does not directly interfere with transcription. This pathway starts with a hormone binding to a membrane receptor, for example, estradiol on the oestrogen receptor α (ERα) (Chen et al. 2014). The hormone then activates adenylyl cyclase-1, which catalyses the production of the secondary messenger cyclic adenosine monophosphate (cAMP). The latter induces the relocation of acyl-coenzyme A binding domain containing protein (ACBD) 3 from the Golgi apparatus to the mitochondrion, the phosphorylation of protein kinase A (PKA) and its recruitment to ACBD3 (Fan et al. 2010). The ACBD3-PKA complex in turn phosphorylates TSPO, increasing its steroidogenic capacity (see Fig. 11.3). This surge in activity is thought to justify the higher binding observed in hormone-stimulated tissues. This hypothesis is further supported by the observation that ACTH does not induce TSPO transcriptional increase in Leydig cells (Boujrad et al. 1994). However, the PKCε pathway in this cell line is already activated and therefore not inducible, which might not be the case for non-cancerous cell types. In these cell types, phosphorylation by the recruited PKA pools on the mitochondria could cause VDAC1 pore closure inducing, via the non-accumulated Ca^{2+}, a NADPH oxidase (NOX) 5-mediated increase in ROS which, through PKCε, establish a positive feedback loop to increase TSPO expression further (Gatliff J, East D et al. in press). Furthermore, the upregulation could be mediated by compensatory post-translational mechanism, like the cAMP-induced internalisation of TSPO into the OMM by metaxin 1 (Rone et al. 2009).

Negative feedback mechanisms for regulation of TSPO expression have also been identified, involving the cAMP-dependent natural antisense transcript (NAT) (Fan and Papadopoulos 2012) and the peroxisome proliferator-activated receptor (PPAR) α, which are also activated by cAMP a, reducing ROS levels whilst also impairing transcription factor binding (Boujrad et al. 1994; Gazouli et al. 2002; Vega et al. 2000).

Having framed the complexity and also controversy surrounding TSPO via the increased understanding of its structure and expression, light is likely to be shed on how the protein associates with a wide array of cellular functions and pathologies, as described below.

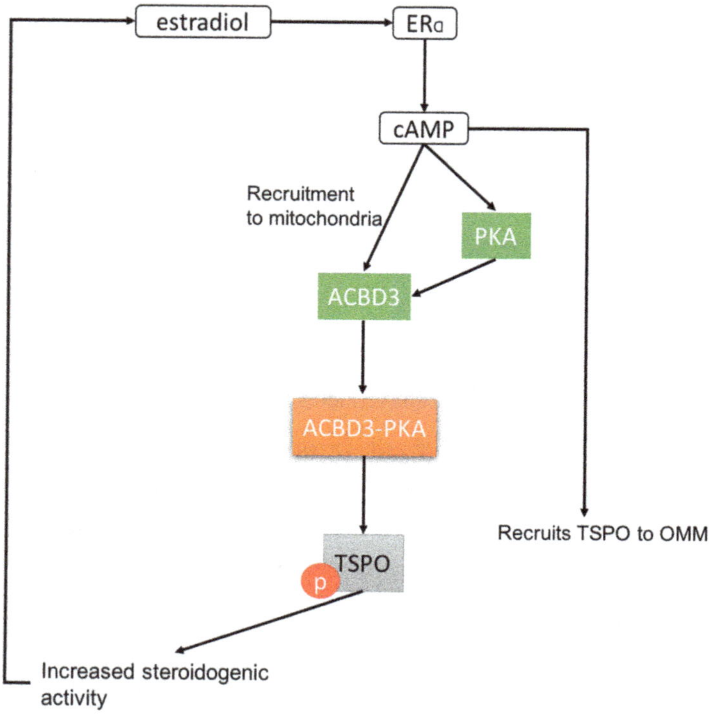

Fig. 11.3 Post-translational modulation of TSPO. As characterised in astrocytes, estradiol can increase cAMP through ERα, which then recruits ACBD3 to the mitochondria where it acts as an adaptor for PKA. The ACBD3-PKA complex can then phosphorylate TSPO. Phosphorylated TSPO displays increased steroidogenic activity, which then feeds back to estradiol, creating a positive feedback loop. cAMP can also control TSPO levels by regulating the internalisation of the protein into the OMM

11.4 The Molecular and Cellular Functions of TSPO

TSPO is involved in a wide variety of different cellular processes revolving around mitochondrial function. A few examples are oxidative stress (Gatliff and Campanella 2015), mitophagy (Gatliff et al. 2014), apoptosis (Werry et al. 2015), porphyrin metabolism (Selvaraj and Stocco 2015) and steroidogenesis (Midzak et al. 2015). This variety of functions is most likely mediated by the interaction with the proteins to which TSPO is known to bind (Fig. 11.4). The most prominent of which are, based on the available literature, the VDAC1 (McEnery et al. 1992), the steroidogenic acute regulatory protein (StAR) (Miller and Auchus 2011), the DBI/ACBD1 and ACBD3 (Fan et al. 2010; Guidotti et al. 1983; Li et al. 2001), the ANT (McEnery et al. 1992) and the 14-3-3γ (Aghazadeh et al. 2012) and PKCε (Aghazadeh et al. 2014). TSPO's vast range of interacting partners, its widespread and variable expression across cell types and species, and the use of varying pharmacological agents in its study have created confusion surrounding its physiological role and

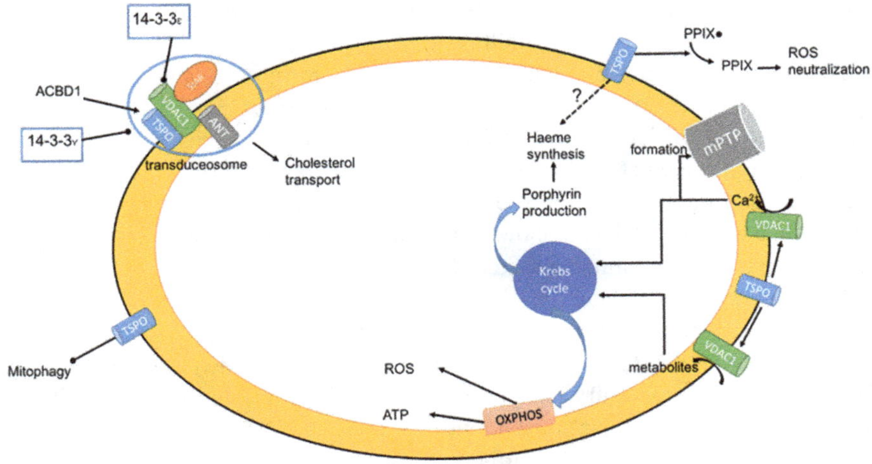

Fig. 11.4 Molecular functions of TSPO. With VDAC1, StAR and ANT, TSPO forms the transduceosome, which is required for putative mitochondrial cholesterol import function associated to TSPO (*arrow*). This complex can be inhibited by 14-3-3Y at different sites (*round arrow*), whilst 14-3-3ℇ binds directly to VDAC1 to do so. ACBD1 can instead enhance TSPO activity by direct binding. The tryptophan residues in the active site of TSPO can mediate free radical transfer to PpIX, leading to ROS neutralization. TSPO can modulate VDAC1 gating, thereby controlling Ca2+ entry into the mitochondria, which can modulate mPTP formation. Additionally, metabolite transport by VDAC1, again under control by TSPO, can modulate the Krebs cycle. The metabolites of the Krebs cycle can either be used for porphyrin production or feed into the oxidative phosphorylation (*OXPHOS*) process, leading to ATP production and ROS formation. TSPO can modulate haeme synthesis in two different ways: by regulating the availability of porphyrin through Krebs-cycle control or indirectly by mediating the porphyrin-dependent ROS neutralisation process. This link has not been fully investigated (*dotted arrow*). Lastly, TSPO can also block mitophagy

suggest that it may hold a highly context-dependent function. However, there are a few molecular mechanisms which the authors hypothesise as underlying TSPO's function and are listed below.

Steroidogenesis. Steroidogenesis begins when cholesterol is initially catalysed into pregnenolone by a specific isoform of cytochrome P450. Different tissue-dependent pathways can then synthesise specific steroids, which in turn exert an effect as hormones on target tissues (Miller and Auchus 2011). This process takes place in a tissue-dependent manner, according to the distribution of the pathway-specific enzymes (Miller and Auchus 2011), with the presence of the precursor in the mitochondria being the rate-limiting factor. Cholesterol is carried from intracellular stores, or low density lipoproteins (LDL), to the OMM by StAR and is then translocated to the inner mitochondrial membrane. Then it is catalysed into pregnenolone, by a translocator complex, or "transduceosome", constituted of StAR, VDAC1, ANT and TSPO (Krueger and Papadopoulos 1990; Liu et al. 2006; Papadopoulos et al. 1997a). Most of the proteins known to interact with TSPO have been shown to regulate its steroidogenic capacity. 14-3-3ε forms a complex with VDAC1, therefore inhibiting steroidogenesis. DBI, later renamed ACBD1 due to its structural motif, binds directly and potentiates the effect of TSPO (Alho et al. 1991; Gray et al. 1986). Additionally, fragments of DBI, or endozepines, have been shown to physiologically act in a similar fashion (Besman et al. 1989; Papadopoulos et al. 1991). As already mentioned, ACBD3 mediates a PKA-induced activation of TSPO. 14-3-3γ binds to the transduceosome to inhibit steroid formation (Aghazadeh et al. 2012).

Despite the intimate relationship between TSPO and steroidogenesis, global genetic mice ablation of TSPO showed no genetic lethality and only minor phenotypic differences in the immune system (Banati et al. 2014; Tu et al. 2014). These results raised the fundamental question whether TSPO is actually necessary for the functions associated with it or whether previous results were due to the unspecific action of its ligands. While this might be true for cases in which the ligands were used at micromolar concentrations, original studies have shown that the affinity of PK11195 reaches nanomolar concentration (Blahos et al. 1995; Campanella et al. 2008; Papadopoulos et al. 1990). Furthermore, Banati et al. (2014) showed the PET-specific ligand dosages did not result in any unspecific binding in TSPO KO mouse tissues. However, Tu et al. (2014) reported basal and ACTH-stimulated steroid production to be unmodified by the absence of the TSPO. This was supported by another publication which showed TSPO KO MA-10 Leydig cells were as responsive to PK11195 treatment as the wild-type cells (WT) (Tu et al. 2015). However these results have been recently contradicted by data on mice with conditional KO in steroidogenic cells driven by the nuclear receptor 5 subfamily a 1 (Nr5a1) transcription factor, which display a defect in ACTH-induced steroid production (Fan et al. 2015). In support of this model, we have found the first publication reporting genetic disruption of TSPO in steroidogenesis, carried out in an R2C Leydig tumour cell line (Papadopoulos et al. 1997b). Interestingly, in this model the mutation is not homozygous, as the cell type in question is polyploidic, containing more than two homologous sets of chromosomes. In this model the absence of a complete set of TSPO genes would create a partial loss of function, but a homozygous mutation would trigger a redundant pathway to fulfil the function normally carried out by TSPO, therefore explaining these apparently contradicting results. This hypothesis is supported by the observation that the A147T human polymorphism on the TSPO gene is responsible for a spectrum of anxiety-related disorders by interfering with microglial neurosteroid production (Colasanti et al. 2013; Costa et al. 2009a, b; Owen et al. 2012). This microglial-specific impairment caused by mutations in TSPO is additionally supported by the KO mice model published by Banati et al. (2014), in which microglia is the only cell type reported to show aberrant behaviour. Furthermore, a role for TSPO in psychiatric disorders is strengthened by the finding of a new generation highly specific ligands successfully used to produce anxiolytic effects by increasing the production of neurosteroids (Rupprecht et al. 2009).

Another possible explanation for the absence of predicted phenotype in the TSPO KO mouse associates with its function as a stress-response protein. The mouse genetic KO studies did not contain analysis of the efficiency of the TSPO-void system against stress; this will most definitely be the next step as the current wealth of data points to a role for this protein in stress physiology. Lastly, it is important to keep in mind that genetic KO models in mammalian systems are not always suitable to test mitochondrial proteins. In contrast to murine studies, genetic inactivation of TSPO in *Drosophila melanogaster* has shown to be protective against apoptotic stimuli, neurodegeneration and ageing (Lin et al. 2014). This same result occurred in the search for a function for the PTEN-induced putative kinase 1 (PINK1), which is a mitochondrial protein responsible for initiating stress-induced mitophagy (Lazarou et al. 2015). Whilst the importance of PINK1 for mitochondrial physiology is confirmed, mouse KO models of PINK1 showed only a mild phenotype (Gautier et al. 2008) in comparison to the *D. melanogaster* model (Clark et al. 2006). Therefore, mouse models of genetic KO provide an excellent control for unspecificity of ligands; however they are not sufficient to refute previous studies on mitochondrial functions.

Porphyrin metabolism. Porphyrins are a class of heterocyclic macrocycle organic compounds, which are precursors to a wide variety of molecules, like chlorophyll and haemoglobin. Porphyrins are also highly reactive and induce oxidative stress when accumulated in the cell. Porphyrin is an endogenous ligand of TSPO whose relationship has been conserved from bacteria to metazoans. In the aforementioned structural biology study with *Bc*TspO, it was shown that the tryptophan residues in the active site of the protein could mediate a free radical transfer to PpIX (Guo et al. 2015). This ROS neutralisation effect was found to be fundamental in counteracting porphyrin toxicity in prokaryotes, as well as in the human liver, haematopoietic system and osteoblasts (Bloomer 1998; Carayon et al. 1996; Ginter et al. 2013; Rosenberg et al. 2014). This function acquires paramount importance in plant systems, with studies on *Arabidopsis thaliana*, and the moss *Physcomitrella patens*, providing evidence regarding TSPO's role in redox homeostasis and stress adaptation (Frank et al. 2007; Lehtonen et al. 2012; Vanhee et al. 2011). *P. patens* upregulates TSPO in response to fungal infection, with a concomitant increase in the expression of ROS-producing proteins as part of the antimicrobial oxidative burst. In TSPO KOs of *P. patents*, oxidative stress is observed at resting condition, with a constitutive upregulation of ROS-producing proteins, excluding NADPH oxidase (NOX). The latter is an important group of cytoplasmic ROS producers to which this study provides one of the few published links between TSPO and NOX to date, with fundamental ramifications on the mechanism of ROS regulation that will be discussed later. In *A. thaliana*, abscisic acid-induced stress also stimulates TSPO expression, which results in degradation of accumulated porphyrins via the classical eukaryotic autophagy pathway (Vanhee et al. 2011). Lastly, genetic interference of TSPO expression in *Danio rerio* and *Gallus gallus domesticus* demonstrates its essential role in haeme production, and erythropoiesis, which is not observed in mice (Nakazawa et al. 2009; Rampon et al. 2009; Zhao et al. 2015).

The novel TSPO-associated processes. TSPO has in fact been linked to various other aspects of mitochondrial physiology, like autophagy, respiration, oxidative stress and Ca^{2+} transport. A prominent hypothesis which has recently arisen from the literature relates these functions to the interaction of TSPO and VDAC1. The latter is a large OMM channel, with a prominent β-barrel structure and a small internal helix working as a plug that can control the travel of metabolites and Ca^{2+} across the mitochondrial membranes. By controlling the travel of Krebs cycle substrates and Ca^{2+}, VDAC1 can simultaneously regulate different processes of mitochondrial physiology, such as ROS production, ATP production, mitophagy and apoptosis (Lemasters and Holmuhamedov 2006; Rizzuto et al. 2012; Shoshan-Barmatz et al. 2015; Ujwal et al. 2008).

VDAC1 conductance is also modulated by kinases (Sheldon et al. 2011), an action thought to be exploited by TSPO as a mean to control its gating. TSPO is believed to control VDAC1 gating by differential expression, with a high TSPO to VDAC ratio facilitating a closed state. When abundant, the 18 kDa protein allows higher phosphorylation of VDAC by PKA by providing a larger docking area for the kinase, a process mentioned in Sect. 3. Indeed, VDAC1 conductance of Ca^{2+} has been shown to be modulated by a TSPO ligand in rat heart mitochondria (Ct et al. 2008); unpublished work within our research group has shown that genetic alteration of TSPO levels controls VDAC1-mediated Ca^{2+} fluxes (Gatliff J, East D et al. in press). The TSPO-VDAC interaction is likely relevant to the search of the protein's function for different reasons. Firstly, the existence of a highly inducible protein that can control Ca^{2+} entry into the mitochondria provides evidence for the dynamic state of mitochondrial Ca^{2+} regulation. Moreover, this regulatory function unravels the intimate relation between TSPO and stress response. Electron microscopy experiments show that genetic overexpression of TSPO, in the absence of other stressors, damages the ultrastructure of the mitochondrion and decreases cristae number. Furthermore, it was shown to increase basal levels of ROS and impairing FCCP-induced mitophagy, which "recycles" damaged mitochondria (Gatliff et al. 2014). This work brings to light the nature of the TSPO, which appears to function as a main regulatory switch for the cell to block physiological mitochondrial functions and redirect the organelle to stress-directed metabolism. Indeed, this hypothesis can potentially explain the apparent contradictory conclusions of the protein's importance for its associated processes. Therefore, to better comprehend the functions described below, the interaction between TSPO and VDAC1 and the stress-induced nature of the function must be taken into account.

The mPTP culprit. A vast quantity of literature links TSPO with the mitochondrial permeability transition pore (mPTP). The latter is a large structure spanning both mitochondrial membranes that form under conditions of mitochondrial Ca^{2+} overload, mitochondrial oxidative stress, adenine nucleotide depletion and elevated phosphate concentrations, allowing for the release of cytochrome c and thereby initiating the apoptotic cascade (Halestrap 2009). Structurally, the pore has not yet been completely defined, as it could be potentially made up from opportunistically formed complexes; however it is known to be closed by the protein cyclophilin-D and the pharmacological agent cyclosporin A. TSPO was initially hypothesised to form a pore in complex with VDAC1 and ANT, which is supported by a vast number of pharmacological studies (Veenman et al. 2007). Furthermore, the high expression of the protein in different cancers, for example, breast cancers and glioblastomas, also credits this hypothesis (Hardwick et al. 1999; Werry et al. 2015). However, studies performed with ligands used high micromolar concentrations, which can cause TSPO-independent effects – an aspect to be revisited in the following section.

Moreover, independent studies showed that ANT was not part of the mPTP but rather that dimers of the mitochondrial ATP synthase constitute the structural basis of the pore (Giorgio et al. 2013; Kokoszka et al. 2004; Sileikyte et al. 2011). Therefore, the current consensus is that TSPO and VDAC1 and ANT are not structurally part of the mPTP but can regulate its opening via control of Ca^{2+} and adenine nucleotide channelling.

Production and regulation of ROS. ROS are group of molecules containing an oxygen atom which either already harbours an unpaired electron or can easily obtain one. The most commonly known ROS are the superoxide anion, O_2^-, hydrogen peroxide, H_2O_2, and hydroxyl radical, OH. Within the cell, ROS can react with macromolecules of many kinds to generate dysfunctional products, thus inducing stress (Court and Coleman 2012; Dias et al. 2013; Ghosh et al. 2011). There is general agreement within the scientific community that ROS is mainly produced within mitochondria as part of the leakage process within the electron transport chain (ETC) that can lead to incomplete reduction of molecular oxygen (Murphy 2009). ROS production at low physiological levels is used by the mitochondrion to modulate cell function (Dröge 2002), and healthy levels of ROS are maintained by a variety of antioxidant systems, such as peroxiredoxins and glutathione (GSH) (Dias et al. 2014; Murphy 2009). Furthermore, ROS is also physiologically produced in the cytoplasm, by either NOX (Bedard and Krause 2007) or xanthine oxidases (Harrison 2002), even though they are believed to contribute less than mitochondria towards total cellular ROS levels. TSPO regulates ROS in different ways. By decreasing VDAC1 gating, the protein can decrease the rate of mitochondrial respiration and the consequent ETC-mediated ROS production. As mentioned previously, through participating in the porphyrin metabolism, TSPO can also neutralise the ROS produced by PpIX, although data suggests that stress-induced TSPO upregulation decreases binding rather inducing further neutralisation (Lehtonen et al. 2012; Vanhee et al. 2011). A spike in NOX upregulation is also observed upon TSPO upregulation in *P. patens*, as a tool to damage invading fungal pathogens. NOX is an important class of cytoplasmic ROS producers used by the mammalian immune system for a variety of purposes, including host defence (Bedard and Krause 2007). NOX5 is an isoform that is activated by a cytoplasmic Ca^{2+} via calmodulin-dependent kinase II (Pandey et al. 2011). Work by the Campanella group highlights the presence of this mechanism in neuronal excitotoxicity. Overexpression of TSPO can decrease VDAC1-mediated Ca^{2+} uptake into the mitochondrion, thereby activating NOX5 and inducing oxidative stress (Gatliff J, East D et al. in press). This dichotomy between oxidative bursts in host defence and oxidative stress in neurons provides a hint as to the low TSPO expression levels in neurons at physiological conditions.

Mitochondria-targeted autophagy. The link between TSPO and autophagy has also recently been unveiled. The latter is a process in which portions of the cytoplasm, or entire organelles, are sequestered by a double membrane and transported to the lysosome for degradation (Lemasters 2005; Matic et al. 2015). Autophagy can be unselective, to sustain cell metabolism during starvation, or targeted to specific organelles, like mitochondria, or even exogenous microorganisms. Nevertheless, upstream initiator proteins, like microtubule-associated protein light chain 3 (LC3) (Tanida et al. 2004), are conserved across eukaryotes and between different types of autophagy (Randow and Youle 2014). As mentioned above, mitophagy induced by FCCP, an ionophore that shuttles protons across the membrane and causes instant membrane depolarisation, is impaired by TSPO in different mammalian cell lines via a ROS and VDAC1-mediated mechanism (Gatliff et al. 2014). In *A. thaliana*, TspO bound to porphyrin is degraded by autophagy via a putative direct interaction with autophagy-related protein 8 (Atg8), an ortholog of LC3, through a putative Atg8-interacting motif (Vanhee et al. 2011). In natural killer cells, TSPO expression has been demonstrated to disrupt human immunodeficiency virus (HIV) replication by simultaneously inducing a low oxidative state in the ER and the ER-associated degradation (ERAD) pathway, which together impair folding and induce degradation of a fundamental protein of the HIV structural scaffold (Zhou et al. 2014).

TSPO in the homeostasis of cellular energy. First of all, VDAC1 gating can control a switch from respiratory ATP production to glycolysis, named Warburg effect in cancer cells (Maldonado and Lemasters 2012). Consequently, TSPO upregulation can be hypothesised to induce a proliferative state with higher glucose consumption and usage of glucose metabolites for anabolic functions. Furthermore, a whole organism zebrafish embryo screening for activators of gluconeogenesis has shown that TSPO ligands, especially PK11195, are potent "hits" (Gut et al. 2013). PK11195 was seen to upregulate different key inducers of the gluconeogenic fasting response, like phosphoenolpyruvate carboxykinase (PEPCK), as well as 5′-aminolevulinate synthase 1 (ALAS-1), the rate-limiting enzyme in the production of haeme. The mechanism via which TSPO can be linked to these processes requires an insight into how haeme can control metabolic pathways. Gluconeogenesis and haeme synthesis compete with aerobic respiration for Krebs cycle intermediates: PEPCK mediates the rate-limiting step in gluconeogenesis, i.e. the catalysis of a Krebs cycle intermediate to phosphoenolpyruvate (Gut et al. 2013), and ALAS-1 regulates haeme biosynthesis by catalysing succinyl CoA (Ponka 1999). These pathways are positively regulated during fasting by PPAR-gamma coactivator-1α (PGC-1α), a master regulator of mitochondrial biogenesis (Handschin et al. 2005; Puigserver et al. 2003). On the other hand, in fed conditions, haeme levels oscillate according to circadian rhythms and control Rev-Erbα, a transcription factor which suppresses gluconeogenesis, both functionally and physically dependent on the molecule (Yin et al. 2007). The putative mechanism with which PK11195 could induce gluconeogenesis would be as follows. PK11195, as described by Guo et al. (2015) in the aforementioned structural biology study, shows competitive binding with porphyrins on the binding site for TSPO. Additionally, the latter seems to be required for haeme biosynthesis, at least in zebrafish and chicken. If PK11195 displaces porphyrin binding or causes haeme synthesis to be defective, the action of Rev-Erbα would be inhibited, inducing a fasting state. Other publications have re-enforced the involvement of TSPO in energy regulation: in TSPO KO Leydig cells, a shift in fatty acid oxidation was observed (Tu et al. 2016), while TSPO ligands, oppositely to genetic TSPO downregulation, were shown to increase glucose uptake, transcription of PGC-1α and release of anti-inflammatory adipokines in adipocytes (Li and Papadopoulos 2015).

TSPO and its interplay with inflammation. Inflammation is defined as an adaptive response that is triggered by noxious stimuli and conditions, presumably developed with aim of restoring homeostasis (Medzhitov 2008). Inflammatory diseases are in fact thought to develop following misregulation of inflammatory pathways, with consequential untargeted effects and prolongation of a non-homeostatic state. Perhaps, this concept is key to understanding the tight relationship between the protein, disease and wide ranging of positive effects produced by the ligands. Nevertheless, despite overwhelming evidence supporting a link between the inflammation and TSPO, a general consensus is still missing on whether the protein is inflammatory or purely functioning as a marker for the process. The strong upregulation associated with neuroinflammation, and the influence ROS levels have on TSPO transcription, highlights the role this protein plays as a consequence to injury. More detailed investigations showed that in a monocytic cell line, Ro5-4846 attenuates ATP-induced formation of the nod-like receptor family, pyrin domain containing 3 (NLRP3) inflammasome (Lee et al. 2016). The TSPO binding ligands were also found to inhibit pro-inflammatory cytokine release in BV2 cells, a murine microglial cell line, and in vivo in mice. Most importantly, the same study showed that TSPO expression negatively correlates with markers of neuroinflammation (Bae et al. 2014). Furthermore, XBD173, a highly selective ligand used to demonstrate the GABAergic synaptic inhibition of neurosteroid production (Rupprecht et al. 2009), was also shown to downregulate retinal microglial inflammation as well as its phagocytic capacity (Karlstetter et al. 2014). Despite the work of Bae et al., it is still unclear whether these functions are dependent on the anti-inflammatory effect of steroids or on the other functions associated with the protein. Nevertheless it is important to take into consideration that from the point of view of the mitochondrion, upregulation of TSPO is indeed an inflammatory event, as it promotes a switch to stress-related, and therefore non-homeostatic, processes like gluconeogenesis, glycolytic metabolism and various other functions elaborated above.

11.5 The Role of TSPO in Molecular and Cellular Dysfunctions

The molecular functions described above should now enable the reader to more easily comprehend the connection between TSPO and pathological conditions. While the literature presents a wide range of explanations for the protective actions of TSPO-targeting pharmacological ligands, we will attempt to only refer to those functions mentioned above.

TSPO is expressed in a range of different cancers, such as breast cancer (Hardwick et al. 1999), glioblastoma (Werry et al. 2015), prostate cancer (Galiègue et al. 2004), endometrial carcinoma (Batra and Iosif 2000) and oesophageal cancer (Sutter et al. 2002).

TSPO may not only serve as a biomarker in these cancers, with its expression 15-fold higher in glioblastomas compared to normal brain tissue, for example, but levels of TSPO expression within the different cancer types have also been correlated with tumorogenicity (Galiègue et al. 2004; Hardwick et al. 1999; Janczar et al. 2015; Veenman et al. 2004). Studies investigating TSPO as a potential target for therapeutic treatment have focused on breast cancer and glioblastomas finding that apoptosis induced by cobalt chloride or erucylphosphohomocholine, are inhibited by ligand treatment (Veenman et al. 2014; Zeno et al. 2009). As cancer cells upregulate anti-apoptotic proteins such as B-cell lymphoma 2 (Bcl-2), interventions have focused on initiating cell death. One mechanism through which that can be achieved is the release of cytochrome c, an inducer of apoptosis, from the mitochondria in response to the permeabilisation of its membrane. By facilitating mPTP opening in tumour cells, which overexpress TSPO, TSPO-inactivating ligands can specifically target these cells, exemplifying the targeted therapy mechanism. It is worth mentioning that different ligands can have opposing effects on the cancer phenotype. This is not only caused by the fact that different ligands bind to different sites within the protein, e.g. Ro5-4846 in comparison to the PK11195 binding described above (Farges et al. 1994), but also by unspecific binding. PK11195 has been observed to exert its effect independently of TSPO knockdown in HeLa cells (Gonzalez-Polo et al. 2005), with a potential mechanism mediated via the F_1F_o-ATPase (Seneviratne et al. 2012). Regardless, TSPO targeting produces an anti-apoptotic effect also by acting on proliferation. Studies in C6 glioma cells have shown that TSPO knockdown causes a decrease in proliferation, whilst overexpressing TSPO has the opposite effect, potentiating proliferation (Rechichi et al. 2008). One of the key features of cancer cell metabolism is in fact the focus of ATP production on the glycolytic pathway. The latter is less effective in energetic terms compared to the mitochondrial Krebs cycle and electron transport chain but can function anaerobically and provide carbon-based compounds as biomass for proliferation (Lemasters and Holmuhamedov 2006; Shoshan-Barmatz et al. 2015). Alteration of the TSPO-VDAC interactions could potentially revert the Warburg effect by promoting mitochondrial respiration.

TSPO is also likely to be involved in neurodegenerative conditions. As mentioned above, expression of TSPO is increased in glia cells following a wide range of brain injuries. These can include progressive disorders such as MS, amyotrophic lateral sclerosis, Alzheimer's disease (AD), PD and Huntington's disease but also acute traumas such as stroke, traumatic brain injury, gliomas and HIV encephalitis (Cagnin et al. 2007; Cosenza-Nashat et al. 2009; Gavish et al. 1999; Rupprecht et al. 2010). In a majority of these disorders, neuroinflammation plays a role, believed to lead to the symptoms observed in the patients. Whilst in AD, for example, neurodegeneration seems to predispose to neuroinflammation, diseases like MS are characterised by an initial inflammatory event initiating neurodegeneration (Naegele and Martin 2014). The relationship between neuroinflammation and neurodegeneration is not fully characterised as of yet. However, there is strong evidence for a positive feedback mechanism between the two (Rupprecht et al. 2010). Not only has

TSPO upregulation been observed in vivo in these diseases, but TSPO ligands have also been shown to be widely protective against the underlying pathology in a wide number of experimental models of CNS diseases (Da Pozzo et al. 2015), even though the exact mechanisms of action have not been fully elucidated yet. The role in steroidogenesis could suggest an anti-inflammatory effect of steroids on microglia. Additionally, since TSPO has been observed in vivo in injured neurons (Cosenza-Nashat et al. 2009; Palzur et al. 2016), its wider role in neurodegeneration, by the improvement of mitochondrial physiology, can be imagined. Work within the Campanella group is elucidating the role of the various molecular mechanisms in an in vitro model of PD: inhibition of mitophagy and induction of oxidative stress appear potential mechanisms for the neuroprotective action of the TSPO ligands.

A wide number of studies have also linked TSPO with cardiovascular diseases (Morin et al. 2016; Qi et al. 2012; Veenman and Gavish 2006). Firstly, TSPO ligands have shown to be protective against ischemia and reperfusion (I/R). I/R damage is caused when cells are insufficiently supplied with oxygen leading to a disruption of ATP synthesis and subsequently cell membrane depolarisation. Once the tissue is reperfused, however, a burst of mitochondrial ROS production is caused as oxygen becomes available once again. The formation of the mPTP following this oxidative burst successively activates the apoptotic pathway. TSPO ligands, by impairing mPTP formation, can block this process, therefore conferring protection. I/R can occur in cardiac cells during myocardial infarction, which is responsible for the largest part of cardiovascular mortality, or during stroke. TSPO ligands have also been found to reduce infarct size and other cardiovascular malfunctions like arrhythmia and cardiac hypertrophy (Qi et al. 2012). A link between TSPO and diabetes, or other disorders of energy regulation, is provided by the mechanistic link with gluconeogenesis and cellular energy regulation. By inducing gluconeogenesis, TSPO can revert insulin insensitivity and glucose tolerance, as demonstrated by a zebrafish study (Gut et al. 2013). Furthermore, in vitro work in adipocytes has shown that TSPO ligands can improve the inflammatory status and induce similar outcomes (Li and Papadopoulos 2015). Lastly, work carried out in TSPO KO Leydig cells has shown a shift to fatty acid oxidation, which, if fine-tuned, could be used by ligand treatment for disorders of energy regulation (Tu et al. 2016).

11.6 Conclusions

By concomitantly pinpointing the past notions and informing on the most recent findings, this chapter provides the reader with an organic overview on the available knowledge on this fascinating protein. Standing as a prime element in cellular and systemic functions as well as dysfunctions, TSPO has been historically exploited as a target for novel biochemically designed diagnostics and therapeutics. Its high degree of conservation, the peculiarity of the structure and role in cholesterol trafficking depict TSPO as a core element in mitochondrial homeostasis and signalling.

Acknowledgements The research activities on TSPO led by M.C. are supported by the following funders gratefully acknowledged: Biotechnology and Biological Sciences Research Council [grant number BB/M010384/1]; the Medical Research Council [grant number G1100809/2]; Bloomsbury Colleges Consortium PhD Studentship Scheme; the London Interdisciplinary Biosciences Consortium; the Petplan Charitable Trust; Umberto Veronesi Foundation, Marie Curie Actions, the LAM-Bighi Grant Initiative and the Italian Ministry of Health [IFO14/01/R/52].

References

Aghazadeh Y, Rone MB, Blonder J, Ye X, Veenstra TD, Hales DB, Culty M, Papadopoulos V (2012) Hormone-induced 14-3-3γ adaptor protein regulates steroidogenic acute regulatory protein activity and steroid biosynthesis in MA-10 Leydig cells. J Biol Chem 287:15380–15394

Aghazadeh Y, Martinez-Arguelles DB, Fan J, Culty M, Papadopoulos V (2014) Induction of androgen formation in the male by a TAT-VDAC1 fusion peptide blocking 14-3-3ε protein adaptor and mitochondrial VDAC1 interactions. Mol Ther J Am Soc Gene Ther 22:1779–1791

Alho H, Costa E, Ferrero P, Fujimoto M, Cosenza-Murphy D, Guidotti A (1985) Diazepam-binding inhibitor: a neuropeptide located in selected neuronal populations of rat brain. Science 229:179–182

Alho H, Harjuntausta T, Schultz R, Pelto-Huikko M, Bovolin P (1991) Immunohistochemistry of diazepam binding inhibitor (DBI) in the central nervous system and peripheral organs: its possible role as an endogenous regulator of different types of benzodiazepine receptors. Neuropharmacology 30:1381–1386

Amiri Z, Weizman R, Katz Y, Burstein O, Edoute Y, Lochner A, Gavish M (1991) Testosterone and cyproterone acetate modulate peripheral but not central benzodiazepine receptors in rats. Brain Res 553:155–158

Anholt RR, Murphy KM, Mack GE, Snyder SH (1984) Peripheral-type benzodiazepine receptors in the central nervous system: localization to olfactory nerves. J Neurosci Off J Soc Neurosci 4:593–603

Anholt RR, De Souza EB, Oster-Granite ML, Snyder SH (1985) Peripheral-type benzodiazepine receptors: autoradiographic localization in whole-body sections of neonatal rats. J Pharmacol Exp Ther 233:517–526

Anholt RR, Pedersen PL, De Souza EB, Snyder SH (1986) The peripheral-type benzodiazepine receptor. Localization to the mitochondrial outer membrane. J Biol Chem 261:576–583

Austin CJD, Kahlert J, Kassiou M, Rendina LM (2013) The translocator protein (TSPO): a novel target for cancer chemotherapy. Int J Biochem Cell Biol 45:1212–1216

Bae K-R, Shim H-J, Balu D, Kim SR, Yu S-W (2014) Translocator protein 18 kDa negatively regulates inflammation in microglia. J Neuroimmune Pharmacol Off J Soc NeuroImmune Pharmacol 9:424–437

Banati RB (2002) Visualising microglial activation in vivo. Glia 40:206–217

Banati RB, Middleton RJ, Chan R, Hatty CR, Kam WW-Y, Quin C, Graeber MB, Parmar A, Zahra D, Callaghan P et al (2014) Positron emission tomography and functional characterization of a complete PBR/TSPO knockout. Nat Commun 5:5452

Bar-Ami S, Fares F, Gavish M (1989) Effect of hypophysectomy and hormone treatment on the induction of peripheral-type benzodiazepine binding sites in female rat genital axis. Horm Metab Res Horm Stoffwechselforschung Horm Métabolisme 21:106–107

Batarseh A, Giatzakis C, Papadopoulos V (2008) Phorbol-12-myristate 13-acetate acting through protein kinase Cepsilon induces translocator protein (18-kDa) TSPO gene expression. Biochemistry (Mosc) 47:12886–12899

Batarseh A, Li J, Papadopoulos V (2010) Protein kinase C epsilon regulation of translocator protein (18 kDa) Tspo gene expression is mediated through a MAPK pathway targeting STAT3 and c-Jun transcription factors. Biochemistry (Mosc) 49:4766–4778

Batarseh A, Barlow KD, Martinez-Arguelles DB, Papadopoulos V (2012) Functional characterization of the human translocator protein (18kDa) gene promoter in human breast cancer cell lines. Biochim Biophys Acta 1819:38–56

Batra S, Iosif CS (2000) Peripheral benzodiazepine receptor in human endometrium and endometrial carcinoma. Anticancer Res 20:463–466

Bedard K, Krause K-H (2007) The NOX family of ROS-generating NADPH oxidases: physiology and pathophysiology. Physiol Rev 87:245–313

Benavides J, Malgouris C, Imbault F, Begassat F, Uzan A, Renault C, Dubroeucq MC, Gueremy C, Le Fur G (1983a) "Peripheral type" benzodiazepine binding sites in rat adrenals: binding studies with [3H]PK 11195 and autoradiographic localization. Arch Int Pharmacodyn Thérapie 266:38–49

Benavides J, Quarteronet D, Imbault F, Malgouris C, Uzan A, Renault C, Dubroeucq MC, Gueremy C, Le Fur G (1983b) Labelling of "peripheral-type" benzodiazepine binding sites in the rat brain by using [3H]PK 11195, an isoquinoline carboxamide derivative: kinetic studies and autoradiographic localization. J Neurochem 41:1744–1750

Benavides J, Menager J, Burgevin MC, Ferris O, Uzan A, Gueremy C, Renault C, Le Fur G (1985) Characterization of solubilized "peripheral type" benzodiazepine binding sites from rat adrenals by using [3H]PK 11195, an isoquinoline carboxamide derivative. Biochem Pharmacol 34:167–170

Benavides J, Fage D, Carter C, Scatton B (1987) Peripheral type benzodiazepine binding sites are a sensitive indirect index of neuronal damage. Brain Res 421:167–172

Bernassau JM, Reversat JL, Ferrara P, Caput D, Lefur G (1993) A 3D model of the peripheral benzodiazepine receptor and its implication in intra mitochondrial cholesterol transport. J Mol Graph 11(236–244):235

Besman MJ, Yanagibashi K, Lee TD, Kawamura M, Hall PF, Shively JE (1989) Identification of des-(Gly-Ile)-endozepine as an effector of corticotropin-dependent adrenal steroidogenesis: stimulation of cholesterol delivery is mediated by the peripheral benzodiazepine receptor. Proc Natl Acad Sci U S A 86:4897–4901

Bitran D, Carlson D, Leschiner S, Gavish M (1998) Ovarian steroids and stress produce changes in peripheral benzodiazepine receptor density. Eur J Pharmacol 361:235–242

Blahos J, Whalin ME, Krueger KE (1995) Identification and purification of a 10-kilodalton protein associated with mitochondrial benzodiazepine receptors. J Biol Chem 270:20285–20291

Bloomer JR (1998) Liver metabolism of porphyrins and haem. J Gastroenterol Hepatol 13:324–329

Boujrad N, Gaillard JL, Garnier M, Papadopoulos V (1994) Acute action of choriogonadotropin on Leydig tumor cells: induction of a higher affinity benzodiazepine-binding site related to steroid biosynthesis. Endocrinology 135:1576–1583

Braestrup C, Squires RF (1977) Specific benzodiazepine receptors in rat brain characterized by high-affinity (3H)diazepam binding. Proc Natl Acad Sci U S A 74:3805–3809

Braestrup C, Albrechtsen R, Squires RF (1977) High densities of benzodiazepine receptors in human cortical areas. Nature 269:702–704

Cagnin A, Kassiou M, Meikle SR, Banati RB (2007) Positron emission tomography imaging of neuroinflammation. Neurother J Am Soc Exp Neurother 4:443–452

Calogero AE, Kamilaris TC, Bernardini R, Johnson EO, Chrousos GP, Gold PW (1990) Effects of peripheral benzodiazepine receptor ligands on hypothalamic-pituitary-adrenal axis function in the rat. J Pharmacol Exp Ther 253:729–737

Campanella M, Szabadkai G, Rizzuto R (2008) Modulation of intracellular Ca2+ signalling in HeLa cells by the apoptotic cell death enhancer PK11195. Biochem Pharmacol 76:1628–1636

Carayon P, Portier M, Dussossoy D, Bord A, Petitprêtre G, Canat X, Le Fur G, Casellas P (1996) Involvement of peripheral benzodiazepine receptors in the protection of hematopoietic cells against oxygen radical damage. Blood 87:3170–3178

Chelli B, Falleni A, Salvetti F, Gremigni V, Lucacchini A, Martini C (2001) Peripheral-type benzodiazepine receptor ligands: mitochondrial permeability transition induction in rat cardiac tissue. Biochem Pharmacol 61:695–705

Chen C, Kuo J, Wong A, Micevych P (2014) Estradiol modulates translocator protein (TSPO) and steroid acute regulatory protein (StAR) via protein kinase A (PKA) signaling in hypothalamic astrocytes. Endocrinology 155:2976–2985

Clark IE, Dodson MW, Jiang C, Cao JH, Huh JR, Seol JH, Yoo SJ, Hay BA, Guo M (2006) Drosophila pink1 is required for mitochondrial function and interacts genetically with parkin. Nature 441:1162–1166

Colasanti A, Owen DR, Grozeva D, Rabiner EA, Matthews PM, Craddock N, Young AH (2013) Bipolar disorder is associated with the rs6971 polymorphism in the gene encoding 18 kDa translocator protein (TSPO). Psychoneuroendocrinology 38:2826–2829

Cosenza-Nashat M, Zhao M-L, Suh H-S, Morgan J, Natividad R, Morgello S, Lee SC (2009) Expression of the translocator protein of 18 kDa by microglia, macrophages and astrocytes based on immunohistochemical localization in abnormal human brain. Neuropathol Appl Neurobiol 35:306–328

Costa B, Pini S, Gabelloni P, Da Pozzo E, Abelli M, Lari L, Preve M, Lucacchini A, Cassano GB, Martini C (2009a) The spontaneous Ala147Thr amino acid substitution within the translocator protein influences pregnenolone production in lymphomonocytes of healthy individuals. Endocrinology 150:5438–5445

Costa B, Pini S, Martini C, Abelli M, Gabelloni P, Landi S, Muti M, Gesi C, Lari L, Cardini A et al (2009b) Ala147Thr substitution in translocator protein is associated with adult separation anxiety in patients with depression. Psychiatr Genet 19:110–111

Court FA, Coleman MP (2012) Mitochondria as a central sensor for axonal degenerative stimuli. Trends Neurosci 35:364–372

Da Pozzo E, Giacomelli C, Barresi E, Costa B, Taliani S, Passetti FDS, Martini C (2015) Targeting the 18-kDa translocator protein: recent perspectives for neuroprotection. Biochem Soc Trans 43:559–565

De Souza EB, Anholt RR, Murphy KM, Snyder SH, Kuhar MJ (1985) Peripheral-type benzodiazepine receptors in endocrine organs: autoradiographic localization in rat pituitary, adrenal, and testis. Endocrinology 116:567–573

Delavoie F, Li H, Hardwick M, Robert J-C, Giatzakis C, Péranzi G, Yao Z-X, Maccario J, Lacapère J-J, Papadopoulos V (2003) In vivo and in vitro peripheral-type benzodiazepine receptor polymerization: functional significance in drug ligand and cholesterol binding. Biochemistry (Mosc) 42:4506–4519

Dias V, Junn E, Mouradian MM (2013) The role of oxidative stress in Parkinson's disease. J Park Dis 3:461–491

Dias IHK, Mistry J, Fell S, Reis A, Spickett CM, Polidori MC, Lip GYH, Griffiths HR (2014) Oxidized LDL lipids increase ?-amyloid production by SH-SY5Y cells through glutathione depletion and lipid raft formation. Free Radic Biol Med 75:48–59

Dröge W (2002) Free radicals in the physiological control of cell function. Physiol Rev 82:47–95

Dubois A, Bénavidès J, Peny B, Duverger D, Fage D, Gotti B, MacKenzie ET, Scatton B (1988) Imaging of primary and remote ischaemic and excitotoxic brain lesions. An autoradiographic study of peripheral type benzodiazepine binding sites in the rat and cat. Brain Res 445:77–90

Fan J, Papadopoulos V (2012) Transcriptional regulation of translocator protein (Tspo) via a SINE B2-mediated natural antisense transcript in MA-10 Leydig cells. Biol Reprod 86:147

Fan J, Rone MB, Papadopoulos V (2009) Translocator protein 2 is involved in cholesterol redistribution during erythropoiesis. J Biol Chem 284:30484–30497

Fan J, Liu J, Culty M, Papadopoulos V (2010) Acyl-coenzyme A binding domain containing 3 (ACBD3; PAP7; GCP60): an emerging signaling molecule. Prog Lipid Res 49:218–234

Fan J, Campioli E, Midzak A, Culty M, Papadopoulos V (2015) Conditional steroidogenic cell-targeted deletion of TSPO unveils a crucial role in viability and hormone-dependent steroid formation. Proc Natl Acad Sci U S A 112:7261–7266

Fantini J, Di Scala C, Evans LS, Williamson PTF, Barrantes FJ (2016) A mirror code for protein-cholesterol interactions in the two leaflets of biological membranes. Sci Rep 6:213–222

Fares F, Bar-Ami S, Brandes JM, Gavish M (1987) Gonadotropin- and estrogen-induced increase of peripheral-type benzodiazepine binding sites in the hypophyseal-genital axis of rats. Eur J Pharmacol 133:97–102

Fares F, Bar-Ami S, Brandes JM, Gavish M (1988) Changes in the density of peripheral benzodiazepine binding sites in genital organs of the female rat during the oestrous cycle. J Reprod Fertil 83:619–625

Farges R, Joseph-Liauzun E, Shire D, Caput D, Fur GL, Ferrara P (1994) Site-directed mutagenesis of the peripheral benzodiazepine receptor: identification of amino acids implicated in the binding site of Ro5-4864. Mol Pharmacol 46:1160–1167

Fay J, Varoga D, Wruck CJ, Kurz B, Goldring MB, Pufe T (2006) Reactive oxygen species induce expression of vascular endothelial growth factor in chondrocytes and human articular cartilage explants. Arthritis Res Ther 8:R189

Frank W, Baar K-M, Qudeimat E, Woriedh M, Alawady A, Ratnadewi D, Gremillon L, Grimm B, Reski R (2007) A mitochondrial protein homologous to the mammalian peripheral-type benzodiazepine receptor is essential for stress adaptation in plants. Plant J 51:1004–1018

Galiègue S, Casellas P, Kramar A, Tinel N, Simony-Lafontaine J (2004) Immunohistochemical assessment of the peripheral benzodiazepine receptor in breast cancer and its relationship with survival. Clin Cancer Res Off J Am Assoc Cancer Res 10:2058–2064

Garnier M, Dimchev AB, Boujrad N, Price JM, Musto NA, Papadopoulos V (1994) In vitro reconstitution of a functional peripheral-type benzodiazepine receptor from mouse Leydig tumor cells. Mol Pharmacol 45:201–211

Gatliff J, Campanella M (2015) TSPO is a REDOX regulator of cell mitophagy. Biochem Soc Trans 43:543–552

Gatliff J, Campanella M (2016) TSPO: kaleidoscopic 18-kDa amid biochemical pharmacology, control and targeting of mitochondria. Biochem J 473:107–121

Gatliff J, East D, Crosby J, Abeti R, Harvey R, Craigen W, Parker P, Campanella M (2014) TSPO interacts with VDAC1 and triggers a ROS-mediated inhibition of mitochondrial quality control. Autophagy 10(12):2279–2296. doi: 10.4161/15548627.2014.991665.

Gautier CA, Kitada T, Shen J (2008) Loss of PINK1 causes mitochondrial functional defects and increased sensitivity to oxidative stress. Proc Natl Acad Sci U S A 105:11364–11369

Gavish M, Bachman I, Shoukrun R, Katz Y, Veenman L, Weisinger G, Weizman A (1999) Enigma of the peripheral benzodiazepine receptor. Pharmacol Rev 51:629–650

Gazouli M, Yao Z-X, Boujrad N, Corton JC, Culty M, Papadopoulos V (2002) Effect of peroxisome proliferators on Leydig cell peripheral-type benzodiazepine receptor gene expression, hormone-stimulated cholesterol transport, and steroidogenesis: role of the peroxisome proliferator-activator receptor alpha. Endocrinology 143:2571–2583

Ghosh N, Ghosh R, Mandal SC (2011) Antioxidant protection: a promising therapeutic intervention in neurodegenerative disease. Free Radic Res 45:888–905

Giatzakis C, Papadopoulos V (2004) Differential utilization of the promoter of peripheral-type benzodiazepine receptor by steroidogenic versus nonsteroidogenic cell lines and the role of Sp1 and Sp3 in the regulation of basal activity. Endocrinology 145:1113–1123

Giatzakis C, Batarseh A, Dettin L, Papadopoulos V (2007) The role of Ets transcription factors in the basal transcription of the translocator protein (18 kDa). Biochemistry (Mosc) 46:4763–4774

Ginter C, Kiburu I, Boudker O (2013) Chemical catalysis by the translocator protein (18 kDa). Biochemistry (Mosc) 52:3609–3611

Giorgio V, von Stockum S, Antoniel M, Fabbro A, Fogolari F, Forte M, Glick GD, Petronilli V, Zoratti M, Szabó I et al (2013) Dimers of mitochondrial ATP synthase form the permeability transition pore. Proc Natl Acad Sci 110:5887–5892

Gomez-Nicola D, Perry VH (2015) Microglial dynamics and role in the healthy and diseased brain: a paradigm of functional plasticity. Neuroscientist 21(2):169–184.

Gonzalez-Polo R-A, Carvalho G, Braun T, Decaudin D, Fabre C, Larochette N, Perfettini J-L, Djavaheri-Mergny M, Youlyouz-Marfak I, Codogno P et al (2005) PK11195 potently sensitizes to apoptosis induction independently from the peripheral benzodiazepin receptor. Oncogene 24:7503–7513

Graeber MB (2010) Changing face of microglia. Science 330:783–788

Gray PW, Glaister D, Seeburg PH, Guidotti A, Costa E (1986) Cloning and expression of cDNA for human diazepam binding inhibitor, a natural ligand of an allosteric regulatory site of the gamma-aminobutyric acid type A receptor. Proc Natl Acad Sci U S A 83:7547–7551

Guidotti A, Forchetti CM, Corda MG, Konkel D, Bennett CD, Costa E (1983) Isolation, characterization, and purification to homogeneity of an endogenous polypeptide with agonistic action on benzodiazepine receptors. Proc Natl Acad Sci U S A 80:3531–3535

Guo Y, Kalathur RC, Liu Q, Kloss B, Bruni R, Ginter C, Kloppmann E, Rost B, Hendrickson WA (2015) Structure and activity of tryptophan-rich TSPO proteins. Science 347:551–555

Gut P, Baeza-Raja B, Andersson O, Hasenkamp L, Hsiao J, Hesselson D, Akassoglou K, Verdin E, Hirschey MD, Stainier DYR (2013) Whole-organism screening for gluconeogenesis identifies activators of fasting metabolism. Nat Chem Biol 9:97–104

Gut P, Zweckstetter M, Banati RB (2015) Lost in translocation: the functions of the 18-kD translocator protein. Trends Endocrinol Metab TEM 26:349–356

Halestrap AP (2009) What is the mitochondrial permeability transition pore? J Mol Cell Cardiol 46:821–831

Handschin C, Lin J, Rhee J, Peyer A-K, Chin S, Wu P-H, Meyer UA, Spiegelman BM (2005) Nutritional regulation of hepatic heme biosynthesis and porphyria through PGC-1α. Cell 122:505–515

Hardwick M, Fertikh D, Culty M, Li H, Vidic B, Papadopoulos V (1999) Peripheral-type benzodiazepine receptor (PBR) in human breast cancer. Cancer Res 59:831–842

Harrison R (2002) Structure and function of xanthine oxidoreductase: where are we now? Free Radic Biol Med 33:774–797

Hirsch T, Decaudin D, Susin SA, Marchetti P, Larochette N, Resche-Rigon M, Kroemer G (1998) PK11195, a ligand of the mitochondrial benzodiazepine receptor, facilitates the induction of apoptosis and reverses Bcl-2-mediated cytoprotection. Exp Cell Res 241:426–434

Huang R, Zhao L, Chen H, Yin R-H, Li C-Y, Zhan Y-Q, Zhang J-H, Ge C, Yu M, Yang X-M (2014) Megakaryocytic differentiation of K562 cells induced by PMA reduced the activity of respiratory chain complex IV. PLoS One 9:e96246

Janczar K, Su Z, Raccagni I, Anfosso A, Kelly C, Durrenberger PF, Gerhard A, Roncaroli F (2015) The 18-kDa mitochondrial translocator protein in gliomas: from the bench to bedside. Biochem Soc Trans 43:579–585

Jaremko L, Jaremko M, Giller K, Becker S, Zweckstetter M (2014) Structure of the mitochondrial translocator protein in complex with a diagnostic ligand. Science 343:1363–1366

Jaremko Ł, Jaremko M, Giller K, Becker S, Zweckstetter M (2015a) Conformational flexibility in the transmembrane protein TSPO. Chem Weinh Bergstr Ger 21:16555–16563

Jaremko M, Jaremko Ł, Giller K, Becker S, Zweckstetter M (2015b) Structural integrity of the A147T polymorph of mammalian TSPO. Chembiochem Eur J Chem Biol 16:1483–1489

Jayakumar AR, Panickar KS, Norenberg MD (2002) Effects on free radical generation by ligands of the peripheral benzodiazepine receptor in cultured neural cells. J Neurochem 83:1226–1234

Jordà EG, Jiménez A, Verdaguer E, Canudas AM, Folch J, Sureda FX, Camins A, Pallàs M (2005) Evidence in favour of a role for peripheral-type benzodiazepine receptor ligands in amplification of neuronal apoptosis. Apoptosis Int J Program Cell Death 10:91–104

Joseph-Liauzun E, Delmas P, Shire D, Ferrara P (1998) Topological analysis of the peripheral benzodiazepine receptor in yeast mitochondrial membranes supports a five-transmembrane structure. J Biol Chem 273:2146–2152

Junck L, Olson JM, Ciliax BJ, Koeppe RA, Watkins GL, Jewett DM, McKeever PE, Wieland DM, Kilbourn MR, Starosta-Rubinstein S (1989) PET imaging of human gliomas with ligands for the peripheral benzodiazepine binding site. Ann Neurol 26:752–758

Karchewski LA, Bloechlinger S, Woolf CJ (2004) Axonal injury-dependent induction of the peripheral benzodiazepine receptor in small-diameter adult rat primary sensory neurons. Eur J Neurosci 20:671–683

Karlstetter M, Nothdurfter C, Aslanidis A, Moeller K, Horn F, Scholz R, Neumann H, Weber BHF, Rupprecht R, Langmann T (2014) Translocator protein (18 kDa) (TSPO) is expressed in reactive retinal microglia and modulates microglial inflammation and phagocytosis. J Neuroinflammation 11:3

Keshari RS, Verma A, Barthwal MK, Dikshit M (2013) Reactive oxygen species-induced activation of ERK and p38 MAPK mediates PMA-induced NETs release from human neutrophils. J Cell Biochem 114:532–540

Kokoszka JE, Waymire KG, Levy SE, Sligh JE, Cai J, Jones DP, MacGregor GR, Wallace DC (2004) The ADP/ATP translocator is not essential for the mitochondrial permeability transition pore. Nature 427:461–465

Kruczek C, Görg B, Keitel V, Pirev E, Kröncke KD, Schliess F, Häussinger D (2009) Hypoosmotic swelling affects zinc homeostasis in cultured rat astrocytes. Glia 57:79–92

Krueger KE, Papadopoulos V (1990) Peripheral-type benzodiazepine receptors mediate translocation of cholesterol from outer to inner mitochondrial membranes in adrenocortical cells. J Biol Chem 265:15015–15022

Lacapère JJ, Delavoie F, Li H, Péranzi G, Maccario J, Papadopoulos V, Vidic B (2001) Structural and functional study of reconstituted peripheral benzodiazepine receptor. Biochem Biophys Res Commun 284:536–541

Lazarou M, Sliter DA, Kane LA, Sarraf SA, Wang C, Burman JL, Sideris DP, Fogel AI, Youle RJ (2015) The ubiquitin kinase PINK1 recruits autophagy receptors to induce mitophagy. Nature 524:309–314

Lee D-H, Steinacker P, Seubert S, Turnescu T, Melms A, Manzel A, Otto M, Linker RA (2015) Role of glial 14-3-3 gamma protein in autoimmune demyelination. J Neuroinflammation 12:187

Lee J-W, Kim LE, Shim H-J, Kim E-K, Hwang WC, Min DS, Yu S-W (2016) A translocator protein 18 kDa ligand, Ro5-4864, inhibits ATP-induced NLRP3 inflammasome activation. Biochem Biophys Res Commun 474:587–593

Lehtonen MT, Akita M, Frank W, Reski R, Valkonen JPT (2012) Involvement of a class III peroxidase and the mitochondrial protein TSPO in oxidative burst upon treatment of moss plants with a fungal elicitor. Mol Plant-Microbe Interact MPMI 25:363–371

Lemasters JJ (2005) Selective mitochondrial autophagy, or mitophagy, as a targeted defense against oxidative stress, mitochondrial dysfunction, and aging. Rejuvenation Res 8:3–5

Lemasters JJ, Holmuhamedov E (2006) Voltage-dependent anion channel (VDAC) as mitochondrial governator – thinking outside the box. Biochim Biophys Acta 1762:181–190

Li H, Papadopoulos V (1998) Peripheral-type benzodiazepine receptor function in cholesterol transport. Identification of a putative cholesterol recognition/interaction amino acid sequence and consensus pattern. Endocrinology 139:4991–4997

Li J, Papadopoulos V (2015) Translocator protein (18 kDa) as a pharmacological target in adipocytes to regulate glucose homeostasis. Biochem Pharmacol 97:99–110

Li H, Degenhardt B, Tobin D, Yao ZX, Tasken K, Papadopoulos V (2001) Identification, localization, and function in steroidogenesis of PAP7: a peripheral-type benzodiazepine receptor- and PKA (RIalpha)-associated protein. Mol Endocrinol Baltim Md 15:2211–2228

Li F, Liu J, Valls L, Hiser C, Ferguson-Miller S (2015a) Identification of a key cholesterol binding enhancement motif in translocator protein 18 kDa. Biochemistry (Mosc) 54:1441–1443

Li F, Liu J, Zheng Y, Garavito RM, Ferguson-Miller S (2015b) Crystal structures of translocator protein (TSPO) and mutant mimic of a human polymorphism. Science 347:555–558

Lin R, Angelin A, Da Settimo F, Martini C, Taliani S, Zhu S, Wallace DC (2014) Genetic analysis of dTSPO, an outer mitochondrial membrane protein, reveals its functions in apoptosis, longevity, and Ab42-induced neurodegeneration. Aging Cell 13:507–518

Liu J, Rone MB, Papadopoulos V (2006) Protein-protein interactions mediate mitochondrial cholesterol transport and steroid biosynthesis. J Biol Chem 281:38879–38893

Maldonado EN, Lemasters JJ (2012) Warburg revisited: regulation of mitochondrial metabolism by voltage-dependent anion channels in cancer cells. J Pharmacol Exp Ther 342:637–641

Matic I, Strobbe D, Frison M, Campanella M (2015) Controlled and impaired mitochondrial quality in neurons: molecular physiology and prospective pharmacology. Pharmacol Res Off J Ital Pharmacol Soc

Mazurika C, Veenman L, Weizman R, Bidder M, Leschiner S, Golani I, Spanier I, Weisinger G, Gavish M (2009) Estradiol modulates uterine 18 kDa translocator protein gene expression in uterus and kidney of rats. Mol Cell Endocrinol 307:43–49

McEnery MW, Snowman AM, Trifiletti RR, Snyder SH (1992) Isolation of the mitochondrial benzodiazepine receptor: association with the voltage-dependent anion channel and the adenine nucleotide carrier. Proc Natl Acad Sci U S A 89:3170–3174

Medzhitov R (2008) Origin and physiological roles of inflammation. Nature 454:428–435

Mercer KA, Weizman R, Gavish M (1992) Ontogenesis of peripheral benzodiazepine receptors: demonstration of selective up-regulation in rat testis as a function of maturation. J Recept Res 12:413–425

Midzak A, Zirkin B, Papadopoulos V (2015) Translocator protein: pharmacology and steroidogenesis. Biochem Soc Trans 43:572–578

Miller WL, Auchus RJ (2011) The molecular biology, biochemistry, and physiology of human steroidogenesis and its disorders. Endocr Rev 32:81–151

Morin D, Musman J, Pons S, Berdeaux A, Ghaleh B (2016) Mitochondrial translocator protein (TSPO): from physiology to cardioprotection. Biochem Pharmacol 105:1–13

Murail S, Robert J-C, Coïc Y-M, Neumann J-M, Ostuni MA, Yao Z-X, Papadopoulos V, Jamin N, Lacapère J-J (2008) Secondary and tertiary structures of the transmembrane domains of the translocator protein TSPO determined by NMR. Stabilization of the TSPO tertiary fold upon ligand binding. Biochim Biophys Acta 1778:1375–1381

Murphy MP (2009) How mitochondria produce reactive oxygen species. Biochem J 417:1–13

Naegele M, Martin R (2014) The good and the bad of neuroinflammation in multiple sclerosis. Handb Clin Neurol 122:59–87

Nakazawa F, Alev C, Shin M, Nakaya Y, Jakt LM, Sheng G (2009) PBRL, a putative peripheral benzodiazepine receptor, in primitive erythropoiesis. Gene Expr Patterns GEP 9:114–121

Olson JMM, Ciliax BJ, Mancini WR, Young AB (1988) Presence of peripheral-type benzodiazepine binding sites on human erythrocyte membranes. Eur J Pharmacol 152:47–53

Owen DRJ, Lewis AJM, Reynolds R, Rupprecht R, Eser D, Wilkins MR, Bennacef I, Nutt DJ, Parker CA (2011) Variation in binding affinity of the novel anxiolytic XBD173 for the 18 kDa translocator protein in human brain. Synap N Y N 65:257–259

Owen DR, Yeo AJ, Gunn RN, Song K, Wadsworth G, Lewis A, Rhodes C, Pulford DJ, Bennacef I, Parker CA et al (2012) An 18-kDa translocator protein (TSPO) polymorphism explains differences in binding affinity of the PET radioligand PBR28. J Cereb Blood Flow Metab Off J Int Soc Cereb Blood Flow Metab 32:1–5

Palzur E, Sharon A, Shehadeh M, Soustiel JF (2016) Investigation of the mechanisms of neuroprotection mediated by Ro5-4864 in brain injury. Neuroscience 329:162–170

Pandey D, Gratton J-P, Rafikov R, Black SM, Fulton DJR (2011) Calcium/calmodulin-dependent kinase II mediates the phosphorylation and activation of NADPH oxidase 5. Mol Pharmacol 80:407–415

Papadopoulos V, Mukhin AG, Costa E, Krueger KE (1990) The peripheral-type benzodiazepine receptor is functionally linked to Leydig cell steroidogenesis. J Biol Chem 265:3772–3779

Papadopoulos V, Berkovich A, Krueger KE, Costa E, Guidotti A (1991) Diazepam binding inhibitor and its processing products stimulate mitochondrial steroid biosynthesis via an interaction with mitochondrial benzodiazepine receptors. Endocrinology 129:1481–1488

Papadopoulos V, Boujrad N, Ikonomovic MD, Ferrara P, Vidic B (1994) Topography of the Leydig cell mitochondrial peripheral-type benzodiazepine receptor. Mol Cell Endocrinol 104:R5–R9

Papadopoulos V, Amri H, Boujrad N, Cascio C, Culty M, Garnier M, Hardwick M, Li H, Vidic B, Brown AS et al (1997a) Peripheral benzodiazepine receptor in cholesterol transport and steroidogenesis. Steroids 62:21–28

Papadopoulos V, Amri H, Li H, Boujrad N, Vidic B, Garnier M (1997b) Targeted disruption of the peripheral-type benzodiazepine receptor gene inhibits steroidogenesis in the R2C Leydig tumor cell line. J Biol Chem 272:32129–32135

Papadopoulos V, Baraldi M, Guilarte TR, Knudsen TB, Lacapère J-J, Lindemann P, Norenberg MD, Nutt D, Weizman A, Zhang M-R et al (2006) Translocator protein (18kDa): new

nomenclature for the peripheral-type benzodiazepine receptor based on its structure and molecular function. Trends Pharmacol Sci 27:402–409
Ponka P (1999) Cell biology of heme. Am J Med Sci 318:241–256
Puigserver P, Rhee J, Donovan J, Walkey CJ, Yoon JC, Oriente F, Kitamura Y, Altomonte J, Dong H, Accili D et al (2003) Insulin-regulated hepatic gluconeogenesis through FOXO1–PGC-1α interaction. Nature 423:550–555
Qi X, Xu J, Wang F, Xiao J (2012) Translocator protein (18 kDa): a promising therapeutic target and diagnostic tool for cardiovascular diseases. Oxid Med Cell Longev 2012:162934
Rampon C, Bouzaffour M, Ostuni MA, Dufourcq P, Girard C, Freyssinet JM, Lacapere J-J, Schweizer-Groyer G, Vriz S (2009) Translocator protein (18 kDa) is involved in primitive erythropoiesis in zebrafish. FASEB J Off Publ Fed Am Soc Exp Biol 23:4181–4192
Randow F, Youle RJ (2014) Self and nonself: how autophagy targets mitochondria and bacteria. Cell Host Microbe 15:403–411
Ransohoff RM, Perry VH (2009) Microglial physiology: unique stimuli, specialized responses. Annu Rev Immunol 27:119–145
Rechichi M, Salvetti A, Chelli B, Costa B, Da Pozzo E, Spinetti F, Lena A, Evangelista M, Rainaldi G, Martini C et al (2008) TSPO over-expression increases motility, transmigration and proliferation properties of C6 rat glioma cells. Biochim Biophys Acta 1782:118–125
Rizzuto R, De Stefani D, Raffaello A, Mammucari C (2012) Mitochondria as sensors and regulators of calcium signalling. Nat Rev Mol Cell Biol 13:566–578
Rogers J, Mastroeni D, Leonard B, Joyce J, Grover A (2007) Neuroinflammation in Alzheimer's disease and Parkinson's disease: are microglia pathogenic in either disorder? Int Rev Neurobiol 82:235–246
Rone M, Liu J, Blonder J, Ye X, Veenstra TD, Young JC, Papadopoulos V (2009) Targeting and insertion of the cholesterol-binding translocator protein into the outer mitochondrial membrane. Biochemistry (Mosc) 48:6909–6920
Rosenberg N, Rosenberg O, Weizman A, Veenman L, Gavish M (2014) In vitro effect of FGIN-1-27, a ligand to 18 kDa mitochondrial translocator protein, in human osteoblast-like cells. J Bioenerg Biomembr 46:197–204
Rupprecht R, Rammes G, Eser D, Baghai TC, Schüle C, Nothdurfter C, Troxler T, Gentsch C, Kalkman HO, Chaperon F et al (2009) Translocator protein (18 kD) as target for anxiolytics without benzodiazepine-like side effects. Science 325:490–493
Rupprecht R, Papadopoulos V, Rammes G, Baghai TC, Fan J, Akula N, Groyer G, Adams D, Schumacher M (2010) Translocator protein (18 kDa) (TSPO) as a therapeutic target for neurological and psychiatric disorders. Nat Rev Drug Discov 9:971–988
Saari TI, Uusi-Oukari M, Ahonen J, Olkkola KT (2011) Enhancement of GABAergic activity: neuropharmacological effects of benzodiazepines and therapeutic use in anesthesiology. Pharmacol Rev 63:243–267
Selvaraj V, Stocco DM (2015) The changing landscape in translocator protein (TSPO) function. Trends Endocrinol Metab 26:341–348
Seneviratne MSD, Faccenda D, De Biase V, Campanella M (2012) PK11195 inhibits mitophagy targeting the F1Fo-ATPsynthase in Bcl-2 knock-down cells. Curr Mol Med 12:476–482
Sheldon KL, Maldonado EN, Lemasters JJ, Rostovtseva TK, Bezrukov SM (2011) Phosphorylation of voltage-dependent anion channel by serine/threonine kinases governs its interaction with tubulin. PLoS One 6:e25539
Shoshan-Barmatz V, Ben-Hail D, Admoni L, Krelin Y, Tripathi SS (2015) The mitochondrial voltage-dependent anion channel 1 in tumor cells. Biochim Biophys Acta BBA – Biomembr 1848:2547–2575
Sileikyte J, Petronilli V, Zulian A, Dabbeni-Sala F, Tognon G, Nikolov P, Bernardi P, Ricchelli F (2011) Regulation of the inner membrane mitochondrial permeability transition by the outer membrane translocator protein (peripheral benzodiazepine receptor). J Biol Chem 286:1046–1053
Smith SM, Vale WW (2006) The role of the hypothalamic-pituitary-adrenal axis in neuroendocrine responses to stress. Dialogues Clin Neurosci 8:383–395

Sutter AP, Maaser K, Höpfner M, Barthel B, Grabowski P, Faiss S, Carayon P, Zeitz M, Scherübl H (2002) Specific ligands of the peripheral benzodiazepine receptor induce apoptosis and cell cycle arrest in human esophageal cancer cells. Int J Cancer 102:318–327

Tamse CT, Lu X, Mortel EG, Cabrales E, Feng W, Schaefer S (2008) The peripheral benzodiazepine receptor modulates Ca^{2+} transport through the VDAC in rat heart mitochondria. J Clin Basic Cardiol 11(1-4):24–29

Tanida I, Ueno T, Kominami E (2004) LC3 conjugation system in mammalian autophagy. Int J Biochem Cell Biol 36:2503–2518

Trincavelli ML, Marselli L, Falleni A, Gremigni V, Ragge E, Dotta F, Santangelo C, Marchetti P, Lucacchini A, Martini C (2002) Upregulation of mitochondrial peripheral benzodiazepine receptor expression by cytokine-induced damage of human pancreatic islets. J Cell Biochem 84:636–644

Tu LN, Morohaku K, Manna PR, Pelton SH, Butler WR, Stocco DM, Selvaraj V (2014) Peripheral benzodiazepine receptor/translocator protein global knock-out mice are viable with no effects on steroid hormone biosynthesis. J Biol Chem 289:27444–27454

Tu LN, Zhao AH, Stocco DM, Selvaraj V (2015) PK11195 effect on steroidogenesis is not mediated through the translocator protein (TSPO). Endocrinology 156:1033–1039

Tu LN, Zhao AH, Hussein M, Selvaraj V (2016) Mitochondrial translocator protein (TSPO) is a regulator of fatty acid metabolism. FASEB J 30:850.8–850.8

Turkheimer FE, Rizzo G, Bloomfield PS, Howes O, Zanotti-Fregonara P, Bertoldo A, Veronese M (2015) The methodology of TSPO imaging with positron emission tomography. Biochem Soc Trans 43:586–592

Ujwal R, Cascio D, Colletier J-P, Faham S, Zhang J, Toro L, Ping P, Abramson J (2008) The crystal structure of mouse VDAC1 at 2.3 Å resolution reveals mechanistic insights into metabolite gating. Proc Natl Acad Sci 105:17742–17747

Vanhee C, Zapotoczny G, Masquelier D, Ghislain M, Batoko H (2011) The arabidopsis multistress regulator TSPO is a heme binding membrane protein and a potential scavenger of porphyrins via an autophagy-dependent degradation mechanism. Plant Cell 23:785–805

Varga B, Markó K, Hádinger N, Jelitai M, Demeter K, Tihanyi K, Vas A, Madarász E (2009) Translocator protein (TSPO 18kDa) is expressed by neural stem and neuronal precursor cells. Neurosci Lett 462:257–262

Veenman L, Gavish M (2006) The peripheral-type benzodiazepine receptor and the cardiovascular system. Implications for drug development. Pharmacol Ther 110:503–524

Veenman L, Levin E, Weisinger G, Leschiner S, Spanier I, Snyder SH, Weizman A, Gavish M (2004) Peripheral-type benzodiazepine receptor density and in vitro tumorigenicity of glioma cell lines. Biochem Pharmacol 68:689–698

Veenman L, Papadopoulos V, Gavish M (2007) Channel-like functions of the 18-kDa translocator protein (TSPO): regulation of apoptosis and steroidogenesis as part of the host-defense response. Curr Pharm Des 13:2385–2405

Veenman L, Gavish M, Kugler W (2014) Apoptosis induction by erucylphosphohomocholine via the 18 kDa mitochondrial translocator protein: implications for cancer treatment. Anticancer Agents Med Chem 14:559–577

Vega RB, Huss JM, Kelly DP (2000) The coactivator PGC-1 cooperates with peroxisome proliferator-activated receptor alpha in transcriptional control of nuclear genes encoding mitochondrial fatty acid oxidation enzymes. Mol Cell Biol 20:1868–1876

Verma A, Nye JS, Snyder SH (1987) Porphyrins are endogenous ligands for the mitochondrial (peripheral-type) benzodiazepine receptor. Proc Natl Acad Sci U S A 84:2256–2260

Weizman A, Amiri Z, Katz Y, Snyder SH, Gavish M (1992) Testosterone prevents castration-induced reduction in peripheral benzodiazepine receptors in Cowper's gland and adrenal. Brain Res 572:72–75

Werry EL, Barron ML, Kassiou M (2015) TSPO as a target for glioblastoma therapeutics. Biochem Soc Trans 43:531–536

Wu Y, Kazumura K, Maruyama W, Osawa T, Naoi M (2015) Rasagiline and selegiline suppress calcium efflux from mitochondria by PK11195-induced opening of mitochondrial permeability transition pore: a novel anti-apoptotic function for neuroprotection. J Neural Transm Vienna Austria 122:1399–14071996

Yeliseev AA, Kaplan S (2000) TspO of rhodobacter sphaeroides. A structural and functional model for the mammalian peripheral benzodiazepine receptor. J Biol Chem 275:5657–5667

Yeliseev AA, Krueger KE, Kaplan S (1997) A mammalian mitochondrial drug receptor functions as a bacterial "oxygen" sensor. Proc Natl Acad Sci 94:5101–5106

Yin L, Wu N, Curtin JC, Qatanani M, Szwergold NR, Reid RA, Waitt GM, Parks DJ, Pearce KH, Wisely GB et al (2007) Rev-erbα, a heme sensor that coordinates metabolic and circadian pathways. Science 318:1786–1789

Zeno S, Zaaroor M, Leschiner S, Veenman L, Gavish M (2009) CoCl(2) induces apoptosis via the 18 kDa translocator protein in U118MG human glioblastoma cells. Biochemistry (Mosc) 48:4652–4661

Zhao AH, Tu LN, Mukai C, Sirivelu MP, Pillai VV, Morohaku K, Cohen R, Selvaraj V (2015) Mitochondrial translocator protein (TSPO) function is not essential for heme biosynthesis. J Biol Chem 291:1591–1603

Zhou T, Dang Y, Zheng Y-H (2014) The mitochondrial translocator protein, TSPO, inhibits HIV-1 envelope glycoprotein biosynthesis via the endoplasmic reticulum-associated protein degradation pathway. J Virol 88:3474–3484

Chapter 12
Protein Import Channels in the Crossroads of Mitochondrial Function

Ma Su Su Aung, Ruth Hartke, Stephen Madamba, Oygul Mirzalieva, and Pablo M. Peixoto

12.1 Introduction

Exchange of information is a central theme in biology because it underlies the evolution of symbiotic interactions. Conceivably, the most relevant of such interactions was the eukaryotic merger that led to the origin of mitochondria – commonly referred to as the power plants of almost every eukaryotic cell (Sagan 1967; Schwartz and Dayhoff 1978; Gray et al. 1999; Gray 2015). Information exchange between mitochondria and the cell is in the form of ions, metabolites, amino acids, nucleic acids, and proteins. The benefits of this particular exchange include regulation of metabolism, stress response, division, differentiation, survival, and death.

Without exception, exchanged molecules need to cross two physical boundaries, the inner and the outer mitochondrial membranes. Interestingly, the lipid-protein composition of the outer mitochondrial membrane resembles that of related prokaryotic ancestors. For example, a homolog protein present in the plasma membrane of virtually all prokaryotes is the most abundant protein in the outer membrane of mitochondria. Pioneer investigators in the field continue to name it the mitochondrial porin, modernly known as the voltage-dependent anion channel (VDAC). Channels like VDAC form gateways for exchange of signaling molecules between mitochondrial and the other cellular compartments. The proteinaceous components

M.S.S. Aung
Macaulay Honors College at Baruch College of City University of New York, New York, NY, USA

R. Hartke • S. Madamba • O. Mirzalieva • P.M. Peixoto (✉)
Graduate Center and Baruch College of City University of New York, New York, NY 10010, USA
e-mail: pablo.peixoto@baruch.cuny.edu

O. Mirzalieva
Baruch College of City University of New York, New York, NY, USA

T.K. Rostovtseva (ed.), *Molecular Basis for Mitochondrial Signaling*, Biological and Medical Physics, Biomedical Engineering,
DOI 10.1007/978-3-319-55539-3_12

of these channels, like over 98% of all mitochondrial proteins, have to be imported from the cytoplasm. We present in this chapter an updated review of the protein import pathways and of how they enable mitochondrial biogenesis and control of cellular metabolism.

12.2 Protein Import Pathways

With striking homology to bacterial transport systems, the mitochondrial protein import machinery is highly conserved in evolution (Hoogenraad et al. 2002; Dolezal et al. 2006). However, the machinery grew more complex after the eukaryotic merger – it is estimated that approximately 40 different proteins contribute to protein import. This number is likely to grow: we estimate that ~17% of the proteins recently identified in mitochondrial fractions, many of which were confirmed by fluorescence microscopy (Pagliarini et al. 2008; Calvo et al. 2016), have unknown functions. Nonetheless, several import complexes mediate the transfer of nuclear-encoded proteins to their final destination within mitochondria (Neupert and Herrmann 2007; Chacinska et al. 2009; Schmidt et al. 2010) (Fig. 12.1). The first import gateway to be crossed by virtually all proteins entering mitochondria is the translocase of the outer membrane (TOM complex). Downstream import pathways diverge from this point. (1) Once in the intermembrane space, proteins harboring β-barrel transmembrane domains are chaperoned (by Tim9-Tim10 and Tim8-Tim13) toward the sorting and assembly machinery (SAM complex) in the outer membrane. (2) Matrix-destined proteins are freighted across the outer and inner membranes in a dynamic interlock between the TOM complex and the translocase

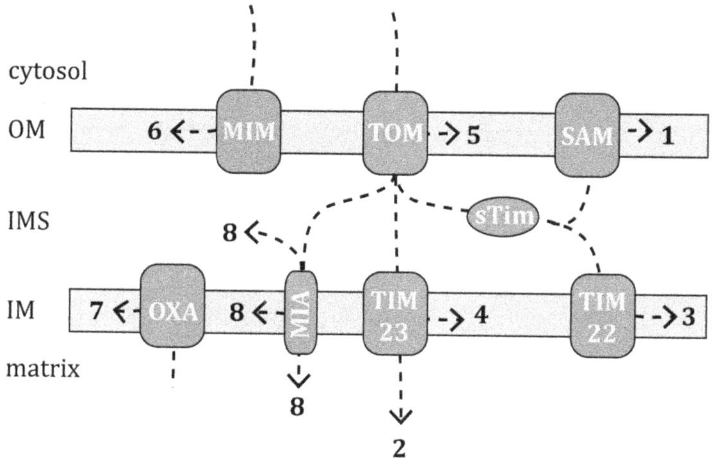

Fig. 12.1 Protein import pathways into mitochondria. The *diagram* indicates the protein import pathways indicated with *dotted lines* into each one of the mitochondrial subcompartments: the outer membrane (*OM*), the intermembrane space (*IMS*), the inner membrane (*IM*), or the matrix. The *numbers* represent the distinct pathways further described in the text

of the inner membrane TIM23. The process requires binding of the presequence translocase-associated motor (PAM) to the matrix side of TIM23. (3) Alternatively, the small Tim chaperones escort the transfer of a different subset of proteins to the translocase of the inner membrane TIM22. This particular translocase releases multi-spanning proteins into the inner membrane. The mechanism of such release remains to be determined (Peixoto et al. 2007). (4) In some cases, the transport through TOM-TIM23 is disrupted, and TIM23 instead releases the protein into the inner membrane (Botelho et al. 2011). (5) In a similar fashion, a study suggests that the TOM complex is competent for lateral release of proteins into the outer membrane (Harner et al. 2011), although it remains to be demonstrated if this alternative pathway is undertaken under physiological conditions by outer membrane proteins. (6) The mitochondrial import machinery (MIM) facilitates the insertion of outer membrane proteins containing multiple α-helical transmembrane segments (Becker et al. 2011; Dimmer et al. 2012). In addition, Mim1, a component of the MIM complex, participates in the biogenesis of some outer membrane proteins anchored via a single N-terminus α-helix (Becker et al. 2008; Hulett et al. 2008; Popov-Celeketic et al. 2008). (7) In a different sorting route, the oxidase and assembly (OXA) complex aids insertion into the inner membrane of proteins synthesized in the matrix (Hell et al. 1998). (8) Finally, the mitochondrial intermembrane space transport and assembly machinery (MIA complex) aids the proper folding of proteins with cysteine motifs that are either destined to the intermembrane space (Chacinska et al. 2004), the inner membrane (Darshi et al. 2012; Wrobel et al. 2013), or the matrix (Zhuang et al. 2013; Longen et al. 2014).

12.3 Regulation of Protein Import in Health and Disease

Disease-associated mutations in proteins of the import machinery are rare. The short list includes Mohr-Tranebjaerg syndrome and dilated cardiomyopathy with ataxia, which are associated with mutations in the homolog genes coding for Tim8a and Tim14, respectively (Tranebjaerg et al. 1995; Koehler et al. 1999). This scarceness suggests that most mutations are lethal early during human development. However, recent studies start to indicate that protein import can be affected by post-translational modifications and cellular stress. For example, a mutant variant of the mitochondrial superoxide dismutase 1 (SOD1) impairs protein import in spinal cord mitochondria. This mutation is involved in the etiology of amyotrophic lateral sclerosis (ALS) (Li et al. 2010). Similar examples in which protein import seems to be involved in the pathophysiology of age-related diseases includes Alzheimer's, Huntington's, and Parkinson's disease (Gottschalk et al. 2014; Di Maio et al. 2016).

The regulation of protein import in healthy cells has been more extensively studied in skeletal muscles, where studies have shown that import of matrix-destined proteins, such as mitochondrial transcription factor A (TFAM) and outer membrane proteins such as Tom40 was accelerated as an adaptation to chronic contractile activity (Takahashi et al. 1998; Gordon et al. 2001; Joseph et al. 2010). In contrast, contractile inactivity impaired import of mitochondrial matrix-destined proteins

(Singh and Hood 2011; Tryon et al. 2015), coinciding with a loss of mitochondrial content. These effects were associated with depleted ATP levels and reduced mitochondrial membrane potential, underscoring the adaptive plasticity of protein import and assembly in response to changes in skeletal muscle contractile activity. This is a concept that may open possibility for therapeutic intervention. In a recent study, a novel role for the apoptosis-associated outer membrane proteins Bax and Bak in the regulation of mitochondrial protein import was identified (Zhang et al. 2013). A double KO model of these two proteins had reduced mitochondrial protein import and reduced expression of components of the protein import machinery. Interestingly, this impairment was reversed following an endurance training protocol. Pharmacological induction is thus a promising approach, as studies have shown an increase in biogenesis and protein import upon treatment with thyroid hormone (Hood et al. 1992; Craig et al. 1998). More recently, a study suggested that the Bax activator protein Bim interacts with TOM subunits Tom20, Tom22, Tom40, and Tom70, but the physiological relevance of such interaction remains to be studied (Frank et al. 2015). Finally, the recent discovery of specific protein import inhibitors may help shed light into the mechanisms of distinct import pathways in healthy and diseased cells (Hasson et al. 2010; Dabir et al. 2013).

Studies on the mechanisms of protein import regulation are emerging as a hot topic in the field. The glucose-sensitive casein kinases 1 and 2 (CK1 and CK2) and protein kinase A (PKA) were shown to phosphorylate the TOM subunits Tom20, Tom22, and Tom40 (Schmidt et al. 2011; Rao et al. 2012; Gerbeth et al. 2013). While the CKs promoted assembly of the TOM subunits, PKA inhibited import of Tom40 (Rao et al. 2012). Another recent study has suggested that assembly of the TOM complex can be enhanced by the cyclin-dependent kinase via phosphorylation of the Tom6, which also promoted mitochondrial respiration (Harbauer et al. 2014a). However, the mechanisms described above pertain to regulation of assembly, but not of functioning of the TOM complex. Mechanistic data aside from the classic regulation by mitochondrial targeting sequences (MTS) is scarce. A few studies suggest that phosphorylation of certain proteins with dual cellular localization can either enhance or inhibit binding to the TOM chaperone Hsp70 (Robin et al. 2002; Avadhani et al. 2011). Another example of regulation of dual targeting involves proteolytic modification of the cytochrome p450 (Boopathi et al. 2008).

12.4 Import of Mitochondrial DNA Transcription Regulators, Nucleic Acids, and Steroidogenic Proteins

The process of mitochondrial biogenesis is essential to cell viability, as mitochondria cannot be produced de novo. As descendants of ancient proteobacteria, mammalian mitochondria house their own genome (mtDNA), which is circular, double stranded, consisting of 16,569 base pairs. The mtDNA is organized into DNA-protein complexes called mitochondrial nucleoids and 37 genes coding. 22 tRNAs, 2 mitochondrial rRNAs, and 13 protein subunits for electron transport chain

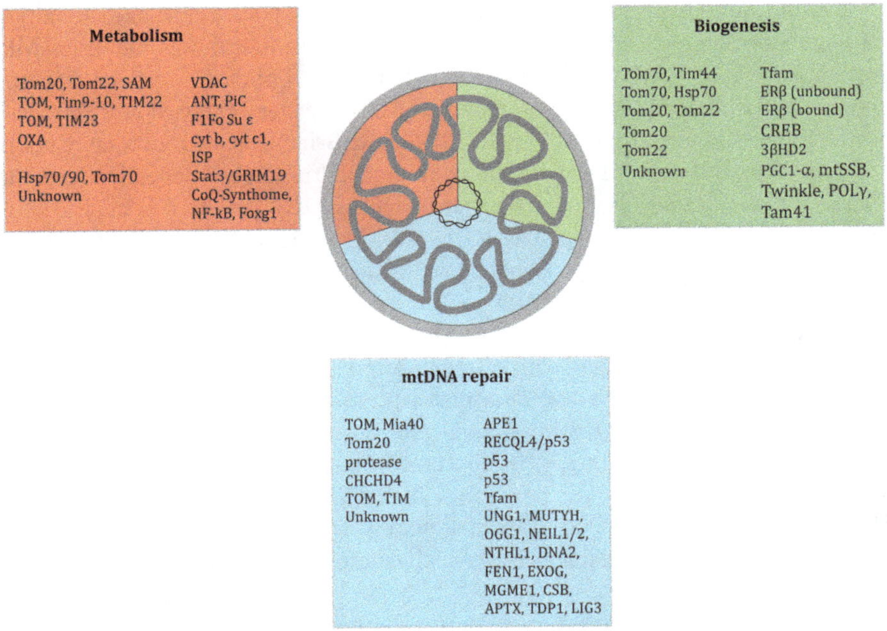

Fig. 12.2 Summary of known interactions between mitochondrial protein import components and regulators of mitochondrial biogenesis, DNA repair, and metabolism. Each *colored rectangle* includes a paired list of known interactions between import components and imported proteins involved in metabolism (*red*), biogenesis (*green*), and mtDNA repair (*blue*)

complexes except for Complex II (Akhmedov and Marin-Garcia 2015; DeBalsi et al. 2016). Along with mtDNA, protein machinery for mtDNA packaging, replication, transcription, and repair co-localizes in mitochondrial nucleoids. Among the major nucleoid-associated proteins are transcription factor A of mitochondria (Tfam), mitochondrial single-stranded DNA-binding protein (mtSSB), mtDNA helicase Twinkle, mtDNA polymerase γ (POL γ), and mtRNA polymerase (POLRMT) (Garrido et al. 2003; Garstka et al. 2003; Akhmedov and Marin-Garcia 2015). Import of nuclear-encoded mitochondrial proteins is necessary for maintaining mitochondrial function as well as allowing for the organelles to grow and divide. Figure 12.2 presents a current list of known interactions between protein import components and regulators of mitochondrial biogenesis, DNA repair, and metabolism.

Like the nuclear genome, mitochondrial DNA undergoes transcription and replication regulated by nuclear genes. A key regulator of mitochondrial biogenesis is peroxisome proliferator-activated receptor-γ coactivator-1α (PGC-1α), a transcriptional coactivator of nuclear transcription factors, including nuclear respiratory factors 1 and 2 (NRF1/NRF2) (Wu et al. 1999). The NRFs promote expression of mitochondrial transcription factor A (Tfam), a crucial activator of mtDNA transcription and replication (Larsson et al. 1998).

A number of other proteins, which localize to both the nucleus and mitochondria, can influence mitochondrial gene expression. The transcription factor cAMP response element-binding protein (CREB) localizes to both the nucleus and the mitochondrial matrix, where it promotes cAMP-dependent expression of respiratory complex subunits (De Rasmo et al. 2009). CREB-induced mtDNA transcription requires phosphorylation by cytosolic or mitochondrial cAMP-dependent protein kinase (PKA) and translocation to the matrix, mediated by Tom20 (Ryu et al. 2005). Estrogen receptor β (ERβ) localizes to the mitochondria and is targeted to import receptors depending on its ligand-bound status; when unbound, it targets to Tom70 at Hsp70, and when bound, it targets to Tom20/Tom22 at one of its three LXXLL motifs (Simpkins et al. 2008). It is found in the matrix as well, where it may exert its effect through interactions with estrogen response element sequences found in mtDNA or through pathways involving transcription factors like CREB. Additionally, the orphan receptor estrogen-related receptor α, which has been thought to be involved in estrogen signaling, is required for PGC-1α-induced mitochondrial biogenesis and is able to induce biogenesis in SAOS2 osteosarcoma cells lacking PGC-1α (Schreiber et al. 2004). Replication of mtDNA must also be closely coordinated with fission and fusion events, which determine mitochondrial number and morphology.

12.4.1 PGC-1α

Regarded as the master regulator of mitochondrial biogenesis, PGC-1α was initially described as an activator of the nuclear receptor PPARγ peroxisome proliferator-activated receptor gamma (Puigserver et al. 1998). Expression of PGC-1α is greatly increased upon exposure to cold, elevating PPARγ activity, indicating a role in adaptive thermogenesis. PGC-1α is found in tissues where mitochondria are abundant, including skeletal muscle, liver, heart, and brown adipose tissue. Since its discovery, PGC-1α has been found to induce expression of proteins affecting mitochondrial function, including uncoupling proteins and those involved in oxidative phosphorylation. Notably, PGC-1α regulates mitochondrial biogenesis primarily by inducing Tfam expression through interactions with the transcription factors NRF1 and NRF2. Wu et al. demonstrated the physical interactions between PGC-1 and NRF-1 and showed that binding site on PGC-1 overlaps with the region required for its interaction with PPARγ (Wu et al. 1999). Its role outside of mitochondrial biogenesis has been reviewed elsewhere (Knutti and Kralli 2001; Liang and Ward 2006).

Investigations into the subcellular localization of PGC-1α have found that its cytoplasmic and nuclear distribution is dynamic, e.g., favoring the nucleus in response to conditions like oxidative stress (Anderson et al. 2008). The same study showed that the NAD-dependent deacetylase sirtuin 1 (SIRT1), which deacetylases and influences PGC-1α transcriptional activity, co-localizes with PGC-1α in the nucleus after oxidative stress. This is unsurprising, considering that PGC-1α has also been found to powerfully induce expression of the antioxidant system in several mitochondria-rich cell types. During myogenesis, PGC-1α was shown to reduce levels of ROS, and when downregulated, transcription of mitophagic genes LC3 and PINK1 was increased following ROS-dependent translocation of the transcription factor FOXO1 to the nucleus (Baldelli et al. 2014). A protective effect of PGC-1α has also been demonstrated in neurons, in which ROS induces expression of PGC-1α and the closely related PGC-1β, causing increased expression levels of antioxidant enzymes (St-Pierre et al. 2006). While greater PGC-1α nuclear activity may lead to an increase in mitochondria number and subsequent ROS production, these results nevertheless suggested a strong antioxidant response.

The function of PGC-1α as a nuclear coactivator has been well studied, but recently it has been shown that the protein also localizes to mitochondria. Aquilano et al. found that both PGC-1α and SIRT1 localize to mitochondria and that SIRT1 associates with Tfam, a downstream target of PGC-1α (Aquilano et al. 2010). They also showed that PGC-1α and SIRT1 form complexes with Tfam, suggesting a possible direct role in mitochondrial biogenesis.

Although the presence of PGC-1α in mitochondria has been demonstrated, the mechanism by which it is imported is not understood. Conditions like elevated ROS induce PGC-1α translocation to the nucleus, and it has been shown that PGC-1α can accumulate in the mitochondria in response to oxidative stress as caused by acute exercise (Safdar et al. 2011). Further study of ROS-induced shifts in subcellular distribution of PGC-1α may determine if the downstream result of nuclear (antioxidant upregulation) or mitochondrial (Tfam interaction) accumulation is favored.

PGC-1α is one of the numerous mitochondrial proteins that lacks the classical mitochondrial targeting amino acid sequence (MTS). As Safdar et al. speculate, PGC-1α may be imported into mitochondria by a mechanism requiring posttranslational modifications, similar to other MTS-lacking proteins, or by association with a protein that does carry an MTS. One such mechanism may involve PGC-1α phosphorylation by the kinase p38 MAPK (Akimoto et al. 2005). However, it has been revealed that in mouse hippocampal tissue and purified mitochondria, import of a small PGC-1α isoform (35 kDa) is dependent on the outer membrane protein VDAC and mitochondrial membrane potential; it was also shown that PGC-1α physically interacts with both VDAC and PINK1, suggesting a direct role in mitophagy (Choi et al. 2013). It is thought that the varied isoforms are due to posttranslational modifications. Whether these modifications regulate binding of specific PGC-1α isoforms to proteins like Tfam, VDAC, or PINK1 has yet to be determined.

12.4.2 Tfam

Unlike PGC-1α, Tfam contains a typical mitochondrial targeting sequence, and its interactions with the protein import machinery have been recently characterized. As mentioned above, Tfam forms nucleoid structures by binding mtDNA in the mitochondrial matrix, possibly aiding to protect mtDNA from oxidative damage. It was reported that in bovine retinal epithelial cells, Tfam binds to both the Tom70 subunit of the TOM complex and the Tim44 subunit of the TIM23 complex (Santos and Kowluru 2013). Under diabetic conditions, accumulation of TIM23 and TOM complex subunits (Tim23, Tom40, Tom70) in the mitochondria decreased alongside decreases in Tfam binding to each subunit. Expression of the subunits did not change except for Tim44, which also experienced lowered mitochondrial accumulation and Tfam binding.

Tfam binds to mtDNA in a non-sequence-specific manner and promotes transcription through interactions with mitochondrial transcription factor B2 (McCulloch and Shadel 2003). Consequently, reduced import of Tfam would be expected to affect mtDNA copy number in a cell as well as disrupt ATP production due to changes in expression of respiratory complex complexes. However, even small changes in Tfam concentration enough to alter the Tfam:mtDNA ratio can greatly affect DNA compaction and mtDNA replication. The in vivo ratio is estimated to be 1 Tfam molecule per 15–18 base pairs (bp). In a rolling-circle DNA replication assay, increasing the ratio to 1 Tfam per 8 bp inhibited mtDNA replication (Kukat et al. 2011; Farge et al. 2014). It was proposed that at high Tfam concentrations and thus greater mtDNA coverage, peeling of mtDNA strands is blocked, decreasing transcription and replication. A recent study found reduced PGC-1α and Tfam expression in Alzheimer's hippocampal tissue, implicating impaired mitochondrial biogenesis in Alzheimer's disease (Sheng et al. 2012). Impaired mitochondrial biogenesis has been identified in a number of diseases, and it may be useful to establish if it is caused in part by dysregulation of the Tfam:mtDNA ratio (Uittenbogaard and Chiaramello 2014). As reported earlier, PGC-1α only binds to mitochondrial Tfam bound to the D-loop region of mtDNA, but the functional nature of this relationship is not well understood. Does PGC-1α actually directly activate Tfam to promote mtDNA transcription and replication? Other factors to consider include acetylation status of Tfam, the role of the protein import machinery in regulating the Tfam:mtDNA ratio, and the role of PGC-1α/SIRT1-Tfam complexes in mtDNA maintenance.

12.4.3 Import of Nucleic Acids

Interestingly, plant, mammalian, and yeast mitochondria are capable of importing DNA (Koulintchenko et al. 2003, 2006; Weber-Lotfi et al. 2009). The mechanism of DNA import has not yet been demonstrated, but accumulating evidence suggests VDAC is the key import channel. Uptake of DNA was inhibited in isolated mitochondria, both plant and mammalian, when exposed to the VDAC blockers DIDS and heparin (Weber-Lotfi et al. 2009). A recent study by Weber-Lotfi et al. has revealed other proteins that may be involved in DNA import (Weber-Lotfi et al. 2015). The precursor of the ATP synthase β subunit associates with the outer membrane, where it may cooperate with VDAC by binding to DNA. A copper-binding protein of Complex I in the inner membrane they term CuBPp was shown to be required for import of long DNA fragments by acting as a receptor in the intermembrane space. Translocation of DNA into the matrix remains poorly understood and may involve different mechanisms in plants and mammals. Inhibitors of the inner membrane protein adenine nucleotide translocator (ANT) block DNA import in plant, but not in mammalian mitochondria (Koulintchenko et al. 2006). Mitochondria also import RNA, chiefly tRNAs, and VDAC has again been identified as a key regulator, although TOM and TIM23 have been shown to be involved (Campo et al. 2016). Nucleic acid import into mitochondria is an emerging subject of research, and continued work will be necessary to understand the mechanisms and functions of this process.

12.4.4 Protein Import and Lipidogenesis

Mitochondrial biogenesis requires not only the import and synthesis of proteins but also of lipids, as the organelle is a membrane-bound structure. Several lipids can be made by mitochondria, including phosphatidylethanolamine (PE), phosphatidylglycerol (PG), and cardiolipin (CL), but all other lipids must be imported. Lipid production in the cell occurs primarily in the endoplasmic reticulum (ER), whose membrane associates with the OMM at sites called the mitochondria-associated membrane (MAM) (Vance 1990). These sites are thought to be where one of the main mechanisms of lipid transport into mitochondria occurs; other proposed mechanisms include vesicular traffic and lipid transfer proteins (Horvath and Daum 2013).

While both the outer and inner membranes contain phospholipids present in all cellular membranes, their composition differs especially in regard to CL, which is synthesized in the inner membrane. Most of the mitochondrial (and cellular) CL is found in the inner membrane, where it interacts with numerous proteins, including each of the respiratory complexes and cytochrome c, suggesting a key role of CL in mitochondrial function (Flis and Daum 2013). One of the key proteins required for CL biosynthesis is translocator and maintenance protein 41 (Tam41), a matrix-facing inner membrane protein first identified in yeast as having a role in protein import by maintaining TIM23 integrity (Gallas et al. 2006; Tamura et al. 2006). Tam41 was later shown in yeast to be involved in the CL synthesis pathway as a CDP-diacylglycerol (CDP-DAG) synthase, indicating coordination between protein import and lipid biosynthesis (Tamura et al. 2013). A recent study has shown that CL also directly interacts with mtDNA and in CL-deficient yeast experiencing thermal stress, mtDNA segregation is diminished (Luévano-Martínez et al. 2015). Interestingly, another recent study revealed a new role for a familiar mitochondrial biogenesis protein; in hearts of adult mice lacking both PGC-1α and PGC-1β, CL levels are reduced due to decreased expression of Cds1, one of several CDP-DAG synthases (Lai et al. 2014). Finally, Tom22 has been recently shown to interact with the steroidogenic enzyme 3β-hydroxysteroid dehydrogenase 2, implicating the TOM subunit in the process of conversion of pregnenolone to progesterone in gonadal and adrenal tissues (Rajapaksha et al. 2016).

12.5 Mitochondrial DNA and Damage

Various sources of both endogenous and exogenous stresses can damage mtDNA. Reactive oxygen species (ROS) generated by oxidative phosphorylation is the major source of endogenous mtDNA damage. The lesions generated by ROS include base modifications, abasic (AP) sites, DNA strand breaks, and DNA crosslinks. Additionally, ROS can damage the sugar-phosphate backbone in mtDNA. Errors in mtDNA replication machinery can also cause point mutations and deletions. Due to the close proximity of mitochondrial genome to the ROS emission site on the inner mitochondrial membrane and the lack of protective histones, mtDNA is more susceptible to oxidative damage than nuclear DNA (nDNA). For the above reasons, damage to mtDNA by exogenous genotoxic agents such as industrial byproducts, UV, environmental toxins, and alkylating agents is also more extensive (Akhmedov and Marin-Garcia 2015).

As key players in regulating cellular processes from energy metabolism, generation of ROS, and regulation of cytosolic Ca^{2+} to stress response, and determination of the fate of cells, dysfunctional mitochondria may lead to a wide spectrum of pathogenesis including neuromuscular disorders, neurodegenerative diseases, cardiovascular diseases, and cancer. Thus, the maintenance of mitochondrial genomic integrity is crucial for mitochondrial function and viability of cells. In the following sections, we will present various mtDNA repair pathways and the import of crucial proteins involving in the mtDNA repair mechanism.

12.6 mtDNA Repair

It is currently accepted that mammalian mtDNA possesses almost all nDNA repair mechanisms, including base excision repair (BER) (primary repair pathway), single-strand DNA breaks (SSBs) repair, mismatch repair, and, possibly, homologous recombination (HR) and nonhomologous end joining (NHEJ). All of the proteins involved in mtDNA repair are encoded by nDNA and therefore must be translocated into the mitochondrial matrix through TOM and subsequently through one of the inner membrane channels TIM22 or TIM23. The most well-studied proteins involved in the repair pathways include apurinic/apyrimidinic endonuclease 1 (APE1), Tfam, mtSSB, POL γ, p53, and RECQL4.

12.6.1 Base Excision Repair (BER)

As mentioned previously, BER is the primary mtDNA repair pathway, which is used for excision and repair of oxidized, deaminated and alkylated bases, as well as SSBs repair. BER is also the first and most extensively characterized repair pathway in mitochondria (Takao et al. 1998; Larsen et al. 2005; Prakash and Doublie 2015).

The first step in BER is the removal of the damaged base by cleavage of an N-glycosidic bond between the nitrogenous base and its deoxyribose. This cleavage is catalyzed by DNA glycosylases, the highly conserved nuclear-encoded enzymes that contain the mitochondrial targeting sequence (MTS) that allows them to translocate to mitochondria. Depending on whether DNA glycosylase possesses the intrinsic lyase activity, it can be mechanistically characterized as monofunctional (lacks a lyase activity) or bifunctional (possesses a lyase activity). Monofunctional DNA glycosylases rely on AP endonuclease (APE1) to hydrolyze the phosphate backbone, after the AP site is created. The terminal DNA ends then undergo dRP lyase reaction and gap filling by DNA polymerase. This process of replacing a single nucleotide is known as short-patch BER (SP-BER) (Demple and Sung 2005).

12.6.1.1 APE1

Because APE1 is the only protein that can hydrolyze AP sites generated by DNA glycosylases, its function directly influences mitochondrial genome stability. APE1 contains MTS at residues 289–318 that allow the protein to be translocated to mitochondrial inner membrane in response to oxidative stress. The mechanism of APE1 translocation into the matrix is not yet understood. Mutations at Lys299 and Arg301 abolish the import of APE1 to mitochondria under oxidative stress, which suggests that these are the critical sites. It has been shown that APE1 passes through TOM, although the mechanism is still not clear. Upon passage through TOM, APE1 interacts with the mitochondrial import and assembly protein Mia40, the main component of the MIA import pathway, which catalyzes the formation of a disulfide bond between APE1's Cys93 and Mia40's Cys55 residues. After the import, another component of the MIA pathway, augmenter of liver regeneration (ALR), reoxidizes Mia40. Although APE1 is normally localized in the inner membrane space, some of it is imported into the matrix via the presently unknown mechanism. Expression levels of Mia40 affect translocation of APE1 and, as a result, mitochondrial genome stability. Additionally, abnormal expression and localization of APE1 was found to be associated with tumor aggressiveness and chemoresistance (Barchiesi et al. 2015).

12.6.1.2 UNG1

Human uracil DNA glycosylase 1 (UDG1 or UNG1) is a monofunctional glycosylase, which is a mitochondrial isoform of the enzyme produced both by alternative splicing and transcription from a different start site. The active site, an Asp145 residue, initiates catalytic cleavage of the N-glycosidic bond on single- (ss) or double-stranded (ds) DNA. To aid the binding of uracil to the active site, another UNG1 residue, Leu272, is inserted into the minor groove of DNA and causes a local disruption (Prakash and Doublie 2015). Although it is known that UNG1 preprotein contains a MTS, which gets cleaved upon entry to the inner membrane, and that residues 1–28 are sufficient to ensure import, no research investigating the mechanism of such import is currently available (Otterlei et al. 1998).

12.6.1.3 MUTYH

A human homolog of the bacterial mutY. MUTYH recognizes and excises the undamaged adenine opposite to 7,8-dihydro-8-oxo-guanine (8-oxo-G) lesion as means of preventing C:G to A:T transversion mutations. POL γ then inserts the cytosine to correctly pair the bases. These transversion mutations are among the most often found in several prevalent cancers (Markkanen et al. 2012; Prakash and Doublie 2015). Despite containing an MTS and playing a key role in the repair of one of the most abundant and highly mutagenic kinds of oxidative damage, 8-oxo-G, MUTYH import into mitochondria has not been studied (Markkanen et al. 2012; Prakash and Doublie 2015).

12.6.1.4 OGG1

Another kind of enzyme responsible for 8-oxo-G lesion repair is 8-oxo-G DNA glycosylase 1 (OGG1), a bifunctional glycosylase and a human homolog of the bacterial mutM. Several isoforms of OGG1 produced by alternative splicing have been described, all of them containing the same N-terminal MTS but varying C-terminus. Literature suggests that import of OGG1 into mitochondrial matrix improves with aerobic exercise and declines in an age-dependent manner (Szczesny et al. 2003; Radak et al. 2009).

In addition to OGG1, three other bifunctional DNA glycosylases have been documented in mammalian mitochondria, NTHL1, NEIL1, and NEIL2. The first one, NTHL1, is responsible for the excision of oxidized pyrimidine lesions. NEIL1 exhibits the preference for guanidinohydantoin (Gh), spiroiminodihydantoin (Sp), thymine glycol (Tg), 5-hydroxyuracil (5-OHU), dihydrouracil (DHU), and ring-opened 2,6-diamino-4-hydroxy-5-formamidopyrimidine (Fapy) damaged bases in dsDNA, and NEIL2 excises 5-OHU lesions in ssDNA (Brooks et al. 2013; Akhmedov and Marin-Garcia 2015). To our knowledge, the mechanism of import of these three enzymes has not yet been proposed.

12.6.1.5 p53

The p53 protein is mostly known as a tumor suppressor and a transcription factor, but p53 also regulates various cellular processes such as DNA repair, cell cycle, apoptosis, redox homeostasis, and metabolism. Not surprisingly, p53 is commonly referred to as the “guardian of the genome” (Park et al. 2016a). p53 participates in almost all nDNA repair pathways with the exception of nucleotide excision repair (NER). The roles of p53 in mtDNA repair include 3–5′ exonuclease activity, involvement in mtDNA replication, and ensuring the accuracy of mtDNA synthesis by excising mispaired nucleotide bases (Bakhanashvili et al. 2008).

At least three distinct mechanisms of p53 translocation into mitochondria have been discovered in unstressed normal cells. The first import pathway involves the physical interaction of p53 with RecQ helicase-like protein 4 (RECQL4), a nuclear DNA helicase containing 1,208 amino acid residues. The amino-terminal region of RECQL4 contains both MTS and NLS (nuclear localizing signals) (NLS) and MTS. The MTS of RECQL4 allows it to interact with Tom20 and to likely cross the outer membrane via the TOM complex. The NLS region is positioned at amino acids 270–400 of RECQL4 and 293–362 of p53. The interaction of RECQL4 and p53 at the aforementioned positions masks their respective NLSs, allowing the translocation of the complex into the mitochondrial matrix (Croteau et al. 2012; De et al. 2012; Park et al. 2016a). However, the precise translocation pathway across the inner membrane has yet to be identified.

The biological functions of interaction between RECQL4 and p53 in mitochondria are not known. In unstressed normal cells, RECQL4 co-localizes with p53 in mitochondria in addition to the nucleoplasm and participates in mtDNA maintenance. Its specific involvement in mtDNA replication and repair is not clearly understood (Croteau et al. 2012). RECQL4-deficient cells exhibit mitochondrial bioenergetic defects and dysfunction involving reduced mtDNA copy number, increased ROS production, elevated mitochondrial fragmentation, and diminished mtDNA repair capacity after oxidative stress, highlighting its role in mitochondrial function and genomic stability (Chi et al. 2012).

The second pathway of p53 import involves the activation of cryptic MTS by a cytosolic protease, which recognizes serine protease consensus sites present in mammalian p53 (Park et al. 2016a). The proteolytic cleavage of p53 results in a ~40 kDa fragment, which is imported through the membranes of mitochondria (Boopathi et al. 2008). However, additional research is necessary to represent the nature of the signals and/or mechanism of this pathway.

The third p53 import pathway contains a disulfide relay system consisting of the import receptor CHCHD4 and the FAD-dependent sulfhydryl oxidase (GFER) in the intermembrane space (IMS). CHCHD4, coiled-coil-helix-coiled-coil-helix domain containing four proteins, is the mammalian homolog of Mia40 and can form intermolecular disulfide bonds with cysteine residue in proteins targeted for translocation into the mitochondria (Zhuang et al. 2013). The protein p53 possesses cysteine-rich motifs (Cys-135/Cys-141 and Cys-275/Cys-277), which can form two intramolecular disulfide bonds. Thus, in a respiration-dependent manner, CHCHD4 can covalently bind p53 at the Cys-135 residue (disulfide Cys-135/Cys-141 pair) upon its translocation across the TOM complex (Park et al. 2016a). This mitochondrial disulfide relay system needs the regeneration of oxidized CHCHD4 through its oxidase GFER, which successively transfers the electrons to oxidized cytochrome c. As a result, p53 import via this process requires the reoxidation of reduced cytochrome c by respiratory Complex IV (Zhuang et al. 2013). Other substrates of Mia40 such as TIM22 and mitochondrial ribosome Mrp10 are required to translocate to the inner membrane and matrix, respectively. Nevertheless, it is not clear how p53 via its interaction with CHCHD4 is imported into the matrix space, where mtDNA resides (Park et al. 2016a). Clearly, CHCHD4-mediated p53 import plays a role in mtDNA repair. The CHCHD4-overexpressed cells elevate the recovery of mtDNA integrity compared with the control. In contrast, CHCHD4-depleted cells have decreased mtDNA integrity (Zhuang et al. 2013).

Although the exact function of p53 in mitochondria requires more research, p53 is indeed essential for mtDNA repair and accuracy of DNA synthesis. Studies show that p53 physically interacts with mtDNA and POL γ, acting as an external proof-reader for DNA replication. In Addition, p53 elevates the binding of Tfam to cyto-toxic-damaged DNA, and enhances BER through direct interaction with the repair complex in the inner membrane (Bakhanashvili et al. 2008).

12.6.1.6 Tfam in mtDNA Repair

Besides taking roles in mtDNA transcription, replication, regulation of mtDNA copy number, and organization of the mitochondrial nucleoid, as previously mentioned in section 12.4.2, Tfam participates in mtDNA repair and modulates BER enzymatic activities. Recent study by Canugovi et al. found that Tfam reduced the activities of the steps of BER involving OGG1, UDG, APE1, and POL γ in vitro by binding to mtDNA. The purpose of Tfam by doing so is not likely to inhibit DNA repair or mtDNA metabolic processes rather to modulate the access of those proteins to mtDNA. The interacting proteins including p53 can regulate Tfam/mtDNA affinity and promote BER in mtDNA, suggesting a role of Tfam as an important regulator of BER. Furthermore, Tfam knockdown cells express higher mutation frequency and mtDNA damage (Canugovi et al. 2010). The import pathway of Tfam into mitochondria is described in Mitochondrial Biogenesis section.

12.6.2 Single-Strand DNA Breaks (SSBR)

SSBR is often regarded as a sub-pathway of BER due to its different end processing events to restore phosphate (5′-P) and hydroxyl (3′-OH) before ligation proceeds. There are other instances in which end processing at a SSB may take place without being preceded by the other BER steps involving removal of a damaged base and cleavage at an abasic (AP) site. In those instances, two enzymes, tyrosyl-DNA-phosphodiesterase 1 (TDP1) and aprataxin (APTX), are directly involved (Meagher and Lightowlers 2014).

12.6.2.1 TDP1 and APTX

TDP1 and APTX are recently discovered enzymes of the mtDNA repair network of SSBR. TDP1 can repair 3′ lesions induced by chain terminating nucleoside analogues. APTX is required to remove the covalent attachment of adenine monophosphate (AMP) to the 5′ end of mtDNA (5′-AMP), caused by abortive ligase activity next to an existing lesion. To this date, it is not known exactly how TDP1 is imported to mitochondria lacking N-terminal MTS (Das et al. 2010). In contrast, APTX possesses an alternatively spliced isoform containing 14 aa N-terminal sequence, which is possibly an MTS. The MTS is cleaved upon entry into the mitochondria by a mitochondrial processing peptidase (MPP) (Meagher and Lightowlers 2014). Further research is required to describe how TDP1 and APTX are translocated to the mtDNA damage sites. Mutations in the genes encoding TDP1 and APTX result in failure to repair 3' lesions and to remove covalent attachment, respectively. Such failure reduces mtDNA stability, since SSBs can further result in double-strand breaks. Oxidative mtDNA damage accumulates in cells lacking TDP1, suggesting a role in ROS-induced mtDNA damage repair. Furthermore, mutations in TDP1 and APTX can cause ataxia, which is often associated with mitochondrial disease (Das et al. 2010; Meagher and Lightowlers 2014).

12.6.3 Mismatch Repair (MMR)

As the name suggests, MMR repairs base-base mismatches and small nucleotide insertion/deletion mispairs. The presence of the nuclear MMR proteins in the mitochondria has been controversial (Martin 2011). de Souza-Pinto et al., in 2009, did not observe the localization of the nuclear MMR proteins, MutS homolog 3 and 6 (MSH3 and MSH6, respectively), and MutL homolog 1 (MLH1) in human mitochondria (de Souza-Pinto et al. 2009). However, Martin et al. and others have detected the presence of MLH1 in the mitochondria and found a role for MLH1 in oxidative mtDNA repair. Deficiency of MLH1 in addition to silent mitochondrial genes, POLG and PINK1, shows an increase 8-oxo-G lesions (Martin 2011). A human homolog, hMSH5, directly interacts with mtDNA, Twinkle, and POL γ. Oxidation-induced lesions are more efficiently repaired (Bannwarth et al. 2012) in cells that overexpress hMSH5. Furthermore, de Souza-Pinto et al. identified the Y-box-binding protein (YB-1) as an a mtDNA mismatch-binding and mismatch repair factor in human mitochondria. It is shown that YB-1 directly binds to mtDNA and participates in the repair (de Souza-Pinto et al. 2009). To date the extent of mismatch repair activity and the import pathways of MMR proteins into mitochondria are not clearly known and required further research (Martin 2011; Bannwarth et al. 2012).

12.6.3.1 DNA Ligase III (LIG3)

LIG3 is the only DNA ligase known in mitochondria. The absence of LIG3 in vitro is accompanied by the loss of mtDNA without losing cell viability (Spadafora et al. 2016). In 1999, Lakshimpathy et al. demonstrated that the upstream 5′ end of LIG3 cDNA coded for an MTS (Lakshmipathy and Campbell 1999). When entering mitochondria, the resultant MTS is cleaved producing a mature protein with similar molecular weight as the nuclear variant (Meagher and Lightowlers 2014). Mitochondrial LIG3 is mainly involved in BER and in replication of mtDNA. In addition, LIG3 plays a role in an alternative pathway of DNA double-strand break repair that supports NHEJ (Kukshal et al. 2015). The exact translocation pathway of LIG3 to the mitochondrial matrix has yet to be identified.

12.6.3.2 mtSSB

mtSSB is the only known mitochondrial single-stranded DNA-binding protein. It is essential in mtDNA replication and could have a functional role similar to that of the nuclear ssDNA-binding replication protein A (RPA) in managing mitochondrial replication-associated DNA repair. Little is known about a role of mtSSB in the context of mtDNA repair (Wollen Steen et al. 2012). The import pathway of mtSSB in mammalian mitochondria is not well studied. However, in the plant, *Arabidopsis thaliana*, a gene encoding a mitochondrially targeted SSB (MTSSB) was identified. The At4g11060 gene codes for a protein of 201 aa, including a 84-bp region coding 28-residue presumed N-terminal mitochondrial targeting transit peptide (Edmondson et al. 2005). There is a high similarity between the human, bacetrial, and plant ortologs of mtSSB, suggesting a highly conserved function in plasmidial DNA repair (Edmondson et al. 2005) (Akhmedov and Marin-Garcia 2015).

12.6.3.3 POL γ

As the only known polymerase found in mammalian mitochondria, POL γ promotes mtDNA replication, recombination, and repair (Graziewicz et al. 2006). The extent of import of this enzyme into the mitochondria has not been studied. POL γ are accountable for high accuracy of mtDNA replication through 3–5′ exonucleolytic proofreading activity and nucleotide selectivity (Akhmedov and Marin-Garcia 2015). Mutations in POL γ can cause point mutations and deletions in mtDNA and can be the source of mitochondrial diseases such as progressive external ophthalmoplegia (PEO), Parkinson's disease, and Alpers syndrome (Akhmedov and Marin-Garcia 2015; DeBalsi et al. 2016). POL γ fills single-nucleotide gaps in the presence of a 5′ terminal deoxyribose phosphate (dRP) flap (Graziewicz et al. 2006). The process is aided by flap endonuclease (FEN1) and DNA nuclease/helicase (DNA2). Interestingly, DNA2 does not contain a canonical MTS, but its translocation to mitochondria requires a specific sequence between 734 and 829 amino

acid residues. By interacting with POL γ, DNA2 activates the polymerase (Zheng et al. 2008; Akhmedov and Marin-Garcia 2015; Ding and Liu 2015). In addition, the mitochondrial 5′ exo/endonuclease G (EXOG) is capable of excising displaced 5′ flaps, while depletion of EXOG affects mitochondrial genome and leads to mitochondrial dysfunction in different kinds of human cells (Zhuang et al. 2013; Akhmedov and Marin-Garcia 2015). Finally, a newly discovered 5–3′ exonuclease MGME1 has been shown to create ligatable DNA ends in combination with POL γ. Loss of MGME1 function results in multisystemic mitochondrial disease in humans (Akhmedov and Marin-Garcia 2015; Uhler et al. 2016).

12.7 Protein Import and Metabolism

Mitochondria are the major metabolic hub in eukaryotic organisms, which has granted these organelles the nickname "powerhouse of the cell." The major metabolic function of mitochondria is carbohydrate metabolism, involving production of ATP by way of the tricarboxylic acid cycle and oxidative phosphorylation. Additionally, two other major metabolic processes occurring within mitochondria include energy production through fatty acid oxidation as well as citrulline production in the urea cycle (Nakagawa and Guarente 2009; Cheng and Ristow 2013).

12.7.1 Import of Metabolic Exchangers

Mitochondrial metabolism relies on import of its protein machinery. Import pathways vary depending on the nature and final destination of such proteins. One component of the metabolic machinery is the voltage-dependent anion channel (VDAC), which allows transport of ions, metabolites, and nucleic acids across the outer membrane. In its highest conductance state, VDAC is involved in the exchange of anionic metabolites pyruvate, ATP, ADP, P_i, and nucleotides and is additionally capable of transporting non-electrolyte substances of up to 5 kDa in size (Pavlov et al. 2005; Lemasters and Holmuhamedov 2006; Mannella and Kinnally 2008; Rostovtseva 2012). Anion gating occurs with a 50–60% reduction in conductance; consequently, due to the electrostatic nature of the VDAC channel walls, flow of small cations such as Ca^{2+} is facilitated (Colombini 2004). The import and assembly of VDAC is driven by an MTS that directs it to Tom20 and Tom22 receptors within the TOM complex. Once imported through TOM, the SAM complex integrates VDAC to the OMM (Kozjak-Pavlovic et al. 2007).

VDAC interacts with several metabolite transporters located in the inner membrane. The inorganic phosphate carrier or PiC is imported in an ATP/$\Delta\Psi$-dependent manner to the IM and is responsible for the direction of P_i to the matrix (Pratt et al. 1991). The pyrimidine nucleotide carrier −1 (PNC-1), also referred to as SLC25A33, is induced by insulin like growth factor (IGF-I) and is responsible for the import of nucleotides for biogenesis of mtDNA (Favre et al. 2010; Di Noia et al. 2014). PNC aids in maintaining an appropriate ratio of mtDNA to nDNA genes and is responsible for cell maintenance but can support tumor growth (Favre et al. 2010). VDAC's interaction with the adenine nucleotide translocase (ANT) allows for exchange of ATP and ADP between the matrix and the cytosol via VDAC (Gavalda-Navarro et al. 2014). The proteins that form this metabolite pathway are imported through TOM and delivered to TIM22 by Tim9/Tim10 chaperone proteins (Endres et al. 1999) In addition to ATP/ADP exchange, ANT is also involved in the uncoupling of fatty acids prior to β-oxidation. Deficiency or malfunction of ANT leads to oxidative damage and has been linked to cardiac arrhythmias, lowered production of erythrocytes and B-lymphocytes, as well as a possible link to type 2 diabetes (Ciapaite et al. 2006; Cho et al. 2015; Roussel et al. 2015).

As mentioned above, mitochondrial metabolism and the import of its protein machinery are inherently codependent. The 13 proteins coded by mtDNA are incorporated into the inner membrane as subunits in Complexes I, III, IV, and V of the electron transport chain (ETC), while all other protein components involved in oxidative phosphorylation are nuclear encoded and must be imported to the matrix or inner membrane (Bentinger et al. 2010; Alcazar-Fabra et al. 2016). During import of matrix-targeted proteins, positively charged presequences are electrophoretically driven to the matrix because of the $\Delta\Psi$ created by the ETC (Harbauer et al. 2014b). This codependency is further exemplified by the ATP requirement of chaperone interactions with the proteins en route to the mitochondrial matrix (Wachter et al. 1994) and by the effect of $\Delta\Psi$ depletion on steady-state levels of Tom20 and Tim23 (Joseph et al. 2004).

Mitochondrial import pathways employ highly conserved protein complexes, which are present in *Saccharomyces cerevisiae*, resulting in myriad studies performed on this readily accessible organism (Campo et al. 2016). In the respiratory Complex III, or the cytochrome bc_1 complex, all subunits are nuclear encoded with the exception of the mtDNA-encoded respiratory subunit cytochrome b. The Rieske iron-sulfur protein (ISP), one of three respiratory subunits of Complex III involved in redox, has been shown in multiple studies, to contain a 30 amino acid-long MTS at the N-terminal portion of the precursor protein, which is subsequently cleaved prior to incorporation into Complex III (Brandt et al. 1993; Conte and Zara 2011). Interestingly, in this case, the cleaved MTS of ISP is retained to form subunit 9 of Complex III (Brandt et al. 1993). The second nuclear-encoded respiratory subunit (4) of Complex III, cytochrome c_1, has been experimentally shown to follow well-understood import pathways to the IM. In addition to being $\Delta\Psi$ dependent for successful import, cytochrome c_1 possesses a bipartite presequence containing two hydrophobic domains (Rodiger et al. 2011). Evidence indicates that hydrophobic areas of the MTS do not signal a stop transfer to incite lateral release. Initially, cytochrome c_1 imports to the matrix and subsequently exports to the inner membrane for integration to Complex III, which involves dual cleavage of its MTS (Stuart et al. 1990). This suggests the probable TIM23 – OXA pathway of inner membrane import (Campo et al. 2016).

Import of respiratory subunits, such as the above examples of Complex III subunits, demonstrates the direct and most obvious effect protein import has on metabolic function; that is, protein import supplies the metabolic machinery necessary for oxidative phosphorylation to occur. Examining the function of these protein subunits, in turn, demonstrates the converse effect metabolic function has on import. Such is the case with ISP, as it has been shown to not only facilitate electron flow through Complex III but also to function as a gating protein responsible for the release of 4 H^+ with every two rotations of the Q cycle (Gurung et al. 2005; Jafari et al. 2016). The proton-pumping abilities of Complex III contribute to the $\Delta\Psi$ vital for the import of metabolically supportive proteins.

12.7.2 Import of Metabolic Regulators

12.7.2.1 Coenzyme Q

Coenzyme Q (CoQ), also known as ubiquinone or Complex II, is an isoprenylated compound with two major biological functions within the ETC: (1) as a single electron acceptor, it forms semiquinone, a semi-reduced conformation with antioxidant properties allowing it to absorb free radicals and reduce emission of ROS, and (2) as an acceptor of two electrons, CoQ drives the ETC by then reducing Complex III (Gonzalez-Mariscal et al. 2014). Additionally, CoQ is an electron carrier involved in pyrimidine synthesis and β-oxidation (Allan et al. 2015). CoQ is a unique complex (of the ETC) in that it does not involve proteins encoded by either nDNA or mtDNA. Instead, CoQ is assembled within the matrix in the mevalonate pathway by a complex of proteins referred to as the coenzyme Q synthome or CoQ-synthome (Alcazar-Fabra et al. 2016). This conglomeration of proteins includes 11 subunits named Coq 1–11 as well as ferredoxin (Yah1) and ferredoxin reductase (Arh1) all of which are nuclear encoded (Gonzalez-Mariscal et al. 2014; Allan et al. 2015; Jenkins et al. 2016). The CoQ-synthome is an example of an indirect but vital relationship between metabolism and the import of Coq-synthome protein subunits. Twelve of the subunits are imported and assembled in the matrix, with only the subunit Coq2p being inserted into the IM. Coq2p is thought to anchor the synthome to the IM (Gonzalez-Mariscal et al. 2014). The pathway of import has logically been assumed to utilize the matrix-targeted TOM-TIM23 pathway (Gonzalez-Mariscal et al. 2014). The exact mechanism of import for these proteins, however, has yet to be specifically investigated. While this import pathway is understood to be the method of import for all matrix-targeted proteins, the mechanism of import of CoQ-synthome merits investigation. *Saccharomyces cerevisiae* contains CoQ(6), a CoQ compound analogous to the *Homo sapiens* form CoQ(10) that differs in its number of isoprenyl subunits. It has been experimentally shown that strains grown in the absence of $\Delta\Psi$ were able to continue to synthesize CoQ(6) at normal levels (Santos-Ocana et al. 1998; Gonzalez-Mariscal et al. 2014). This observation elicits many questions: Are Coq-synthome proteins continuing to be imported in the absence of $\Delta\Psi$? Does this suggest a yet undiscovered mechanism of protein import to the matrix? Is the same result able to be replicated with CoQ(10)? Not only is the method of import in question, but new subunits of the CoQ-synthome are continuing to be discovered, making this complex an important topic of continued research.

12.7.2.2 Stat3/GRIM-19

The *g*ene associated with *r*etinoid-*i*nterferon-induced cell *m*ortality 19 (GRIM-19) and signal transduction and activator of transcription 3 (STAT3) have an important relationship that influences metabolic function in mitochondria and introduces a novel mechanism of import. GRIM-19 is a nuclear-encoded protein imported to the IM to be integrated as a component of Complex I (Tammineni et al. 2013) Stat3 is a nuclear-encoded transcription factor protein with both nuclear and mitochondrial functions. The nuclear function involves the LIF/JAK/Stat pathway, while in mitochondria Stat3 not only is integrated with Complex I but also serves as a transcription factor (Wegrzyn et al. 2009).

The mechanism of import for GRIM-19 is $\Delta\Psi$ dependent. Import of Stat3, on the other hand, involves a less conventionally studied mechanism. Phosphorylation of two sites on Stat3, Tyr-705 and Ser-727, determines a nuclear localization. When only Ser-727 is phosphorylated, Stat3 becomes associated with GRIM-19 in the cytosol and co-localizes to the inner membrane of mitochondria. It is thought that phosphorylation of Ser-727 allows for a binding site between the two proteins. GRIM-19 not only behaves as a chaperone for Stat3 but aids in the assembly of Complex I (Tammineni et al. 2013). The co-localization of these two inner membrane proteins may indicate a novel mechanism for import. Prior studies have shown inner membrane-bound proteins to be chaperoned to Tom70 by Hsp70 and Hsp90. Once in the inner membrane, they would be chaperoned to TIM22 by Tim9-Tim10 and Tim8-Tim13 complexes. However, the precise import pathways of Stat3 and GRIM-19 have yet to be studied (Young et al. 2003; Hasson et al. 2010).

The role of mitochondrial Stat3 in metabolism and its physiological consequences are a topic of current research. GRIM-19 knockouts in mice models were shown to be lethal at the embryonic stage, while Stat3 knockouts showed decreased levels of function in Complexes I and II (Wegrzyn et al. 2009). As mentioned above, while Stat3 physically associates with Complex I, it also binds mtDNA and promotes the transcription of ETC proteins (Park et al. 2016b). The co-localization of GRIM-19 and Stat3 in concert with the circumstances under which each is active may suggest these proteins actually have antagonistic roles. GRIM-19, though necessary for the function of Complex I, is an inducer of apoptosis and an inhibitor of Stat3. Conversely, the Stat3 ability to upregulate mtDNA translation may cause anti-apoptotic effects leading to tumorigenesis if left unchecked (Zhang et al. 2003). In addition to the potentially pathological effect mitochondrial Stat3 could produce, recent studies have found potentially vital benefits of the transcription factor. The proliferation increased expression of ETC machinery proteins induced by Stat3 has been seen to support the energy-expensive process of maintaining embryonic stem cells (ESC) in a naïve pluripotent state. In fact, with enhanced oxidative phosphorylation, primed stem cells can return to a naïve pluripotency stage (Carbognin et al. 2016). Stat3 has also been found to be upregulated in neuronal cells surrounding an injury site to the spinal cord, which might suggest a role of Stat3 in cell survival (Park et al. 2016b).

12.7.2.3 F1Fo-ATP Synthase Subunit ε

The F_0F_1-ATP synthase is a two-sectioned multi-subunit complex responsible for physically combining ADP and P_i to produce ATP. The F_1 region of ATP synthase, located in the matrix, contains five protein subunits α, β, γ, δ, and ε (Jonckheere et al. 2012). Subunit ε of region F_1 is a small 5.8 kDa protein which is nuclear encoded by the ATP5E gene. It is important to make the distinction that the F_1 subunit ε of bacteria and chloroplast cells is homologous to the subunit δ of mammalian ATP synthase; in addition, bacteria and chloroplasts do not possess a protein homologous to the mammalian subunit ε (Havlickova et al. 2010). *Saccharomyces cerevisiae* do however possess an F_1 Su ε homologous in structure and function to the mammalian ortholog and is therefore a convenient model for investigation (Lai-Zhang and Mueller 2000).

Unlike other matrix-targeted proteins, Su ε does not possess a cleavable MTS (Havlickova et al. 2010). This structural feature presents a conundrum when considering the mechanism of import. Recently, it was experimentally shown that the subunit ε is imported through TIM2, but independently of $\Delta\Psi$ and without positively charged residues at the N-terminus (Turakhiya et al. 2016). When the positively charged amino acid sequences were added on the N-terminal end, import became dependent of $\Delta\Psi$ (Turakhiya et al. 2016). These results indicate the possibility of a yet undefined mechanism of protein import into mitochondria.

12.7.2.4 NF-kB

The NF-kB family of transcription factors controls the expression of genes involved in inflammation, metabolism, cancer, and development. Aside from its canonical regulation of mitochondrial metabolism by regulating gene expression, NF-kB was shown to localize to the intermembrane space and to interact with the ATP-ADP translocator-1 (Bottero et al. 2001). This interaction promotes the mitochondrial recruitment of NF-kB with a concomitant decrease in its nuclear activity. This result is further corroborated with decreased expression of known nuclear anti-apoptotic NF-kB target genes, Bcl-xL and c-IAP-2 (Zamora et al. 2004). Another NF-kB family member, RelA, is also present in the mitochondria, where it binds to mitochondrial DNA and inhibits the expression of cytochrome c oxidase III (Cogswell et al. 2003). Interestingly, mitochondrial p53 levels negatively correlate with RelA levels. This potential interplay between mitochondrial p53 and RelA levels is further supported by the fact that overexpression of p53 mitigates the inhibitory effect of RelA on mitochondrial gene expression (Johnson et al. 2011). It has been proposed that these antagonistic functions of p53 and RelA on mitochondrial respiration influence the metabolic switch from oxidative to anaerobic metabolism (Johnson et al. 2011). Importantly, we found no studies into the mechanisms of mitochondrial translocation of this widely studied family of transcription factors.

12.7.2.5 Foxg1

Forkhead box g1 or Foxg1 is a transcription factor that functions in both the nucleus and mitochondria. Its function, which is necessary for the formation of the mature mammalian cerebral cortex structure, may play a key role in the signaling pathways between nucleus and mitochondria. Foxg1 aids in proliferation and differentiation of cell tissues that lead to growth and maturation of the telencephalon and mesencephalon (Ahlgren et al. 2003). Nucleus-targeted Foxg1 acts as a repressor for transcription, specifically of fibroblast growth factors. Overexpression of mtFoxg1 results in larger mitochondria (often longer than 4 μm) as well as an increase in cell differentiation. Not only does mt-Foxg1 promote differentiation, but also it supports mitochondrial fusion. Conversely, full-length Foxg1 is involved with mitochondrial fission. The dual roles within matrix-targeted Foxg1 suggest it is a regulatory factor, which aids in producing the correct ratio of cells for cerebral cortex formation. Import of mt-Foxg1 involves a MTS at aa 277–302, a location farther from the N-terminus than classically observed. The $\Delta\Psi$-dependent mechanism of import for mt-Foxg1 is not specifically defined, but it has been suggested that the full-length protein is imported and subsequently broken into smaller fragments as varied functions are favored (Pancrazi et al. 2015). Foxg1 mutations can lead to congenital encephalopathies including Rett syndrome (Kumakura et al. 2014), as well as facial dysmorphisms, and epilepsy (Cellini et al. 2016).

12.8 Concluding Remarks

With the elucidation of the mitochondrial proteome in several organisms including humans, the development of tracing techniques, and the emerging studies on regulatory mechanisms, the field of mitochondrial protein import is gaining renovated interest. Nuclear transcription and metabolic factors are now known to regulate mitochondrial function and dysfunction in cancer, neurodegeneration, and cardiac injury. Understanding the mechanism of mitochondrial recruitment will support development of therapeutic interventions. It has also become evident that the protein import machinery may function as a rheostat that adapts to match changes in cellular metabolic needs. Targeting this machinery with exercise or pharmacological compounds is thus becoming an attractive approach to modulate mitochondrial mass in healthy and diseased cells. In addition to identifying distinct regulatory mechanisms, future studies should aim at defining the interactions between different machineries, signaling networks, bioenergetic complexes, mitochondrial dynamics, and apoptosis.

Acknowledgments This work was supported partially by CUNY CIRG grant 80209 and Eugene Lang Foundation grant 78592 to PMP.

References

Ahlgren S, Vogt P, Bronner-Fraser M (2003) Excess FoxG1 causes overgrowth of the neural tube. J Neurobiol 57(3):337–349

Akhmedov AT, Marin-Garcia J (2015) Mitochondrial DNA maintenance: an appraisal. Mol Cell Biochem 409(1–2):283–305

Akimoto T, Pohnert SC, Li P, Zhang M, Gumbs C, Rosenberg PB, Williams RS, Yan Z (2005) Exercise stimulates Pgc-1α transcription in skeletal muscle through activation of the p38 MAPK pathway. J Biol Chem 280(20):19587–19593

Alcazar-Fabra M, Navas P, Brea-Calvo G (2016) Coenzyme Q biosynthesis and its role in the respiratory chain structure. Biochim Biophys Acta 1857(8):1073–1078

Allan CM, Awad AM, Johnson JS, Shirasaki DI, Wang C, Blaby-Haas CE, Merchant SS, Loo JA, Clarke CF (2015) Identification of Coq11, a new coenzyme Q biosynthetic protein in the CoQ-synthome in *Saccharomyces cerevisiae*. J Biol Chem 290(12):7517–7534

Anderson RM, Barger JL, Edwards MG, Braun KH, O'Connor CE, Prolla TA, Weindruch R (2008) Dynamic regulation of PGC-1α localization and turnover implicates mitochondrial adaptation in calorie restriction and the stress response. Aging Cell 7(1):101–111

Aquilano K, Vigilanza P, Baldelli S, Pagliei B, Rotilio G, Ciriolo MR (2010) Peroxisome proliferator-activated receptor gamma co-activator 1alpha (PGC-1alpha) and sirtuin 1 (SIRT1) reside in mitochondria: possible direct function in mitochondrial biogenesis. J Biol Chem 285(28):21590–21599

Avadhani NG, Sangar MC, Bansal S, Bajpai P (2011) Bimodal targeting of cytochrome P450s to endoplasmic reticulum and mitochondria: the concept of chimeric signals. FEBS J 278(22): 4218–4229

Bakhanashvili M, Grinberg S, Bonda E, Simon AJ, Moshitch-Moshkovitz S, Rahav G (2008) p53 in mitochondria enhances the accuracy of DNA synthesis. Cell Death Differ 15(12):1865–1874

Baldelli S, Aquilano K, Ciriolo MR (2014) PGC-1[alpha] buffers ROS-mediated removal of mitochondria during myogenesis. Cell Death Dis 5:e1515

Bannwarth S, Figueroa A, Fragaki K, Destroismaisons L, Lacas-Gervais S, Lespinasse F, Vandenbos F, Pradelli LA, Ricci JE, Rotig A, Michiels JF, Vande Velde C, Paquis-Flucklinger V (2012) The human MSH5 (MutSHomolog 5) protein localizes to mitochondria and protects the mitochondrial genome from oxidative damage. Mitochondrion 12(6):654–665

Barchiesi A, Wasilewski M, Chacinska A, Tell G, Vascotto C (2015) Mitochondrial translocation of APE1 relies on the MIA pathway. Nucleic Acids Res 43(11):5451–5464

Becker T, Pfannschmidt S, Guiard B, Stojanovski D, Milenkovic D, Kutik S, Pfanner N, Meisinger C, Wiedemann N (2008) Biogenesis of the mitochondrial TOM complex: Mim1 promotes insertion and assembly of signal-anchored receptors. J Biol Chem 283(1):120–127

Becker T, Wenz LS, Kruger V, Lehmann W, Muller JM, Goroncy L, Zufall N, Lithgow T, Guiard B, Chacinska A, Wagner R, Meisinger C, Pfanner N (2011) The mitochondrial import protein Mim1 promotes biogenesis of multispanning outer membrane proteins. J Cell Biol 194(3):387–395

Bentinger M, Tekle M, Dallner G (2010) Coenzyme Q – biosynthesis and functions. Biochem Biophys Res Commun 396(1):74–79

Boopathi E, Srinivasan S, Fang JK, Avadhani NG (2008) Bimodal protein targeting through activation of cryptic mitochondrial targeting signals by an inducible cytosolic endoprotease. Mol Cell 32(1):32–42

Botelho SC, Osterberg M, Reichert AS, Yamano K, Bjorkholm P, Endo T, von Heijne G, Kim H (2011) TIM23-mediated insertion of transmembrane alpha-helices into the mitochondrial inner membrane. EMBO J 30(6):1003–1011

Bottero V, Rossi F, Samson M, Mari M, Hofman P, Peyron JF (2001) Ikappa b-alpha, the NF-kappa B inhibitory subunit, interacts with ANT, the mitochondrial ATP/ADP translocator. J Biol Chem 276(24):21317–21324

Brandt U, Yu L, Yu CA, Trumpower BL (1993) The mitochondrial targeting presequence of the Rieske iron-sulfur protein is processed in a single step after insertion into the cytochrome bc1 complex in mammals and retained as a subunit in the complex. J Biol Chem 268(12):8387–8390

Brooks SC, Adhikary S, Rubinson EH, Eichman BF (2013) Recent advances in the structural mechanisms of DNA glycosylases. Biochim Biophys Acta 1834(1):247–271

Calvo SE, Clauser KR, Mootha VK (2016) MitoCarta2.0: an updated inventory of mammalian mitochondrial proteins. Nucleic Acids Res 44(D1):D1251–D1257

Campo ML, Peixoto PM, Martinez-Caballero S (2016) Revisiting trends on mitochondrial mega-channels for the import of proteins and nucleic acids. J Bioenerg Biomembr 49:75–99

Canugovi C, Maynard S, Bayne AC, Sykora P, Tian J, de Souza-Pinto NC, Croteau DL, Bohr VA (2010) The mitochondrial transcription factor A functions in mitochondrial base excision repair. DNA Repair (Amst) 9(10):1080–1089

Carbognin E, Betto RM, Soriano ME, Smith AG, Martello G (2016) Stat3 promotes mitochondrial transcription and oxidative respiration during maintenance and induction of naive pluripotency. EMBO J 35(6):618–634

Cellini E, Vignoli A, Pisano T, Falchi M, Molinaro A, Accorsi P, Bontacchio A, Pinelli L, Giordano L, Guerrini R, F. S. S. Group (2016) The hyperkinetic movement disorder of FOXG1-related epileptic-dyskinetic encephalopathy. Dev Med Child Neurol 58(1):93–97

Chacinska A, Pfannschmidt S, Wiedemann N, Kozjak V, Sanjuan Szklarz LK, Schulze-Specking A, Truscott KN, Guiard B, Meisinger C, Pfanner N (2004) Essential role of Mia40 in import and assembly of mitochondrial intermembrane space proteins. EMBO J 23(19):3735–3746

Chacinska A, Koehler CM, Milenkovic D, Lithgow T, Pfanner N (2009) Importing mitochondrial proteins: machineries and mechanisms. Cell 138(4):628–644

Cheng Z, Ristow M (2013) Mitochondria and metabolic homeostasis. Antioxid Redox Signal 19(3):240–242

Chi Z, Nie L, Peng Z, Yang Q, Yang K, Tao J, Mi Y, Fang X, Balajee AS, Zhao Y (2012) RecQL4 cytoplasmic localization: implications in mitochondrial DNA oxidative damage repair. Int J Biochem Cell Biol 44(11):1942–1951

Cho J, Seo J, Lim CH, Yang L, Shiratsuchi T, Lee MH, Chowdhury RR, Kasahara H, Kim JS, Oh SP, Lee YJ, Terada N (2015) Mitochondrial ATP transporter Ant2 depletion impairs erythropoiesis and B lymphopoiesis. Cell Death Differ 22(9):1437–1450

Choi J, Batchu VVK, Schubert M, Castellani RJ, Russell JW (2013) A novel PGC-1α isoform in brain localizes to mitochondria and associates with PINK1 and VDAC. Biochem Biophys Res Commun 435(4):671–677

Ciapaite J, Bakker SJ, Diamant M, van Eikenhorst G, Heine RJ, Westerhoff HV, Krab K (2006) Metabolic control of mitochondrial properties by adenine nucleotide translocator determines palmitoyl-CoA effects. Implications for a mechanism linking obesity and type 2 diabetes. FEBS J 273(23):5288–5302

Cogswell PC, Kashatus DF, Keifer JA, Guttridge DC, Reuther JY, Bristow C, Roy S, Nicholson DW, Baldwin AS Jr (2003) NF-kappa B and I kappa B alpha are found in the mitochondria. Evidence for regulation of mitochondrial gene expression by NF-kappa B. J Biol Chem 278(5):2963–2968

Colombini M (2004) VDAC: the channel at the interface between mitochondria and the cytosol. Mol Cell Biochem 256–257(1–2):107–115
Conte L, Zara V (2011) The Rieske iron-sulfur protein: import and assembly into the cytochrome bc(1) complex of yeast mitochondria. Bioinorg Chem Appl 2011:363941
Craig EE, Chesley A, Hood DA (1998) Thyroid hormone modifies mitochondrial phenotype by increasing protein import without altering degradation. Am J Phys 275(6 Pt 1):C1508–C1515
Croteau DL, Singh DK, Hoh Ferrarelli L, Lu H, Bohr VA (2012) RECQL4 in genomic instability and aging. Trends Genet 28(12):624–631
Dabir DV, Hasson SA, Setoguchi K, Johnson ME, Wongkongkathep P, Douglas CJ, Zimmerman J, Damoiseaux R, Teitell MA, Koehler CM (2013) A small molecule inhibitor of redox-regulated protein translocation into mitochondria. Dev Cell 25(1):81–92
Darshi M, Trinh KN, Murphy AN, Taylor SS (2012) Targeting and import mechanism of coiled-coil helix coiled-coil helix domain-containing protein 3 (ChChd3) into the mitochondrial intermembrane space. J Biol Chem 287(47):39480–39491
Das BB, Dexheimer TS, Maddali K, Pommier Y (2010) Role of tyrosyl-DNA phosphodiesterase (TDP1) in mitochondria. Proc Natl Acad Sci U S A 107(46):19790–19795
De S, Kumari J, Mudgal R, Modi P, Gupta S, Futami K, Goto H, Lindor NM, Furuichi Y, Mohanty D, Sengupta S (2012) RECQL4 is essential for the transport of p53 to mitochondria in normal human cells in the absence of exogenous stress. J Cell Sci 125(Pt 10):2509–2522
De Rasmo D, Signorile A, Roca E, Papa S (2009) cAMP response element-binding protein (CREB) is imported into mitochondria and promotes protein synthesis. FEBS J 276(16):4325–4333
de Souza-Pinto NC, Mason PA, Hashiguchi K, Weissman L, Tian J, Guay D, Lebel M, Stevnsner TV, Rasmussen LJ, Bohr VA (2009) Novel DNA mismatch-repair activity involving YB-1 in human mitochondria. DNA Repair (Amst) 8(6):704–719
DeBalsi KL, Hoff KE, Copeland WC (2016) Role of the mitochondrial DNA replication machinery in mitochondrial DNA mutagenesis, aging and age-related diseases. Ageing Res Rev 33:89
Demple B, Sung JS (2005) Molecular and biological roles of Ape1 protein in mammalian base excision repair. DNA Repair (Amst) 4(12):1442–1449
Di Maio R, Barrett PJ, Hoffman EK, Barrett CW, Zharikov A, Borah A, Hu X, McCoy J, Chu CT, Burton EA, Hastings TG, Greenamyre JT (2016) Alpha-synuclein binds to TOM20 and inhibits mitochondrial protein import in Parkinson's disease. Sci Transl Med 8(342):342ra378
Di Noia MA, Todisco S, Cirigliano A, Rinaldi T, Agrimi G, Iacobazzi V, Palmieri F (2014) The human SLC25A33 and SLC25A36 genes of solute carrier family 25 encode two mitochondrial pyrimidine nucleotide transporters. J Biol Chem 289(48):33137–33148
Dimmer KS, Papic D, Schumann B, Sperl D, Krumpe K, Walther DM, Rapaport D (2012) A crucial role for Mim2 in the biogenesis of mitochondrial outer membrane proteins. J Cell Sci 125(Pt 14):3464–3473
Ding L, Liu Y (2015) Borrowing nuclear DNA helicases to protect mitochondrial DNA. Int J Mol Sci 16(5):10870–10887
Dolezal P, Likic V, Tachezy J, Lithgow T (2006) Evolution of the molecular machines for protein import into mitochondria. Science 313(5785):314–318
Edmondson AC, Song D, Alvarez LA, Wall MK, Almond D, McClellan DA, Maxwell A, Nielsen BL (2005) Characterization of a mitochondrially targeted single-stranded DNA-binding protein in *Arabidopsis thaliana*. Mol Gen Genomics 273(2):115–122
Endres M, Neupert W, Brunner M (1999) Transport of the ADP/ATP carrier of mitochondria from the TOM complex to the TIM22.54 complex. EMBO J 18(12):3214–3221
Farge G, Mehmedovic M, Baclayon M, van den Wildenberg SMJL, Roos WH, Gustafsson CM, Wuite GJL, Falkenberg M (2014) In vitro-reconstituted nucleoids can block mitochondrial DNA replication and transcription. Cell Rep 8(1):66–74
Favre C, Zhdanov A, Leahy M, Papkovsky D, O'Connor R (2010) Mitochondrial pyrimidine nucleotide carrier (PNC1) regulates mitochondrial biogenesis and the invasive phenotype of cancer cells. Oncogene 29(27):3964–3976
Flis VV, Daum G (2013) Lipid transport between the endoplasmic reticulum and mitochondria. Cold Spring Harb Perspect Biol 5(6):a013235

Floyd S, Favre C, Lasorsa FM, Leahy M, Trigiante G, Stroebel P, Marx A, Loughran G, O'Callaghan K, Marobbio CM, Slotboom DJ, Kunji ER, Palmieri F, O'Connor R (2007) The insulin-like growth factor-I-mTOR signaling pathway induces the mitochondrial pyrimidine nucleotide carrier to promote cell growth. Mol Biol Cell 18(9):3545–3555

Frank DO, Dengjel J, Wilfling F, Kozjak-Pavlovic V, Hacker G, Weber A (2015) The pro-apoptotic BH3-only protein Bim interacts with components of the translocase of the outer mitochondrial membrane (TOM). PLoS ONE 10(4):e0123341

Gallas MR, Dienhart MK, Stuart RA, Long RM (2006) Characterization of Mmp37p, a *Saccharomyces cerevisiae* mitochondrial matrix protein with a role in mitochondrial protein import. Mol Biol Cell 17(9):4051–4062

Garrido N, Griparic L, Jokitalo E, Wartiovaara J, van der Bliek AM, Spelbrink JN (2003) Composition and dynamics of human mitochondrial nucleoids. Mol Biol Cell 14(4): 1583–1596

Garstka HL, Schmitt WE, Schultz J, Sogl B, Silakowski B, Perez-Martos A, Montoya J, Wiesner RJ (2003) Import of mitochondrial transcription factor A (TFAM) into rat liver mitochondria stimulates transcription of mitochondrial DNA. Nucleic Acids Res 31(17):5039–5047

Gavalda-Navarro A, Villena JA, Planavila A, Vinas O, Mampel T (2014) Expression of adenine nucleotide translocase (ANT) isoform genes is controlled by PGC-1alpha through different transcription factors. J Cell Physiol 229(12):2126–2136

Gerbeth C, Schmidt O, Rao S, Harbauer AB, Mikropoulou D, Opalinska M, Guiard B, Pfanner N, Meisinger C (2013) Glucose-induced regulation of protein import receptor Tom22 by cytosolic and mitochondria-bound kinases. Cell Metab 18(4):578–587

Gonzalez-Mariscal I, Garcia-Teston E, Padilla S, Martin-Montalvo A, Pomares-Viciana T, Vazquez-Fonseca L, Gandolfo-Dominguez P, Santos-Ocana C (2014) Regulation of coenzyme Q biosynthesis in yeast: a new complex in the block. IUBMB Life 66(2):63–70

Gordon JW, Rungi AA, Inagaki H, Hood DA (2001) Effects of contractile activity on mitochondrial transcription factor A expression in skeletal muscle. J Appl Physiol (1985) 90(1):389–396

Gottschalk WK, Lutz MW, He YT, Saunders AM, Burns DK, Roses AD, Chiba-Falek O (2014) The broad impact of TOM40 on neurodegenerative diseases in aging. J Park Dis Alzheimers Dis 1(1):pii: 12

Gray MW (2015) Mosaic nature of the mitochondrial proteome: implications for the origin and evolution of mitochondria. Proc Natl Acad Sci U S A 112(33):10133–10138

Gray MW, Burger G, Lang BF (1999) Mitochondrial evolution. Science 283(5407):1476–1481

Graziewicz MA, Longley MJ, Copeland WC (2006) DNA polymerase gamma in mitochondrial DNA replication and repair. Chem Rev 106(2):383–405

Gurung B, Yu L, Xia D, Yu CA (2005) The iron-sulfur cluster of the Rieske iron-sulfur protein functions as a proton-exiting gate in the cytochrome bc(1) complex. J Biol Chem 280(26):24895–24902

Harbauer AB, Opalinska M, Gerbeth C, Herman JS, Rao S, Schonfisch B, Guiard B, Schmidt O, Pfanner N, Meisinger C (2014a) Mitochondria. Cell cycle-dependent regulation of mitochondrial preprotein translocase. Science 346(6213):1109–1113

Harbauer AB, Zahedi RP, Sickmann A, Pfanner N, Meisinger C (2014b) The protein import machinery of mitochondria-a regulatory hub in metabolism, stress, and disease. Cell Metab 19(3):357–372

Harner M, Neupert W, Deponte M (2011) Lateral release of proteins from the TOM complex into the outer membrane of mitochondria. EMBO J 30(16):3232–3241

Hasson SA, Damoiseaux R, Glavin JD, Dabir DV, Walker SS, Koehler CM (2010) Substrate specificity of the TIM22 mitochondrial import pathway revealed with small molecule inhibitor of protein translocation. Proc Natl Acad Sci U S A 107(21):9578–9583

Havlickova V, Kaplanova V, Nuskova H, Drahota Z, Houstek J (2010) Knockdown of F1 epsilon subunit decreases mitochondrial content of ATP synthase and leads to accumulation of subunit c. Biochim Biophys Acta 1797(6–7):1124–1129

Hell K, Herrmann JM, Pratje E, Neupert W, Stuart RA (1998) Oxa1p, an essential component of the N-tail protein export machinery in mitochondria. Proc Natl Acad Sci U S A 95(5):2250–2255

Hood DA, Simoneau JA, Kelly AM, Pette D (1992) Effect of thyroid status on the expression of metabolic enzymes during chronic stimulation. Am J Phys 263(4 Pt 1):C788–C793

Hoogenraad NJ, Ward LA, Ryan MT (2002) Import and assembly of proteins into mitochondria of mammalian cells. Biochim Biophys Acta 1592(1):97–105

Horvath SE, Daum G (2013) Lipids of mitochondria. Prog Lipid Res 52(4):590–614

Hulett JM, Lueder F, Chan NC, Perry AJ, Wolynec P, Likic VA, Gooley PR, Lithgow T (2008) The transmembrane segment of Tom20 is recognized by Mim1 for docking to the mitochondrial TOM complex. J Mol Biol 376(3):694–704

Jafari G, Wasko BM, Kaeberlein M, Crofts AR (2016) New functional and biophysical insights into the mitochondrial Rieske iron-sulfur protein from genetic suppressor analysis in *C. elegans*. WormBook 5(2):e1174803

Jenkins BJ, Daly TM, Morrisey JM, Mather MW, Vaidya AB, Bergman LW (2016) Characterization of a plasmodium falciparum orthologue of the yeast ubiquinone-binding protein, Coq10p. PLoS ONE 11(3):e0152197

Johnson RF, Witzel II, Perkins ND (2011) p53-dependent regulation of mitochondrial energy production by the RelA subunit of NF-kappaB. Cancer Res 71(16):5588–5597

Jonckheere AI, Smeitink JA, Rodenburg RJ (2012) Mitochondrial ATP synthase: architecture, function and pathology. J Inherit Metab Dis 35(2):211–225

Joseph AM, Rungi AA, Robinson BH, Hood DA (2004) Compensatory responses of protein import and transcription factor expression in mitochondrial DNA defects. Am J Physiol Cell Physiol 286(4):C867–C875

Joseph AM, Ljubicic V, Adhihetty PJ, Hood DA (2010) Biogenesis of the mitochondrial Tom40 channel in skeletal muscle from aged animals and its adaptability to chronic contractile activity. Am J Physiol Cell Physiol 298(6):C1308–C1314

Knutti D, Kralli A (2001) PGC-1, a versatile coactivator. Trends Endocrinol Metab 12(8):360–365

Koehler CM, Leuenberger D, Merchant S, Renold A, Junne T, Schatz G (1999) Human deafness dystonia syndrome is a mitochondrial disease. Proc Natl Acad Sci 96(5):2141–2146

Koulintchenko M, Konstantinov Y, Dietrich A (2003) Plant mitochondria actively import DNA via the permeability transition pore complex. EMBO J 22(6):1245–1254

Koulintchenko M, Temperley RJ, Mason PA, Dietrich A, Lightowlers RN (2006) Natural competence of mammalian mitochondria allows the molecular investigation of mitochondrial gene expression. Hum Mol Genet 15(1):143–154

Kozjak-Pavlovic V, Ross K, Benlasfer N, Kimmig S, Karlas A, Rudel T (2007) Conserved roles of Sam50 and metaxins in VDAC biogenesis. EMBO Rep 8(6):576–582

Kukat C, Wurm CA, Spahr H, Falkenberg M, Larsson NG, Jakobs S (2011) Super-resolution microscopy reveals that mammalian mitochondrial nucleoids have a uniform size and frequently contain a single copy of mtDNA. Proc Natl Acad Sci U S A 108(33):13534–13539

Kukshal V, Kim IK, Hura GL, Tomkinson AE, Tainer JA, Ellenberger T (2015) Human DNA ligase III bridges two DNA ends to promote specific intermolecular DNA end joining. Nucleic Acids Res 43(14):7021–7031

Kumakura A, Takahashi S, Okajima K, Hata D (2014) A haploinsufficiency of FOXG1 identified in a boy with congenital variant of Rett syndrome. Brain and Development 36(8):725–729

Lai L, Wang M, Martin OJ, Leone TC, Vega RB, Han X, Kelly DP (2014) A role for peroxisome proliferator-activated receptor γ coactivator 1 (PGC-1) in the regulation of cardiac mitochondrial phospholipid biosynthesis. J Biol Chem 289(4):2250–2259

Lai-Zhang J, Mueller DM (2000) Complementation of deletion mutants in the genes encoding the F1-ATPase by expression of the corresponding bovine subunits in yeast *S. cerevisiae*. Eur J Biochem 267(8):2409–2418

Lakshmipathy U, Campbell C (1999) The human DNA ligase III gene encodes nuclear and mitochondrial proteins. Mol Cell Biol 19(5):3869–3876

Larsen NB, Rasmussen M, Rasmussen LJ (2005) Nuclear and mitochondrial DNA repair: similar pathways? Mitochondrion 5(2):89–108

Larsson NG, Wang J, Wilhelmsson H, Oldfors A, Rustin P, Lewandoski M, Barsh GS, Clayton DA (1998) Mitochondrial transcription factor A is necessary for mtDNA maintenance and embryogenesis in mice. Nat Genet 18(3):231–236

Lemasters JJ, Holmuhamedov E (2006) Voltage-dependent anion channel (VDAC) as mitochondrial governator – thinking outside the box. Biochim Biophys Acta 1762(2):181–190

Li Q, Vande Velde C, Israelson A, Xie J, Bailey AO, Dong MQ, Chun SJ, Roy T, Winer L, Yates JR, Capaldi RA, Cleveland DW, Miller TM (2010) ALS-linked mutant superoxide dismutase 1 (SOD1) alters mitochondrial protein composition and decreases protein import. Proc Natl Acad Sci U S A 107(49):21146–21151

Liang H, Ward WF (2006) PGC-1alpha: a key regulator of energy metabolism. Adv Physiol Educ 30(4):145–151

Longen S, Woellhaf MW, Petrungaro C, Riemer J, Herrmann JM (2014) The disulfide relay of the intermembrane space oxidizes the ribosomal subunit mrp10 on its transit into the mitochondrial matrix. Dev Cell 28(1):30–42

Luévano-Martínez LA, Forni MF, dos Santos VT, Souza-Pinto NC, Kowaltowski AJ (2015) Cardiolipin is a key determinant for mtDNA stability and segregation during mitochondrial stress. Biochim Biophys Acta (BBA) Bioenerg 1847(6–7):587–598

Mannella CA, Kinnally KW (2008) Reflections on VDAC as a voltage-gated channel and a mitochondrial regulator. J Bioenerg Biomembr 40(3):149–155

Markkanen E, Hubscher U, van Loon B (2012) Regulation of oxidative DNA damage repair: the adenine:8-oxo-guanine problem. Cell Cycle 11(6):1070–1075

Martin SA (2011) Mitochondrial DNA repair. In: Storici DF (ed) DNA repair â€" on the pathways to fixing DNA damage and errors. Edited by Francesca Storici, ISBN 978-953-307-649-2, Published: September 9, 2011 under CC BY-NC-SA 3.0 license

McCulloch V, Shadel GS (2003) Human mitochondrial transcription factor B1 interacts with the C-terminal activation region of h-mtTFA and stimulates transcription independently of its RNA methyltransferase activity. Mol Cell Biol 23(16):5816–5824

Meagher M, Lightowlers RN (2014) The role of TDP1 and APTX in mitochondrial DNA repair. Biochimie 100:121–124

Nakagawa T, Guarente L (2009) Urea cycle regulation by mitochondrial sirtuin, SIRT5. Aging (Albany NY) 1(6):578–581

Neupert W, Herrmann JM (2007) Translocation of proteins into mitochondria. Annu Rev Biochem 76:723–749

Otterlei M, Haug T, Nagelhus TA, Slupphaug G, Lindmo T, Krokan HE (1998) Nuclear and mitochondrial splice forms of human uracil-DNA glycosylase contain a complex nuclear localisation signal and a strong classical mitochondrial localisation signal, respectively. Nucleic Acids Res 26(20):4611–4617

Pagliarini DJ, Calvo SE, Chang B, Sheth SA, Vafai SB, Ong SE, Walford GA, Sugiana C, Boneh A, Chen WK, Hill DE, Vidal M, Evans JG, Thorburn DR, Carr SA, Mootha VK (2008) A mitochondrial protein compendium elucidates complex I disease biology. Cell 134(1):112–123

Pancrazi L, Di Benedetto G, Colombaioni L, Della Sala G, Testa G, Olimpico F, Reyes A, Zeviani M, Pozzan T, Costa M (2015) Foxg1 localizes to mitochondria and coordinates cell differentiation and bioenergetics. Proc Natl Acad Sci U S A 112(45):13910–13915

Park JH, Zhuang J, Li J, Hwang PM (2016a) p53 as guardian of the mitochondrial genome. FEBS Lett 590(7):924–934

Park KW, Lin CY, Benveniste EN, Lee YS (2016b) Mitochondrial STAT3 is negatively regulated by SOCS3 and upregulated after spinal cord injury. Exp Neurol 284(Pt A):98–105

Pavlov E, Grigoriev SM, Dejean LM, Zweihorn CL, Mannella CA, Kinnally KW (2005) The mitochondrial channel VDAC has a cation-selective open state. Biochim Biophys Acta 1710(2–3):96–102

Peixoto PM, Grana F, Roy TJ, Dunn CD, Flores M, Jensen RE, Campo ML (2007) Awaking TIM22, a dynamic ligand-gated channel for protein insertion in the mitochondrial inner membrane. J Biol Chem 282(26):18694–18701

Popov-Celeketic J, Waizenegger T, Rapaport D (2008) Mim1 functions in an oligomeric form to facilitate the integration of Tom20 into the mitochondrial outer membrane. J Mol Biol 376(3):671–680

Prakash A, Doublie S (2015) Base excision repair in the mitochondria. J Cell Biochem 116(8):1490–1499

Pratt RD, Ferreira GC, Pedersen PL (1991) Mitochondrial phosphate transport. Import of the H+/Pi symporter and role of the presequence. J Biol Chem 266(2):1276–1280

Puigserver P, Wu Z, Park CW, Graves R, Wright M, Spiegelman BM (1998) A cold-inducible coactivator of nuclear receptors linked to adaptive thermogenesis. Cell 92(6):829–839

Radak Z, Atalay M, Jakus J, Boldogh I, Davies K, Goto S (2009) Exercise improves import of 8-oxoguanine DNA glycosylase into the mitochondrial matrix of skeletal muscle and enhances the relative activity. Free Radic Biol Med 46(2):238–243

Rajapaksha M, Kaur J, Prasad M, Pawlak KJ, Marshall B, Perry EW, Whittal RM, Bose HS (2016) An outer mitochondrial translocase, Tom22, is crucial for inner mitochondrial steroidogenic regulation in adrenal and gonadal tissues. Mol Cell Biol 36(6):1032–1047

Rao S, Schmidt O, Harbauer AB, Schonfisch B, Guiard B, Pfanner N, Meisinger C (2012) Biogenesis of the preprotein translocase of the outer mitochondrial membrane: protein kinase A phosphorylates the precursor of Tom40 and impairs its import. Mol Biol Cell 23(9):1618–1627

Robin MA, Anandatheerthavarada HK, Biswas G, Sepuri NB, Gordon DM, Pain D, Avadhani NG (2002) Bimodal targeting of microsomal CYP2E1 to mitochondria through activation of an N-terminal chimeric signal by cAMP-mediated phosphorylation. J Biol Chem 277(43):40583–40593

Rodiger A, Baudisch B, Langner U, Klosgen RB (2011) Dual targeting of a mitochondrial protein: the case study of cytochrome c1. Mol Plant 4(4):679–687

Rostovtseva TK (2012) VDAC structure, function, and regulation of mitochondrial and cellular metabolism. Biochim Biophys Acta 1818(6):1437

Roussel J, Thireau J, Brenner C, Saint N, Scheuermann V, Lacampagne A, Le Guennec JY, Fauconnier J (2015) Palmitoyl-carnitine increases RyR2 oxidation and sarcoplasmic reticulum Ca2+ leak in cardiomyocytes: role of adenine nucleotide translocase. Biochim Biophys Acta 1852(5):749–758

Ryu H, Lee J, Impey S, Ratan RR, Ferrante RJ (2005) Antioxidants modulate mitochondrial PKA and increase CREB binding to D-loop DNA of the mitochondrial genome in neurons. Proc Natl Acad Sci U S A 102(39):13915–13920

Safdar A, Little JP, Stokl AJ, Hettinga BP, Akhtar M, Tarnopolsky MA (2011) Exercise increases mitochondrial PGC-1α content and promotes nuclear-mitochondrial cross-talk to coordinate mitochondrial biogenesis. J Biol Chem 286(12):10605–10617

Sagan L (1967) On the origin of mitosing cells. J Theor Biol 14(3):255–274

Santos JM, Kowluru RA (2013) Impaired transport of mitochondrial transcription factor A (TFAM) and the metabolic memory phenomenon associated with the progression of diabetic retinopathy. Diabetes Metab Res Rev 29(3):204–213

Santos-Ocana C, Cordoba F, Crane FL, Clarke CF, Navas P (1998) Coenzyme Q6 and iron reduction are responsible for the extracellular ascorbate stabilization at the plasma membrane of *Saccharomyces cerevisiae*. J Biol Chem 273(14):8099–8105

Schmidt O, Pfanner N, Meisinger C (2010) Mitochondrial protein import: from proteomics to functional mechanisms. Nat Rev Mol Cell Biol 11(9):655–667

Schmidt O, Harbauer AB, Rao S, Eyrich B, Zahedi RP, Stojanovski D, Schonfisch B, Guiard B, Sickmann A, Pfanner N, Meisinger C (2011) Regulation of mitochondrial protein import by cytosolic kinases. Cell 144(2):227–239

Schreiber SN, Emter R, Hock MB, Knutti D, Cardenas J, Podvinec M, Oakeley EJ, Kralli A (2004) The estrogen-related receptor α (ERRα) functions in PPARγ coactivator 1α (PGC-1α)-induced mitochondrial biogenesis. Proc Natl Acad Sci U S A 101(17):6472–6477

Schwartz RM, Dayhoff MO (1978) Origins of prokaryotes, eukaryotes, mitochondria, and chloroplasts. Science 199(4327):395–403

Sheng B, Wang X, Su B, Lee H-g, Casadesus G, Perry G, Zhu X (2012) Impaired mitochondrial biogenesis contributes to mitochondrial dysfunction in Alzheimer's disease. J Neurochem 120(3):419–429

Simpkins JW, Yang S, Sarkar SN, Pearce V (2008) Estrogen actions on mitochondria-physiological and pathological implications. Mol Cell Endocrinol 290(1–2):51–59

Singh K, Hood DA (2011) Effect of denervation-induced muscle disuse on mitochondrial protein import. Am J Physiol Cell Physiol 300(1):C138–C145

Spadafora D, Kozhukhar N, Alexeyev MF (2016) Presequence-independent mitochondrial import of DNA ligase facilitates establishment of cell lines with reduced mtDNA copy number. PLoS ONE 11(3):e0152705

St-Pierre J, Drori S, Uldry M, Silvaggi JM, Rhee J, Jäger S, Handschin C, Zheng K, Lin J, Yang W, Simon DK, Bachoo R, Spiegelman BM (2006) Suppression of reactive oxygen species and neurodegeneration by the PGC-1 transcriptional coactivators. Cell 127(2):397–408

Stuart RA, Nicholson DW, Wienhues U, Neupert W (1990) Import of apocytochrome c into the mitochondrial intermembrane space along a cytochrome c1 sorting pathway. J Biol Chem 265(33):20210–20219

Szczesny B, Hazra TK, Papaconstantinou J, Mitra S, Boldogh I (2003) Age-dependent deficiency in import of mitochondrial DNA glycosylases required for repair of oxidatively damaged bases. Proc Natl Acad Sci U S A 100(19):10670–10675

Takahashi M, Chesley A, Freyssenet D, Hood DA (1998) Contractile activity-induced adaptations in the mitochondrial protein import system. Am J Phys 274(5 Pt 1):C1380–C1387

Takao M, Aburatani H, Kobayashi K, Yasui A (1998) Mitochondrial targeting of human DNA glycosylases for repair of oxidative DNA damage. Nucleic Acids Res 26(12):2917–2922

Tammineni P, Anugula C, Mohammed F, Anjaneyulu M, Larner AC, Sepuri NB (2013) The import of the transcription factor STAT3 into mitochondria depends on GRIM-19, a component of the electron transport chain. J Biol Chem 288(7):4723–4732

Tamura Y, Harada Y, Yamano K, Watanabe K, Ishikawa D, Ohshima C, Nishikawa S, Yamamoto H, Endo T (2006) Identification of Tam41 maintaining integrity of the TIM23 protein translocator complex in mitochondria. J Cell Biol 174(5):631–637

Tamura Y, Harada Y, Nishikawa S-I, Yamano K, Kamiya M, Shiota T, Kuroda T, Kuge O, Sesaki H, Imai K, Tomii K, Endo T (2013) Tam41 is a CDP-diacylglycerol synthase required for cardiolipin biosynthesis in mitochondria. Cell Metab 17(5):709–718

Tranebjaerg L, Schwartz C, Eriksen H, Andreasson S, Ponjavic V, Dahl A, Stevenson RE, May M, Arena F, Barker D (1995) A new X linked recessive deafness syndrome with blindness, dystonia, fractures, and mental deficiency is linked to Xq22. J Med Genet 32(4):257–263

Tryon LD, Crilly MJ, Hood DA (2015) Effect of denervation on the regulation of mitochondrial transcription factor A expression in skeletal muscle. Am J Physiol Cell Physiol 309(4):C228–C238

Turakhiya U, von der Malsburg K, Gold VA, Guiard B, Chacinska A, van der Laan M, Ieva R (2016) Protein import by the mitochondrial presequence translocase in the absence of a membrane potential. J Mol Biol 428(6):1041–1052

Uhler JP, Thorn C, Nicholls TJ, Matic S, Milenkovic D, Gustafsson CM, Falkenberg M (2016) MGME1 processes flaps into ligatable nicks in concert with DNA polymerase gamma during mtDNA replication. Nucleic Acids Res 44(12):5861–5871

Uittenbogaard M, Chiaramello A (2014) Mitochondrial biogenesis: a therapeutic target for neurodevelopmental disorders and neurodegenerative diseases. Curr Pharm Des 20(35):5574–5593

Vance JE (1990) Phospholipid synthesis in a membrane fraction associated with mitochondria. J Biol Chem 265(13):7248–7256

Wachter C, Schatz G, Glick BS (1994) Protein import into mitochondria: the requirement for external ATP is precursor-specific whereas intramitochondrial ATP is universally needed for translocation into the matrix. Mol Biol Cell 5(4):465–474

Weber-Lotfi F, Ibrahim N, Boesch P, Cosset A, Konstantinov Y, Lightowlers RN, Dietrich A (2009) Developing a genetic approach to investigate the mechanism of mitochondrial competence for DNA import. Biochim Biophys Acta (BBA) Bioenerg 1787(5):320–327

Weber-Lotfi F, Koulintchenko MV, Ibrahim N, Hammann P, Mileshina DV, Konstantinov YM, Dietrich A (2015) Nucleic acid import into mitochondria: new insights into the translocation pathways. Biochim Biophys Acta (BBA) Mol Cell Res 1853(12):3165–3181

Wegrzyn J, Potla R, Chwae YJ, Sepuri NB, Zhang Q, Koeck T, Derecka M, Szczepanek K, Szelag M, Gornicka A, Moh A, Moghaddas S, Chen Q, Bobbili S, Cichy J, Dulak J, Baker DP, Wolfman A, Stuehr D, Hassan MO, Fu XY, Avadhani N, Drake JI, Fawcett P, Lesnefsky EJ, Larner AC (2009) Function of mitochondrial Stat3 in cellular respiration. Science 323(5915):793–797

Wollen Steen K, Doseth B, P. Westbye M, Akbari M, Kang D, Falkenberg M, Slupphaug G (2012) mtSSB may sequester UNG1 at mitochondrial ssDNA and delay uracil processing until the dsDNA conformation is restored. DNA Repair (Amst) 11(1):82–91

Wrobel L, Trojanowska A, Sztolsztener ME, Chacinska A (2013) Mitochondrial protein import: Mia40 facilitates Tim22 translocation into the inner membrane of mitochondria. Mol Biol Cell 24(5):543–554

Wu Z, Puigserver P, Andersson U, Zhang C, Adelmant G, Mootha V, Troy A, Cinti S, Lowell B, Scarpulla RC, Spiegelman BM (1999) Mechanisms controlling mitochondrial biogenesis and respiration through the thermogenic coactivator PGC-1. Cell 98(1):115–124

Young JC, Hoogenraad NJ, Hartl FU (2003) Molecular chaperones Hsp90 and Hsp70 deliver preproteins to the mitochondrial import receptor Tom70. Cell 112(1):41–50

Zamora M, Merono C, Vinas O, Mampel T (2004) Recruitment of NF-kappaB into mitochondria is involved in adenine nucleotide translocase 1 (ANT1)-induced apoptosis. J Biol Chem 279(37):38415–38423

Zhang J, Yang J, Roy SK, Tininini S, Hu J, Bromberg JF, Poli V, Stark GR, Kalvakolanu DV (2003) The cell death regulator GRIM-19 is an inhibitor of signal transducer and activator of transcription 3. Proc Natl Acad Sci U S A 100(16):9342–9347

Zhang Y, Iqbal S, O'Leary MF, Menzies KJ, Saleem A, Ding S, Hood DA (2013) Altered mitochondrial morphology and defective protein import reveal novel roles for Bax and/or Bak in skeletal muscle. Am J Physiol Cell Physiol 305(5):C502–C511

Zheng L, Zhou M, Guo Z, Lu H, Qian L, Dai H, Qiu J, Yakubovskaya E, Bogenhagen DF, Demple B, Shen B (2008) Human DNA2 is a mitochondrial nuclease/helicase for efficient processing of DNA replication and repair intermediates. Mol Cell 32(3):325–336

Zhuang J, Wang PY, Huang X, Chen X, Kang JG, Hwang PM (2013) Mitochondrial disulfide relay mediates translocation of p53 and partitions its subcellular activity. Proc Natl Acad Sci U S A 110(43):17356–17361

Chapter 13
Substrate Selection and Its Impact on Mitochondrial Respiration and Redox

Sonia Cortassa, Steven J. Sollott, and Miguel A. Aon

13.1 Introduction

We increasingly recognize that complex gene-environment and gene-nutrient interactions underlie an organism's response to physiological or pathophysiological stimuli. An individual genomic variant (Mendelian-type inborn error) can affect subtly one primary metabolite flux, without evidence of clinical disease; however, in a complex disease, state variations perturbing a network of metabolite fluxes may attain the clinical threshold for disease, either alone or in combination with environmental factors (Lanpher et al. 2006). The capacity to comprehensively assess gene, protein, transcript, and metabolite profiles, including posttranslational modifications, through high-throughput "omics" studies, has opened new avenues of research possibilities, among them the characterization of metabolic remodeling associated with disease, e.g., diabetes, cancer, aging, or normal physiology such as observed in the fast-feed transition or caloric restriction (Kelley and Mandarino 2000; Mitchell et al. 2016).

Multiple environmental (e.g., nutrition) and genetic interactions produce different phenotypic patterns of which metabolites and metabolic fluxes are main functional readouts. Recently, the concept of *metabolism-epigenome-genome axis* was proposed to account for the dynamic and reciprocal feedback loops between the epigenome and the genome, in turn, driven by metabolic-elicited modifications, e.g., histones acetylation and deacetylation, DNA methylation, and the feedback from the genome

S. Cortassa • S.J. Sollott • M.A. Aon (✉)
Laboratory of Cardiovascular Science, National Institute on Aging,
National Institutes of Health, Baltimore, MD, USA
e-mail: miguel.aon@nih.gov

T.K. Rostovtseva (ed.), *Molecular Basis for Mitochondrial Signaling*,
Biological and Medical Physics, Biomedical Engineering,
DOI 10.1007/978-3-319-55539-3_13

to the epigenome, exemplified by the expression level of histone acetyltransferases, histone deacetylases, histone methyltransferases, and demethylating enzymes (Aon et al. 2016). Hypothetically, a metabolism-epigenome-genome axis provides a comprehensive framework for analyzing the modulation of genotype-phenotype interactions in response to different nutritional environments. According to this concept, the epigenome represents an interface between metabolism and the gene expression machinery of nuclear and mitochondrial DNAs (Donohoe and Bultman 2012; Keating and El-Osta 2015; Mcknight 2010; Wallace 2010; Wallace and Fan 2010), whereas its direct dynamic readout is embodied by the fluxome, defined as the ensemble of metabolic fluxes resulting from genes expressed and proteins translated including their posttranslational modifications (Aon 2013; Cascante and Marin 2008; Cortassa et al. 2015). The change in fluxome dynamics under different cellular functions or in response to nutritional status such as caloric restriction (Mitchell et al. 2016), starvation, or hypoxia has repercussions at both epigenetic and genetic levels thus retro-influencing metabolism. Mechanistically, the tricarboxylic acid cycle (TCA) metabolite citrate is important for acetyl-CoA (AcCoA) generation required for lipogenesis as well as for the acetylation of histones in nuclei (Salminen et al. 2014; Wallace and Fan 2010). Epigenetic modifications via histone acetylation and cellular energy metabolism are linked by ATP citrate lyase (ACLY) in a glucose-dependent manner (Wellen et al. 2009). In mitochondria, AcCoA is generated from pyruvate by PDH, and subsequently citrate synthase catalyzes the conversion of AcCoA and oxaloacetate into citrate in the TCA cycle. Citrate can be transported from mitochondria, via the citrate carrier, into cytoplasm, where ACLY generates AcCoA from citrate (Choudhary et al. 2014).

13.2 Mitochondrial Energy-Redox Functions

The interrelationship between mitochondrial energy and driving forces such as phosphorylation (ATP), electrochemical ($\Delta\Psi_m$, ΔpH), and redox (NAD(P)H, GSH, ROS) is able to quickly change in response to substrate and ADP levels. These energetic changes determine intra- and extramitochondrial redox environments (RE) involving antioxidants level (e.g., glutathione peroxidase, peroxiredoxin), post-translational modifications (glutathionylation, oxidation), and H_2O_2 emission, thus influencing the cytoplasmic redox status (Cortassa et al. 2014; Dey et al. 2016; Jones and Sies 2015; Kembro et al. 2013; Sies 2015; Swain et al. 2016). The energetic status responds to fluctuations in ADP availability modulated by nutrients (e.g., substrate type, abundance) and energy demand (e.g., sedentarism, physical activity) and may, in turn, generate rhythmically changing levels of mitochondrial H_2O_2 emission (Cortassa et al. 2014) thereby conveying the energetic status to ROS signaling, both converging to tune mitochondrial and cellular function.

The oxidative potential of the TCA cycle, via NADH, influences both respiration and the provision of NAD(P)H, to restoring the antioxidant systems. As a matter of fact, the mitochondrial redox potential of thioredoxin2 (Trx2) decreases (becomes more reducing) from −322 mV at baseline to −350 mV in state 4 and state 3 respiration, during the transition from non-energized to energized (Stanley et al. 2011). This corresponds to a tenfold decrease in the ratio of oxidized to reduced Trx2 and a 2.4-fold increase in the percentage of the Trx2 pool in the reduced form. Under these conditions, after glutamate/malate addition, $Trx(SH)_2$ rises in parallel with $\Delta\Psi_m$, and NAD(P)H, as well as GSH in the mitochondrial matrix (Stanley et al. 2011). As another example, in isolated heart mitochondria, a lower respiratory flux is observed under oxidative stress than in its absence, at similar $\Delta\Psi_m$ under both conditions (Cortassa et al. 2014). These results suggested that the NADH-electron donor capacity to respiration might diminish under stress likely due to redirection of electrons to the antioxidant systems. Indeed, the relationship between respiration and ROS is altered by oxidative stress, resulting in decreased mitochondrial energetic performance and higher levels of ROS emission (Cortassa et al. 2014). The intramitochondrial RE is highly influenced by the type of substrate that in the case of lipids is in the form of reducing equivalents, e.g., NADH and $FADH_2$.

As a dynamic metric, the RE is a function of the different redox couples accounting for both their redox potential and the concentration of the reduced species (Jones 2002; Kembro et al. 2013; Schafer and Buettner 2001). Although specific for each subcellular compartment (Dey et al. 2016; Jones and Go 2010; Kaludercic et al. 2014; Swain et al. 2016), the RE dynamics between compartments (e.g., mitochondria, cytoplasm) is interdependent, mediated by exchange of ROS and redox-related components such as GSH and H_2O_2 (Dey et al. 2016; Jones and Go 2010; Kembro et al. 2014b). In the case of mitochondria, regeneration of glutathione (GSH) from its oxidized form (GSSG) requires glutathione reductase harnessing the more negative reduction potential of NADPH, which, in turn, will be regenerated by the transhydrogenase coupling hydride transfer between NADH and NADP to the proton motive force (Aon et al. 2007; Hoek and Rydstrom 1988; Nickel et al. 2015; Rydstrom 2006). Since mitochondria cannot synthesize GSH, and the fact that GSSG cannot cross the membrane, the reduction of the latter strictly depends on compartmentalized mitochondrial NADPH generation, a crucial event in the ROS scavenging capacity by antioxidant systems (Aon et al. 2007; Dey et al. 2016; Swain et al. 2016).

The significant role of compartmentation in controlling ROS levels, the RE, and dynamic behavior also depends on the concerted and continuous function of the ROS scavenging systems, e.g., glutathione/thioredoxin, to keep low rates of H_2O_2 emission from mitochondria (Aon et al. 2012; Stanley et al. 2011). Therefore, the duplication of antioxidant defense systems in multiple compartments appears as a natural and efficient salvage mechanism to avoid or to reduce oxidative bursts (Kembro et al. 2013, 2014b). Failure to maintain NAD(P)H supply during oxidative stress or increased work is a key contributor to ROS overload which may lead to reperfusion-related arrhythmias after ischemic injury (Akar et al. 2005; Aon et al. 2009; Brown et al. 2010; Swain et al. 2016; Xie et al. 2013) or heart failure, the latter due, in part, to impaired mitochondrial Ca^{2+} signaling to the TCA cycle (Liu et al. 2014).

Current wisdom suggests that under high energy demand, e.g., exercise, in the absence of additional oxidative stress, mitochondria will function at relatively more reduced RE (Aon et al. 2015; Hafstad et al. 2015). However, a shift will happen under pathological conditions, displacing mitochondrial function toward a more oxidized RE (Alleman et al. 2014, 2016; Aon et al. 2014; Tocchetti et al. 2012, 2015). From this perspective, the TCA cycle function in response to nutrient availability can be viewed as a signaling task via its effects on NAD(P)H provision to the antioxidant systems and direct modulation of enzyme function via ROS, in addition to its well-known role as a source of metabolite precursors and NADH. For example, alpha ketoglutarate dehydrogenase and PDH are both deactivated by H_2O_2 (Crane et al. 1983; Mailloux 2015; Nulton-Persson et al. 2003) or influenced by the triggering of increased expression of antioxidant enzymes via the Nrf2 antioxidant response element (Nguyen et al. 2009).

13.3 Fatty Acids, Mitochondrial Function, and Oxidative Stress

FAs are main metabolic fuels, and β-oxidation represents their main degradation pathway, for example, in heart and skeletal muscle (Eaton 2002). FA beta-oxidation is a major pathway of energy metabolism providing ~80% of the ATP required for the liver and the heart (Eaton et al. 1996). The rate of β-oxidation is led by demand since an increase in work rate and ATP utilization leads to faster oxidative phosphorylation (OxPhos) and TCA cycle activity. In turn, the decrease in NADH and AcCoA levels leads to an increase in the β-oxidation flux (Eaton 2002; Eaton et al. 1996; Lopaschuk et al. 2010; Neely et al. 1969; Oram et al. 1973).

The FAs released during triacylglyceride (TAG) catabolism are mainly used for β-oxidation and subsequent ATP synthesis via OxPhos in mitochondria. In oxidative tissues such as the heart, TAG-derived FAs are utilized as an energy source, but they also serve as signaling molecules as well as building blocks for membranes and complex lipids. Hepatocytes, heart and skeletal myocytes, adrenocortical cells, enterocytes, adipocytes, and macrophages may all contain large amounts of lipid droplets (LDs). Excessive LD accumulation is a hallmark of T2DM, obesity, atherosclerosis, hepatic steatosis, and other metabolic diseases (Aon et al. 2014; Singh and Cuervo 2012; Singh et al. 2009; Walther and Farese 2009; Walther and Farese 2012).

As a major energy source, FAs may provide up to two thirds of ATP synthesized via reducing equivalents derived from β-oxidation in mitochondria. The saturated FA palmitate (16:0) supplies about three times higher energy equivalents than glucose in the form of reducing power [7 NADH, 7 FADH2 plus 8 AcCoA] with a net yield of 106 moles of ATP accounting for the energetic cost of activating the FA (-2 ATP), whereas unsaturated FA oleate (18:1) supplies 8 NADH, 7 FADH2 plus 9 AcCoA with a net ATP yield of 109 moles, in both cases assuming that OxPhos generates 1.5 ATP per FADH2 and 2.5 ATP per NADH oxidized (Nelson and Cox 2013). The reducing equivalents are not only able to contribute electrons to the respiratory/energetic machinery but also to the antioxidant systems via mitochondrial transhydrogenase that converts NADH to NADPH, the latter being a major electron donor to the glutathione and thioredoxin systems from mitochondria (Hoek and Rydstrom 1988).

Preservation of the intracellular RE is crucial for vital functions such as division, differentiation, contractile work, and survival, among others (Aggarwal and Makielski 2013; Aon et al. 2007, 2009; Brown et al. 2010; Fisher-Wellman and Neufer 2012; Jeong et al. 2012; Juhaszova et al. 2004; Lloyd et al. 2012; Muoio and Neufer 2012; Schafer and Buettner 2001). Mitochondria are main drivers of the intracellular RE (Alleman et al. 2014; Aon et al. 2015; Nickel et al. 2014) and together with peroxisomes constitute the main subcellular compartments where lipid degradation occurs. ROS imbalance can be transduced into redox-mediated posttranslational modifications and signaling via H_2O_2, a mild oxidant reacting with cysteine residues in proteins, affecting, e.g., protein traffic, enzyme, and receptor transcription factor activity, throughout compartmentalized cellular redox circuits (D'autreaux and Toledano 2007; Gauthier et al. 2013; Jones and Go 2010; Kaludercic et al. 2014; Kembro et al. 2013). The ability of H_2O_2 to freely diffuse throughout cellular compartments enables propagation of intracellular physiological and pathophysiological signals (Aon et al. 2004; Jeong et al. 2012; Juhaszova et al. 2004; Zhou et al. 2010).

A proper cellular/mitochondrial RE is also vital for optimal excitation-contraction (EC) coupling as well as energy supply in the heart (Burgoyne et al. 2012; Christians and Benjamin 2012). Mitochondrial lipid oxidation is a major determinant of the intracellular RE affecting, among other functions, Ca^{2+} handling by interfering with a wide range of proteins implicated in EC coupling (Fauconnier et al. 2007) including the SR Ca^{2+} release channels [the ryanodine receptors], the SR Ca^{2+} pumps, and the sarcolemmal Na^+/Ca^{2+} exchanger (Dedkova and Blatter 2008; Zima and Blatter 2006). In this context, it becomes crucial to know about the impact of ROS on redox balance as a function of substrate oxidation.

The local balance between the ROS-generating and ROS-scavenging capacities in the dense and highly connected mitochondrial network of cardiac cells determines mitochondrion behavior. For instance, mitochondria oscillate when a threshold of ROS is attained, and their collective behavior is tuned via phase and frequency synchronization (Kurz et al. 2010). In turn, the synchronization process is influenced by the size of mitochondrial clusters: large clusters take longer to synchronize resulting in a lower common frequency compared to smaller clusters (Kurz et al. 2016). Importantly, mitochondrial cluster dynamics in cardiomyocytes can be altered by metabolic substrates (glucose, pyruvate, lactate, β-hydroxybutyrate) influencing the synchronization of mitochondrial dynamics, producing a larger frequency distribution and an inverse relation between cluster frequency and size implying a dynamic heterogeneity and functional fragmentation of the mitochondrial population into several localized, smaller clusters (Kurz et al. 2010, 2014, 2015, 2016).

In agreement with the prominent role of lipids on the intracellular redox status, it was shown that with palmitate as a fuel source, a transition from oxidized to reduced cellular redox status in cardiomyocytes from type 2 diabetic (*db/db*) hearts was determined, drastically abating ROS levels (Tocchetti et al. 2012). This effect was coupled to a marked GSH rise both in wild-type and *db/db* myocytes from mice. As a consequence of its favorable effect on cellular redox balance, Palm significantly improved ISO-induced contractile reserve in *db/db*, type 2 (Tocchetti et al. 2015) and type 1 (Tocchetti et al. 2015) diabetic cardiomyocytes, from mice and guinea pig, respectively, and heart trabeculae from Zucker diabetic fatty rat, type 2 diabetes animal model (Bhatt et al. 2015).

Beyond mitochondria, lipids exert a considerable impact on other cellular processes, influencing the functional status of several organs such as the liver, skeletal, and cardiac muscles (Lee et al. 2015; Muoio and Neufer 2012; Roul and Recchia 2015; Singh and Cuervo 2012; Sung et al. 2015). The impact of lipids on mitochondrial redox status and ROS emission, and their links to energetics, is not fully elucidated. At a most basic level, our knowledge remains quite incomplete about the action of lipids on mitochondrial energetic and redox functions. Lipids can act both as uncouplers and OxPhos inhibitors (Wojtczak and Schonfeld 1993), and the consequences of these counteracting effects on mitochondrial energetic, redox, and signaling functions are just starting to be unraveled (Aon et al. 2014; Kienesberger et al. 2013; Schonfeld and Wojtczak 2008). Recent data indicate that there is a concentration effect of lipids on the redox and energetic response of mitochondria; under the threshold concentration lipids can have beneficial actions as opposed to deleterious ones depending on the threshold levels achieved (Cortassa, Sollott, Aon, unpublished).

13.4 Glucose Metabolism, Pyruvate Transport, and Pyruvate Dehydrogenase Complex Regulation

Cytoplasmic pyruvate is derived from multiple sources in the cytosol, namely, glycolysis, and precursors lactate and alanine. Pyruvate diffuses freely across the outer mitochondrial membrane through nonselective pores but, like other charged molecules, requires specialized transport across the inner membrane. The mitochondrial pyruvate carrier (MPC) conducts pyruvate across the inner mitochondrial membrane to the matrix and thereby occupies a critical link between cytosolic and mitochondrial metabolisms. The mammalian MPC protein complex comprises two obligate, paralogous subunits, designated MPC1 and MPC2, which are encoded by the MPC1 and MPC2 genes and highly conserved across eukaryotes (Bricker et al. 2012). In liver mitochondria, besides the TCA cycle, pyruvate can be channeled toward gluconeogenesis by carboxylation to oxaloacetate by the enzyme pyruvate carboxylase. This reaction regulates oxaloacetate supply to phosphoenolpyruvate kinase and, therefore, the overall gluconeogenic rate (Gray et al. 2015). In type 2 diabetes, elevated hepatic β-oxidation drives gluconeogenesis by raising mitochondrial levels of reducing equivalents and AcCoA which allosterically activates pyruvate carboxylase.

Metabolic flexibility denotes the capacity of a system to adjust fuel selection, primarily glucose and FAs, depending on nutrient availability (Kelley and Mandarino 2000; Zhang et al. 2014). Immediately downstream the MPC, sitting at the crossroad – utilization pathways of glucose-linked substrates (as sources of oxidative energy or as precursors of lipogenesis), or FAs (as preferred substrates for supplying

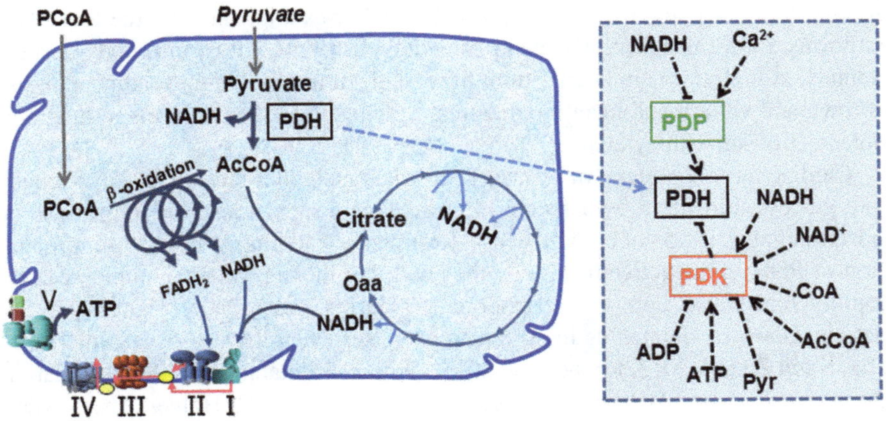

Fig. 13.1 Scheme of the model used to test the influence of PDH regulation on mitochondrial substrate selection. The model encompasses (i) the uptake of pyruvate via the pyruvate carrier and its initial oxidation by pyruvate dehydrogenase (PDH) and the TCA cycle; (ii) the uptake of palmitoyl-CoA (PCoA) into mitochondria followed by its oxidation via β-oxidation, based on the model of van Eunen and collaborators (Van Eunen et al. 2013); Acetyl-CoA (AcCoA) is at the branching point in which glucose and fatty acids degradation come together before entering the TCA cycle for complete oxidation. The model also includes (iii) oxidative phosphorylation, (iv) ionic transport for H^+, Ca^{2+}, and inorganic phosphate, (v) ATP/ADP exchange through the adenine nucleotide translocator, and (vi) the generation and scavenging of ROS as described in Kembro et al. (2013). Pyruvate dehydrogenase (PDH) complex regulation accounts for pyruvate dehydrogenase kinase (PDK) and pyruvate dehydrogenase phosphatase (PDP) modulation by all known effectors, as modeled by the factor α_{PDH} described in the text (see Eqs. 13.1 and 13.2). The *right box* depicts the effectors (*positive* indicated with *arrow heads*, *negative* by *blunt lines*) and their targets PDK and PDP

AcCoA and NADH) – the PDH complex plays a critical role in the use of either carbohydrate or fat as fuel. In animals, the PDH complex together with Mg^{2+}, thiamin pyrophosphate, CoA, and NAD^+ catalyzes the oxidative decarboxylation of pyruvate into AcCoA, CO_2, and NADH, by an operationally nonreversible reaction (forward rate constant ~10^7 times higher than in the reverse direction, at pH 7) (Randle 1986). The PDH reaction has a role in ATP synthesis and in the biosynthesis of FAs and TCA cycle intermediates from glucose.

Regulation of the mammalian PDH complex proceeds via specific kinases (pyruvate dehydrogenase kinase, PDK) and phosphatases (pyruvate dehydrogenase phosphatase, PDP) that render the enzymatic complex phosphorylated (inactive) or dephosphorylated (active) (Fig. 13.1) (see Mailloux 2015; Roche et al. 2001 for reviews). The E1 component of the PDH complex is interconverted between active and inactive forms, resulting in activity proportional to the fraction of E1 tetramers that are not phosphorylated.

The crucial role of PDH manifests differently depending upon the physiological condition or the body organ where tissue-selective control of this reaction is indicated by the distinct patterns of kinase isoform expression and the highly conserved primary structures of the different PDK isoforms (Holness et al. 2000;

Roche et al. 2001; Roche and Hiromasa 2007). For example, among the four PDK isoforms, PDK2 is apparently the most widely distributed in various body tissues, namely, at high levels in heart, brain, liver, skeletal muscle, kidney, adipose tissue (brown and white), and lactating mammary gland, and at lower levels in lung and spleen (Bowker-Kinley et al. 1998; Wu et al. 1998, 2000).

Cardiac muscle preferentially uses FAs or ketone bodies but can rapidly upregulate glucose utilization by activating PDH during a rapid transition to exercise or postprandial elevation of blood glucose. Resting skeletal muscle has a lower proportion of PDH in the active form than the heart, but this enzymatic complex can be rapidly reactivated upon dephosphorylation during initiation of exercise, or its activity down-modulated as in sustained exercise, when the use of FAs increases (Egan and Zierath 2013, Roche et al. 2001). Within skeletal muscle, PDH regulation differs in slow- and fast-twitch muscles (Holness et al. 2000; Holness and Sugden 1990; Sugden et al. 1997, 2000), the latter exhibiting a greater reliance on glucose thus maintaining a greater portion of PDH in the active form. In slow-twitch muscle, the enhanced FA oxidation after feeding high-fat diets is mainly attributed to the upregulation of pyruvate dehydrogenase kinase 4, PDK4 (Zhang et al. 2014). During starvation, the need to protect glucose stores can be fulfilled by the overexpression of one or more PDK kinase isoforms as in liver (Denyer et al. 1986; Jones et al. 1992; Marchington et al. 1987; Sugden et al. 1996, 1998), kidney (Sugden et al. 1999), and lactating mammary gland (Baxter and Coore 1978) resulting in a shutdown or acute reduction of PDH activity in these tissues (Roche et al. 2001). In the liver, PDH activity is down-modulated immediately after feeding, even following a rise in insulin levels, until liver glycogen is replenished (Holness et al. 1988; Sugden et al. 1998). In fed animals, PDH activity in liver increases in response to the availability of mobile forms of glucose-linked substrates (e.g., glucose, alanine) rather than insulin-enhancing pyruvate dehydrogenase phosphatase (PDP) activity, leading to FA synthesis. In the brain, the PDH reaction plays a crucial role in the complete oxidation of glucose (Malloch et al. 1986).

13.4.1 General Effector Regulation of PDK and PDP Activities That Modulate Mammalian PDH

The PDH complex is negatively regulated allosterically by the enzymatic products AcCoA and NADH and activated by NAD^+, ADP, and CoA (Fig. 13.1). These allosteric effectors modulate the kinase PDK and phosphatase PDP. The dedicated PDK/PDP system responds to metabolite and hormone signals to vary PDH activity in response to changes in nutritional state (Roche and Hiromasa 2007).

Among the PDK isoforms, only PDK2 exhibits strong sensitivity for all regulatory responses shown for PDK. PDK2 activity is greatly stimulated by NADH and AcCoA (Bao et al. 2004a, b), the products resulting from the PDH reaction and β-oxidation. Thus, elevation of PDK2 activity plays an important role in suppressing PDH activity to favor use of fat over carbohydrate as an oxidative fuel. ADP and pyruvate act synergistically to decrease PDK2 activity (Bao et al. 2004b).

PDP activity requires Mg^{2+} and effectors such as Ca^{2+}, to decrease the Km of one or more PDP isoforms for Mg^{2+}, and polyamines, an insulin second messenger (Roche et al. 2001). Ca^{2+} acts to directly upregulate the portion of active PDH by enhancing PDP1 activity up to tenfold (Denton et al. 1996; Thomas and Denton 1986; Yan et al. 1996). The possible regulatory role of Mg^{2+} is suggested by the sensitivity of the activation constant of PDP1 to Ca^{2+} depending on Mg^{2+} levels (~0.8 μM Ca^{2+} in the presence of saturating Mg^{2+}, increasing to 2 μM at 1.0 mM Mg^{2+}) (Roche et al. 2001). The PDP isoform activities are regulated by effector-altered sensitivities to Mg^{2+} level through very different mechanisms. With PDP1, its regulatory subunit plays a key role in modulating PDP1 catalytic site in response to Mg^{2+} level. PDP1 activity has been shown to occur in a wide range of tissues including the heart, skeletal muscle, kidney, brain, and liver. PDP2 was elevated in liver and adipose tissue suggesting its importance in fat-synthesizing tissues (Huang et al. 1998). PDP2 is probably the target for insulin activation of PDH in adipose tissue.

Hormonal signaling cascades such as the insulin signaling pathway also play a part in modulating PDH activity in response to whole-body changes in nutrition and energy state. Insulin enhances PDH activity in fat-synthesizing tissues by producing a second messenger that enhances PDP activity by lowering the Km of a phosphatase for Mg^{2+} (Denton et al. 1986; Thomas et al. 1986). Hormonal signals can also alter the short-term control of PDH activity by altering kinase activity. Signals that increase the production of pyruvate from glucose (e.g., adrenalin in skeletal muscle) will enhance pyruvate inhibition of PDK activity (Randle 1998; Sugden and Holness 1994). Hormone signals that enhance triglyceride breakdown and therefore FA oxidation indirectly stimulate kinase activity due to the resulting elevation of the intra-mitochondrial levels of NADH and AcCoA (Randle 1998; Roche et al. 2001; Roche and Hiromasa 2007; Sugden and Holness 1994).

13.5 Computational Modeling of Mammalian PDH

13.5.1 Background

The critical role of PDH in redirecting catabolism toward the utilization of glucose or FAs is driven by the negative feedback of its own reaction products, AcCoA and NADH (also supplied by FAs, ketone bodies, and the degradation of several amino acids), and can enhance kinase activity up to fourfold (Cate and Roche 1978; Pettit et al. 1975). Figure 13.1 displays a scheme with the strategic location of PDH in the metabolic network, highlighting its dependence on multiple regulatory effectors from mitochondria.

Decreased PDH activity restricts carbohydrate consumption as a result of the increase in mitochondrial $NADH/NAD^+$ and AcCoA/CoA ratios that stimulate PDH inactivation (Batenburg and Olson 1976; Hansford 1976). Stimulation of responsive PDK isoforms is produced by the use of these reactants to increase the proportion of reduced and acetylated lipoyl groups within the complex (Ravindran et al. 1996; Yang et al. 1998).

AcCoA inhibits PDH, potentially competing with CoA ($Ki = 5–10\ \mu M$) (Quinlan et al. 2014). Reduced availability of AcCoA decreases malonyl-CoA, an inhibitor of lipid utilization, thus forcing β-oxidation which is facilitated by upregulation of PDK4 (Foster 2012; Sugden et al. 2000; Zhang et al. 2014). In isolated mitochondria, AcCoA can be removed by at least two mechanisms: the condensation of oxaloacetate with AcCoA to generate citrate through citrate synthase and, upon addition of carnitine, the conversion of AcCoA to acetylcarnitine, catalyzed by carnitine acetyltransferase. Removal of AcCoA by either pathway should promote flux through the PDH complex.

The effects of NADH/NAD^+ and AcCoA/CoA are mediated by the oxidation, reduction, and acetylation state of the lipoyl group, an 80-amino acid, free-folding domain in the N-terminal region of E2 (Roche et al. 2001). Elevation of the NADH/NAD^+ and AcCoA/CoA ratios facilitates the stimulation of the activity of certain kinase isoforms, including PDK2, through covalent changes in the E2 component. PDK2 binds to the lipoyl domain depending on the redox and acetylation status of the latter, determining the enzymatic activity (Roche and Hiromasa 2007; Steussy et al. 2001). The regulatory response of PDK2 is also positively dependent on K^+ ions in the presence of physiologic levels of chloride and phosphate anions. It seems likely that, mechanistically, this product stimulation of PDK2 activity results from speeding up the rate of dissociation of ADP, which is reduced in the presence of elevated ions (Bao et al. 2004a, b). Potent synergistic inhibition of PDK2 activity by elevated ADP and pyruvate requires both K^+ and Pi. The marked reduction in binding of PDK2 to the L2 domain of E2 due to binding of ADP and pyruvate (aided by K^+ and phosphate) likely makes a major contribution (beyond slowing ADP dissociation) to the potent inhibition by these effectors (Roche and Hiromasa 2007).

13.5.2 Modular Analysis of PDH Activity and Its Regulation

Our model of PDH accounts for the regulatory influence of multiple effectors on its activity through α_{PDH}, a comprehensive integrative factor that accounts for all well-known regulators targeting either PDK or PDP, including ATP, ADP, Ca^{2+}, pyruvate in addition to AcCoA, CoA, NADH, and NAD (Fig. 13.1):

$$V_{PDH} = \frac{V_{PDH}^{\max}\alpha_{PDH}\left(\frac{NAD^+}{K_{M\_PDH}^{NAD}}\right)\left(\frac{Pyr}{K_{M\_PDH}^{Pyr}}\right)\left(\frac{CoA}{K_{M\_PDH}^{CoA}}\right)}{\left(1+\frac{Pyr}{K_{M\_PDH}^{Pyr}}\right)\left(\frac{1+\frac{CoA}{K_{M\_PDH}^{CoA}}+\frac{CoA_T}{K_{D\_PDH}^{CoA}}\left(1+\frac{CoA}{CoA_T}\right)}{1+\frac{NAD^+}{K_{M\_PDH}^{NAD}}+\frac{NAD_T}{K_{D\_PDH}^{NAD}}\left(1+\frac{NAD^+}{NAD_T}\right)}\right)} \tag{13.1}$$

With

$$\alpha_{PDH} = \frac{\left(1+\frac{Ca_m^{2+}}{K_{DP}^{Ca}}\right)\left(\left(\frac{NAD^+}{\mathrm{NADH}}\right)^{n\_NAD} F_{DP\_NAD}+1\right)}{\left(1+\frac{Ca_m^{2+}}{K_{DP}^{Ca}}\right)\left(\left(\frac{NAD^+}{\mathrm{NADH}}\right)^{n\_NAD} F_{DP\_NAD}+1\right) + \left(1+\frac{ATP}{K_{PDK}^{ATP}}\right)\left(1+\frac{K_{PDK}^{ADP}}{ADP}\right)\left(1+\frac{K_{PDK}^{Pyr}}{Pyr}\right)\left(\left(\frac{\mathrm{AcCoA}}{CoA}\right)^{n\_CoA} F_{PDK\_\mathrm{AcCoA}}+1\right)} \quad (13.2)$$

The symbols used in Eqs. (13.1 and 13.2) are defined in Table 13.1.

The main modulators of the interconversion between the non-phosphorylated and phosphorylated forms of PDH are AcCoA/CoA and NADH/NAD molar ratios. An increase in either ratio augments the proportion of inactive PDH since the activity of the kinase PDK is stimulated by both AcCoA and NADH while it is inhibited by CoA and NAD. NADH renders PDH more inactive via inhibition of the phosphatase PDP, which is reversed by NAD (Pettit et al. 1975).

The 3D plots in Figs. 13.2 and 13.3 show the dependence of PDH activity (Figs. 13.2a and 13.3a) and the factor α_{PDH} (Figs. 13.2b and 13.3b) as a function of its substrate pyruvate or Ca^{2+} and both AcCoA/CoA and NADH/NAD ratios, respectively. PDH activity increases exponentially at low AcCoA/CoA ratio concomitantly with the sensitivity to pyruvate. From the behavior of α_{PDH} (Fig. 13.2b), it can be seen that the modulatory role of AcCoA/CoA appears to dominate the enzyme activity at low rather than high ratios, since at high AcCoA/CoA the activity of PDH is negligible, whereas α_{PDH}, although low, is not, suggesting that other effectors may prevail. Comparatively, the degree of enzyme activation attained by NADH/NAD and Ca^{2+} is about fivefold lower than AcCoA/CoA and pyruvate (compare Figs. 13.2a and 13.3a).

The PDH model was able to simulate the experimental data of Pettit and colleagues (Pettit et al. 1975) obtained in highly purified preparations of PDH complexes and their component enzymes from bovine kidney and heart. The steady-state activity of PDH was modulated throughout a wide range of AcCoA/CoA (Fig. 13.4a) and NADH/NAD (Fig. 13.4b) ratios, both experimentally and computationally. The model correctly predicts a decrease in activity as a function of an increase in the ratios, with is only a minor impact at substantially low ratios (<0.1) (Fig. 13.4a, b, insets).

13.5.3 Integrated Analysis of PDH Activity and Regulation

Next, we analyzed the PDH behavior in an integrated model of mitochondrial metabolism (Cortassa, Sollott, Aon, unpublished) including simultaneous degradation of glucose-derived substrates (pyruvate, Pyr) and FAs (palmitoyl-CoA, PCoA)

Table 13.1 Parameters used in pyruvate dehydrogenase modeling

Symbol	Value	Units	Description
V_{PDH}^{max}	7.5	mM s^{-1}	Maximal rate of pyruvate dehydrogenase (PDH)
α_{PDH}	var	–	Factor integrating PDH effectors through regulation of PDK and PDP
$K_{M\_PDH}^{NAD}$	0.05	mM	Michaelis constant (K_M) of PDH for NAD
$K_{M\_PDH}^{Pyr}$	0.1	mM	K_M of PDH for pyruvate
$K_{M\_PDH}^{CoA}$	0.006	mM	K_M of PDH for coenzyme A (CoA)
$K_{D\_PDH}^{CoA}$	0.03	mM	Dissociation constant for CoA
$K_{D\_PDH}^{NAD}$	0.04	mM	Dissociation constant for NAD
CoA_T	1.0	mM	Total concentration of mitochondrial CoA species
NAD_T	1.0	mM	Total concentration of mitochondrial NAD
K_{DP}^{Ca}	0.001	mM	Activation constant of PDP for Ca2+
K_{PDK}^{ATP}	0.2	mM	Activation constant of PDK for ATP
K_{PDK}^{ADP}	0.05	mM	Inhibition constant of PDK for ADP
K_{PDK}^{Pyr}	0.1	mM	Inhibition constant of PDK for pyruvate
F_{DP_NAD}	6.7	–	NADH inhibitory factor of PDP
F_{PDK_AcCoA}	5.3	–	AcCoA activation factor of PDK
n_NAD	0.2	–	Exponential coefficient of NADH/NAD ratio
n_CoA	0.6	–	Exponential coefficient of AcCoA/CoA ratio

(Fig. 13.1). We sought to understand the impact of substrate selection (glucose-FA) on the regulation of PDH activity via AcCoA/CoA and NADH/NAD ratios, when the enzyme is integrated to mitochondrial metabolism. Figure 13.5 depicts the PDH flux as a function of both ratios, when either glucose (via pyruvate) is changing at constant FA input (via PCoA) or vice versa. Overall, and as expected from the behavior of isolated PDH, its flux decreases as a function of increasing AcCoA/CoA (compare Figs. 13.4a and 13.5a). However, PDH flux as a function of NADH/NAD displays a different behavior than its activity when isolated, i.e., increasing rather than decreasing as a function of the ratio (compare Figs. 13.4b and 13.5b). When integrated, the PDH flux is the result of the instantaneous composition of all regulatory effectors. This interpretation was confirmed by the model's ability to reproduce the trajectory of PDH flux in the integrated system, across the family of PDH activity curves corresponding to the isolated enzyme (Fig. 13.6), which is obtained when using the steady-state values of all regulatory effectors to parameterize the PDH rate expression (e.g., see Eqs. 13.1 and 13.2). When PCoA is changing

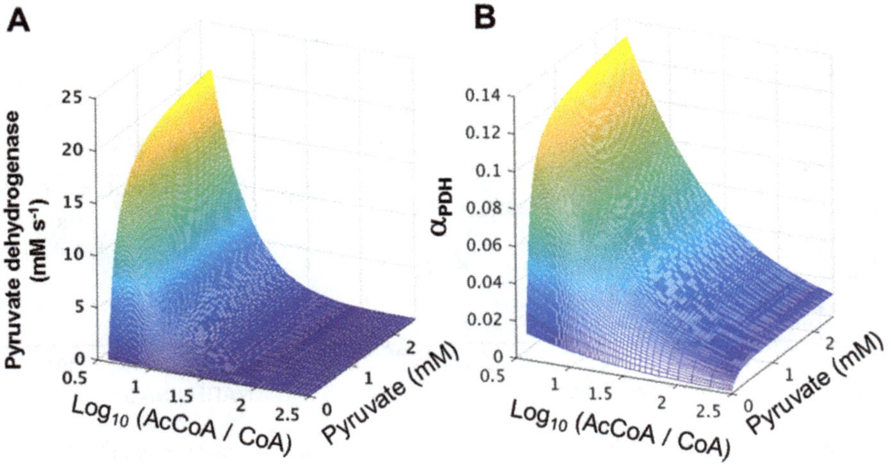

Fig. 13.2 Modular analysis of PDH activity and its modulatory factor α_{PDH} as a function of pyruvate and AcCoA/CoA ratio. The PDH rate expression (Eq. 13.1) was studied as a function of pyruvate concentration in the range 0.01–2.5 mM, while CoA was varied between 5×10^{-4} and 0.05 mM while keeping AcCoA constant at 0.2 mM. Other parameters were as follows: NADH = 0.1 mM; NAD^+ = 0.9 mM; ATP = 1.0 mM; ADP = 0.5 mM; Ca^{2+} = 0.2 µM. Maximal rate and other parameters utilized are indicated in Table 13.1

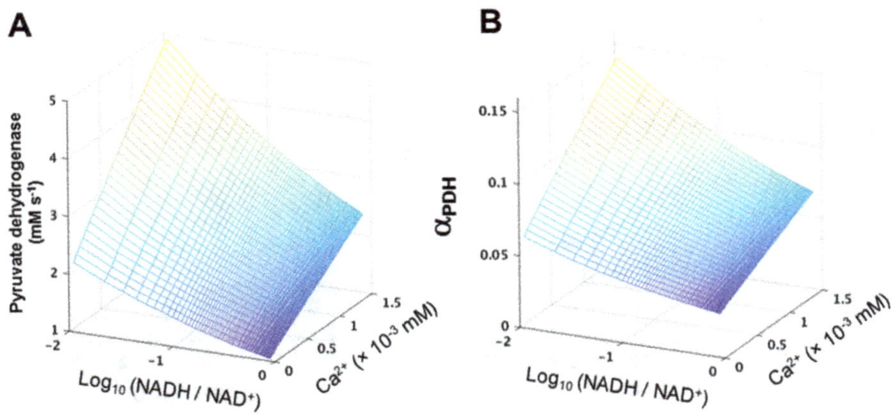

Fig. 13.3 Modular analysis of PDH activity and its modulatory factor α_{PDH} as a function of Ca^{2+} and NADH/NAD ratio. The PDH rate expression (Eq. 13.1) was analyzed as a function of Ca^{2+} (range 1×10^{-5}–1.5×10^{-3} mM) and NADH (range 0.01–0.8 mM) concentrations, while NAD^+ was kept constant at 0.9 mM. Other parameters were as follows: AcCoA = 0.2 mM; CoA = 0.01 mM; ATP = 1.0 mM; ADP = 0.5 mM; Pyr = 0.5 mM. Maximal rate and other parameters utilized are indicated in Table 13.1

at constant Pyr, we observe an overall similar qualitative behavior although for a more restricted range of variation in both ratios and activity (Fig. 13.5). Together, these results suggest that AcCoA/CoA regulation prevails over NADH/NAD in the

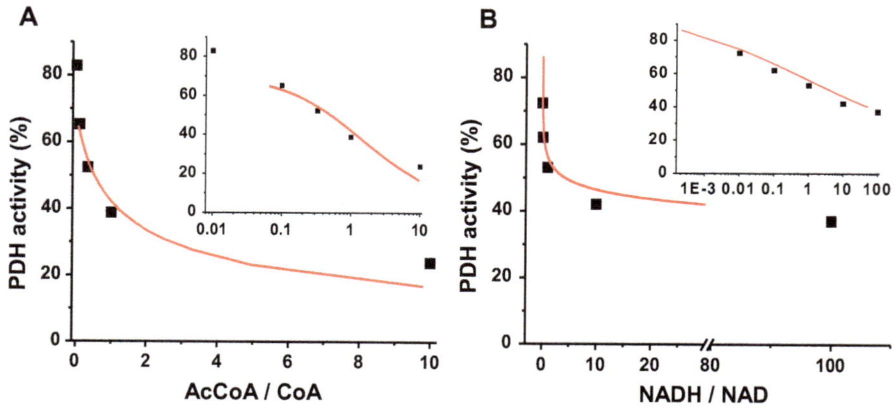

Fig. 13.4 Comparative study of PDH activity as a function of the main modulatory ratios NADH/NAD and AcCoA/CoA. The expression of PDH rate (Eq. 13.1) was evaluated at (mM): $Ca^{2+} = 2\times10^{-4}$, ADP = 0.5, ATP = 1.0, and Pyr = 2.5. In panel **A** NADH was kept constant at 0.1 mM, AcCoA = 0.05, while CoA varied in the range 5×10^{-3} – 0.76 mM. In Panel **B** CoA was constant at 0.1 mM, while NADH changed from 1×10^{-4} to 0.975 and NAD changed in parallel to keep the total pyridine nucleotide pool at 1.0 mM

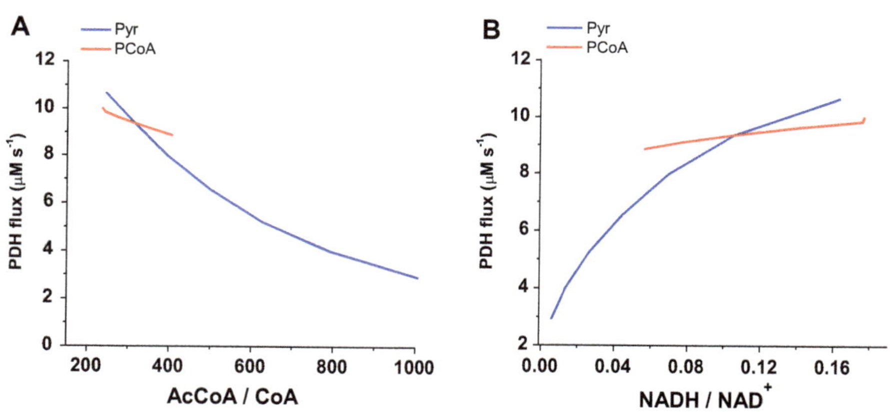

Fig. 13.5 Steady-state PDH flux in the integrated model when either pyruvate (Pyr) or palmitoyl-CoA (PCoA) was varied (PCoA/Pyr ratio). A computed simulation of the complete model was run until all state variables reached steady state (i.e., their time derivatives were $<1\times10^{-10}$). Pyr was adjusted in the range from 4×10^{-3} to 0.01 mM (indicated in *blue lines*) while keeping PCoA constant at 0.04 mM. In the steady states indicated with a *red line* PCoA was varied from 0.01 to 0.06 mM, while Pyr remained constant at 0.009 mM. In these plots the AcCoA/CoA and NADH/NAD ratios are computed from their steady-state values occurring upon variations in the relative proportion of substrates Pyr and PCoA. Consequently, all axes are representing state variables thus explaining the trend observed in the PDH flux value as a function of NADH/NAD, since the values of every regulatory effector of PDH activity are simultaneously changing in these simulations

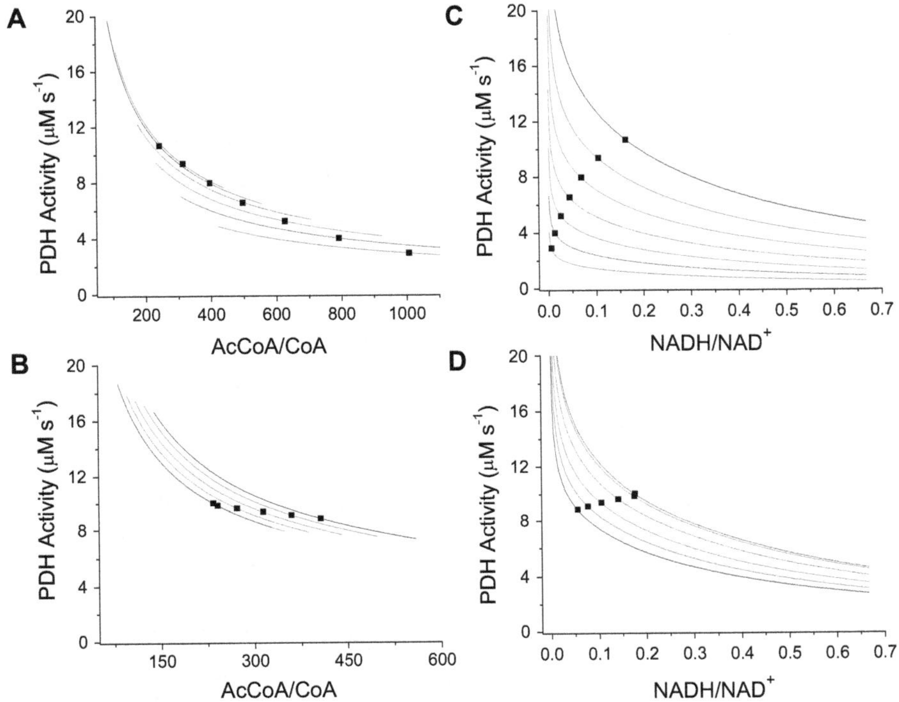

Fig. 13.6 Modular study of the PDH activity under conditions reproducing the steady states obtained with different PCoA/Pyr ratios. The concentrations of substrates and effectors other than the ones in the x-axis were fixed at the values corresponding to the steady states represented in the curves in Fig. 13.5. The PDH activity was calculated as a function of the ratio AcCoA/CoA and NADH/NAD under conditions mimicking variations in Pyr (*top panels*) or PCoA (*bottom panels*). The *squares* indicate the exact conditions for all substrates and effectors that were represented in Fig. 13.5

integrated system, where the flux through the PDH complex as a systemic property results from the instantaneous levels of all regulatory effectors.

Respiration and H_2O_2 emission both increase as a function of PCoA/Pyr ratio when either one of them is changing in constant proportion to the other, suggesting that ROS generation is matched by the ROS scavenging systems under these conditions (Fig. 13.7).

To address the question of substrate selection, we quantified the fluxes driven by Pyr or PCoA as a function of their ratio to simulate changing nutrient availability and the sensitivity of the pathways' flux to both substrates. Figure 13.8a shows the fluxes through β-oxidation via carnitine palmitoyl transferase 1 (CPT1) and PDH as a function of the ratio PCoA/Pyr. When Pyr increases at constant PCoA, i.e., decreasing PCoA/Pyr ratio, the flux through PDH increases 3.6-fold, whereas that of CPT1 decreased 15%. On the other hand, increasing PCoA at constant Pyr pro-

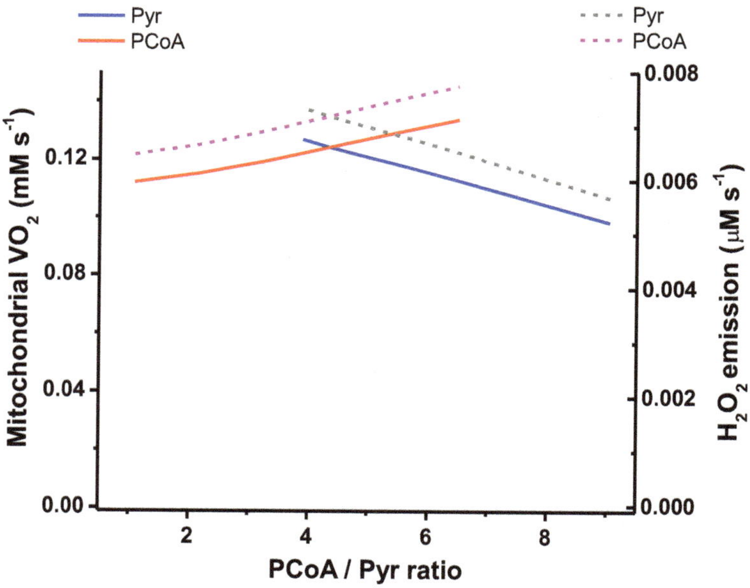

Fig. 13.7 Steady-state respiration and H_2O_2 emission fluxes as a function of PCoA/Pyr ratios. The O_2 consumption (*solid lines*) and H_2O_2 emission (*dashed lines*) fluxes were obtained in the same simulations presented in Fig. 13.5 when Pyr (in *blue* or *gray*) or PCoA (in *red* or *magenta*) was changed. All simulation conditions correspond to those specified in the legend of Fig. 13.5

duced ~40% increase in flux through CPT1 as compared to ~11% decrease through PDH. Together, these results suggest that the flux from Pyr is much more sensitive to nutrient availability than from PCoA, thus making substrate selection toward Pyr more sensitive, at least under these conditions (Fig. 13.8b). Quantitatively speaking, the flux values through PDH and CPT1 determined with our model agree very well with published results, respectively, of glucose oxidation in whole heart (3–10 vs. 5–7 μM s^{-1}) (Buchanan et al. 2005; Cortassa et al. 2015; Kashiwaya et al. 1994) and with palmitate oxidation in cardiomyocytes (0.4–0.6 vs. 0.4 μM s^{-1}) (Luiken et al. 2009) or respiration in the presence of both palmitate and glucose in cardiomyocytes (0.12 vs. 0.4 μM s^{-1}) (Wang et al. 2011).

Taken together, the results obtained agree with the idea that the crucial regulatory role played by PDH in substrate selection in the short term depends on glucose and FAs' availability and on the resulting composition of metabolite levels that activate or inhibit the kinases and phosphatases from the PDH complex. Within this complex regulatory picture, our simulations also indicate that, under these conditions, the phosphorylated form of PDH predominates over the non-phosphorylated one, as can be judged by the low α_{PDH} values (~1.2×10^{-3}, when either pyruvate or PCoA is varying; see also Figs. 13.2, 13.4, and 13.5).

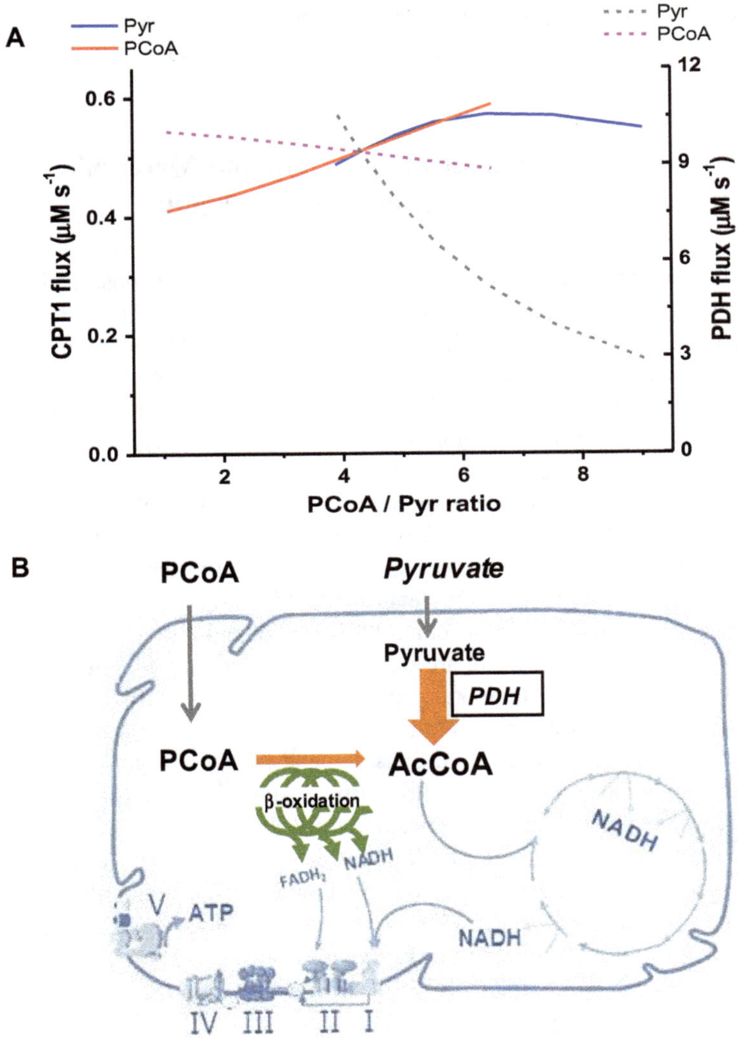

Fig. 13.8 Steady-state fluxes through PDH and CPT1 upon changing proportions between pyruvate and PCoA. The fluxes through the transport step of FAs CPT1 (*solid lines*) and regulatory PDH (*dashed lines*) obtained in the simulations in Fig. 13.5 were represented as a function of the ratio of PCoA over pyruvate. *Blue* and *gray* lines indicate when Pyr was varied at constant PCoA, whereas *red* and *magenta* ones represent changes in PCoA at constant Pyr. The scheme in panel **B** is a graphical representation of the sensitivity of flux regulation through the PDH complex and CPT1 in response to changing proportion of glucose- and FA-derived substrates

Accordingly, a higher response of pyruvate flux through PDH as compared to FAs through CPT1 was determined, indicating a higher sensitivity of selection toward carbohydrates (Fig. 13.8). Judging from the agreement between the PDH behaviors whether isolated from or integrated with mitochondrial metabolism, simulations also show a predominant regulatory impact by the AcCoA/CoA ratio as compared to NADH/NAD (compare Figs. 13.4 and 13.5).

13.6 Modulation of Substrate Selection and Metabolic Remodeling in Health, Disease, and Aging

(Patho)physiological situations involve differential selection of substrate fuel which can exert a significant impact on the organism or cell behavior. For instance, obese and type 2 diabetic patients exhibit greater rates of FA oxidation and insulin resistance unlike lean healthy individuals (the latter whom, under insulin stimulation, are able to switch from predominantly FA oxidation to elevation of glucose uptake, oxidation, and storage). This capacity to adjust fuel selection as a function of nutrient availability has been termed *metabolic flexibility* (Kelley and Mandarino 2000), in which the mitochondrial PDH complex plays a crucial regulatory role (Randle 1986; Roche et al. 2001; Roche and Hiromasa 2007; Sugden et al. 1998; Zhang et al. 2014). PDH regulation involves short-term (e.g., allosteric inhibition/activation) as well as long-term (e.g., gene expression, transcriptional, posttranslational) mechanisms (Randle 1998; Roche et al. 2001; Roche and Hiromasa 2007; Sugden et al. 1997, 1998) which, in turn, are subjected to circadian regulation (Bellet and Sassone-Corsi 2010). The temporal regulatory dimension of PDH activity is of great importance because both acute (e.g., fast-feed transition) and chronic (e.g., metabolic disorder) nutritional conditions demand flexible, or generate inflexible, metabolic responses. For example, long-term consumption of a high-saturated fat diet may cause hyperglycemia, hyperinsulinemia, glucose intolerance, and obesity. In skeletal muscle, consumption of a high-fat diet leads to the use of lipid-derived fuels as respiratory substrates, a switch modulated, in part, by upregulation of PDK, the kinase activity associated with PDH (Zhang et al. 2014). Another level of regulation is given by nutrient availability-sensitive posttranslational modifications in the presence or absence of sirtuin3 (SIRT3) that reshape the mitochondrial acetylome, potentially affecting multiple enzymatic activities (Finkel 2015; Foster et al. 2013).

Glucose stores are important for their utilization as preferred substrate by the central nervous system. Inactivation of PDH along with greatly diminished glucose oxidation may happen in muscle, liver, and fat cells due to diabetes, starvation, long-term feeding of high-fat diet, or obesity. As the studies of Randle and coworkers first demonstrated (Randle 1986; Randle et al. 1963), during starvation and the diabetic state, the acute decreases in the PDH activity of these tissues are engendered by a marked induction of kinase activity (Randle 1998; Roche et al. 2001). In starvation and hibernation, enhanced PDK activity is beneficial for preventing loss of body carbohydrate while favoring the use of the more abundant lipid fuels. Along with increased kinase levels, hormonally controlled increase in FA and ketone body oxidation elevates AcCoA and NADH, which, in turn, stimulate the activity of certain PDK isoforms.

Severe limitation of glucose use by PDH inactivation raises blood glucose in the diabetic state, while glucose clearance is impaired making a major contribution to the pathology of diabetes. High blood glucose damages vascular cells (Brownlee 2001; Choi et al. 2008; Giacco and Brownlee 2010; Laakso 1999; Lasker 1993) and the myocardium leading to progressive vascular damage and heart dysfunction in both type 1 (insulin-deficient) and type 2 (insulin-resistant) diabetes (Aon et al. 2015; Bhatt et al. 2015; Tocchetti et al. 2012, 2015), which amount to ~4% and 96%, respectively, of the diabetic population according to the 2012 American Diabetes Association statistics. Diabetes is characterized by increased circulating concentrations of glucose and FAs. Irrespective of hyperglycemia, the heart from diabetics relies heavily on FA utilization with a concomitant decrease in glucose oxidation (Boudina and Abel 2010; Carley and Severson 2005). Historically, the glucose-FA cycle, also known as the Randle cycle (Randle 1998; Randle et al. 1963), has played a relevant role as the biochemical mechanism explaining the control and functional impact of fuel selection, FA over glucose oxidation (Hue and Taegtmeyer 2009). However, on its own, hyperglycemia induces cellular damage that involves the increase of the flux of glucose and other sugars through the polyol, hexosamine, advanced glycation end products (AGEs), and diacylglycerol (DAG) pathways, the latter leading to protein kinase C (PKC) activation (Brownlee 1995, 2001; Giacco and Brownlee 2010). These changes are also implicated in the hyperglycemia-mediated modifications and impairments of cell redox assets (Aon et al. 2015; Brownlee 2001; Tocchetti et al. 2012, 2015; Williamson et al. 1993).

Under starvation and diabetic conditions, the PDK4 isoform (less sensitive to pyruvate inhibition vs. other isoforms) is overexpressed in several tissues, particularly in heart and skeletal muscle under conditions of limited consumption of carbohydrate (Holness et al. 2000; Roche et al. 2001; Sugden et al. 1998, 1999, 2000; Wu et al. 1998, 2000). Besides prolonged starvation, feeding a high-fat, low-carbohydrate, diet increases PDK4 in both slow- and fast-twitch muscles. Refeeding or insulin treatment reverses the effects of starvation or diabetes, respectively. In rat liver, PDK2 and PDK4 are both overexpressed under conditions of starvation, diabetes, or feeding a high-fat, low-carbohydrate diet or in response to artificial elevation of cAMP or 3,5,3′-triiodothyronine (Denyer et al. 1986; Jones et al. 1992; Sugden et al. 1998; Sugden and Holness 1994; Wu et al. 2000). These effects can be also reversed by insulin or refeeding with a carbohydrate-rich diet (Holness et al. 1988). Selective elevation of PDK2, following maintenance on a high-fat diet, could be prevented or reversed by a diet supplemented with long-chain ω-3-fatty acids (Sugden et al. 1998).

A healthy heart can utilize various substrates (glucose, FAs, ketones, lactate) to satisfy its continuous energy requirements, although under postabsorptive or fasting conditions it preferentially uses FAs (Kolwicz and Tian 2009; Lionetti et al. 2011). The heart's high adaptability can also be found throughout its life cycle where at the fetal stage it relies on carbohydrate substrates, whereas at more mature stages it predominantly consumes FAs as fuel. The FA dependence of the heart can be enhanced by diabetes (Belke et al. 1999; Lopaschuk 2002). In general, aging and disease drive substantial metabolic remodeling that includes changes in mitochondrial function, impaired metabolic flexibility, and reduced insulin sensitivity (Finkel 2015). In the aged heart, the capacity for glucose utilization prevails over FA oxidation (Hansford 1983; Lesnefsky et al. 2016; Van Bilsen et al. 2009), although tending to be insufficient for sustaining energy supply under stress (Kolwicz and Tian 2009). Unlike the reported decline of protein levels from mitochondrial metabolism (including respiratory complexes, TCA cycle, FA, and amino acid metabolisms), a significant increase in glycolytic and extracellular structural proteins happens with age (Tocchi et al. 2015). Aging impairs mitochondrial OxPhos, particularly so in interfibrillar mitochondria, affecting the activity of complexes III and IV, which accounts in large measure for the known decrease in respiration (Lesnefsky et al. 2016). Phenotypically, the metabolic profile of the aging heart bears some similarity to that of heart failure. Pathological hypertrophy is associated with reversion to a fetal gene expression pattern and an increased reliance on carbohydrate fuel rather than FAs which in turn are less consumed (Kolwicz and Tian 2009).

13.7 Concluding Remarks

Mitochondrial metabolism and PDH activity are central players in substrate selection, a process that underlies metabolic remodeling and flexibility in healthy, diseased, and aged states. Within the complex regulatory picture of the PDH activity, with dedicated kinases and phosphatases as key targets, the mitochondrial AcCoA/CoA ratio assumes a predominant role compared to NADH/NAD. Under conditions of different nutrient availability, the higher sensitivity of PDH toward glucose-derived substrates (i.e., pyruvate) is in agreement with the crucial role played by this enzymatic complex in preserving glucose stores for brain function that, in turn, determines the overall balance between nutrient utilization, storage, and turnover in the organism's function. By accounting for all major effectors of the PDH complex, modeling and results presented herein now enable a more realistic and detailed understanding of the regulation of selection between glucose- and FA-derived substrates in the presence of all major redox-energetic mitochondrial functions.

Acknowledgments This work was supported by the Intramural Research Program of the National Institutes of Health, National Institute on Aging.

References

Aggarwal NT, Makielski JC (2013) Redox control of cardiac excitability. Antioxid Redox Signal 18:432–468

Akar FG, Aon MA, Tomaselli GF, O'Rourke B (2005) The mitochondrial origin of postischemic arrhythmias. J Clin Invest 115:3527–3535

Alleman RJ, Katunga LA, Nelson MA, Brown DA, Anderson EJ (2014) The "goldilocks zone" from a redox perspective-adaptive vs. deleterious responses to oxidative stress in striated muscle. Front Physiol 5:358

Alleman RJ, Tsang AM, Ryan TE, Patteson DJ, Mcclung JM, Spangenburg EE, Shaikh SR, Neufer PD, Brown DA (2016) Exercise-induced protection against reperfusion arrhythmia involves stabilization of mitochondrial energetics. Am J Physiol Heart Circ Physiol 310:H1360–H1370

Aon MA (2013) Complex systems biology of networks: the riddle and the challenge. In: Aon MA, Saks V, Schlattner U (eds) Systems biology of metabolic and signaling networks. energy, mass and information transfer, 1st edn. Springer-Verlag Berlin Heidelberg, Heidelberg

Aon MA, Cortassa S, O'rourke B (2004) Percolation and criticality in a mitochondrial network. Proc Natl Acad Sci U S A 101:4447–4452

Aon MA, Cortassa S, Maack C, O'Rourke B (2007) Sequential opening of mitochondrial ion channels as a function of glutathione redox thiol status. J Biol Chem 282:21889–21900

Aon MA, Cortassa S, Akar FG, Brown DA, Zhou L, O'rourke B (2009) From mitochondrial dynamics to arrhythmias. Int J Biochem Cell Biol 41:1940–1948

Aon MA, Stanley BA, Sivakumaran V, Kembro JM, O'Rourke B, Paolocci N, Cortassa S (2012) Glutathione/thioredoxin systems modulate mitochondrial H2O2 emission: an experimental-computational study. J Gen Physiol 139:479–491

Aon MA, Bhatt N, Cortassa S (2014) Mitochondrial and cellular mechanisms for managing lipid excess. Front Physiol 5:1–13

Aon MA, Tocchetti CG, Bhatt N, Paolocci N, Cortassa S (2015) Protective mechanisms of mitochondria and heart function in diabetes. Antioxid Redox Signal 22:1563–1586

Aon MA, Cortassa S, Juhaszova M, Sollott SJ (2016) Mitochondrial health, the epigenome and healthspan. Clin Sci (Lond) 130:1285–1305

Bao H, Kasten SA, Yan X, Hiromasa Y, Roche TE (2004a) Pyruvate dehydrogenase kinase isoform 2 activity stimulated by speeding up the rate of dissociation of ADP. Biochemistry 43: 13442–13451

Bao H, Kasten SA, Yan X, Roche TE (2004b) Pyruvate dehydrogenase kinase isoform 2 activity limited and further inhibited by slowing down the rate of dissociation of ADP. Biochemistry 43:13432–13441

Batenburg JJ, Olson MS (1976) Regulation of pyruvate dehydrogenase by fatty acid in isolated rat liver mitochondria. J Biol Chem 251:1364–1370

Baxter MA, Coore HG (1978) The mode of regulation of pyruvate dehydrogenase of lactating rat mammary gland. Effects of starvation and insulin. Biochem J 174:553–561

Belke DD, Larsen TS, Lopaschuk GD, Severson DL (1999) Glucose and fatty acid metabolism in the isolated working mouse heart. Am J Phys 277:R1210–R1217

Bellet MM, Sassone-Corsi P (2010) Mammalian circadian clock and metabolism – the epigenetic link. J Cell Sci 123:3837–3848

Bhatt NM, Aon MA, Tocchetti CG, Shen X, Dey S, Ramirez-Correa G, O'Rourke B, Gao WD, Cortassa S (2015) Restoring redox balance enhances contractility in heart trabeculae from type 2 diabetic rats exposed to high glucose. Am J Physiol Heart Circ Physiol 308:H291–H302

Boudina S, Abel ED (2010) Diabetic cardiomyopathy, causes and effects. Rev Endocr Metab Disord 11:31–39

Bowker-Kinley MM, Davis WI, Wu P, Harris. RA, Popov KM (1998) Evidence for existence of tissue-specific regulation of the mammalian pyruvate dehydrogenase complex. Biochem J 329(Pt 1):191–196

Bricker DK, Taylor EB, Schell JC, Orsak T, Boutron A, Chen YC, Cox JE, Cardon CM, Van Vranken JG, Dephoure N, Redin C, Boudina S, Gygi SP, Brivet M, Thummel CS, Rutter J (2012) A mitochondrial pyruvate carrier required for pyruvate uptake in yeast, Drosophila, and humans. Science 337:96–100

Brown DA, Aon MA, Frasier CR, Sloan RC, Maloney AH, Anderson EJ, O'Rourke B (2010) Cardiac arrhythmias induced by glutathione oxidation can be inhibited by preventing mitochondrial depolarization. J Mol Cell Cardiol 48:673–679

Brownlee M (1995) Advanced protein glycosylation in diabetes and aging. Annu Rev Med 46:223–234

Brownlee M (2001) Biochemistry and molecular cell biology of diabetic complications. Nature 414:813–820

Buchanan J, Mazumder PK, Hu P, Chakrabarti G, Roberts MW, Yun UJ, Cooksey RC, Litwin SE, Abel ED (2005) Reduced cardiac efficiency and altered substrate metabolism precedes the onset of hyperglycemia and contractile dysfunction in two mouse models of insulin resistance and obesity. Endocrinology 146:5341–5349

Burgoyne JR, Mongue-Din H, Eaton P, Shah AM (2012) Redox signaling in cardiac physiology and pathology. Circ Res 111:1091–1106

Carley AN, Severson DL (2005) Fatty acid metabolism is enhanced in type 2 diabetic hearts. Biochim Biophys Acta 1734:112–126

Cascante M, Marin S (2008) Metabolomics and fluxomics approaches. Essays Biochem 45:67–81

Cate RL, Roche TE (1978) A unifying mechanism for stimulation of mammalian pyruvate dehydrogenase(a) kinase by reduced nicotinamide adenine dinucleotide, dihydrolipoamide, acetyl coenzyme A, or pyruvate. J Biol Chem 253:496–503

Choi SW, Benzie IF, Ma SW, Strain JJ, Hannigan BM (2008) Acute hyperglycemia and oxidative stress: direct cause and effect? Free Radic Biol Med 44:1217–1231

Choudhary C, Weinert BT, Nishida Y, Verdin E, Mann M (2014) The growing landscape of lysine acetylation links metabolism and cell signalling. Nat Rev Mol Cell Biol 15:536–550

Christians ES, Benjamin IJ (2012) Proteostasis and REDOX state in the heart. Am J Physiol Heart Circ Physiol 302:H24–H37

Cortassa S, O'Rourke B, Aon MA (2014) Redox-optimized ROS balance and the relationship between mitochondrial respiration and ROS. Biochim Biophys Acta 1837:287–295

Cortassa S, Caceres V, Bell LN, O'Rourke B, Paolocci N, Aon MA (2015) From metabolomics to fluxomics: a computational procedure to translate metabolite profiles into metabolic fluxes. Biophys J 108:163–172

Crane D, Haussinger D, Graf P, Sies H (1983) Decreased flux through pyruvate dehydrogenase by thiol oxidation during t-butyl hydroperoxide metabolism in perfused rat liver. Hoppe Seylers Z Physiol Chem 364:977–987

D'autreaux B, Toledano MB (2007) ROS as signalling molecules: mechanisms that generate specificity in ROS homeostasis. Nat Rev Mol Cell Biol 8:813–824

Dedkova EN, Blatter LA (2008) Mitochondrial Ca2+ and the heart. Cell Calcium 44:77–91

Denton RM, Mccormack JG, Thomas AP (1986) Mechanisms whereby insulin and other hormones binding to cell surface receptors influence metabolic pathways within the inner membrane of mitochondria. Ann N Y Acad Sci 488:370–384

Denton RM, Mccormack JG, Rutter GA, Burnett P, Edgell NJ, Moule SK, Diggle TA (1996) The hormonal regulation of pyruvate dehydrogenase complex. Adv Enzym Regul 36:183–198

Denyer GS, Kerbey AL, Randle PJ (1986) Kinase activator protein mediates longer-term effects of starvation on activity of pyruvate dehydrogenase kinase in rat liver mitochondria. Biochem J 239:347–354

Dey S, Sidor A, O'Rourke B (2016) Compartment-specific control of reactive oxygen species scavenging by antioxidant pathway enzymes. J Biol Chem 291:11185–11197

Donohoe DR, Bultman SJ (2012) Metaboloepigenetics: interrelationships between energy metabolism and epigenetic control of gene expression. J Cell Physiol 227:3169–3177

Eaton S (2002) Control of mitochondrial beta-oxidation flux. Prog Lipid Res 41:197–239

Eaton S, Bartlett K, Pourfarzam M (1996) Mammalian mitochondrial beta-oxidation. Biochem J 320(Pt 2):345–357

Egan B, Zierath JR (2013) Exercise metabolism and the molecular regulation of skeletal muscle adaptation. Cell Metab 17:162–184

Fauconnier J, Andersson DC, Zhang SJ, Lanner JT, Wibom R, Katz A, Bruton JD, Westerblad H (2007) Effects of palmitate on Ca(2+) handling in adult control and ob/ob cardiomyocytes: impact of mitochondrial reactive oxygen species. Diabetes 56:1136–1142

Finkel T (2015) The metabolic regulation of aging. Nat Med 21:1416–1423

Fisher-Wellman KH, Neufer PD (2012) Linking mitochondrial bioenergetics to insulin resistance via redox biology. Trends Endocrinol Metab 23:142–153

Foster DW (2012) Malonyl-CoA: the regulator of fatty acid synthesis and oxidation. J Clin Invest 122:1958–1959

Foster DB, Liu T, Rucker J, O'Meally RN, Devine LR, Cole RN, O'Rourke B (2013) The cardiac acetyl-lysine proteome. PLoS One 8:e67513

Gauthier LD, Greenstein JL, Cortassa S, O'Rourke B, Winslow RL (2013) A computational model of reactive oxygen species and redox balance in cardiac mitochondria. Biophys J 105:1045–1056

Giacco F, Brownlee M (2010) Oxidative stress and diabetic complications. Circ Res 107:1058–1070

Gray LR, Sultana MR, Rauckhorst AJ, Oonthonpan L, Tompkins SC, Sharma A, Fu X, Miao R, Pewa AD, Brown KS, Lane EE, Dohlman A, Zepeda-Orozco D, Xie J, Rutter J, Norris AW, Cox JE, Burgess SC, Potthoff MJ, Taylor EB (2015) Hepatic mitochondrial pyruvate carrier 1 is required for efficient regulation of gluconeogenesis and whole-body glucose homeostasis. Cell Metab 22:669–681

Hafstad AD, Boardman N, Aasum E (2015) How exercise may amend metabolic disturbances in diabetic cardiomyopathy. Antioxid Redox Signal 22:1587–1605

Hansford RG (1976) Studies on the effects of coenzyme A-SH: acetyl coenzyme A, nicotinamide adenine dinucleotide: reduced nicotinamide adenine dinucleotide, and adenosine diphosphate: adenosine triphosphate ratios on the interconversion of active and inactive pyruvate dehydrogenase in isolated rat heart mitochondria. J Biol Chem 251:5483–5489

Hansford RG (1983) Bioenergetics in aging. Biochim Biophys Acta 726:41–80

Hoek JB, Rydstrom J (1988) Physiological roles of nicotinamide nucleotide transhydrogenase. Biochem J 254:1–10

Holness MJ, Sugden MC (1990) Glucose utilization in heart, diaphragm and skeletal muscle during the fed-to-starved transition. Biochem J 270:245–249

Holness MJ, Maclennan PA, Palmer TN, Sugden MC (1988) The disposition of carbohydrate between glycogenesis, lipogenesis and oxidation in liver during the starved-to-fed transition. Biochem J 252:325–330

Holness MJ, Kraus A, Harris RA, Sugden MC (2000) Targeted upregulation of pyruvate dehydrogenase kinase (PDK)-4 in slow-twitch skeletal muscle underlies the stable modification of the regulatory characteristics of PDK induced by high-fat feeding. Diabetes 49:775–781

Huang B, Gudi R, Wu P, Harris RA, Hamilton J, Popov KM (1998) Isoenzymes of pyruvate dehydrogenase phosphatase. DNA-derived amino acid sequences, expression, and regulation. J Biol Chem 273:17680–17688

Hue L, Taegtmeyer H (2009) The Randle cycle revisited: a new head for an old hat. Am J Physiol Endocrinol Metab 297:E578–E591

Jeong EM, Liu M, Sturdy M, Gao G, Varghese ST, Sovari AA, Dudley SC Jr (2012) Metabolic stress, reactive oxygen species, and arrhythmia. J Mol Cell Cardiol 52:454–463

Jones DP (2002) Redox potential of GSH/GSSG couple: assay and biological significance. Methods Enzymol 348:93–112

Jones DP, Go YM (2010) Redox compartmentalization and cellular stress. Diabetes Obes Metab 12(Suppl 2):116–125

Jones DP, Sies H (2015) The redox code. Antioxid Redox Signal 23:734–746

Jones BS, Yeaman SJ, Sugden MC, Holness MJ (1992) Hepatic pyruvate dehydrogenase kinase activities during the starved-to-fed transition. Biochim Biophys Acta 1134:164–168

Juhaszova M, Zorov DB, Kim SH, Pepe S, Fu Q, Fishbein KW, Ziman BD, Wang S, Ytrehus K, Antos CL, Olson EN, Sollott SJ (2004) Glycogen synthase kinase-3beta mediates convergence of protection signaling to inhibit the mitochondrial permeability transition pore. J Clin Invest 113:1535–1549

Kaludercic N, Deshwal S, Di Lisa F (2014) Reactive oxygen species and redox compartmentalization. Front Physiol 5:285

Kashiwaya Y, Sato K, Tsuchiya N, Thomas S, Fell DA, Veech RL, Passonneau JV (1994) Control of glucose utilization in working perfused rat heart. J Biol Chem 269:25502–25514

Keating ST, El-Osta A (2015) Epigenetics and metabolism. Circ Res 116:715–736

Kelley DE, Mandarino LJ (2000) Fuel selection in human skeletal muscle in insulin resistance: a reexamination. Diabetes 49:677–683

Kembro JM, Aon MA, Winslow RL, O'Rourke B, Cortassa S (2013) Integrating mitochondrial energetics, redox and ROS metabolic networks: a two-compartment model. Biophys J 104:332–343

Kembro JM, Cortassa S, Aon MA (2014) Complex oscillatory redox dynamics with signaling potential at the edge between normal and pathological mitochondrial function. Front Physiol 5:257

Kienesberger PC, Pulinilkunnil T, Nagendran J, Dyck JR (2013) Myocardial triacylglycerol metabolism. J Mol Cell Cardiol 55:101–110

Kolwicz SC Jr, Tian R (2009) Metabolic therapy at the crossroad: how to optimize myocardial substrate utilization? Trends Cardiovasc Med 19:201–207

Kurz FT, Aon MA, O'Rourke B, Armoundas AA (2010) Spatio-temporal oscillations of individual mitochondria in cardiac myocytes reveal modulation of synchronized mitochondrial clusters. Proc Natl Acad Sci U S A 107:14315–14320

Kurz FT, Aon MA, O'Rourke B, Armoundas AA (2014) Cardiac mitochondria exhibit dynamic functional clustering. Front Physiol 5:329

Kurz FT, Derungs T, Aon MA, O'Rourke B, Armoundas AA (2015) Mitochondrial networks in cardiac myocytes reveal dynamic coupling behavior. Biophys J 108:1922–1933

Kurz FT, Kembro JM, Flesia AG, Armoundas AA, Cortassa S, Aon MA, Lloyd D (2016) Network dynamics: quantitative analysis of complex behavior in metabolism, organelles and cells, from experiments to models and back. Wiley Interdiscip Rev Syst Biol Med. doi:10.1002/wsbm.1352

Laakso M (1999) Hyperglycemia and cardiovascular disease in type 2 diabetes. Diabetes 48:937–942

Lanpher B, Brunetti-Pierri N, Lee B (2006) Inborn errors of metabolism: the flux from Mendelian to complex diseases. Nat Rev Genet 7:449–460

Lasker RD (1993) The diabetes control and complications trial. Implications for policy and practice. N Engl J Med 329:1035–1036

Lee J, Homma T, Kurahashi T, Kang ES, Fujii J (2015) Oxidative stress triggers lipid droplet accumulation in primary cultured hepatocytes by activating fatty acid synthesis. Biochem Biophys Res Commun 464:229–235

Lesnefsky EJ, Chen Q, Hoppel CL (2016) Mitochondrial metabolism in aging heart. Circ Res 118:1593–1611

Lionetti V, Stanley WC, Recchia FA (2011) Modulating fatty acid oxidation in heart failure. Cardiovasc Res 90:202–209

Liu T, Takimoto E, Dimaano VL, Demazumder D, Kettlewell S, Smith G, Sidor A, Abraham TP, O'Rourke B (2014) Inhibiting mitochondrial Na+/Ca2+ exchange prevents sudden death in a Guinea pig model of heart failure. Circ Res 115:44–54

Lloyd D, Cortassa S, O'Rourke B, Aon MA (2012) What yeast and cardiomyocytes share: ultradian oscillatory redox mechanisms of cellular coherence and survival. Integr Biol (Camb) 4:65–74

Lopaschuk GD (2002) Metabolic abnormalities in the diabetic heart. Heart Fail Rev 7:149–159

Lopaschuk GD, Ussher JR, Folmes CD, Jaswal JS, Stanley WC (2010) Myocardial fatty acid metabolism in health and disease. Physiol Rev 90:207–258

Luiken JJ, Niessen HE, Coort SL, Hoebers N, Coumans WA, Schwenk RW, Bonen A, Glatz JF (2009) Etomoxir-induced partial carnitine palmitoyltransferase-I (CPT-I) inhibition in vivo does not alter cardiac long-chain fatty acid uptake and oxidation rates. Biochem J 419:447–455

Mailloux RJ (2015) Still at the center of it all: novel functions of the oxidative krebs cycle. Bioenergetics 4:122

Malloch GD, Munday LA, Olson MS, Clark JB (1986) Comparative development of the pyruvate dehydrogenase complex and citrate synthase in rat brain mitochondria. Biochem J 238:729–736

Marchington DR, Kerbey AL, Jones AE, Randle PJ (1987) Insulin reverses effects of starvation on the activity of pyruvate dehydrogenase kinase in cultured hepatocytes. Biochem J 246: 233–236

Mcknight SL (2010) On getting there from here. Science 330:1338–1339
Mitchell SJ, Madrigal-Matute J, Scheibye-Knudsen M, Fang E, Aon M, Gonzalez-Reyes JA, Cortassa S, Kaushik S, Gonzalez-Freire M, Patel B, Wahl D, Ali A, Calvo-Rubio M, Buron MI, Guiterrez V, Ward TM, Palacios HH, Cai H, Frederick DW, Hine C, Broeskamp F, Habering L, Dawson J, Beasley TM, Wan J, Ikeno Y, Hubbard G, Becker KG, Zhang Y, Bohr VA, Longo DL, Navas P, Ferrucci L, Sinclair DA, Cohen P, Egan JM, Mitchell JR, Baur JA, Allison DB, Anson RM, Villalba JM, Madeo F, Cuervo AM, Pearson KJ, Ingram DK, Bernier M, De Cabo R (2016) Effects of sex, strain, and energy intake on hallmarks of aging in mice. Cell Metab 23:1093–1112
Muoio DM, Neufer PD (2012) Lipid-induced mitochondrial stress and insulin action in muscle. Cell Metab 15:595–605
Neely JR, Bowman RH, Morgan HE (1969) Effects of ventricular pressure development and palmitate on glucose transport. Am J Phys 216:804–811
Nelson DL, Cox MM (2013) Lehninger principles of biochemistry. W. H. Freeman and Company, New York
Nguyen T, Nioi P, Pickett CB (2009) The Nrf2-antioxidant response element signaling pathway and its activation by oxidative stress. J Biol Chem 284:13291–13295
Nickel A, Kohlhaas M, Maack C (2014) Mitochondrial reactive oxygen species production and elimination. J Mol Cell Cardiol 73:26–33
Nickel AG, von Hardenberg A, Hohl M, Loffler JR, Kohlhaas M, Becker J, Reil JC, Kazakov A, Bonnekoh J, Stadelmaier M, Puhl SL, Wagner M, Bogeski I, Cortassa S, Kappl R, Pasieka B, Lafontaine M, Lancaster CR, Blacker TS, Hall AR, Duchen MR, Kastner L, Lipp P, Zeller T, Muller C, Knopp A, Laufs U, Bohm M, Hoth M, Maack C (2015) Reversal of mitochondrial transhydrogenase causes oxidative stress in heart failure. Cell Metab 22:472–484
Nulton-Persson AC, Starke DW, Mieyal JJ, Szweda LI (2003) Reversible inactivation of alpha-ketoglutarate dehydrogenase in response to alterations in the mitochondrial glutathione status. Biochemistry 42:4235–4242
Oram JF, Bennetch SL, Neely JR (1973) Regulation of fatty acid utilization in isolated perfused rat hearts. J Biol Chem 248:5299–5309
Pettit FH, Pelley JW, Reed LJ (1975) Regulation of pyruvate dehydrogenase kinase and phosphatase by acetyl-CoA/CoA and NADH/NAD ratios. Biochem Biophys Res Commun 65:575–582
Quinlan CL, Goncalves RL, Hey-Mogensen M, Yadava N, Bunik VI, Brand MD (2014) The 2-oxoacid dehydrogenase complexes in mitochondria can produce superoxide/hydrogen peroxide at much higher rates than complex I. J Biol Chem 289:8312–8325
Randle PJ (1986) Fuel selection in animals. Biochem Soc Trans 14:799–806
Randle PJ (1998) Regulatory interactions between lipids and carbohydrates: the glucose fatty acid cycle after 35 years. Diabetes Metab Rev 14:263–283
Randle PJ, Garland PB, Hales CN, Newsholme EA (1963) The glucose fatty-acid cycle Its role in insulin sensitivity and the metabolic disturbances of diabetes mellitus. Lancet 1:785–789
Ravindran S, Radke GA, Guest JR, Roche TE (1996) Lipoyl domain-based mechanism for the integrated feedback control of the pyruvate dehydrogenase complex by enhancement of pyruvate dehydrogenase kinase activity. J Biol Chem 271:653–662
Roche TE, Hiromasa Y (2007) Pyruvate dehydrogenase kinase regulatory mechanisms and inhibition in treating diabetes, heart ischemia, and cancer. Cell Mol Life Sci 64:830–849
Roche TE, Baker JC, Yan X, Hiromasa Y, Gong X, Peng T, Dong J, Turkan A, Kasten SA (2001) Distinct regulatory properties of pyruvate dehydrogenase kinase and phosphatase isoforms. Prog Nucleic Acid Res Mol Biol 70:33–75
Roul D, Recchia FA (2015) Metabolic alterations induce oxidative stress in diabetic and failing hearts: different pathways, same outcome. Antioxid Redox Signal 22:1502–1514
Rydstrom J (2006) Mitochondrial NADPH, transhydrogenase and disease. Biochim Biophys Acta 1757:721–726
Salminen A, Kaarniranta K, Hiltunen M, Kauppinen A (2014) Krebs cycle dysfunction shapes epigenetic landscape of chromatin: novel insights into mitochondrial regulation of aging process. Cell Signal 26:1598–1603

Schafer FQ, Buettner GR (2001) Redox environment of the cell as viewed through the redox state of the glutathione disulfide/glutathione couple. Free Radic Biol Med 30:1191–1212
Schonfeld P, Wojtczak L (2008) Fatty acids as modulators of the cellular production of reactive oxygen species. Free Radic Biol Med 45:231–241
Sies H (2015) Oxidative stress: a concept in redox biology and medicine. Redox Biol 4:180–183
Singh R, Cuervo AM (2012) Lipophagy: connecting autophagy and lipid metabolism. Int J Cell Biol 2012:282041
Singh R, Kaushik S, Wang Y, Xiang Y, Novak I, Komatsu M, Tanaka K, Cuervo AM, Czaja MJ (2009) Autophagy regulates lipid metabolism. Nature 458:1131–1135
Stanley BA, Sivakumaran V, Shi S, Mcdonald I, Lloyd D, Watson WH, Aon MA, Paolocci N (2011) Thioredoxin reductase-2 is essential for keeping low levels of H(2)O(2) emission from isolated heart mitochondria. J Biol Chem 286:33669–33677
Steussy CN, Popov KM, Bowker-Kinley MM, Sloan RB Jr, Harris RA, Hamilton JA (2001) Structure of pyruvate dehydrogenase kinase. Novel folding pattern for a serine protein kinase. J Biol Chem 276:37443–37450
Sugden MC, Holness MJ (1994) Interactive regulation of the pyruvate dehydrogenase complex and the carnitine palmitoyltransferase system. FASEB J 8:54–61
Sugden MC, Fryer LG, Holness MJ (1996) Regulation of hepatic pyruvate dehydrogenase kinase by insulin and dietary manipulation in vivo. Studies with the euglycaemic-hyperinsulinaemic clamp. Biochim Biophys Acta 1316:114–120
Sugden MC, Orfali KA, Fryer LG, Holness MJ, Priestman DA (1997) Molecular mechanisms underlying the long-term impact of dietary fat to increase cardiac pyruvate dehydrogenase kinase: regulation by insulin, cyclic AMP and pyruvate. J Mol Cell Cardiol 29:1867–1875
Sugden MC, Fryer LG, Orfali KA, Priestman DA, Donald E, Holness MJ (1998) Studies of the long-term regulation of hepatic pyruvate dehydrogenase kinase. Biochem J 329(Pt 1):89–94
Sugden MC, Holness MJ, Donald E, Lall H (1999) Substrate interactions in the short- and long-term regulation of renal glucose oxidation. Metabolism 48:707–715
Sugden MC, Kraus A, Harris RA, Holness MJ (2000) Fibre-type specific modification of the activity and regulation of skeletal muscle pyruvate dehydrogenase kinase (PDK) by prolonged starvation and refeeding is associated with targeted regulation of PDK isoenzyme 4 expression. Biochem J 346(Pt 3):651–657
Sung MM, Hamza SM, Dyck JR (2015) Myocardial metabolism in diabetic cardiomyopathy: potential therapeutic targets. Antioxid Redox Signal 22:1606–1630
Swain L, Kesemeyer A, Meyer-Roxlau S, Vettel C, Zieseniss A, Guntsch A, Jatho A, Becker A, Nanadikar MS, Morgan B, Dennerlein S, Shah AM, El-Armouche A, Nikolaev VO, Katschinski DM (2016) Redox imaging using cardiac myocyte specific transgenic biosensor mice. Circ Res 119:1004–1016
Thomas AP, Denton RM (1986) Use of toluene-permeabilized mitochondria to study the regulation of adipose tissue pyruvate dehydrogenase in situ. Further evidence that insulin acts through stimulation of pyruvate dehydrogenase phosphate phosphatase. Biochem J 238:93–101
Thomas AP, Diggle TA, Denton RM (1986) Sensitivity of pyruvate dehydrogenase phosphate phosphatase to magnesium ions. Similar effects of spermine and insulin. Biochem J 238:83–91
Tocchetti CG, Caceres V, Stanley BA, Xie C, Shi S, Watson WH, O'rourke B, Spadari-Bratfisch RC, Cortassa S, Akar FG, Paolocci N, Aon MA (2012) GSH or palmitate preserves mitochondrial energetic/redox balance, preventing mechanical dysfunction in metabolically challenged myocytes/hearts from type 2 diabetic mice. Diabetes 61:3094–3105
Tocchetti CG, Stanley BA, Sivakumaran V, Bedja D, O'Rourke B, Paolocci N, Cortassa S, Aon MA (2015) Impaired mitochondrial energy supply coupled to increased H2O2 emission under energy/redox stress leads to myocardial dysfunction during Type I diabetes. Clin Sci (Lond) 129:561–574
Tocchi A, Quarles EK, Basisty N, Gitari L, Rabinovitch PS (2015) Mitochondrial dysfunction in cardiac aging. Biochim Biophys Acta 1847:1424–1433

Van Bilsen M, Van Nieuwenhoven FA, Van Der Vusse GJ (2009) Metabolic remodelling of the failing heart: beneficial or detrimental? Cardiovasc Res 81:420–428

Van Eunen K, Simons SM, Gerding A, Bleeker A, Den Besten G, Touw CM, Houten SM, Groen BK, Krab K, Reijngoud DJ, Bakker BM (2013) Biochemical competition makes fatty-acid beta-oxidation vulnerable to substrate overload. PLoS Comput Biol 9:e1003186

Wallace DC (2010) The epigenome and the mitochondrion: bioenergetics and the environment [corrected]. Genes Dev 24:1571–1573

Wallace DC, Fan W (2010) Energetics, epigenetics, mitochondrial genetics. Mitochondrion 10:12–31

Walther TC, Farese RV Jr (2009) The life of lipid droplets. Biochim Biophys Acta 1791:459–466

Walther TC, Farese RV Jr (2012) Lipid droplets and cellular lipid metabolism. Annu Rev Biochem 81:687–714

Wang H, Sreenevasan U, Hu H, Saladino A, Polster BM, Lund LM, Gong DW, Stanley WC, Sztalryd C (2011) Perilipin 5, a lipid droplet-associated protein, provides physical and metabolic linkage to mitochondria. J Lipid Res 52:2159–2168

Wellen KE, Hatzivassiliou G, Sachdeva UM, Bui TV, Cross JR, Thompson CB (2009) ATP-citrate lyase links cellular metabolism to histone acetylation. Science 324:1076–1080

Williamson JR, Chang K, Frangos M, Hasan KS, Ido Y, Kawamura T, Nyengaard JR, Van Den Enden M, KILO C, Tilton RG (1993) Hyperglycemic pseudohypoxia and diabetic complications. Diabetes 42:801–813

Wojtczak L, Schonfeld P (1993) Effect of fatty acids on energy coupling processes in mitochondria. Biochim Biophys Acta 1183:41–57

Wu P, Sato J, Zhao Y, Jaskiewicz J, Popov KM, Harris RA (1998) Starvation and diabetes increase the amount of pyruvate dehydrogenase kinase isoenzyme 4 in rat heart. Biochem J 329(Pt 1):197–201

Wu P, Blair PV, Sato J, Jaskiewicz J, Popov KM, Harris RA (2000) Starvation increases the amount of pyruvate dehydrogenase kinase in several mammalian tissues. Arch Biochem Biophys 381:1–7

Xie C, Biary N, Tocchetti CG, Aon MA, Paolocci N, Kauffman J, Akar FG (2013) Glutathione oxidation unmasks proarrhythmic vulnerability of chronically hyperglycemic guinea pigs. Am J Physiol Heart Circ Physiol 304:H916–H926

Yan J, Lawson JE, Reed LJ (1996) Role of the regulatory subunit of bovine pyruvate dehydrogenase phosphatase. Proc Natl Acad Sci U S A 93:4953–4956

Yang D, Gong X, Yakhnin A, Roche TE (1998) Requirements for the adaptor protein role of dihydrolipoyl acetyltransferase in the up-regulated function of the pyruvate dehydrogenase kinase and pyruvate dehydrogenase phosphatase. J Biol Chem 273:14130–14137

Zhang S, Hulver MW, Mcmillan RP, Cline MA, Gilbert ER (2014) The pivotal role of pyruvate dehydrogenase kinases in metabolic flexibility. Nutr Metab (Lond) 11:10

Zhou L, Aon MA, Almas T, Cortassa S, Winslow RL, O'Rourke B (2010) A reaction-diffusion model of ROS-induced ROS release in a mitochondrial network. PLoS Comput Biol 6:e1000657

Zima AV, Blatter LA (2006) Redox regulation of cardiac calcium channels and transporters. Cardiovasc Res 71:310–321

Index

T.K. Rostovtseva (ed.), *Molecular Basis for Mitochondrial Signaling*, Biological and Medical Physics, Biomedical Engineering,
DOI 10.1007/978-3-319-55539-3

R

S

V

W

Y

www.ingramcontent.com/pod-product-compliance
Lightning Source LLC
Chambersburg PA
CBHW050232230625
28565CB00001B/19
* 9 7 8 3 3 1 9 5 5 5 3 7 9 *